T0226423

NIELS BOHR
COLLECTED WORKS
VOLUME 11

AAGE BOHR AND NIELS BOHR, ISRAEL 1953. (PHOTOGRAPH BY STEFAN ROZENTAL)

NIELS BOHR

COLLECTED WORKS

GENERAL EDITOR

FINN AASERUD

THE NIELS BOHR ARCHIVE, COPENHAGEN

VOLUME 11
THE POLITICAL ARENA
(1934–1961)

EDITED BY

FINN AASERUD

2005

ELSEVIER

AMSTERDAM · BOSTON · HEIDELBERG · LONDON · NEW YORK · OXFORD · PARIS
SAN DIEGO · SAN FRANCISCO · SINGAPORE · SYDNEY · TOKYO

ELSEVIER B.V.
Radarweg 29
P.O. Box 211, 1000 AE Amsterdam
The Netherlands

Library of Congress Catalog Card Number: 70-126498
ISBN Collected Works: 0 7204 1800 3
ISBN Volume 11: 0 444 51336 1

⊛ The paper used in this publication meets the requirements of ANSI/NISO Z39.48-1992 (Permanence of Paper).

Transferred to digital print 2007

Printed and bound by CPI Antony Rowe, Eastbourne

FOREWORD

The present book is the first of the last two volumes of the Niels Bohr Collected Works. The last volume – Volume 12, "Popularization and People (1911–1962)" – will be published shortly. The two volumes were originally projected as one volume, but the amount of material proved too large to be accommodated between two covers.

Part I of Volume 11 is devoted to Bohr's self-appointed mission to promote an "open world" between nations, which took up a large part of his time and concentration from the autumn of 1943, when he understood that the atomic bomb was on the way to becoming a reality, until he died in November 1962. However, his public writings on the topic are not nearly in proportion to the effort he put into it. As explained in the introduction to Part I, this has called for the inclusion of more unpublished material than is usual in the Bohr Collected Works, which as a rule are centred on Bohr's publications. Some of this archival material is presented together with Bohr's publications, whereas the rest, which serves to place the publications in context, is to be found in the appendix to Part I.

The majority of the unpublished material stems from the Niels Bohr Archive (NBA), notably the Niels Bohr Political Papers, whereas other material originates from research abroad. In the early 1990s I visited several archival repositories in the United States. My residence as a Carlsberg Fellow at Churchill College in Cambridge, England, in the spring term of 1996, made possible extensive studies in British archives, notably at the Public Records Office in London and the Churchill Archives Centre in Cambridge.

Most of Bohr's publications in Part II of the present volume, as well as in Volume 12, may best be termed occasional writings. As such, they differ substantially from the publications in the first ten volumes of the Collected Works, which represent Bohr's writings relating to physics (Volumes 1 to 9) and his most important philosophical writings (Volume 10). It testifies to the effort that Bohr put into everything he wrote, as well as to the consistency of his thinking, that far from constituting a miscellany of not quite focused contributions, all publications reproduced in Volumes 11 and 12 show an impressive quality of expression and unity of thought.

Part II of the present volume documents Bohr's involvement in political and social activities other than those laid out in Part I. In accordance with the standard used in previous volumes of the Collected Works, this involvement is documented only in so far as it resulted in publications on Bohr's part. The publications can be divided into two major categories, the first of which consists of Bohr's writings in his capacity as a leader of specific institutions for the advancement of science, in particular the Royal Danish Academy of Sciences and Letters. The second category comprises lectures or writings prepared for a variety of special events.

The introduction to Part II seeks to place Bohr's contributions in the context in which they were presented. Especially in the second category of writings, this context is often mentioned only in passing or not at all, as Bohr tended to concentrate on one or another topic of particular interest. These are the same topics as those dealt with in publications in other volumes of the Collected Works – especially in Volume 10, in Part I of the present volume and in the writings presented under the heading of "Popularization" in Volume 12. However, no two of Bohr's contributions are identical, and he always added some new perspective to the topic, adapted to the audience for his presentation. Indeed, having a broader orientation than most others of Bohr's publications, they are particularly well suited to present him to the general public, both as a thinker and a person. A few letters to and from Bohr, illustrating the published material in Part II, are reproduced in an appendix.

* * *

In such a long and complicated project it is impossible to thank everybody who has contributed. Some people, however, must be mentioned in particular. I extend my apologies to those who have offered their valuable services, but who are not mentioned here.

When I started my employment as Director of the NBA in 1989 (which includes the general editorship of the Niels Bohr Collected Works), my predecessor Erik Rüdinger had already prepared a preliminary table of contents for what was then conceived to be the last volume in the series. Erik's preparatory work has been of inestimable value for continuing the project. Although the organization of the material has been changed considerably since then and much unpublished material has been included, only a few publications have been added to Erik's original list. I thank Erik for enabling me to make such a head start and for his continued interest and help throughout the project.

Before my stay in Cambridge, I conducted a series of interviews with Niels Bohr's son, Aage Bohr, who was with his father during the critical years of exile from Denmark beginning in the autumn of 1943 and lasting until the late summer of 1945. Aage continued as Niels Bohr's main confidant in political as well as other questions until his father's death. The interviews were the beginning of a close working and personal relationship, which has added immeasurably to the present volume. Aage has been instrumental in helping with the interpretation of the archival material for Part I and providing the recollections which I use freely in the introductions to Part I and Part II. In addition, he has taken an active part in writing and structuring in particular the introduction to Part I, without ever seeking to impose his views. Aage has also made himself available whenever advice was needed regarding any aspect of the work. Besides improving the quality of this volume, the collaboration with Aage has been a uniquely rewarding personal experience for me.

During my stay in England, Richard Ponman at the Historical and Records Section of the Cabinet Office in London was particularly helpful in guiding me through the maze of archival material. After my return to Denmark, Jørgen Grunnet at the Danish Embassy in Washington was helpful locating and re-trieving further documents from the U.S. government archives. More recently, Milton O. Gustafson of the U.S. National Archives has provided similar invalu-able assistance.

In addition to the few footnotes in Bohr's original texts, several numbered explanatory footnotes have been prepared by the editor in order to place the writings in context. Particularly useful help in this regard was provided by Knud Max Møller (1922–2004), biochemist and a close friend of the NBA. Others who have provided helpful information on one or more particular subjects include Libeta Chernobrow (head, interlibrary loan, Weizmann Insti-tute of Science, Israel), Ernest D. Courant (emeritus professor of physics at Brookhaven National Laboratory, U.S.A.), Ole Farver (chemist at the Danish University of Pharmaceutical Sciences and President of the Danish Weizmann Society), Paul Josephson (historian at Colby College, Maine, U.S.A.), Jørgen Kalckar (editor of Volumes 6 and 7 of the Collected Works), Henrik Knudsen (historian of science at the University of Aarhus), Alexei Kojevnikov (historian of science at the University of Georgia, Athens, U.S.A), Niels Nielsen (Di-rector of the Nordic Region of the Conference Board), Povl Ølgaard (reactor physicist at the Risø National Laboratory from the time of its establishment), Erik Rüdinger, Arne Sell (retired physician and historian of the Danish radium stations), Stephen Twigge (Head of the Research and Planning Unit at The National Archives, London) and Johnny Wøllekær (archivist, Odense City Archives). I thank them all.

Translating Niels Bohr's writings has constituted a special challenge, especially as the last two volumes of the Collected Works include more material originally written in Danish than did any of the earlier ones. As was also the case for Volumes 7 and 10, the translations have been a joint effort between Felicity Pors and the editor. Others have provided substantial help in the process. During the early period of preparations Helle Bonaparte, the secretary at the NBA until 1991, made a few preliminary translations. Hilde Levi (1909–2003) – who made her career in Denmark in physics as well as in the application of physics to biology after escaping Nazi Germany in 1934 – made an invaluable effort, which was unfortunately cut short by her illness and subsequent death. Erik Rüdinger and Jørgen Kalckar, as well as my son Andreas Næs Aaserud, have been of great assistance in the last stages of the translation process. As always, a main challenge has been to retain Bohr's style, including his generally long sentences, while making the English idiomatic. Arbitrary misspellings have been corrected, whereas typical ones have been retained. Misspellings of names have been kept in transcripts of original texts, but have been corrected in explanatory footnotes and translations. Whenever a non-English contribution is reproduced in facsimile, any explanatory footnotes appear in the translation only. When both the translation and the original version of a document have been transcribed, explanatory footnotes are placed in the original, with the footnote numbers repeated in the translation.

A nearly complete version of Part I was submitted to several readers for comment. This provoked particularly incisive reactions from the historian of science John L. Heilbron as well as Jørgen Kalckar and Erik Rüdinger. Useful comments were also provided by Andrew D. Jackson (physicist and chairman of the NBA board of directors), Helge Kragh (historian of science at the University of Aarhus); David Favrholdt (philosopher at the University of Odense and editor of Volume 10 of the Collected Works), Vilhelm Bohr (Department Chair at the U.S. National Institutes of Health and member of the NBA board of directors) and Henry Nielsen (historian of science at the University of Aarhus). John Heilbron has made an equally thorough reading of the introduction to Part II, as has Hans von Bülow (staff member of the secretariat of the Danish Atomic Energy Commission from its first year of operation in 1956 and its head from 1965 until the abolition of the Commission in 1976), Vilhelm Bohr and Henry Nielsen. I am grateful to them all for their valuable input.

For a publication such as this, it is essential to prepare an extensive and useful subject index. For the early stage of this work I was fortunate to secure the help of Helle Bonaparte. Christina Olausson, a student of physics at the Niels Bohr Institute, picked up where Helle left off. Both have contributed greatly to the work.

My closest collaborators continue to be the staff at the NBA, Felicity Pors and Anne Lis Rasmussen. Apart from the considerable task of translating Bohr's words, Felicity has contributed to all aspects of preparing the volume, from identifying specific articles and locating photographs through commenting on the introductions to taking part in structuring the volume. Lis has also contributed fruitfully to discussions of all aspects of the volume, in addition to typing all non-facsimile material and dealing with the many complex issues involved in formatting the book, including the generation of the index. I thank Felicity and Lis for their great dedication to the work.

In addition to the grant from the Carlsberg Foundation covering my stay in Cambridge, financial support for the preparation of Volumes 11 and 12 has been provided by the Lounsbery Foundation, New York. Furthermore, Hilde Levi bequeathed a significant sum of money to the NBA, which has been used for a variety of tasks connected with the Collected Works. I thank Hilde and the two foundations for this financial help.

Elsevier, represented by Publisher Carl Schwarz, has shown considerable understanding for the several delays in relation to carefully planned deadlines. As was the case for Volume 10, the practical work on the layout of the volume has been handled brilliantly by Betsy Lightfoot at Elsevier.

Finally, I would like to thank the NBA's board of directors for its constant trust and encouragement, and my wife, Gro Synnøve Næs, for her patience and never-failing help and support.

Finn Aaserud
The Niels Bohr Archive
October 2005

CONTENTS

PART I: AN "OPEN WORLD"

PART II: OTHER POLITICAL AND SOCIAL INVOLVEMENTS

INVENTORY OF RELEVANT MANUSCRIPTS
IN THE NIELS BOHR ARCHIVE

INDEX

EARLIER VOLUMES OF THE
NIELS BOHR COLLECTED WORKS

The General Editors of the *Niels Bohr Collected Works* have been: Léon Rosenfeld (1904–1974) (Volumes 1 to 3), Erik Rüdinger (Volumes 5 to 9, Volume 7 jointly with Finn Aaserud) and Finn Aaserud (Volume 10). All volumes are published by North-Holland/Elsevier.

Vol. 1, *Early Work (1905–1911)* (ed. J. Rud Nielsen), 1972.

Vol. 2, *Work on Atomic Physics (1912–1917)* (ed. Ulrich Hoyer), 1981.

Vol. 3, *The Correspondence Principle (1918–1923)* (ed. J. Rud Nielsen), 1976.

Vol. 4, *The Periodic System (1920–1923)* (ed. J. Rud Nielsen), 1977.

Vol. 5, *The Emergence of Quantum Mechanics (mainly 1924–1926)* (ed. Klaus Stolzenburg), 1984.

Vol. 6, *Foundations of Quantum Physics I (1926–1932)* (ed. Jørgen Kalckar), 1985.

Vol. 7, *Foundations of Quantum Physics II (1933–1958)* (ed. Jørgen Kalckar), 1996.

Vol. 8, *The Penetration of Charged Particles Through Matter (1912–1954)* (ed. Jens Thorsen), 1987.

Vol. 9, *Nuclear Physics (1929–1952)* (ed. Rudolf Peierls), 1986.

Vol. 10, *Complementarity Beyond Physics (1928–1962)* (ed. David Favrholdt), 1999.

ABBREVIATED TITLES OF PERIODICALS

Fys. Tidsskr.	Fysisk Tidsskrift (Copenhagen)
Ned. T. Natuurk.	Nederlandsch Tijdschrift voor Natuurkunde (Amsterdam)
Overs. Dan. Vidensk. Selsk. Virks.	Oversigt over Det Kongelige Danske Videnskabernes Selskabs Virksomhed (Copenhagen)
Proc. Amer. Phil. Soc.	Proceedings of the American Philosophical Society (Philadelphia)

OTHER ABBREVIATIONS

AB	Akademisk Boldklub (Academic Ball Club), Copenhagen
AEC	Atomic Energy Commission, U.S.A.
AHQP	Archive for History of Quantum Physics
AIP	American Institute of Physics, College Park, Maryland
ATV	Akademiet for de tekniske Videnskaber (Academy of Technical Sciences), Denmark
BMSS $(x.y)$	Bohr Manuscripts, AHQP, section y on mf. x
BSC $(x.y)$	Bohr Scientific Correspondence, AHQP, section y on mf. x
BSC-Supp.	Bohr Scientific Correspondence, Supplement, NBA
BPP $x.y$	Bohr Political Papers, NBA, folder x, item y
CAB x/y	Records of the Cabinet Office, PRO, series x, folder y
CERN	Conseil Européen pour la Recherche Nucléaire, Geneva
CHAD	Chadwick Papers, Churchill Archives Centre, Cambridge, England
FRUS	Foreign Relations of the United States, printed series published by the U.S. Department of State
IUPAP	International Union for Pure and Applied Physics
mf.	microfilm
NATO	North Atlantic Treaty Organization
NBA	Niels Bohr Archive, Copenhagen
NORDITA	Nordic Institute for Theoretical Physics, Copenhagen
PREM	Records of the Premier's Office, PRO
PRO	Public Records Office, The National Archives, London, England

SNU	Selskabet for Naturlærens Udbredelse (Society for the Dissemination of Natural Science), Denmark
UNAEC	United Nations Atomic Energy Commission
UNESCO	United Nations Educational, Scientific and Cultural Organization

ACKNOWLEDGEMENTS

For some of the published work in this volume, the editors and the publisher were unfortunately unable to trace the copyright holders and thus to enter a formal request for permission to reproduce the material. These works were nevertheless considered sufficiently important to be reprinted without further delay. The effort to identify the copyright holders will be continued.

N. Bohr, "Science and Civilization", The Times, 11 August 1945, is reprinted by permission of The Times, London.

N. Bohr, "A Challenge to Civilization", Science **102** (1945) 363–364, is reprinted by permission of the American Association for the Advancement of Science.

N. Bohr, "Atomic Physics and International Cooperation", Proc. Amer. Phil. Soc. **91** (1947) 137–138, is reprinted by permission of the American Philosophical Society.

The following contributions in Politiken: "Niels Bohr kom i Gaar hjem fra Rusland", 28 May 1934; H. Iacobæus, "Niels Bohr og Sovjetrusland" (extract), 30 May 1934; "Prof. Niels Bohrs Svar", 30 May 1934; N. Bohr, "Universitetet og Forskningen", 3 June 1941; N. Bohr, "Menneskehedens Valg mellem Katastrofe og lykkeligere Kaar", 1 January 1946; "Niels Bohrs dybt alvorlige appel", 19 January 1951, are reprinted by permission of Politiken.

The following contributions in Berlingske Aftenavis: V. Pürschel, "Professor Niels Bohrs Udtalelse", 30 May 1934; "En Udtalelse af Professor Niels Bohr", 31 May 1934, are reprinted by permission of Berlingske Tidende.

The following contributions by N. Bohr from Overs. Dan. Vidensk. Selsk. Virks.: "Mødet den 20. Oktober 1939", Juni 1939 – Maj 1940, pp. 25–26; "Mødet den 15. Marts [1940]", Juni 1939 – Maj 1940, pp. 40–41; "Mødet den 20. Septbr. 1940", Juni 1940 – Maj 1941, pp. 25–26; "Mødet den 30. Januar 1942", Juni 1941 – Maj 1942, pp. 32–34; "Mødet den 13. November 1942 paa 200-Aarsdagen for Selskabets Stiftelse", Juni 1942 – Maj 1943, pp. 26–28, 31–32, 36, 40–41, 44–48; "Mødet den 19. Oktober 1945", Juni 1945 – Maj 1946, pp. 29–31; "Mødet den 25. April 1947", Juni 1946 – Maj 1947, pp. 53–54; "Mødet den 17. Oktober 1947 til Minde om Kong Christian X", Juni 1947 – Maj 1948, pp. 26–29; "Mødet den 11. Marts 1949", Juni 1948 – Maj 1949, pp. 45–46; "Mødet den 2. Februar 1951", Juni 1950 – Maj 1951, pp. 42–45; "Mødet den 19. Oktober 1951", Juni 1951 – Maj 1952, pp. 33–34; "Mødet den 16. November 1951", Juni 1951 – Maj 1952, p. 39; "Mødet den 14. Oktober 1960", Juni 1960 – Maj 1961, pp. 39–41, are reprinted by permission of the Royal Danish Academy of Sciences and Letters.

The following contributions by N. Bohr in Fys. Tidsskr.: "Ottende Uddeling af H.C. Ørsted Medaillen", **39** (1941) 175–177, 192–193; "Niende Uddeling af H.C. Ørsted Medaillen", **51** (1953) 65–67, 80; "Tiende og ellevte Uddeling af H.C. Ørsted Medaillen", **57** (1959) 145, 158; "Mindeaften for Kirstine Meyer i Selskabet for Naturlærens Udbredelse", **40** (1942) 173–175, are reprinted by permission of Selskabet for Naturlærens Udbredelse.

N. Bohr, "Science and its International Significance", Danish Foreign Office Journal, No. 208 (May 1938) 61–63, is reprinted by permission of the Danish Foreign Office.

The publisher's introductory remarks as well as N. Bohr, "Preface", … fra Thrige **6** (No. 1, February 1953) 2–4, are reprinted by permission of the Thomas B. Thrige Foundation.

The pages introducing Bohr's session as well as his lecture, "Greater International Cooperation is Needed for Peace and Survival", Atomic Energy in Industry: Minutes of 3rd Conference October 13–15, 1954, National Industrial Conference Board, Inc., New York 1955, pp. 18–26, are reprinted by permission of the Conference Board.

N. Bohr, "Det fysiske grundlag for industriel udnyttelse af atomkerne-energien", Tidsskrift for Industri, nr. 7–8 (1955) 168–179, is reprinted by permission of Dansk Industri.

N. Bohr, "Atomerne og Samfundet", Den liberale venstrealmanak, ASAs Forlag, Copenhagen 1956, pp. 25–32, is reprinted by permission of Venstres Landsorganisation.

"Hilsen til udstillingen fra professor Niels Bohr", Elektroteknikeren **53** (1957) 363; "Professor Niels Bohr om Risø", Elektroteknikeren **54** (1958) 238–239, are reprinted by permission of Dansk Energi.

"The Presentation of the first Atoms for Peace Award to Niels Henrik David Bohr, October 24, 1957", National Academy of Sciences, Washington, D.C. 1957, pp. 1, 5, 18–22, is reprinted by permission of the MIT Archives & Special Collections.

N. Bohr, "Om Maalingsproblemet i Atomfysikken", Festskrift til N.E. Nørlund i Anledning af hans 60 Aars Fødselsdag den 26. Oktober 1945, Ejnar Munksgaard, Copenhagen 1946, pp. 163–167, is reprinted by permission of the publishers and Dansk Matematisk Forening.

N. Bohr, "Mathematics and Natural Philosophy", The Scientific Monthly **82** (1956) 85–88, is reprinted by permission of New York University.

"Niels Bohr's Hilsen", Akademisk Boldklub 1939–1949, Copenhagen 1949, pp. 7–8, is reprinted by permission of Akademisk Boldklub.

N. Bohr, "Tale ved Statsradiofoniens udsendelse den 16. oktober 1953", Det kongelige Teater, Forestillingen lørdag den 17. oktober 1953, pp. 4–6, is reprinted by permission of Det kongelige Teater.

N. Bohr, "Israels genopbygning: Et æventyr af ejendommelig art", Israel **7** (No. 2, 1954) 14–17, is reprinted by permission of Dansk Zionistforbund.

N. Bohr, "Kampens mål: At vi i frihed kan se hen til en lysere fremtid", Ti år efter, Kammeraternes Hjælpefond, Copenhagen 1955, is reprinted by permission of Kammeraternes Hjælpefond.

PART I

AN "OPEN WORLD"

INTRODUCTION

by

FINN AASERUD

During the last two decades of his life, Niels Bohr devoted a large part of his manifold efforts and interests to developing and realizing his idea of an "open world" between nations. In essence, the effort constituted a response to the challenge to civilization posed by the advent of the atomic bomb. It started in October 1943 when Bohr learned about the advanced stage of the bomb project upon being forced to escape to England from Nazi-occupied Denmark. It lasted, without interruption, until his death in November 1962. During these twenty years, Bohr acted as a statesman of science, with the doggedness and determination that characterized his efforts in all fields.

However, Bohr's publications dealing directly with these questions are scarce, and their number is not nearly in proportion to his effort. In fact, no more than six of Bohr's published papers deal centrally with these issues. Moreover, all but one of these papers – his famous "Open Letter to the United Nations" – are brief and deal with general perspectives rather than with immediate questions of politics.[1]

This state of affairs presents a dilemma for the editor of the Niels Bohr Collected Works, which on the one hand are intended to provide as balanced a view as possible of Bohr's total activities, yet on the other hand are based on Bohr's own words as expressed in his publications. Indeed, the first ten volumes in the series claim completeness only with regard to Bohr's publications proper, for which editors' introductions and archival material provide context, illustration and explanation. Thus, whereas a strict allegiance to the policy of focusing on the published papers would leave Bohr's central concern during

---

[1] N. Bohr, *Open Letter to the United Nations, June 9th, 1950*, J.H. Schultz, Copenhagen 1950, reproduced in this volume on pp. [171] ff.

[3]

the last two decades of his life severely under-documented, a departure from it would involve an awkward imbalance in the criteria for the selection of material to be included in the series. How can this dilemma be resolved?

The nature of Bohr's "open world" activities, and the documentation they have left, suggest a compromise that not only makes evident the centrality of this idea in Bohr's thinking and doing, but also provides an understanding of the development of the idea itself and the way that Bohr chose to promote it. The Open Letter gives an important clue in this regard, consisting as it does of comments newly written in 1950 and a selection of previously unpublished documents from Bohr's hand. In 1950, then, Bohr considered that documents he had written long before without the intention of being published, were indeed quite publishable.

The previously unpublished documents reproduced by Bohr in the Open Letter are memoranda constructed just as carefully as any of his publications proper. Instead of being intended for publication, however, Bohr wrote them as background documents for the confidential discussions he carried on with statesmen. Such confidentiality was, of course, required in the climate of secrecy during World War II. Bohr continued this approach after the war, and it was only when he felt a need to involve wider circles that he decided to publish selected parts of the documents originally prepared for President Franklin D. Roosevelt, as well as a document written for the American Secretary of State George C. Marshall three years after the war, in his Open Letter.

The compromise chosen in this last volume of the Collected Works to document Bohr's ideas and activities for an "open world" is the approach chosen by Bohr himself in the Open Letter. To the six published papers already mentioned and complete versions of the memoranda first published in the Open Letter has been added material never before published, according to two criteria: first, they are "publishable" in the sense that Bohr put as much work into them as he did for a publication proper; second, they are particularly important for documenting the development of Bohr's ideas. These unpublished documents comprise memoranda on political matters written for various statesmen both during and after World War II. They are divided into three separate sets, according to the periods in which they were written. A fourth set of documents consists of Bohr's unpublished statements for a Danish audience following his homecoming in 1945.

To a greater extent than his other publications, Bohr's writings on political issues can only be understood in relation to the dramatic story of the background for their creation. Section 1 of this introduction describes how Bohr, in connection with his efforts to promote international cooperation in science, was gradually confronted with political issues up to the early 1940s. Section

[4]

2 deals with Bohr's activities and writings during the war, his most intensive period as a political actor, when he came closest to affecting political decisions of the nations controlling the atomic bomb.

After the bombing of Hiroshima Bohr published his views, primarily for British, American and Danish audiences, but he also tried to reach physicists in the Soviet Union. In order to keep open his option for confidential contact on the political level, he restricted his publications to a general account of the implications of the atomic bomb for future relations among nations, carefully avoiding any mention of his confidential approaches to statesmen. While Section 3 deals with the context of these public activities, Section 4 describes Bohr's simultaneous confidential approaches, most notably his meeting with Marshall in 1948.

The "Open Letter to the United Nations" constituted a merging of Bohr's previously separate confidential and public approaches, in that Bohr here made public not only his general viewpoint, but also his confidential contacts with statesmen. Section 5 describes how the Open Letter came to be, touches upon its reception and briefly outlines the developments after 1950. The Open Letter signified a change in Bohr's political activities in that a larger circle became involved in the promotion of his ideas. Although he was as committed as ever to his political cause, from 1950 until his death in 1962 Bohr produced only one published or "publishable" document devoted centrally to his broad vision of an open world.[2] With the exception of this document – a brief sequel to the Open Letter in 1956 – Bohr's political statements during this period therefore lie outside the scope of the Collected Works.

In this introduction references to quotations from Bohr's publications are given in footnotes, whereas quotations from his "publishable" writings are indicated in boldfaced margin notes. Both kinds of documents are reproduced immediately following this introduction.[3] In addition, the account is based on a considerable number of other memoranda, letters and notes, which are reproduced in an appendix.[4] The majority of these documents are part of the Bohr Political Papers (BPP), which are now available to scholars at the Niels Bohr Archive. However, several documents stem from other archives, notably the Public Records Office in London. Whenever such a document is quoted in this introduction, it is identified in the margin with normal typeface, with a reference to the page number in the Appendix where it is reproduced in full.

---

[2] Open letter from Bohr to Hammarskjöld, 9 November 1956, privately printed. Reproduced on pp. [191] ff.
[3] See pp. [85]–[192], below.
[4] See pp. [193]–[333], below.

Complete references, as well as editor's notes (when required), are provided immediately before the documents themselves.

1. POLITICS AND INTERNATIONAL SCIENTIFIC COOPERATION UNTIL 1943

To Bohr science always constituted, by its very nature, a truly international enterprise. In the years between the two world wars, the responsibility he felt for advancing international cooperation in science was to bring him gradually into action on the political scene and to prepare him for his subsequent activities during the war.

Bohr was exposed at an early age to the ideal and practice of international research. His father, the prominent physiologist Christian Bohr, directed a laboratory at the University of Copenhagen which was in fruitful contact with research and colleagues in other countries. Upon completing his Ph.D. at the University, Bohr was inspired by the environment he experienced in the laboratory in Manchester of Ernest Rutherford, whose personality and example would provide guidance for the rest of his life. Besides preparing the ground for his crucial contributions to the understanding of the atom, Bohr's Manchester experience led to several lasting contacts with fellow physicists from other countries.[5]

Bohr's ambition to build his own institute for theoretical physics at the University of Copenhagen was realized in 1921, and he succeeded from the start in making it an international research centre with young promising physicists from all over the world visiting for limited periods. During the first years of the institute's existence some of the leading physicists of the day, including Max Planck and Albert Einstein, as well as Rutherford, visited Copenhagen. A special style of cooperation developed around Bohr, sometimes referred to as the "Copenhagen Spirit", in which humour, informality and openness were important elements in scientific discussions as well as in human relationships.

Bohr was strongly opposed to the contemporary trend of excluding the Central Powers of World War I from international scientific cooperation. Bohr refused all connections with institutions holding this view, such as the newly established International Union for Pure and Applied Physics (IUPAP). His insistence on including all nations contributed importantly to the success of his institute and provided a basis for its tradition of harmonious scientific cooperation which continues to this day.

---

[5] The early years of Bohr's life are touched upon in Vol. 1, especially the *Biographical Sketch* by the editor, Léon Rosenfeld, pp. XVII–XLVIII, on pp. XVIII–XXI.

Bohr's visit to the Soviet Union in 1934 included a visit to Kharkov, where he attended an international conference on theoretical physics. Left to right: D.D. Ivanenko, L. Tisza, L. Rosenfeld, –, Y.B. Rumer, N. Bohr, J.G. Crowther, L.D. Landau, M.S. Plesset, Y.I. Frenkel, I. Waller, E.J. Williams, V. Gordon, A.V. Fock and I. Tamm.

Bohr's institute was anything but an ivory tower. In order to obtain funding for his many visitors, as well as instruments for their research, Bohr wrote numerous applications for economic support, notably to funding agencies such as the Carlsberg and the Rask–Ørsted Foundations in Denmark, and the International Education Board and (from the 1930s) the Rockefeller Foundation in the United States. In addition, Bohr's direct contact with the international physics community brought contemporary developments of world politics closer to him than to most Danes. He experienced at close range the dramatic developments in the Soviet Union and Germany.

Productive contact with the Soviet Union[6] began with the visits to Copenhagen in the late 1920s of George Gamow and Lev Landau, who were soon to become leading members of the international physics community. In the following years the situation became more difficult, as severe restrictions were imposed on foreign travel for Soviet scientists and as their opportunities for open discussions became strongly limited. Gamow's stay in Copenhagen con-

---

[6] See P.J.E. Kragh, *Niels Bohr and the Soviet Union between the Two World Wars: Resources at the Niels Bohr Archive*, Master's Thesis, Copenhagen University 2003.

stituted one of his visits to several European institutions up to 1931. Upon his return to the Soviet Union, his several applications for permission to visit foreign institutions were rejected until he was allowed to attend the Solvay Conference in Brussels in 1933. Gamow was so worried about his future as a physicist in his home country that he never returned.

Bohr sought to improve scientific relations with the Soviet Union by accepting an invitation from the Academy of Sciences of the U.S.S.R.[7] and Russian universities, visiting in the spring of 1934. He went to several institutions for physics and met Russian officials, especially Nikolai Bukharin, who had played a major role in Soviet politics since the revolution in 1917. Bukharin was editor of the newspaper "Izvestia", which printed an extensive interview with Bohr. While praising the Soviet Union's support of science and its appreciation of the close relationship between theory and practice, Bohr here pleaded for more openness in international relations between scientists:

Manuscript for
Izvestia interview,
12 May 34
German
Full text on p. [196]
Translation on p. [199]

"It is extremely necessary to promote in every way the scientific cooperation between, and the mutual support to, scientists of the various countries. It is important that it is recognized abroad what large exceptional efforts are being made in your country towards developing science and how one here supports even the fields of science that are particularly far from direct practical application. Very important in this regard are the direct contact and mutual personal connections between foreign and Soviet scientists. It is extremely necessary that they pay mutual visits to one another as often as possible and in greater numbers. And this applies in particular to the younger generation of scientists, which is particularly able to pick up what is new. I believe that in science there will and must exist a sense of community and cooperation that one may not be able to create in all fields. Science all over the world must be directed towards the welfare of humanity. In this regard one should not distinguish the social problems from the scientific problems."

In other words, the openness required for scientific work had a broader function than serving science itself. This was a point Bohr would return to in his political campaign during and after World War II.

In the interview Bohr also referred to visits to cultural and industrial institutions in Leningrad (now St. Petersburg) and described many favourable

---

[7] The Academy had changed its name from the Russian Academy of Sciences at its 200th anniversary in 1925, when it also became more closely integrated into the Soviet government system.

impressions. He emphasized, however, "that it is difficult to say to what an extent everything taking place here may be transferred to other countries."

The original transcript of the interview, written in German and endorsed by Bohr, was made available to the Danish news agency Ritzau. The communist daily "Arbejderbladet" published a translation of the entire interview, noting Bohr's "enthusiastic" statements about the Soviet Union.[8] On his return, Bohr gave a briefer interview to the journalist Merete Bonnesen from "Politiken", one of the major Danish newspapers.[9] Bohr's visit, and particularly the "Izvestia" interview, caused critical comments in Letters to the Editor of "Politiken" and "Berlingske Aftenavis". The claim that he naively endorsed the Soviet regime provoked Bohr to explain his position in two further Letters to the Editor.[10]

Only a few months after Bohr's visit and less than a year after Gamow's escape, the Russian physicist Peter Kapitza was retained in the Soviet Union during his summer vacation. For the previous thirteen years Kapitza had been a prominent member of Rutherford's group in Cambridge, where he had even obtained his own laboratory. The international physics community was upset, and in his unsuccessful efforts to seek Kapitza's return, Rutherford asked Bohr whether he "might be in a position either directly or indirectly to express your views on the matter to the proper authorities in Russia." Bohr proceeded to write a personal letter to Bukharin, using the opportunity to thank him for "our lively conversation [during Bohr's recent visit] about the general philosophical issues which are so close to our hearts." With regard to Kapitza, Bohr returned to the arguments he had presented in the "Izvestia" interview:

Rutherford to Bohr,
6 Dec 34
English
Full text on p. [212]

Bohr to Bukharin,
15 Dec 34
German
Full text on p. [214]
Translation on p. [215]

> "Lord Rutherford who, as you know, has always committed himself to the international cooperation among scholars, and to cultivating the particularly friendly relations with the Russian colleagues, does not only consider the unfortunate consequences for the local interests which would result from a possible interruption of Prof. Kapitza's work in Cambridge, but he just as much fears that if such a situation became publicly known, it could have a very regrettable effect on the cooperation between foreign and Russian scientists, which now develops so very promisingly, particularly thanks to the astute and magnificent support of science on the part of the Soviet government."

[8] Arbejderbladet, 27 May 1934.

[9] Politiken, 28 May 1934. Reproduced on p. [203] and translated on p. [204].

[10] The critical comments were printed in Politiken and in Berlingske Aftenavis, both on 30 May. Bohr's response in Politiken was printed immediately after the critical comment, whereas his letter to Berlingske Aftenavis appeared the following day. The Letters are reproduced on p. [206] and translated on p. [208].

Bohr to Rutherford,
15 Dec 34
English
Full text on p. [216]

Sending a copy of the letter to Rutherford, Bohr described Bukharin as "a most attractive and intelligent man." However, Bohr was "not sure that he has still the great influence which he exerted for a long time after the revolution".

Rutherford to Bohr,
25 Dec 34
English
Full text on p. [218]

Ten days later Rutherford responded to Bohr asking for help with a different line of attack. After a discussion with "representatives of the University [of Cambridge] and the Royal Society on the Kapitza question", he enclosed a draft of a "memorial". He asked Bohr to reformulate the memorial as he found appropriate, to distribute it to non-British representatives of the international physics community for their signature, and, finally, to send the result personally to Maxim Litvinov, Stalin's Commissar for Foreign Affairs, who, as Rutherford surmised in his first letter to Bohr, had been "absent when the Kapitza question was officially dealt with". Rutherford specifically encouraged Bohr to obtain advice in the matter from Paul Langevin, a prominent French physicist considered to be on good terms with the Soviet authorities.

Bohr to Rutherford,
7 Jan 35
English
Full text on p. [221]

While agreeing with Rutherford that an unofficial approach on the part of scientists was preferable to a public diplomatic protest, Bohr opposed the suggested collective petition, being "somewhat afraid that any semi-public international appeal might harden the case prematurely and make it a matter of prestige of the Russian government to keep its original decision." Bohr preferred "a more private communication with Litvinoff from some independent individual, and if not yourself, perhaps me or perhaps better Langevin". To obtain a second opinion, Bohr followed Rutherford's advice to involve Langevin in the question, sharing all the material on the matter sent by Rutherford,

Bohr to Langevin,
7 Jan 35
English
Full text on p. [222]

while expressing his reservation that "he was not sure that a more private communication with Litvinoff might not be more effective, and at the same time less dangerous for Kapitza".

In the meanwhile, Kapitza was given his own physics laboratory in Moscow, the Institute for Physical Problems under the auspices of the Soviet Academy. He soon became a major force in Soviet physics. Since Langevin did not answer, evidently finding the situation difficult, Bohr did not pursue the matter further. Nevertheless, the Kapitza incident reflects Bohr's independent political judgement.

Bohr never received a response from Bukharin. As Bohr suspected, his influence was declining; he was executed in the Great Terror of 1937–38. A little later Landau also disappeared. This time Bohr wrote a two-page letter to Stalin himself, stating in part:

Bohr to Stalin,
23 Sep 38
German
Full text on p. [223]
Translation on p. [225]

"In view of the great importance of this matter for science in the USSR, as well as for international scientific cooperation, I turn to you with the urgent plea that an investigation into the fate of prof. Landau ..."

[10]

Kapitza, at whose laboratory Landau had taken up employment in 1937, also intervened forcefully in the matter. While it is unknown whether Bohr's letter played any role for Landau's subsequent release, his action represents another step in his political development and shows that he did not hesitate to turn to a world leader when the situation demanded it.

The rise to power of the Nazis in Germany in 1933, with their racist ideology, was felt by Bohr as a betrayal of civilization itself. He deemed the event as a blow in particular to the ideals and practices of the international science community and saw an urgent need to rescue his persecuted colleagues. He threw himself personally into the fight, using his influence to help Jewish physicists escape from Germany to Denmark. Most often he succeeded in finding permanent work for them elsewhere, especially in the United States. Bohr was a member of the Danish Committee for the Support of Refugee Intellectual Workers, established in October 1933 to help refugees obtain permission to enter Denmark and to seek financial support for them.

Bohr spoke against the racist ideology by arguing that the relationship between different human cultures can be recognized as "complementary" in terms of the concept he had proposed as a basis for understanding quantum mechanics.[11] He presented this argument in a newspaper interview in connection with his fiftieth birthday in 1935,[12] as well as at the Second International Congress of Anthropological and Ethnological Sciences three years later.[13] Although the Congress was held in Denmark, Bohr's lecture was reported in the "New York Times" the day after it was given.[14]

The 1930s saw the rise of nuclear physics, and Bohr was particularly alert to this development. With the support of foundations in Denmark and in the United States he was able to build up facilities for experimental research in nuclear physics, while at the same time making important contributions to the theory of nuclear reactions. The field had arisen in Europe, but came gradually to be dominated by the United States, especially on the experimental side. An increasing number of visiting researchers at the Copenhagen institute came from America, and Bohr made extensive visits to the United States.[15]

---

[11] Vol. 6 is devoted in part to Bohr's introduction of this concept.

[12] P. Vinding, *Conversation with Niels Bohr*, Berlingske Aftenavis, 2 October 1935, pp. 9–11. Reproduced and translated in Vol. 12.

[13] N. Bohr, *Natural philosophy and human cultures*, Nature **143** (1939), 268–272. Reproduced (from the conference proceedings) in Vol. 10, pp. [240]–[249].

[14] The New York Times, 5 August 1938. Reproduced in Vol. 10, p. [238].

[15] Bohr's contributions to nuclear physics are dealt with in Vol. 9. See also F. Aaserud, *Redirecting Science: Niels Bohr, philanthropy and the rise of nuclear physics*, Cambridge University Press, Cambridge 1990.

Bohr's insights were crucial for the further understanding of the process of nuclear fission and hence for developing the atomic bomb. The fission process was discovered at the end of 1938 by the German scientists Otto Hahn and Fritz Strassmann and explained by Otto Robert Frisch and Lise Meitner in the light of Bohr's ideas. Frisch went on to conduct pioneering experiments at Bohr's institute confirming their explanation, while Bohr, together with the American physicist John Wheeler, made essential theoretical contributions in Princeton where Bohr was visiting for the spring term. Bohr was able to draw the crucial conclusion that a chain reaction, with its vast release of energy, cannot be realized in materials found in nature. Hence, a separation of isotopes on a scale previously beyond imagination would be required to create a nuclear explosion. On this basis Bohr wrote in March 1939 to his wife Margrethe about the possibility of building an atomic bomb: "I do not believe in it, but over here one is very occupied by this."[16] On more than one occasion he also publicly expressed his view that it would be infeasible in practice to construct an atomic bomb "with the present technical means".[17]

Although the German occupation of Denmark on 9 April 1940 in effect prevented the presence of foreign visitors at Bohr's institute, research in physics continued without serious interruption. As for developments outside the institute, Bohr agreed to write the introduction to the impressive eight-volume "Denmark's Culture in the Year 1940", a comprehensive overview of all aspects of contemporary Danish culture which came to signify Danish cultural independence in defiance of the German presence. The central theme of Bohr's introduction was the ability of Danish culture to absorb elements of other cultures and make them its own. It concludes:[18]

"What destiny has in store for us and for others is hidden from our eyes, but no matter how far-reaching the consequences, in all areas of human life, may be of the crisis in which the world finds itself today, we are entitled

---

[16] Niels to Margrethe Bohr, 17 March 1939. The Niels Bohr–Margrethe Bohr correspondence is still in the possession of the Bohr family. I am grateful to Aage Bohr for allowing me to quote selected passages in the present introduction.

[17] See, for example, his review lecture given to the Danish Society for the Dissemination of Natural Science on 6 December 1939, *Nyere Undersøgelser over Atomkernernes Omdannelser* [*Recent Investigations of the Transmutations of Atomic Nuclei*], Fys. Tidsskr. **39** (1941) 3–32. Reproduced in Vol. 9, pp. [413]–[442] (Danish original) and [443]–[466] (English translation). The quotation is from p. [466].

[18] N. Bohr, *Dansk Kultur. Nogle indledende Betragtninger* [*Danish Culture. Some Introductory Reflections*] in *Danmarks Kultur ved Aar 1940*, Det Danske Forlag, Copenhagen 1941–1943, Vol. 1, pp. 9–17. Reproduced in Vol. 10, pp. [253]–[261] (Danish original) and [262]–[272] (English translation).

to hope that our people, as long as we can retain freedom to develop the outlook that is so deeply rooted in us, will also in the future be able to serve honourably the cause of mankind."

In consideration of the situation and Bohr's as a rule careful way of expressing himself these words amounted to an encouragement to his fellow countrymen to stand firm against the German occupiers.

The first indication that reached Bohr of an effort to produce an atomic bomb came in September 1941 when his younger German colleague and close friend Werner Heisenberg visited Copenhagen. The official reason for Heisenberg's visit was his participation in an astrophysics symposium at the German Cultural Institute, one of several such institutions established by the German authorities in occupied countries in order to encourage cooperation. Bohr avoided contact with the German authorities, but agreed to a private conversation with Heisenberg. In it, Heisenberg mentioned that he was pre-occupied with problems relating to military applications of atomic energy.[19] Although Heisenberg's information may have caused him to reconsider the question briefly,[20] Bohr remained sceptical with regard to the possibility of producing an atomic bomb, as documented by his correspondence with his British colleague James Chadwick in 1943.

The correspondence started at the end of January with a secret message from Chadwick inviting Bohr to come to England to work on "a particular problem in which your assistance would be of the greatest help," Bohr immediately understood what was referred to, yet refused the invitation on the following grounds:

> "Not only I feel it to be my duty in our desperate situation to help to resist the threat against the freedom of our institutions and to assist in the protection of the exiled scientists, who have sought refuge here. Still neither such duties nor even the dangers of retaliation to my collaborators and relatives might perhaps not carry sufficient weight to detain me here, if I felt, that I could be of real help in other ways, but I do not think that this is probable. Above all I have to the best of my judgment convinced myself, that in spite of all future prospects any immediate use of the latest marvellous discoveries of atomic physics is impracticable."

Bohr–Chadwick letters, early 1943
English
Full text on p. [227]

[19] Note in the handwriting of Bohr's son, Erik, 21 Mar 1954, BPP 1.6. Bohr's unsent letters pertaining to the meeting can be read at the website www.nba.nbi.dk and in the Danish journal *Naturens Verden* **84** (No. 8–9, 2001), pp. I–XL.

[20] S. Rozental, *Niels Bohr: Memoirs of a Working Relationship*, Christian Ejlers, Copenhagen 1998, p. 59.

In September 1943 the political situation in Denmark changed dramatically, with the Nazis planning to deport all Jews in Denmark and send them to concentration camps. The arrest of Bohr as an enemy of the Nazi regime had already been contemplated a month before but was postponed on the grounds that it would attract less notice if made in connection with the large-scale deportation.[21] Information about the imminent arrest was leaked, and Bohr was warned on 29 September. With the help of contacts within the resistance movement, he was able to escape to Sweden that night.

In Stockholm Bohr visited the Swedish Foreign Minister and, later on the same day, Saturday 2 October, the Swedish king, who agreed to make public the Swedish offer to Hitler to accept the Danish Jews in Sweden. At this time, too, the invitation to England was repeated, and now Bohr accepted.

2. THE POLITICAL CAMPAIGN DURING THE WAR EXILE, 1943–1945

Although Bohr had already shown both an aptitude for and interest in political questions, his attempt to influence world politics started only during his wartime exile, prompted by his recognition that the atomic bomb was becoming a reality. This was the period in which his ideas were closest to being realized, both because of the novelty of his argument and because of his access to the statesmen responsible for decisions about the development of the atomic bomb.[22]

---

[21] N. Blaedel, *Harmony and Unity: The Life of Niels Bohr*, Science Tech Publishers, Madison and Springer-Verlag, Berlin 1988, p. 215.

[22] The study of Bohr's wartime political activities was pioneered by the British historian Margaret Gowing in her official history of the British atomic energy project: M. Gowing, *Britain and Atomic Energy 1939–1945*, Macmillan, London 1964. Gowing was able to interview Bohr, who plays a major role in her book, extensively for her project and had full access to the BPP. As government historian she also had access to official papers which were otherwise closed to researchers at the time. Even after the British government papers have entered the public domain, Gowing's work constitutes an invaluable guide for identifying, locating and interpreting them. After the death of Niels Bohr, Aage Bohr published a brief, yet informative, article on his father's wartime political activities in a memorial volume for Bohr: A. Bohr, *The War Years and the Prospects Raised by the Atomic Weapons* in *Niels Bohr: His life and work as seen by his friends and colleagues* (ed. S. Rozental), North–Holland, Amsterdam 1967, pp. 191–214. The present editor's first publication on Bohr's politics, stemming from a conference in Berkeley in January 1998 on "Physicists in the Postwar Political Arena", likewise concentrates on his wartime activities: F. Aaserud, *The scientist and the statesmen: Niels Bohr's political crusade during World War II*, Historical Studies in the Physical and Biological Sciences **30** (1, 1999), 1–47. The present section constitutes a further development of this paper.

THE CAMPAIGN BEGINS

Bohr arrived by bomber in Scotland on 6 October 1943 and was brought to London the following day. As had been arranged before his departure from Stockholm, he was joined on 15 October by his son, Aage, who was 21 at the time and in the middle of his study of physics at the University of Copenhagen. In his work both inside and outside physics Niels was dependent on a "helper" to clarify his thoughts and put them down in writing. Aage was going to play a particularly crucial role in this regard in the course of the travels with his father.

In London Bohr was briefed immediately about the present stage of the development of the atomic bomb by Chadwick, who met him at the airport. He also soon met Sir John Anderson, who from 1941 had been given responsibility at ministerial level for the British atomic bomb project, code-named "Tube Alloys". The British Prime Minister Winston Churchill subsequently appointed Anderson Chancellor of the Exchequer. A chemist by education, Anderson had, as a young man at the beginning of the century, spent a year doing research on uranium in the thoroughly international research environment at the University of Leipzig.[23] Not long before Bohr's arrival, he had helped to prepare the ground for the Quebec Agreement, which resolved the tangled issue of British–American cooperation in the development of the atomic bomb. Prior to the agreement, each side had been suspicious of the other's motivation. Thus, some people in the American leadership were wary that the British involved themselves in the atomic bomb project not only in order to win the war, but also with a view to the prospect of postwar economic advantage.[24] The agreement, signed by Churchill and Roosevelt on 19 August 1943, played a crucial role in alleviating the suspicion. Anderson found it difficult to negotiate with the leaders of the American organization responsible for the technical aspects of the "Manhattan Project" to develop the bomb, Vannevar Bush and James Conant. Anderson's experience in this regard was going to have an impact on Bohr's subsequent activities.

Bohr seems to have concluded very soon after the briefings that not only was the powerful weapon likely to be developed during the war; it might also, precisely because of its power, provide a vehicle to secure international understanding after the war. A few months later he started to write down his thoughts on these matters.

Both the British and the Americans were eager to enroll Bohr for the atomic

---

[23] See J.W. Wheeler–Bennett, *John Anderson, Viscount Waverley*, Macmillan, London 1962.
[24] Gowing, *Britain and Atomic Energy*, ref. 22, p. 167.

bomb project. The formal affiliation was arranged with the British Directorate of Tube Alloys, which was under Anderson's jurisdiction. Bohr became "Scientific Adviser", his son Aage "Junior Scientific Officer." The employment was given on the understanding that Bohr would have full access to the American project and would take part in the work in Los Alamos, New Mexico, where the Americans had recently built a huge research facility – known to insiders as "Y" – to build the bomb.

Before he left for the United States at the end of November, Bohr had several meetings with Anderson, who took the Danish scientist into his confidence about both his anxieties as regards the relationship with the Americans and his concern about the general postwar implications of the atomic bomb. Anderson gave Bohr an introduction to Lord Halifax, who as British ambassador in Washington was the highest British official in the United States. The direct link to Halifax and, through him, back to Anderson, was a strong encouragement to Bohr and would prove crucial for the promotion of his political ideas in the months to come.

Bohr met with Halifax on 21 December 1943, fifteen days after he had arrived in the United States. In a follow-up letter Bohr did not go into the substance of their discussion, but limited his remarks to a justification of his intervention in political matters:

Bohr to Halifax,
28 Dec 43
English
Full text on p. [231]

> "The unique hopes and peculiar dangers involved in this project are, of course, fully appreciated by the responsible statesmen, but nevertheless, it must be the duty of a scientist, to whom opportunity is given to follow the gradual development of the technical aspects of the project, to call to their attention any detail of this development which may have a bearing on the realization of such hopes and on the elimination of such dangers."

In his response two days later, Halifax only acknowledged receipt of Bohr's letter.

After spending Christmas with friends in New York, Bohr joined General Leslie Groves, the head of the Manhattan Project, on the train to Los Alamos, arriving there for the first time on New Years Eve 1943. The very next day a meeting was organized debating the status of the German bomb project on the basis of Bohr's experience. Bohr's first visit to Los Alamos lasted about three weeks.

At Los Alamos Bohr found himself among a great number of old acquaintances, many of whom had worked with him in Copenhagen. He was treated as a member of a family and affectionately called "Uncle Nick" (on the basis of his official codename, Nicholas Baker).

Before a decisive test of the completed weapon could be accomplished, scientists had to rely on a theoretical analysis of each of the several steps in its operation. Bohr's main task at Los Alamos was to form a judgement of the overall functioning of the weapon under construction. His days were hence filled with discussions with staff members engaged in the various phases of the work, ranging from analyses of the nuclear process, through studies of behaviour of materials under extreme conditions, to problems associated with the ignition of the device. At the same time, Bohr made contributions to scientific and technical problems as they arose.

Bohr served as adviser to Groves personally and was a close friend and colleague of J. Robert Oppenheimer, the scientific director at "Y", with whom he discussed all aspects of the work and the operation of the laboratory. In the words of Oppenheimer, Bohr served "as a scientific father confessor to the younger men."[25] One of these "younger men" – Victor Weisskopf, who had already spent substantial time with Bohr in Copenhagen – recalled that Bohr "inspired many of us engaged in the work of war to think about the future and to prepare our minds for the task of peace that lay ahead."[26] However, Bohr felt obliged not to reveal his own confidential contacts with the statesmen.

Upon his return from Los Alamos, Bohr's political views matured further, and he discussed them on several occasions with Lord Halifax. By mid-February he felt ready to share his ideas with Anderson in a letter that constitutes his first attempt at a written presentation of his political outlook based on the prospects of an atomic bomb.

His main ideas, which he would pursue for the rest of the war, were already in place:

> "I know that the question of control has already been considered within your committee,[27] but the more I have learned and thought about the possible development in this new field of science and technique, the more I am convinced that no kind of customary measures will suffice for this purpose and that no real safety can be achieved without a universal agreement based on mutual confidence."

**Letter to Anderson,
16 Feb 44
English
Full text on p. [86]**

Such an agreement, Bohr emphasized, would require complete "openness about

---

[25] R.G. Hewlett and O.E. Anderson, Jr., *The New World: A History of the United States Atomic Energy Commission*, Vol. 1, 1939–1946, Pennsylvania University Press, University Park 1949, p. 310.

[26] V. Weisskopf, *The Joy of Insight: Passions of a Physicist*, Basic Books, New York 1991, p. 144. This is one of several physicists' accounts of their experiences in Los Alamos during the war, including their encounters with Niels Bohr.

[27] The Tube Alloys Consultative Council, John Anderson's policy-making body.

industrial efforts including military applications,'' an openness that could only be achieved by "a compensating guarantee of common security against dangers of unprecedented acuteness." The achievement of such an agreement was so crucial to Bohr that he considered that "it might even be necessary temporarily to renounce the great promises of the project as regards peaceful industrial development" of nuclear energy until "an established control will prevent any illegitimate use of the powerful new materials." Although the details of an agreement still needed to be worked out, Bohr wrote,

> "… the main point of the argument is that the impending realization of the project would not only seem to necessitate, but should also, due to the urgency of confidence, facilitate, a new approach to the problem of international relationship."

Bohr would repeat this point again and again.

As Bohr saw it, a change for the better would require "an early initiative from the side which by good fortune has obtained a lead in the efforts to master forces of nature hitherto beyond human reach." Bohr stressed that the initiative "should in no way impede the importance of the project for immediate military objectives." The aim was "preventing a future competition … by strengthening the confidence within the United Nations [to] turn this unparalleled enterprise to lasting benefit for the common cause."[28] Although he did not yet say so explicitly, the "early initiative" called for Great Britain and the United States to enter into a dialogue with the other United Nations not involved in the development of the atomic bomb, in particular the Soviet Union, about future control measures.

"I am well aware," Bohr wrote, "that I here enter upon questions about which only the responsible statesmen, who alone have insight in the political possibilities, are able to judge." He appreciated "the necessity of separating sharply between the actual work on the project and its possible implications of general political character," recognizing the "serious complications for handling the whole situation" that "might arise from discussions of such implications even within the circle of officials and technicians charged with the development of the project."

---

[28] A preliminary United Nations Charter had been adopted after the United States entered the war in December 1941. Signed by 26 nations fighting the Axis Powers – including the "Big Four", Great Britain, the United States, the Soviet Union and China – it was the direct precursor of the United Nations as we know it today.

Bohr's letter was sent by way of Lord Halifax, who in his covering letter related his experience of conversing with Bohr:

> "Dr. B[ohr] has written to you to tell you of certain ideas which have arisen in his mind as the result of his visit here.
>
> He has discussed them several times with me and I myself think that there may be some hope in what he has in mind. These scientists find it very difficult to make their thought precise on political problems, and [Halifax's Deputy and Minister at the Embassy] Ronnie Campbell and I have had to do a lot of work with B[ohr] to get any clear idea of how his thought worked.
>
> But I think we succeeded in doing it fairly well in the end."

Halifax to Anderson,
18 Feb 44
English
Full text on p. [232]

Halifax found the effort worth the trouble:

> "His thought is, I think, that, as long as no competition has begun (and the situation in this respect may change very rapidly after the surrender of Germany), Great Britain and the United States would seem to have in their hands a card which they could use in their negotiations with others of the United Nations for the improvement of the world situation with a view to the assurance of future peace. This idea of his about the *political* treatment of the project seems sensible, if it can be translated into a practical course of action."

In spite of his favourable view of the substance of Bohr's thoughts, the Ambassador evidently felt that the Danish physicist might not appreciate his forthright comments on the difficulty of communicating with him, for he added as a postscript: "You won't, I hope, ever show B[ohr] this letter!"

Bohr's and Halifax's letters dealt mainly with the substance of Bohr's ideas and not with the equally important question of the appropriate channel for putting them into action, a question that the two must have pondered from their very first meeting. It was clear to Halifax that since the U.S. contribution to developing the bomb was far greater than the British, the political initiative proposed by Bohr ought to come from the Americans. Since the British did not want to be seen by the Americans as standing behind Bohr's views, Anderson and Halifax agreed that the best approach would be for Bohr to find a personal contact who could bring his ideas to the appropriate U.S. officials. They also agreed that Bohr's case constituted high policy and ought to be dealt with by President Roosevelt himself.

[19]

Bohr sent off his letter to Anderson the day after he succeeded in establishing a promising contact to the President. The contact was Felix Frankfurter, whom Bohr had met in London and Washington before the war, when the two discussed their deeply shared concern about international relations and the refugee problem. Frankfurter had been a Justice of the U.S. Supreme Court since 1939 and was a trusted personal adviser to Roosevelt.[29] He was not cleared for the highly secret atomic bomb project, so when Bohr and Frankfurter met again at the Danish Embassy in December 1943, the topic was not touched upon. However, in a subsequent private conversation, which took place at the Supreme Court on 15 February 1944, Frankfurter indicated that he knew about the existence of the project and was concerned about its implications. Then, as Bohr recalled it in an account of his relationship with Frankfurter written the following year:

Bohr on Frankfurter,
6 May 45
English
Full text on p. [291]

"... without entering upon matters of military secrecy B[ohr] ... told F[rankfurter] that in his opinion the accomplishment of the project should offer a unique opportunity for furthering a future harmonious co-operation between nations. On hearing this F[rankfurter] said that, knowing President Roosevelt, he was confident that the President would be very responsive to such ideas as B[ohr] outlined."

Aage Bohr remembers vividly his father telling him about Frankfurter's last words at their meeting: "Let's hope that this will be a memorable day."

Bohr now considered whether it would be advisable to visit England in order to present the latest developments to Anderson personally. For the moment, however, Halifax suggested in the covering letter to Anderson that his Deputy Ronald Campbell

"... come over to tell you what we have got out of B[ohr] as regards his thought on this political side, in greater detail than it is probably either good or possible to put in a letter. For I do believe that B[ohr]'s ideas call for very urgent and deep consideration by the Prime Minister and yourself."

Campbell's briefing seems to have been effective, for shortly thereafter Anderson presented Churchill in writing with the very problems on Bohr's mind. In a richly argued memorandum Anderson wrote that

---

[29] See *Roosevelt and Frankfurter: Their Correspondence 1928–1945* (ed. M. Freedman), Little, Brown and Company, Boston and Toronto 1967.

"... no plans for world organization, which ignore the potentialities of Tube Alloys, can be worth the paper on which they are written. Indeed, it may well be that our thinking on these matters must now be on an entirely new plane. ... one or two Americans close to the White House, who have some knowledge of Tube Alloys, are becoming increasingly concerned about the future and are thinking of urging the President to give the problem of international control his serious and urgent attention."[30]

Anderson to Churchill,
21 Mar 44
English
Full text on p. [233]

The Chancellor drew the conclusion that Roosevelt might soon want to talk to the British Prime Minister about these matters, in which case, Anderson stressed, "it is essential that we should know what we want." One point to be settled with Roosevelt would be

"... whether, and if so when, we should jointly say anything to the Russians. If we jointly decide to work for international control, there is much to be said for communicating to the Russians in the near future the bare fact that we expect, by a given date, to have this devastating weapon; and for inviting them to collaborate with us in preparing a scheme for international control."

When reading this part of the document, Churchill circled the word "collaborate," writing in the margin, "On no account." Anderson's memo continues:

"At the same time, there would seem to be little risk of the Russians, if they chose to be unco-operative, being assisted in the development of their own plans by a communication of the kind suggested."

Here Churchill jotted "No" in the margin.

Meanwhile Frankfurter raised Bohr's arguments with Roosevelt. As he wrote in an account of his relationship with Bohr:

"On this particular occasion I was with the President for about an hour and a half and practically all of it was consumed by this subject. He told me the whole thing 'worried him to death' (I remember the phrase vividly), and he was very eager for all the help he could have in dealing with the problem."

Frankfurter on Bohr,
18 Apr 45
English
Full text on p. [281]

---

[30] The reference to a postwar "world organization" should be seen in the context of the continuing plans to develop the United Nations, as evidenced by the Moscow Conference of the "Big Four" in October 1943 and, subsequently, the high-level talks in Dumbarton Oaks outside Washington, D.C., a year later.

In his account of his relationship with Frankfurter, Bohr recalls that his friend told him about Roosevelt's reaction at the end of March, after Bohr had just returned from another visit to Los Alamos. Roosevelt had authorized Frankfurter to convey with "proper discretion" to Churchill that "it was a matter for Prime Minister Churchill and himself to find the best ways of handling the project to the benefit of all mankind, [and] that he would welcome any suggestion to this purpose from the Prime Minister." Frankfurter proceeded to write out a "formula" to convey to the British. The original document, which Aage Bohr recalls as handwritten, seems to be lost. However, Bohr prepared a typewritten document addressed to Anderson which, in addition to a brief explanation of the background, contains the message that Bohr was authorized to bring from the American President to the British Prime Minister:

"Message" to Anderson,
Apr 44
English
Full text on p. [89]

"I was, in utmost confidence and under the most urgent pledge not to disclose any personal sources, trusted with the information that the President, quite apart from all connections with his technical advisers, in his own mind is deeply occupied with the immense consequences of the project, and that he realizes he shares the responsibility for the handling of the matter to the greatest possible benefit for the common cause solely with the Prime Minister; further I was assured that the President would not only welcome, but is eager from the Prime Minister to receive any suggestions for dealing with this problem, so fraught with danger, but also with great hope for mankind."

Bohr only referred to Frankfurter as the "President's friend", who had been informed about the bomb project "through legitimate sources."

Bohr seems to have been able to relate Frankfurter's news to Halifax on 5 April, a week after his conversation with the former. Bohr and Halifax continued their discussions the following day. Halifax recommended to Anderson that Bohr go to England at the earliest opportunity, to which suggestion Anderson promptly agreed.

At the meetings with Halifax Bohr presented a document on which he had been working for several weeks. The document, dated 2 April 1944, constitutes a substantial expansion of the letter to Anderson from February and is the first extensive written account of Bohr's position.

Memo,
2 Apr 1944
English
Full text on p. [90]

The general structure of Bohr's argument was already in place: first, a survey of the history of the atomic bomb project concentrating on the technical aspects and underscoring "the radical difference between this project and all previous technical efforts"; second, a general presentation of what Bohr perceived as the implications of the project after the war, repeating and building upon formulations already contained in the letter to Anderson from mid-February; and third, arguments for the need to act without delay.

In the third and last section Bohr observed that "the nations united against ag[g]ression will face most serious causes of disagreement due to conflicting attitudes towards social and economic problems". Bohr foresaw that these differences would come to the fore when the war was over. In view of the existence of atomic weapons this was a particularly dangerous development, and Bohr warned that the British and the Americans would not be able to maintain their monopoly in atomic weapons for long. Thus, he wrote that the apparent state of

"... no substantial progress as regards the [atomic bomb] project ... outside the United States of America and the British Commonwealth ... may change as soon as a relief of the strain of war permits other powers like the Soviet Union, where the necessary resources are at disposal, to concentrate on similar efforts."

This prospect strengthened Bohr's call, also expressed in his letter to Anderson, for "an early initiative from the side which by good fortune has achieved a lead in the efforts of mastering mighty forces of nature hitherto beyond human reach."

Bohr then turned to the question of what might be done. Reiterating that "the responsible statesmen alone are in a position to judge the actual political possibilities," Bohr now proposed that they might find "helpful support" in the "world-wide scientific collaborations" developed among physicists before the war. In particular, "personal connections between scientists in confidence of the American and British Governments on the one hand and the Soviet Government on the other might possibly be of help in establishing preliminary and non-committing contact." He concluded:

"Such suggestions imply no underrating of the difficulty and delicacy of the steps to be taken by the Governments in order to obtain an arrangement satisfactory to all concerned, but should only point to some aspects of the situation which may ease such endeavours. Should the efforts be crowned with success, the project will surely have brought about a turning point in history and this wonderful adventure will stand as a symbol of the benefit to mankind, which science can offer when handled in a truly human spirit."

MEETING WITH CHURCHILL

Just before leaving the United States, Bohr had received a message from the Soviet Embassy in London via the Danish Embassy, informing him that a letter from Peter Kapitza, who by then was well settled in the Soviet Union, was awaiting him there. Kapitza had sent the letter to the Soviet Embassy in Stockholm in late October 1943 upon learning that Bohr had escaped to Sweden. Forwarding the letter to their London Embassy, the Russians were evidently informed where Bohr had gone after he had left Stockholm.

Kapitza to Bohr,
28 Oct 43
English
Full text on p. [229]

In his letter Kapitza invited Bohr to come to the Soviet Union, "where everything will be done to give you and your family a shelter and where we now have all the necessary conditions for carrying on scientific work." As for the war, Kapitza considered that "now the worst is well over", reporting that "[w]e scientists have done everything in our power to put our knowledge at the disposal of this war cause." While interpreting Kapitza's invitation as a genuine offer to take part in basic research, Bohr shared the viewpoint of British intelligence that his conversation with the Counsellor at the Soviet Embassy, where he picked up the letter, indicated that the Soviets knew about the efforts in America to develop an atomic bomb and that they might be trying to start a similar project themselves.

Just as he shared Kapitza's letter with British intelligence, so Bohr's response was written in consultation with the appropriate officials, including Anderson. In his answer Bohr was quite non-committal about the prospect of visiting Kapitza in the Soviet Union. Without going further into his present activities,

Bohr to Kapitza,
29 Apr 44
English
Full text on p. [238]

Bohr expressed his satisfaction with "the new promises inspired by the mutual sympathy and respect among the United Nations arisen from the comradeship in the fight for the ideals of freedom and humanity." No doubt, Bohr felt frustrated about stating his goal of international understanding without being able to reveal the means for accomplishing it – the reality of the atomic bomb. Bohr considered Kapitza an obvious contact should his suggestion to inform the Soviet Union about the bomb be accepted.

The Kapitza incident was only a brief interruption for Bohr, whose first priority was to convey the message from the American President to the British Prime Minister. He did report to Anderson immediately upon his arrival on 16 April, showing the Chancellor his typewritten note containing the message, which Anderson welcomed as an important development. He promised to communicate it to Churchill who, however, as Anderson emphasized, was strongly preoccupied with urgent war matters.

Anderson to Churchill,
27 Apr 44
English
Full text on p. [236]

On 27 April Anderson communicated another memo to the Prime Minister in continuation of his approach of 21 March. He urged Churchill "to take the initiative" to discuss with Roosevelt "the question whether arrangements for the control of Tube Alloys can be devised". Anderson enclosed a draft telegram that the British Prime Minister might send to the American President asking "whether any plans for future world security which do not take account of the implications and possibilities of this tremendous development must not be quite unreal." Again Churchill's reaction was totally negative. Anderson did not inform Bohr about his approaches to Churchill. Nor did he bring up the name of Niels Bohr to his Prime Minister.

Churchill to Roosevelt,
draft telegram
27 Apr 44
English
Full text on p. [237]

Meanwhile Bohr was hoping for an early opportunity to meet Churchill. He had high expectations, since not knowing about Anderson's negative experiences he could only imagine that Churchill would respond favourably to Roosevelt's initiative. A conversation with the head of the British Secret Intelligence Service (or MI6), known only as "C",[31] added to Bohr's optimism. "C" had had access to Bohr's documents and appeared impressed. Aage Bohr recalls that during the meeting with Bohr, "C" emphasized how fortunate it was that the country had a Prime Minister with imagination to grasp the far-reaching implications of the atomic bomb project. He had added that he saw Churchill every morning for briefings and offered to bring up the question of a possible meeting with Bohr.

Still not having heard from Anderson about conveying his message to Churchill, Bohr sent a letter to the Chancellor on 9 May restating his mission. He emphasized that his "… purpose was not to urge you to advocate a certain line of policy to the Prime Minister." On the contrary, Bohr explained,

Bohr to Anderson,
9 May 44
English
Full text on p. [239]

"The aim of my delicate mission … was to convey to the Prime Minister the information that the President does not only regard future policy concerning the project as a matter for Mr. Churchill and himself but that he, without discussing the greater issues with his technical advisers, in his own mind is deeply concerned with the unique hope the situation inspires, and eager to receive any suggestion to this effect from Mr. Churchill."

Bohr's task was to encourage Churchill to consider,

"… in his own mind … the best way of turning to the benefit of England and of all mankind the opportunity which thus offers itself.
    I cannot help feeling that the only way in which I can carry out my mission is if I can be given the opportunity to see Mr. Churchill."

The Chancellor agreed to Bohr's bringing up the matter with the President of the Royal Society, the physiologist Sir Henry Dale who, as a member of Anderson's Consultative Council, was privy to Tube Alloys, a circumstance

---

[31] After MI6's founder, Sir Mansfield Cumming. The person known to Bohr as "C" was Stewart Menzies.

Dale to Churchill,
11 May 44
English
Full text on p. [241]

which allowed Bohr to talk to him in detail about his mission. Dale wrote a long letter to Churchill, "on behalf of the scientific community", pleading that Bohr be given "the opportunity of imparting to you a message which he has been charged to deliver to you in person, by a most intimate personal adviser of President Roosevelt." In support of his appeal, Dale noted that a vote among the world's scientists would most likely place Bohr "first among all the men of all countries who are now active in any department of science."

South African Prime Minister Jan Christian Smuts, who with his much-read book, "Holism and Evolution",[32] had contributed to the discussion of the philosophical implications of natural science, was quite taken by Bohr's thinking. When Anderson's wife offered Smuts the opportunity, probably in the autumn of 1943, to meet the Danish scientist, he is reported to have exclaimed: "This is tremendous, as though one was meeting Shakespeare or Napoleon – someone who is changing the history of the world."[33] Most likely, Smuts also had a chance to talk to Bohr at a dinner in the Royal Society in November 1943 to which Bohr had been invited to speak about international scientific relations. Churchill had high regard for Smuts, whom he had known at least since World War I and whom he had made an honorary Field Marshal in the British Army in September 1941. Nevertheless, Smuts had not yet been informed officially about the bomb project and must have recommended Bohr to Churchill simply as a person worth talking to.

A meeting between Bohr and Churchill took place on 16 May 1944. In view of Churchill's preoccupation with preparing D-Day, as well as his attitude as expressed in his reactions to Anderson's earlier approaches, it is hardly surprising that it was not successful. Churchill's personal science adviser Lord Cherwell was present, and as most of the time was taken up by a discussion between Churchill and Cherwell about British–American relations with regard to Tube Alloys, Bohr had little opportunity to convey his message. Asked directly by Churchill about the identity of the "President's friend", Bohr felt obliged to give Frankfurter's name. When Bohr asked Churchill after the meeting whether he could send him a letter, "the Prime Minister replied that he would always be honoured to receive a letter from Professor Bohr but hoped that it would not be about politics."[34]

One can only speculate to what extent the meeting would have been more successful had Bohr been informed about Churchill's negative attitude toward his views, for which he was not prepared. It also appears that the letter from

---

[32] Macmillan, London 1926.
[33] Wheeler–Bennett, *John Anderson*, ref. 23, p. 297.
[34] Gowing, *Britain and Atomic Energy*, ref. 22, p. 355.

Dale, which was handed to Churchill shortly before the meeting, may have complicated the situation. Aage Bohr recalls that Anderson told his father that he regretted not having asked Cherwell to consider arranging the meeting on the basis of Smuts's recommendation alone and to refrain from handing Churchill Dale's letter, which the Prime Minister may have found intrusive and which furthermore contained points that it might have been better for Bohr to have presented himself.

In the face of what must have been a great disappointment, Bohr did not waver. He immediately wrote to Churchill. His letter began by explaining the physical basis of the atomic bomb project and his own experience with the work, praising in particular the "enthusiastic co-operation between the British and American colleagues." He pointed out "that the enterprise, immense as it is, has still proved to demand a smaller effort than might have been anticipated." This circumstance made an early arms race more likely than ever. In Bohr's words, the limited effort required to develop an atomic bomb would "obviously have an important bearing on the question of an eventual competition about the formidable weapon, and on the question of establishing an effective control."

Letter to Churchill,
22 May 44
English
Full text on p. [96]

Only in the last two paragraphs did Bohr refer to their recent meeting, explaining that it had been far from his intention "to venture any comment about the way in which the great joint enterprise has been so happily arranged by the statesmen." On the contrary, Bohr continued, "I wished rather to give expression to the profound conviction I have met everywhere on my journey that the hope for the future lies above all in the most brotherly friendship between the British Commonwealth and the United States." He emphasized that it was "in this spirit of co-operation" that "the President's friend" had entrusted Bohr to convey to Churchill

> "… that the President is deeply occupied in his own mind with the stupendous consequences of the project, in which he sees grave dangers, but also unique opportunities, and that he hopes together with you to find ways of handling the situation to the greatest benefit of all mankind."

Bohr sent the letter to Churchill fully aware of the difficulties involved in getting his thoughts through. It did not help the matter that when Cherwell handed the letter to Churchill, he attached a note stating that it contained hardly anything that was new and that Bohr's most important point was that he did not want Frankfurter's name to get back to America. On the other hand, Aage Bohr recalls that when his father broached the possibility to Anderson that Churchill might not read the letter, Anderson responded, "we will see to it that he does."

During his stay in England, Bohr was able to devote considerable effort to clarifying and elaborating the presentation of his views as set forth in the memorandum of 2 April 1944. The resulting notes (which because of their similarity to the April memo are not reproduced in this volume) were shown to and discussed with people he could confide in.

In a personal note to the Chancellor, possibly written at Anderson's request, Bohr went a step further in commenting on a procedural issue, namely how the question of far-reaching control measures might be taken up in negotiations among the great powers:

Memo,
23 May 44
English
Full text on p. [99]

"The question will surely be raised what kind of arrangement is actually suggested, but I think that it must first be asked what the situation would be if an arrangement about the project, however difficult, cannot be obtained. In particular, it seems to me that unless the question of the concessions which the various nations would be prepared to make was brought up in the negotiations among the great powers, it would never be possible for any one of them to know what their own interests really are and with what intentions from the side of the others they will have to reckon."

In his memo of 2 April, and again in a memo of 19 May not reproduced here, Bohr stated that "it is practically certain that no substantial progress has been achieved by the Axis Powers." Now he considered that even limited progress on the part of the Germans might enable them to assist the Soviets after the war if the relationship of trust between the former Allies was not maintained:

"Not only may preparations for a competition be under way in U.S.S.R., and if matters are left to themselves until the war against Germany and Japan is finished the great opportunity may be lost, but it cannot even be excluded that in Germany, especially if information of the success of the Anglo–American enterprise has leaked out, some initial progress may recently have been made which may bring them into a position to offer substantial help to their present enemies if they could persuade them of the insincerity of intentions of others among the United Nations."

Bohr had received information himself relating to the activities of physicists in Germany. The information was contained in letters from his wife, Margrethe. Margrethe Bohr often met the Austrian physicist Lise Meitner, a close colleague of Bohr as well as a friend of the Bohr family, who in the summer of 1938 had found refuge in Stockholm upon escaping from Berlin.

[28]

Up to the occupation of Denmark, she made extensive visits to Copenhagen, working at the institute and staying at Bohr's home at the Carlsberg Mansion. Margrethe Bohr now related to her husband that Meitner had told her about a letter she had recently received from her former collaborator in Berlin, Otto Hahn, reporting that he and other physicists were moving south. Meitner had expressed anxiety that "the Germans nevertheless did something with uranium, for Heisenberg and his collaborators were eagerly working at an institute in southern Germany."[35]

In a telegram of 21 May 1944 Bohr asked his wife for further information about "where Heisenberg is working, who his collaborators are, why she [Meitner] thinks he is working on Uranium and if so what evidence she has to suggest that there is any prospect of success in these efforts."[36] Upon taking up the matter with Meitner once more, Margrethe Bohr cabled back. She wrote that Meitner's anxiety was not only based on the move of the Berlin physicists to a place in the south, which her correspondents refused to identify, but also on information from an unnamed man, who had visited Stockholm recently, that large amounts of metallic uranium were being produced in Germany.[37]

Bohr also sought to find out what was happening in Germany through a different channel. He thus sent a message to his colleague Christian Møller at the institute in Copenhagen. Bohr reminded Møller that he had been worried about the possibility that the Germans might "utilize nuclear energy as a weapon of war ... ever since Heisenberg's visit to Copenhagen in the autumn of 1941." "For a long time we felt reassured", Bohr continued, "but lately rumours about German activity in this area have again come to the fore". He stressed that "any further information might therefore be significant." Møller responded on 29 September 1944 that Heisenberg and his younger colleague Carl Friedrich von Weizsäcker, who had been with Heisenberg in Copenhagen in 1941, "do not seem to have worked with the question during the last year". At the end of November, Eric Welsh of the British Secret Service, through whom Bohr's messages to and from Scandinivia were communicated, would seriously question this judgement,[38] whereupon Bohr reported to Welsh that a recent letter to Meitner from another German colleague, Max von Laue, might

Bohr to Møller,
May 44
Danish
Full text on p. [243]
Translation on p. [244]

Møller to Bohr,
29 Sep 44
Danish
Full text on p. [264]
Translation on p. [264]

[35] Margrethe to Niels Bohr, 2 and 10 May 1944, Niels Bohr–Margrethe Bohr correspondence, ref. 16. The quotation is from the second letter.

[36] Niels to Margrethe Bohr, 21 May 1944, BPP 3.3.

[37] The telegram to Bohr seems to have been lost, but its contents are repeated in Margrethe to Niels Bohr, 5 June 1944, Niels Bohr–Margrethe Bohr correspondence, ref. 16.

[38] Welsh to Bohr, 28 November 1944, BPP 3.2.

constitute further indication that the Germans were seeking to build an atomic bomb.[39]

Before his return to the United States in mid-June 1944 Bohr brought up the question of whether there was a need to change his formal status in order to maintain his link to the British organization in the uncertain situation after Churchill's negative response. Anderson did not consider a change of status necessary:

Anderson to Campbell,
2 Jun 44
English
Full text on p. [244]

"In my view the most satisfactory arrangement would be for Bohr to continue in the somewhat anomalous and undefined position which he has occupied during the last few months – that is to say that he should continue to be given access to the various sections of the work in the U.S.A., and to be allowed to talk with both the American and British scientists engaged in the work; and that his personal advice should be available to the U.S. authorities and to myself alike."

Anderson wrote to Groves in the same vein. Bohr, who had continued to ponder the matter before and during his voyage to the United States, was reassured on his arrival to learn about Anderson's views and to be handed the letter to Groves, which he was advised to deliver personally. It would turn out that Groves agreed with Anderson that there was no need to change Bohr's status.

MEETING WITH ROOSEVELT

Any anxiety as to how the news about the unsuccessful meeting with Churchill would be received in the U.S. was soon dispelled. Immediately upon his arrival in Washington, Bohr reported to Frankfurter and Halifax. As Bohr subsequently wrote to Anderson, Frankfurter, "learning about the reception of the message in London, had given the President a full account of the matter, which the President not only heartily approved of, but also welcomed as a fortunate development." On 25 June Frankfurter told Bohr that Roosevelt had expressed his wish to see Bohr and suggested that Bohr work out a memorandum for him to read as an introduction to their meeting. The opportunity to present his views in a document to the American President personally was, of course, a great challenge to Bohr.

Bohr to Anderson,
13 Jul 44
English
Full text on p. [250]

On 5 July Bohr sent the result of his efforts to Frankfurter. His covering letter shows how he liked to try out his writing on others:

[39] Bohr to Welsh, 9 December 1944, BPP 3.2, based on information contained in Margrethe to Niels Bohr, 11 November 1944, Niels Bohr–Margrethe Bohr correspondence, ref. 16.

"As you will see, the points are the same as in my previous notes, but I have tried throughout to stress international scientific co-operation as the most natural apology for my interest in the matter. I have omitted some of the more sentimental passages, but left a sentence of such character at the very end. For the case you should think that also this passage should be omitted, I am sending a version of the last page without it."

Bohr to Frankfurter,
5 Jul 44
English
Full text on p. [247]

The passage in question is practically identical to the last sentence in the memo of 2 April. It was retained in the version forwarded to Roosevelt.[40]

A day later Bohr wrote to Frankfurter again, expressing great unease:

"Since I sent the memorandum to you I have had serious anxieties that it may not correspond to your expectations and perhaps not at all be well suited for the purpose. When writing it I felt most strongly that a statement on a matter like this which presents so many sentimental and practical aspects is very difficult unless one knows exactly the background on which it will be received."

Bohr to Frankfurter,
6 Jul 44
English
Full text on p. [248]

He left the final judgement to Frankfurter:

"... I am not only ready to change the form and content of the memorandum entirely, in accordance with any criticism and suggestion, but even prepared that you may find it better to wait with any written statement until the situation has further developed and I perhaps through a personal interview have learned on what points an opinion from me might be requested."

Frankfurter answered reassuringly:

"The memorandum, you must let me say, is just right – precisely what it should be for its purpose. It has gone on its way, with an appropriate covering note, including information of your continuing stay in Washington.
  I cannot forego saying that you are conducting matters of the deepest concern for mankind with a delicacy and wisdom worthy of the enterprise."

Frankfurter to Bohr,
10 Jul 44
English
Full text on p. [248]

The Memorandum constitutes a particularly well-developed expression of Bohr's views. It is structured in the same way as his first memo of 2 April. The first part on the scientific background and technical developments made an

---

[40] Bohr's Memorandum, as reproduced in the published correspondence between Roosevelt and Frankfurter, *Roosevelt and Frankfurter*, ref. 29, pp. 728–735, includes the sentence.

impressive popular introduction to the field. In the second half of the document, four passages are taken from Bohr's letter to Churchill and five more from the April memo.

As for substance, the prospect of a postwar nuclear arms race loomed larger than before, and Bohr gave his most detailed account so far of possible German and Soviet developments. As long as he stayed in occupied Denmark, he wrote, it had been "possible, due to connections originating from regular visits of German physicists to the Institute for Theoretical Physics in Copenhagen in the years between the wars, rather closely to follow the work on such lines which from the very beginning of the war was organized by the German Government." In language that Frankfurter in his covering letter to Roosevelt described as "his able but quite ancient English" Bohr wrote that in spite of a mobilization

"Memorandum",
3 Jul 44
English
Full text on p. [101]

Frankfurter to Roosevelt,
10 Jul 44
English
Full text on p. [249]

> "... of expert knowledge and considerable material resources, it appeared from all information available to us, that at any rate in the initial for Germany so favourable stages of the war, it was never by the Government deemed worth while to attempt the immense and hazardous technical enterprise which an accomplishment of the project would require."

Recently, however, Bohr had "had the opportunity with American and British Intelligence Officers to discuss the latest information, pointing to a feverish German activity on nuclear problems." Bohr suggested that this might be owing to "knowledge of the progress of the work in America." Likewise, Bohr referred to his correspondence with Kapitza and his conversation with "the Counsellor of the Soviet Embassy", who had given Bohr "the impression that the Soviet Officials were very interested in the effort in America about the success of which some rumours may have reached the Soviet Union."

In the second half of the Memorandum Bohr repeated his statement to Churchill that the effort to develop a bomb had "proved far smaller than might have been anticipated". He took the formulation of his main point, which followed from this observation, almost word by word from his first letter to Anderson of 16 February:

> "The prevention of a competition prepared in secrecy will therefore demand such concessions regarding exchange of information and openness about industrial efforts including military preparations as would hardly be conceivable unless at the same time all partners were assured of a compensating guarantee of common security against dangers of unprecedented acuteness.

The establishment of effective control measures will, of course, involve intricate technical and administrative problems, but the main point of the argument is that the accomplishment of the project would not only seem to necessitate but should also, due to the urgency of mutual confidence, facilitate a new approach to the problem of international relationship."

In the final letter in the exchange with Frankfurter about the Memorandum, Bohr expressed his gratitude that the document had now been forwarded and informed the "President's friend" that "I have just sent a copy of it to Anderson to whom I have confidentially conveyed such information as you advised me." In his letter to Anderson, already quoted, Bohr explained, as he had to Frankfurter, that "the repeated allusions to international scientific co-operation are, as you will understand, primarily meant as an apology for the interest of a scientist in such matters." As for the reference in the Memorandum to his contact with Kapitza, Bohr reiterated that this correspondence "has so far not been mentioned to any other American authority." This confirms Bohr's and Anderson's shared view that the matters in question should be discussed only at the very highest level.

Bohr was not able to see the President immediately, as he had to leave Washington for Los Alamos on 15 July. Yet the prospects for interesting the American leadership in his views had improved dramatically since he returned from England. The situation had improved considerably in Great Britain as well, as Bohr learned from Anderson's response to his letter of 13 July. Bohr was still in Los Alamos when he was notified that the letter awaited him in Washington. He sent his son to Washington to pick it up. Aage considered the letter so important that he took the train to Boston to show it to Frankfurter.

Anderson reported the welcome news that "the Prime Minister now fully recognizes that it would be appropriate that he should raise the long-term problem personally with the President when they next meet, which I personally hope will be before very long." Anderson also commended Bohr for his Memorandum, "which seems to me, if I may say so, to present the whole matter in very clear and compact form suitable for the enlightenment of a non-scientific reader."

Anderson had reopened the matter with Churchill in a note written just after Bohr had departed from England. On the same day Smuts too wrote to the Prime Minister on the same subject. Smuts had been taken into confidence about the atomic bomb project soon after the meeting between Bohr and Churchill, which allowed Bohr to discuss his hopes and concerns with him. Smuts had promised to bring the matter up with Churchill.

Bohr to Frankfurter,
14 Jul 44
English
Full text on p. [251]

Bohr to Anderson,
13 Jul 44
English
Full text on p. [250]

Anderson to Bohr,
21 Jul 44
English
Full text on p. [252]

[33]

Smuts to Churchill,
15 Jun 44
English
Full text on p. [245]

After noting his discussions "with the scientific expert who saw you also," i.e., Bohr, Smuts wrote:

"While it may be wise to keep the secret to ourselves for the moment, it will not long remain a secret, and its disclosure after the war may start the most destructive competition in the world. It would therefore be advisable for you and the President once more to consider this matter, and especially the question whether Stalin should be taken into the secret."

Churchill responded to Anderson on 16 July, agreeing that the matter ought to be discussed between himself and the American President. This was the basis for the welcome news that Anderson submitted to Bohr five days later. Bohr had thus not only been able to reach the American President; his mission in England also seemed to have had its intended effect.

On 25 August, a week after Bohr had returned from his stay at Los Alamos, Roosevelt invited him for a conversation the following day. Before the interview Bohr had the opportunity to see Halifax and Campbell, going through the "considerations" that Bohr was going to present "to his highly-placed friend." The discussion with Halifax and Campbell covered several issues, including the consequences of a negative response by the Soviet Union, that seem not to have been taken up in the conversation with Roosevelt. However, the issues would reappear in several of Bohr's subsequent memos.

Campbell to Anderson,
25 Aug 44
English
Full text on p. [253]

For security reasons, Bohr entered through a back door of the White House and the interview was held off the record. Nevertheless, the substance of the conversation can be reconstructed to some extent from the existing sources.[41]

The President was most friendly and was very open in his discussion of the political problems raised by the bomb. He had read the memorandum and "gave NB opportunity to explain the views he had stated". Halifax reported to Anderson immediately after the interview: "As B[ohr] understood[,] these views were fully appreciated. Above all, the friend [the President] agreed that nothing could be lost but very much gained by an early approach to the S[oviet] U[nion]." "After the discussion about the necessity of an understanding and the immense improvement in international relations this would entail", Bohr later recalled, "the President said that he had the best hopes that an understanding with the Russians could be reached" (BP). Indeed, Roosevelt went as far as to

Bohr notes,
undated
English
Full text on p. [288]

Halifax to Anderson,
27 Aug 44
English
Full text on p. [254]

Bohr notes,
3 May 45
English
Full text on p. [290]

[41] These sources are indicated in the margin below at their first quotation. Bohr's personal notes of 3 May 1945, which are quoted several times, are referred to as "BP" in the text. Bohr's notes from the time of the interview itself were destroyed to avoid any criticism of keeping private notes in view of the allegations to be described below.

[34]

state "that an approach to Russia must be tried and that it would open a new era of human history."[42] Bohr was often to quote this impressive sentence in conversation later in life.

Roosevelt described the impression he had received of Stalin from their personal meetings and referred to him as "a realistic thinking and reasonable man" (BP), a statement that Aage Bohr has interpreted to mean that "Stalin was enough of a realist to understand the revolutionary importance of this scientific and technical advance and the consequences it implied."[43] Aage Bohr's account of Roosevelt's statements during the interview continues:[44]

> "He [Roosevelt] mentioned that he had heard how the negotiations with Churchill in London had gone, but added that the latter had often reacted in this way at the first instance. However, Roosevelt said, he and Churchill always managed to reach agreement, and he thought that Churchill would eventually come around to sharing his point of view in the matter. He would discuss the problems with Churchill at their forthcoming meeting and hoped to see my father soon afterwards. Furthermore my father could write to him at any time."

Aage Bohr recalls the content if not the exact wording of his father's first words when they met outside the White House after the interview: "It will be done!"[45] Two weeks later Bohr wrote to Anderson that the interview had been "in every way most satisfactory".

In the course of their conversation Bohr had asked the President "whether he could report to Sir John Anderson" (BP) about the interview, to which the "president answered that he had no secrets at all for Mr. Churchill, and he [the President] next asked (or rather expressed the hope?) that Anderson would attend the meeting" (BP) between Churchill and Roosevelt, which would take place in Quebec three weeks later. When wiring the account of the interview to Anderson the following day, Halifax suggested that the Chancellor invite Bohr along with him to Quebec to discuss matters. However, in his response Anderson reported that a decision had been made that he would not come to Quebec after all. This decision, in fact, owed itself to Roosevelt's unwillingness to discuss financial matters at the conference, which made the participation of

Bohr to Anderson,
12 Sep 44
English
Full text on p. [258]

---

[42] Gowing, *Britain and Atomic Energy*, ref. 22, p. 357. Gowing's observations are based on extensive personal interviews with Bohr.

[43] A. Bohr, *War Years*, ref. 22, p. 207.

[44] *Ibid.*

[45] Personal conversation with Aage Bohr.

[35]

the Chancellor of the Exchequer inappropriate.[46] Instead, Churchill would be accompanied by Cherwell, to whom Anderson had given a full briefing. In this situation, Anderson wrote, "it would we think be dangerous from all points of view for B[ohr] to make any attempt to come to place of meeting." Anderson felt reassured that the British Prime Minister would take the necessary initiative for a discussion with the American President about the postwar implications of the atomic bomb.

Before the meeting in Quebec, Bohr wrote a letter to Roosevelt to thank him for the interview and to reiterate his main arguments, stressing in particular the urgency of the matter: "Any information about the success of the project in America is likely to have caused the utmost intensification of similar projects elsewhere." Moreover, after the defeat of Germany, "expert knowledge and technical experience collected there will presumably become available in equal measure for the great victorious nations." The present moment, Bohr argued, is therefore "most favourable for considerations of the question of control by the friendly governments most concerned lest opportunities be forfeited of forestalling a fateful competition about the new weapon". Bohr commented next on the procedure:

> "Of course, it is wholly outside my province to suggest the procedures appropriate for such delicate problems of statesmanship. But as a scientist it occurs to me that in this unique situation pre-war scientific connections may prove helpful in conveying, with entire regard for security, an understanding of how much would be at stake should the great prospects of atomic physics materialize, and in preparing an adequate realization of the great benefit which would ensue from a whole-hearted co-operation on effective control measures."

Aage Bohr recalls that by emphasizing this point, Bohr was following the advice of Frankfurter that suggestions regarding procedure would be especially useful to the President.

In a subsequent letter to Anderson, Bohr put his letter to Roosevelt in context:

> "The purpose of the letter, of which I enclose a copy, was to remind the President of the main points of the conversation, and the wording of various passages corresponds closely to the way in which the President expressed himself about the political situation."

---

[46] See the appendix to Part I, p. [255], ref. 70.

Anderson to Halifax,
1 Sep 44
English
Full text on p. [256]

Letter to Roosevelt,
7 Sep 44
English
Full text on p. [109]

Bohr to Anderson,
12 Sep 44
English
Full text on p. [258]

The letter to Roosevelt was delivered by Frankfurter on the very evening of the President's departure for Quebec on 8 September. In a covering note Frankfurter summed up Bohr's views: "In a word, the argument is that appropriate candor would risk very little. Withholding, on the other hand, might have grave consequences." Frankfurter gave his full endorsement, stating that "these questions are very serious."

Frankfurter to Roosevelt,
8 Sep 44
English
Full text on p. [257]

MISUNDERSTANDINGS

The meeting between Churchill and Roosevelt went contrary to Bohr's impression from the interview and Anderson's report of Churchill's good intentions. Not only did the two statesmen not discuss the substance of Bohr's views. They even questioned the propriety of his conduct. This occurred after the two men had retired to Roosevelt's private mansion at Hyde Park in upstate New York. In an "Aide-mémoire of Conversation between the President and the Prime Minister at Hyde Park"[47] the two world leaders first stated their shared attitude towards what they seem to have considered Bohr's general idea:

> "The suggestion that the world should be informed regarding tube alloys, with a view to an international agreement regarding its control and use, is not accepted."

Aide-mémoire,
18 Sep 44
English
Full text on p. [259]

The omission of Bohr's central point, namely the use of the atomic bomb project to promote confidence among nations, leaves the impression that it had not been discussed seriously.

After stating in the second paragraph that the bomb might be used against "the Japanese," in the third and last paragraph the aide-mémoire addressed Bohr's conduct:

> "Enquiries should be made regarding the activities of Professor Bohr and steps taken to ensure that he is responsible for no leakage of information particularly to the Russians."

Two days later Churchill expressed his concern in a note to Cherwell. The Prime Minister described Bohr as a "great advocate of publicity", criticizing him for making an "unauthorized disclosure" to Frankfurter and for corresponding with a Russian scientist. He burst out that

Churchill to Cherwell,
20 Sep 44
English
Full text on p. [260]

---

[47] See Gowing, *Britain and Atomic Energy*, ref. 22, p. 447.

"... Bohr ought to be confined or at any rate made to see that he is very near the edge of mortal crimes. I had not visualized any of this before, though I did not like the man when you showed him to me, with his hair all over his head, at Downing Street. Let me have by return your views about this man. I do not like it at all."

Halifax and Campbell protested vigorously,[48] and Cherwell, who was still in the United States after Churchill had returned home, wrote a long note to his superior. He emphasized in particular that the correspondence with Kapitza had taken place in consultation with British authorities and reported that there was "good contemporary evidence at the Embassy here that Frankfurter broached the matter to Bohr (and not Bohr to Frankfurter)". While still finding the substance of Bohr's ideas "rather woolly", Cherwell concluded by clearing the Danish scientist of any suspicion of impropriety:

Cherwell to Churchill,
23 Sep 44
English
Full text on p. [261]

> "I have always found B[ohr] most discreet and conscious of his obligations to England to which he owed a great deal and only the very strongest evidence would induce me to believe that he had done anything improper in this matter."

Cherwell had also had the opportunity to give a similar account to Roosevelt, who had broached the subject in the presence of Vannevar Bush. Bush, as Cherwell reported to Churchill, had in turn confirmed his opinion of Bohr "in all particulars of which he could be expected to have knowledge."

In England Anderson followed the developments by correspondence. He too reacted strongly to the statements about Bohr and his views at Hyde Park. Having read Churchill and Roosevelt's aide-mémoire and Churchill's letter to Cherwell, he wrote at length to the latter. He first sought to clear up the misunderstandings about Bohr's contacts with Kapitza and Frankfurter:

Anderson to Cherwell,
29 Sep 44
English
Full text on p. [262]

> "2. You will remember that correspondence between Baker and Kapitza last April was started by latter, that Baker's reply was vetted by C's people and by myself and that T[ube] A[lloys] is not so much as mentioned in it.
> 3. Halifax and Campbell have full knowledge of what, so far as we are aware, took place between Baker, his contact and his contact's highly placed friend. Accounts I have received do not bear out at all Prime Minister's impression as conveyed in his message to you."

---

[48] *Ibid.*, p. 358.

Anderson went on to point out the severe misrepresentation of Bohr's point of view in the aide-mémoire:

> "4. Beginning of paragraph 1 of aide memoire ... and relevant remarks in Prime Minister's message to you are entirely at variance with what we have always understood to be Baker's ideas, and there is no doubt about the reality of the danger which preoccupies him."

It seems clear that Anderson had not yet seen Cherwell's note to Churchill, as he insisted that "in fairness to Baker, Prime Minister should receive without delay, a correct report of his activities and views so far as they are known to us. ... I hope that you will send one."[49]

Because Roosevelt had expressed his wish to see him again, Bohr postponed his next visit to "Y". As he wrote to Anderson on 12 September, he would stay in Washington in order to "complete a report on my impressions of the technical aspects which General Groves has requested me to write."[50] Bohr, who had no knowledge of what transpired between Churchill and Roosevelt, continued to await Roosevelt's call in Washington after the Hyde Park meeting. While doing so, he stayed in contact with Frankfurter, who also remained uninformed about the outcome of the meeting.

Bohr received a first inkling of what had happened from Chadwick on 6 October. Chadwick, who was head of the British team working on the atomic bomb, had just returned to Washington from London, where he had been briefed about the matter. Chadwick had not previously been informed about the political side of Bohr's efforts. In a note written a few months later Bohr described how Chadwick broke the news to him:[51]

> "... the Chanc[ellor] immediately before his [Chadwick's] departure ... had let him know that complaints about B[ohr]'s talking to unauthorized persons had been expressed during the [Quebec] meeting and that the Chan[cellor] believed it to be a misunderstanding which he was anxious to have completely cleared up.
>
> Chadwick was under the impression that the complaints referred to B[ohr]'s conv[ersation] with the Pres[ident]'s friend".

Bohr report,
30 Dec 44
English
Full text on p. [268]

[49] Anderson did see Cherwell's note eventually, however, as one of the versions of it in the archives has his handwritten annotation "I entirely agree". See p. [262].

[50] This report has not been found, and may never have been completed.

[51] Bohr's note is one of several documents, including his own and Frankfurter's respective accounts of their relationship already quoted (above, pp. [20] and [21], respectively) written in connection with clearing up Churchill's and Roosevelt's misunderstandings regarding Bohr's conduct.

[39]

Thus, Bohr did not learn about the complaints about his contact with Kapitza, possibly because of the obligation Anderson felt to be discreet towards the Prime Minister.

Bohr's immediate reaction is documented in a short note written on the day that he learnt the news:

Bohr memo,
6 Oct 44
English
Full text on p. [265]

"The whole question of a lack of discretion on my part must rest on a misunderstanding from the side of the American organization, which has arisen from my obligation to keep my activities in the delicate matters, reserved for the British and American Governments, completely apart from my participation in the technical work on the project."

Bohr could only imagine that the allegation came from someone among the President's technical advisers who had learnt how Frankfurter had established Bohr's contact with the President. However, the following day Bohr was provided with further information pointing in a different direction. Thus, his notes of 30 December read:

"The next day Chadwick told NB that General Groves in a conversation with him, which he had just had, had told that the complaints about NB's connection with Dr. Frankfurter originated from the President himself."

On learning about the complaints about his contact with Frankfurter, Bohr immediately contacted his friend. Frankfurter reassured him:

"… in the conversation the same evening Frankfurter not only confirmed his previous statements, but assured NB that it must have been a misunderstanding that the President should have uttered such complaints since all their connections had been most scrupulously correct and discrete."

Some days later, after the matter had been cleared up through discussions with American authorities, in particular Bush and Groves, Frankfurter wrote to Bohr:

Frankfurter to Bohr,
14 Oct 44
English
Full text on p. [266]

"Your visit the other day left me with real joy. I was much relieved that needless anxiety was lifted from you. Of course there could not have been any real occasion for concern. Everything you did and said has been done and said with the most fastidious regard for truth and duty. And yet, a disagreeable episode had arisen, and I am very happy all the clouds have cleared away.

You must let me say that you symbolize for me the finest traditions and aspirations of science – science as the pursuit of truth and the true promoter of humanity."

Bohr had yet to be informed about Churchill and Roosevelt's concern about his connections with Kapitza. In his carefully written notes in preparation for a discussion with Anderson in March 1945 Bohr wrote that after the "return to 'Y'", he had "learned from Chadw[ick] who had been travelling with Lord Cherwell, that the question was not a complaint about B[ohr]'s connections with [Frankfurter], but that the main question (issue) had been a blame from the side of the P[rime] M[inister] of B[ohr]'s connections with S[oviet] U[nion]." Thus, Bohr must have learned of Churchill's concern soon after arriving in Los Alamos on 19 October.

Bohr notes,
7 Mar 45
English
Full text on p. [272]

Cherwell's account immediately suggested what might have been the cause of the direction the events had taken. Having not been briefed beforehand about the correspondence with Kapitza, Churchill may have reacted violently when learning from Roosevelt – in conversation or from reading Bohr's Memorandum to the American President – that Bohr was in contact with a Russian colleague. Under the circumstances Roosevelt, who as it appears from Churchill's note to Cherwell came under pressure to explain how Frankfurter had learned about the bomb project, may not have been able to explain the propriety of the Kapitza connection in a satisfactory manner. This is no more than informed guesswork, however, as there are no records of the conversation as it took place in Hyde Park. For his part Anderson, who only a few months before his death was to confide that "he had never been reconciled to the fact that Bohr's counsel had not been followed",[52] may well have regretted that Churchill had not been briefed about Kapitza.

A few days after Bohr had heard about the complaints regarding his discussions with Frankfurter, he "received a message from the Pres[ident] deploring that he, due to a journey, could not receive him, but suggested that he saw Dr. Bush with whom he had talked about the matter discussed during the interview." Thus, Roosevelt suggested a channel that Bohr had avoided owing to his and Anderson's sharp distinction between the political and technical aspects of the bomb project. Notwithstanding his high position as leader of the American organization responsible for utilizing scientific and technological resources for the war effort, with personal accountability to the Chief Executive, Bush's responsibilities were in the technical and administrative, not the

Bohr report,
30 Dec 44
English
Full text on p. [268]

[52] R. Oppenheimer, *Niels Bohr and Atomic Weapons*, The New York Review of Books, 17 December 1964, pp. 6–8, quotation on pp. 6–7.

political, domain. To Bohr and Anderson, the effort to avoid a postwar arms race was not part of the technical task but a political matter at the level of the American President and the British Prime Minister.

The fact that the President himself now encouraged Bohr to speak to Bush changed the situation. A meeting between them took place on or around 11 October. Bohr describes it in his report of 30 December:

> "In the subsequent conversation with Dr. Bush, of which the main topic was the general implications of the project, also B[ohr]'s connection with Dr. F[rankfurter] was discussed, and Dr. Bush expressed great relief in learning that B[ohr] and F[rankfurter] were closely acquainted due to common interest in international cultural co-operation already before the war and that although they had met several times after B[ohr]'s arrival to America, the project had, of course, never been mentioned before F[rankfurter] without being asked told B[ohr] that he was informed about it through legitimate sources and was much concerned about its implications."

While expressing sympathy with Bohr's general views, Bush disagreed on the specifics, as evidenced in Bohr's notes on the meeting prepared for a conversation with Anderson several months later: "[Bush] said, however, that his views differed somewhat as reg[ards] question of timing, [and] [t]hought [that] time [was] not ripe for consultations with Rus[sians]."

Bohr notes,
7 Mar 45
English
Full text on p. [272]

On 17 October Bohr reported to Anderson on his conversation with Bush, as well as on talks with Groves and Chadwick. He concluded: "I believe the misunderstandings have now been cleared up."

Bohr to Anderson,
17 Oct 44
English
Full text on p. [267]

After this intermezzo Bohr spent two months at Los Alamos, where he engaged himself strongly in the new phase the work had entered because of the rapidly increasing availability of active material to be used in a bomb. While there, he pondered how to proceed with the effort to gain understanding for his political views. He felt more acutely than ever the urgency of the matter and the unique opportunity for establishing confidence by means of an early initiative.

Upon his return to Washington in early December Bohr met Halifax and Campbell. They discussed whether to renew the discussions with Bush, the continuation of which Bohr had "not found ... advisable, without directives from the Chancellor". Bohr therefore raised the question of a possible visit to London to discuss matters but found that Anderson did "not think the moment propitious for such a visit." It was arranged for Bohr to send Aage to London for discussions on his father's behalf about practical and technical matters.

Halifax to Anderson,
18 Jan 45
English
Full text on p. [271]

Bohr to Anderson,
18 Jan 45
English
Full text on p. [269]

CONTINUED EFFORTS

During his stay in Washington, Bohr considered what kind of information might be given during preliminary consultations with other nations, as well as how to deal with a negative response. Questions to this effect had been raised in the discussions with Halifax and Campbell just before the interview with Roosevelt. Now Bohr treated them in a new document dated 18 January 1945 entitled "Comments to memorandum of July 3rd 1944."

**"Comments",
18 Jan 45**
English
Full text on p. [111]

The Comments were forwarded via the British Embassy to Anderson along with the letters from Halifax and Bohr of the same date. The letter from Bohr, which reported on the recent progress achieved at Los Alamos, ended with the words:

> "I need not say that I hold myself ready whenever you think that I can in any way be of assistance in the endeavours concerning which you have shown me a confidence which I so deeply appreciate."

Bohr returned to Los Alamos on 20 January, the day before Aage left New York for Montreal en route to London. While in Los Alamos, Bohr made his most important technical contribution during the war, helping to design the "initiator", a component of the bomb which produces a burst of neutrons to set off the explosion when the active material is assembled. Soon, however, Anderson was ready to receive Bohr in London, and after a month in "Y" Bohr was on his way.

He arrived in London in early March and had a long conversation with Anderson a few days later. This conversation is documented in several pages of handwritten notes with the heading "Points for conversation with Chancellor on March 8th 3pm." In these notes Bohr describes in significant detail his activity and experience since their last meeting, including the interview with the President and the subsequent misunderstandings. In addition to shedding unique light on what had transpired since June the year before, the document shows the great care with which Bohr prepared his meetings with Anderson and, by implication, how important he considered it to present all details in a personal conversation. Such presentations were demanding on the listener, and Anderson may well have had this in mind when he wrote to Chadwick a few weeks later: "Discussion with him [Bohr] demands time and patience but is always worth while."

Bohr notes,
7 Mar 45
English
Full text on p. [272]

Anderson to Chadwick,
3 Apr 45
English
Full text on p. [280]

In the ensuing weeks Bohr composed an "Addendum" to the Memorandum as a basis for making another attempt at persuading first of all the American leadership about the advantages to be gained by an early approach to the Soviet Union. In the Addendum Bohr commented on the measures to be taken

"Addendum",
24 Mar 45
English
Full text on p. [113]

"to establish an international control of the manufacture and use of the powerful materials". The task might not be too difficult:

"... the special character of the efforts ... required for the production of the active materials, and the peculiar conditions which govern their use as dangerous explosives, will greatly facilitate such control."

In terms of practical measures, he recommended that

"... a standing expert committee, related to an international security organization, might be charged with keeping account of new scientific and technical developments and with recommending appropriate adjustments of the control measures."

As for "material prepared for armaments," Bohr wrote, it "might ultimately be entrusted to the security organization to be held in readiness for eventual policing purposes."

Bohr now presaged his vision of an "open world" between nations, which after the war would dominate his argument. An international agreement concerning the control of atomic weapons might thus "contribute most favourably towards the settlement of other problems where history and traditions have fostered divergent view-points." In particular,

"... the free access to information, necessary for common security, should have far-reaching effects in removing obstacles barring mutual knowledge about spiritual and material aspects of life in the various countries, without which respect and goodwill between nations can hardly endure."

Bohr also noted the opportunity to "reinforce the intimate bonds which were created in the years before the war between scientists of different nations," and which in the present situation might prove helpful "in connection with the deliberations of the respective governments and the establishment of the control."

In the Addendum Bohr also took up the questions of how an initiative might be presented to other nations and how a possible negative reaction to it could be met. Thus he emphasized that "no information about technical developments of importance for the possible use of the new weapon in the present war, or for the future balance of power, need be exchanged before understanding is reached." If the response was negative, nothing would therefore be lost. On the contrary, "the position of those nations which, with so much foresight, have acquired the lead in the new development would have been greatly strengthened by this proof of sincere intention for an arrangement beneficial to all." However, he considered the prospect for a negative reaction unlikely:

"In all the circumstances it would seem that an understanding could hardly fail to result, when the partners have had a respite for considering the consequences of a refusal to accept the invitation to co-operate, and for convincing themselves of the advantages of an arrangement guaranteeing common security without excluding anyone from participation in the promising utilization of the new sources of material prosperity."

At the end of the Addendum Bohr returned to the question of urgency. First, he emphasized that

"... all such opportunities may, however, be forfeited if an initiative is not taken while the matter can be raised in a spirit of friendly advice. In fact, a postponement ... might ... give the approach the appearance of an attempt at coercion in which no great nations can be expected to acquiesce."

He drew a second point from his Comments of 18 January:

"... it would appear most important that an agreement between the governments be invited before public discussion can arouse sentiments and cause unpredictable complications."

It was in this connection that Bohr now urged "that consultations be initiated sufficiently long before there can be any question of the actual use of the new means of warfare."

The Chancellor agreed that it would be appropriate for Bohr to attempt a renewed approach to Roosevelt on his return to Washington in order to introduce the arguments set forth in the Addendum. However, Bohr was uncertain of how to reach the American President. He thus wrote to Anderson just before leaving for the United States:

"The problem is whether the forwarding through the President's friend of an addendum to the memorandum I prepared last summer might give rise to misunderstandings; or whether it might be preferable this time to approach the President through Dr. Bush.

As you know, the President himself wished me to discuss the matter with Dr. Bush, whose attitude was sympathetic and who actually asked me

Bohr to Anderson, 27 Mar 45 English Full text on p. [278]

to let him have my views in writing so that he could give them closer consideration. It seems clear, therefore, that an approach through him could not be regarded as improper either by the President or by the American organization for the project.

...

Needless to say much will depend on the situation in Washington, where developments may have taken place since my departure; and I would propose, immediately on my arrival, to discuss the whole matter thoroughly with Lord Halifax and the President's friend before decisions are reached."

In a letter to Halifax written just after he had received Bohr's letter, Anderson, as Bohr may well have foreseen, expressed reservations regarding an approach through Bush:

Anderson to Halifax,
28 Mar 45
English
Full text on p. [279]

"This seemed to me dangerous because, apart from the fact that such policy matters seemed to be above Bush's level, he could hardly be expected to keep the thing to himself and if others who are even more security minded got to know of this further activity on B[ohr]'s part, there might be real trouble."

Having informed the ambassador that Bohr had agreed to go no further before consulting Halifax, Anderson added: "I rather think, however, from a remark that he made just before leaving that he would in any case wish to communicate with me again before taking any positive action."

Upon arriving in Washington in early April 1945, Bohr contacted Halifax and Frankfurter to discuss how to proceed. However, the situation changed dramatically on 12 April, when President Roosevelt died. As the church bells announced the event, Halifax and Frankfurter were walking in Lafayette Park just opposite the White House discussing the political questions brought up by Bohr.[53]

Harry S. Truman, the new President, was of course strongly dependent on his advisers. As furthermore Frankfurter was reluctant to contact him directly in the matter, the only remaining option for Bohr seemed to be through Bush. In his first letter to Anderson after Bohr's arrival in Washington, Halifax indicated

Halifax to Anderson,
18 Apr 45
English
Full text on p. [285]

that he was in full agreement with Frankfurter who was "strongly in favour of the suggestion that B[ohr] should discuss the matter with Bush." In so doing, Halifax argued, Bohr "would merely be resuming the discussions where they

---

[53] *Roosevelt and Frankfurter*, ref. 29, p. 726.

were broken off" in October. As the situation had developed, Anderson felt compelled to agree; but, as he wrote to Halifax, it must "be clearly understood that B[ohr] will make no reference whatsoever to views held in the U.K. or to discussions which he has had here."

Anderson to Halifax,
undated
English
Full text on p. [286]

Before seeing Bush, Bohr had another conversation with Halifax, asking permission to bring up "a personal matter" unrelated to the political issues they were discussing. Bohr's question related to the activities of the Danish resistance movement, which he had had the opportunity to follow closely and to which he also owed "the possibility of his escape from Denmark." Bohr explained that the way the Prime Minister and the late President had spoken of the contribution of the Danish resistance "to the common cause" had been a great encouragement to the Danes[54] and that there had "been some expectations that Denmark in some or other way could be represented at the great conference for world security" in San Francisco, where five days later 50 nations would begin their meetings to settle on a final Charter of the United Nations. "That these expectations have so far not been fulfilled", Bohr continued, "has caused some disappointment and anxiety in Denmark". The Ambassador answered that Bohr was fully entitled to bring up the matter and promised to discuss it with his Foreign Secretary, Anthony Eden. Bohr's intervention may or may not have helped Denmark join the conference at a later date and become cosigner on 26 June of the United Nations charter.

Bohr notes,
19 Apr 45
English
Full text on p. [287]

Bohr and Bush met three times – on 23 and 26 (or 27) April and 8 May 1945. Bohr presented the Addendum, as well as the earlier Memorandum, which he was surprised to learn that Bush had not seen. According to Bohr's notes from the meeting, Bush was impressed by the arguments and in particular by the urgency of the matter. He promised to forward the papers immediately to Secretary of War Henry Stimson, who had overall responsibility for the atomic bomb project. Stimson was chairman of the Interim Committee, which Bush was helping to set up and which would consider the wider issues raised by the atomic bomb project. Bush promised to recommend to Stimson that he meet Bohr within a few days. However, Bohr was not going to have direct contact with the Interim Committee.

At his last meeting with Bush Bohr presented yet another memorandum, which is particularly brief and eloquent. He pointed to "the grave responsibility, resting upon our generation" and explained that "the essence of the arguments"

Memo,
8 May 45
English
Full text on p. [118]

[54] Bohr would return to this issue in a talk to former Danish prisoners of war several years later: N. Bohr, *The Goal of the Fight: That we in Freedom May Look Forward to a Brighter Future* in *Ti år efter*, published by Kammeraternes Hjælpefond as a memorial booklet including the programme for a reunion 19–20 March 1955. Reproduced and translated in Part II on pp. [701] ff.

in the Memorandum and Addendum was "that an early invitation to all nations united against aggression ... should afford a unique opportunity for furthering a harmonious relationship between nations." Bohr concluded by stating that such

"... a timely assurance of good-will from the side which has acquired the lead in the new development would seem to form the proper basis for a purposeful policy meeting the requirements of a situation, so fraught with dangers, but at the same time offering the greatest opportunities."

Bush expressed agreement with the views and promised to send the document on.

At about the same time, in early May 1945, Frankfurter began exploring the possibility of approaching the President through Stimson, who was an old friend of his.[55] As Frankfurter subsequently explained it to Bohr, he

Bohr notes,
22 Jun 45
English
Full text on p. [293]

"... felt it his duty to go to Sti[mson] and tell him everything about his connection with R[oosevelt] about the matter and explain B[ohr]'s views. A few days later F[rankfurter] told B[ohr] that St[imson] had been most seriously occupied with the matter and was very sympathetic to such views (F[rankfurter] used the expression that the talk with St[imson] had been as to knock on an open door.)"

In "a further conversation ... St[imson] said that he wished to see B[ohr]."

In preparation for their meeting, Bohr drafted a letter to Stimson dated 20 June, in which he again stressed the valuable help that the governments might find for their endeavours in the worldwide ties between scientists. Because the meeting was called off, the letter was never sent. Nevertheless, Bohr repeated his argument almost verbatim in the two concluding paragraphs of a memo written in England about a month later:

Memo,
12 Jul 45
English
Full text on p. [119]

"If given opportunity, every scientist who has taken part in the basic development will certainly feel it the greatest obligation not only to co-operate loyally in making an eventual control effective, but also to contribute to the utmost of his powers in preparing an understanding of what is at stake and an adequate appreciation of the immense opportunities, which the development presents.

In this connection it may be most essential, however, that the scientists, on whom the governments of every country will depend for advice, from

---

[55] *Roosevelt and Frankfurter*, ref. 29, pp. 304–307.

the very beginning feel assured that the unique situation brought about by so fruitful and promising an exploration of a new domain of knowledge is being handled in a spirit conforming with the ideals of common striving for human progress for which science throughout the ages has stood as a symbol."

Here Bohr allowed himself for the first time not only to recommend the services of his fellow scientists, but also to seek reassurance that, in return, the values of science be employed in the political domain.

Bohr left Washington for London on 26 June 1945, a few days after Stimson had expressed his regret that he would not be able to see Bohr. He spent the rest of the war in London, where he was joined by his wife coming from liberated Denmark and where he continued to be in contact with Anderson. Bohr returned to Denmark on 25 August 1945, after the bombs had fallen and Japan had surrendered.

In his "Open Letter to the United Nations" (1950), Bohr remembered the eventful period he had spent in England and America during the war:[56]

"Looking back on those days, I find it difficult to convey with sufficient vividness the fervent hopes that the progress of science might initiate a new era of harmonious co-operation between nations, and the anxieties lest any opportunity to promote such a development be forfeited."

3. ADDRESSING THE PUBLIC, 1945–1947

As soon as the existence of the atomic bomb was no longer a secret, Bohr was anxious to express his views on its implication in public. He hoped thus to contribute to a broader understanding of the new perspectives presented by the revolutionary technological advance and of the far-reaching measures called for. At the same time, he waited for the atomic powers to begin international consultations, and he kept in contact with the people he had consulted during the war, notably John Anderson. In order not to interfere unduly with the task of the statesmen and to keep open his confidential contacts, he made no reference in his public writings to his political efforts during the war.

WRITING FOR AN INTERNATIONAL AUDIENCE

Having served as cabinet Minister for six and a half years, Anderson returned to Parliament after the surprising landslide victory of the Labour Party,

---

[56] Bohr, *Open Letter*, ref. 1, quotation on p. [180].

which took over government on 27 July 1945. However, the new government under Clement Attlee did make use of the former Chancellor's unequalled expertise and experience by setting up an Advisory Committee on Atomic Energy with Anderson as chair. Thus Anderson continued to play a central role in British questions regarding atomic energy for some time after the end of the war.

Some days before the bomb fell on Hiroshima on 6 August, Bohr, who was still in London, discussed with Anderson his desire to publish an article for the general public. He first considered publishing in the scientific journal "Nature" as an appeal to scientists, but Anderson suggested "The Times". The article appeared on 11 August.

Bohr began by pointing to the "veritable revolution in human resources",[57] which was a result in particular of the advance of science from its roots in early human society through the exploration of the atom and its nucleus, leading ultimately to the possibility of a large-scale release of atomic energy. This development, Bohr wrote, "cannot but raise in the minds of everyone the question of whither the advance of physical science is leading civilization." While the development had "contributed so prolifically to human welfare", a mortal menace could arise

"... unless human society can adjust itself to the exigencies of the situation. ...

Indeed, not only have we left the time behind where each man, for self-protection, could pick up the nearest stone, but we have even reached the stage where the degree of security offered to the citizens of a nation by collective defence measures is entirely insufficient. Against the new destructive powers no defence may be possible, and the issue centres on world-wide cooperation to prevent any use of the new sources of energy which does not serve mankind as a whole. ... It is obvious, however, that no control can be effective without free access to full scientific information and the granting of the opportunity of international supervision of all undertakings which, unless regulated, might become a source of disaster.

Such measures will, of course, demand the abolition of barriers hitherto considered necessary to safeguard national interests but now standing in the way of common security against unprecedented dangers. Certainly the handling of the precarious situation will demand the good will of all nations, but

---

[57] N. Bohr, *Science and Civilization*, The Times, Saturday 11 August 1945; this volume, pp. [121] ff.

it must be recognized that we are dealing with what is potentially a deadly challenge to civilization itself."

For assistance in the great task lying ahead, Bohr pointed to the valuable services that scientists might offer and emphasized the significance of the strong bonds between scientists created by international cooperation, particularly in atomic research. Bohr's final remark also defined his own commitment:

"It need not be added that every scientist who has taken part in laying the foundation for the new development, or has been called upon to participate in work which might have proved decisive in the struggle to preserve a state of civilization where human culture can freely develop, is prepared to assist in any way open to him in bringing about an outcome of the present crisis of humanity worthy of the ideals for which science through the ages has stood."

Soon after his return to Denmark, Bohr was asked to submit an article with similar content to the American journal "Science". Such a statement, "A Challenge to Civilization", was published in the journal's issue of 12 October 1945. Bohr's new contribution was less detailed than the article in "The Times". Yet he expanded upon the significance of openness by stating that[58]

"... the free and open access to information about all scientific and technical progress will in itself go far toward promoting mutual knowledge and understanding of the cultural aspects of life in the various countries, without which respect and good-will between nations can hardly endure."

The importance of openness in its own right was going to figure with increasing prominence in Bohr's political thinking.

In mid-December 1945 Oppenheimer sent Bohr a telegram requesting permission to publish the Times article – which he referred to as "the classic statement of the feelings with which scientists approach the new situation" – "as foreword to book". Bohr wired back the same day that he "gratefully accept proposal". The book in question was "One World Or None: A report to the public on the full meaning of the atomic bomb", published by the Federation of American Scientists.[59] Introduced by physicist Arthur H. Compton, the book consists of fifteen brief statements in addition to Compton's and Bohr's, the ma-

Oppenheimer to Bohr,
telegram
14 Dec 45
English
Full text on p. [306]

Bohr to Oppenheimer,
telegram
14 Dec 45
English
Full text on p. [306]

---

[58] N. Bohr, *A Challenge to Civilization*, Science **102** (1945) 363–364; this volume, pp. [125] ff.
[59] (Eds. D. Masters and K. Way), McGraw-Hill, New York 1946.

jority written by some of the foremost physicists contributing to the war effort. Comprising eighty pages in pamphlet form, the book had a large circulation and must have been considered by Bohr as a particularly welcome vehicle to make his views known to a broader public. It was immediately published in Danish translation.[60]

During his first visit to the United States after the war, Bohr took part in a "Symposium on Present Trends and International Implications of Science" as a representative of the Royal Danish Academy of Sciences and Letters.[61] The meeting took place in Philadelphia and was organized jointly by the National Academy of Sciences and the American Philosophical Society. On this occasion Bohr gave a talk on 25 October 1946 which is particularly interesting in that he presented his views in the context of a broad exposition of the development of atomic physics with its far-reaching conceptual implications. Upon his return to Copenhagen, he submitted an abbreviated written version of the talk, which was published in the Proceedings of the American Philosophical Society.[62]

In his talk, "Atomic Physics and International Cooperation", Bohr began by emphasizing the decisive role of international cooperation in the creation of a whole new branch of science based on the fundamental discoveries of the constituents of the atom and the quantum of action. Thus, the "lines of inquiry pursued in various countries by scientific schools with different traditions and outlooks" had come together through an "international cooperation of an intensity and enthusiasm which has indeed only few counterparts in the history of science." While pointing to the resulting "immense opportunities for a world-wide collaboration on the advance of science and the promotion of human welfare" he stressed, just as he had in his Times article, that "all such perspectives may be overshadowed by deadly menaces to civilization".

Bohr then turned to the philosophical lesson "impressed upon us by the development of physics in our generation" which "is especially suited to further common human understanding", being the result of efforts with "no other aim than to widen the borders of our knowledge and to deepen our understanding of that nature of which we ourselves are a part." In his broadening of the complementarity viewpoint of quantum mechanics, Bohr had observed that complementary relationships also apply in the comparison of human cultures.

---

[60] *Een Verden eller ingen: Udarbejdet af amerikanske Videnskabsmænd med et Forord af Niels Bohr*, P. Haase & Søn, Copenhagen 1946.

[61] *Meeting on 29 November 1946*, Overs. Dan. Vidensk. Selsk. Virks. Juni 1945 – Maj 1946, p. 35.

[62] N. Bohr, *Atomic Physics and International Cooperation*, Proceedings of the American Philosophical Society **91** (1947), 137–138; this volume, pp. [131] ff.

The recognition of this, Bohr wrote, might prove helpful for the "mutual understanding which at present is more needed than ever", thus returning to an issue he had brought up in his lecture in 1938 at the International Congress of Anthropological and Ethnological Sciences.[63] In his Philadelphia talk he illustrated these relationships by asserting that "a full mutual appreciation of the cultural achievements of humanity is indeed irreconcilable with the complacency inherent in any national culture."

Bohr concluded:

"An apology for entering on these general topics in connection with the acute problems facing the world today may perhaps be found in the circumstance that the pursuit of knowledge in a domain of science as remote from human passions as the study of the most elementary physical phenomena has not only forcefully impressed upon us how necessary is openness of mind for the search of truth, but also suggestively reminded us of the common basis of all endeavors for the elevation of human culture. Surely, this circumstance should prove helpful for the strengthening of that spirit which now is incumbent if mankind shall be able jointly to reap the fruits which the progress of science offers."

TRYING TO REACH THE SOVIET UNION

Although he wrote them in English, Bohr did not intend the publications just presented only for an English-speaking audience. In particular, he wanted the Russian public, and especially Russian scientists, to learn about his views.

In his efforts during the war to encourage western statesmen to approach the Soviet Union about the existence of the atomic bomb, he had hoped that his close colleague and friend Peter Kapitza, who was known to have access to the Soviet political leadership, might serve as a middle man.[64] This hope was strengthened by the arrival of Kapitza's letter in April 1944.[65] Thus, when on 21 October 1945 Bohr received a short telegram from Kapitza and his wife Anna "to wish you wellcome on your happy and safe reunion with your family in free Denmark", he wrote to his colleague immediately.

Kapitza to Bohr, telegram
20 Oct 45
English
Full text on p. [295]

---

[63] See p. [11].

[64] On Kapitza's political connections in the Soviet Union, including his correspondence with Stalin, see *Kapitza in Cambridge and Moscow: Life and Letters of a Russian Physicist* (eds. J.W. Boag, P.E. Rubinin and D. Shoenberg), North-Holland, Amsterdam, etc. 1990.

[65] See above, p. [24].

Bohr to Kapitza,
21 Oct 45
English
Full text on p. [296]

Bohr expressed his conviction that after the war international cooperation in science would "contribute more than ever to that understanding between nations in which all peoples put faith and for which so promising a background has been created through the comradeship in the defence of elementary human rights." He situated Kapitza's and his own role by referring to their shared deep admiration for Rutherford, who had died in 1937. Both Bohr and Kapitza had studied and worked under the British scientist, whose wisdom and authority, Bohr wrote, will be "greatly missed in the endeavours to avert new dangers for civilization and to turn the great advance to the lasting benefit for all humanity." He enclosed his articles in "The Times" and "Science", encouraging Kapitza to show them to friends and asking advice about translation into Russian. In these articles, Bohr wrote, "I have tried to give impression for an attitude widely shared among scientists." He expressed eagerness to receive Kapitza's reaction: "I need not add that I shall be most interested to learn what you think yourself about this all-important matter which places so great a responsibility on our whole generation." Bohr was equally interested in "the endeavour to organize scientific research and co-operation on the most effective lines." To him, the establishment of political and scientific collaboration after the war went hand in hand.

Bohr reported on the new developments in a letter to Anderson, the immediate purpose of which was to thank him for the greeting on Bohr's sixtieth birthday published by the former Chancellor in the Danish press.[66] Bohr enclosed his article in "Science" as well as a letter he had recently sent to Bush, in which he had "mentioned certain points of which I have thought in connection with the desirability of inviting the confidence of scientists in other countries".[67] Bohr added that he had just received a most friendly telegram from Kapitza and that he had responded with a letter, of which he enclosed a copy.

Bohr to Anderson,
22 Oct 45
English
Full text on p. [299]

There were good reasons for keeping Anderson informed, for making contact with the Soviet Union was no trivial affair at the time. Only a few days after his homecoming on 25 August 1945 Bohr had been warned by British intelligence that the Soviet Union might have plans to kidnap him; he agreed not to visit the Danish island of Bornholm, where the Soviets still had a military presence. The intelligence authorities were on the alert, ready to protect Bohr. They were assisted in the effort by one of his sons.

---

[66] J. Anderson, *En af de største Aander i vor Generation* [*One of the greatest intellects in our generation*], Politiken, 7 October 1945. The greeting commended Bohr for his international endeavours. The celebration of Bohr's birthday is dealt with below, p. [63].

[67] The letter to Bush has not been found.

Bohr's close relationship with Anderson continued after the war. Here Bohr (left) greets Anderson on his visit to Copenhagen in 1947. Centre: Alec Randall, the British ambassador to Denmark. (Photograph by H. Lund Hansen.)

Just how sensitive the Americans were to the leakage of scientific "secrets" about the atomic bomb is shown by an incident resulting from Bohr's first public lecture in Denmark after his return. The lecture was given in the beginning of October at the first gathering after the war of the Danish Association of Engineers, held in memory of the seventeen Danish engineers who had been killed in the war. Bohr's topic was the advances in nuclear physics leading to the creation of the atomic bomb. The lecture was widely reported in Danish newspapers, most of which concentrated on Bohr's statement that any military defence against the atomic bomb was useless. The newspaper "Politiken" reported on Bohr's lecture under the page-wide heading, "Bohr reveals the secret of the atomic bomb". Bohr felt compelled to respond with a press release. Writing in the third person, he emphasized[68]

[68] The radio talk referred to by Bohr is discussed in the next subsection.

Press release,
4 Oct 45
Danish
Full text on p. [294]
Translation on p. [295]

"... that when reference is made to the secret of the atomic bomb, this is a figure of speech which may give rise to misunderstanding, in that the scientific principles involved in the release of atomic energy were, in the main outlines, public knowledge before the war and their later development is described in detail in the official documents published by the American and British Governments. The talk dealt only with such purely scientific advances and, as the professor on his return to this country has already announced in a statement on the radio, scientists such as himself do not have any precise knowledge of the technical details regarding the production of the active materials or of the strategic problems concerning the potential uses of the new weapons."

Bohr notes,
undated
Danish
Full text on p. [324]
Translation on p. [328]

Groves reacted to the incident by sending "a man over to say that only what was stated in the Smyth Report could be discussed." The Smyth Report was the official American document defining what information about the atomic bomb could be given to the public, which Bohr had indeed already referred to in his press release.[69]

After submitting his telegram Kapitza wrote a letter to his Danish friend, which thus crossed with Bohr's. He sent it via a highly irregular channel. Mogens Fog, who had played a prominent role in the resistance during the war and was now a communist member of the Danish liberation government, told Bohr that a Soviet physicist wanted to arrange a secret meeting at which he would hand him a letter from Kapitza. According to his typed report written in connection with the incident, Bohr responded that he "could not engage in secret arrangements of any kind and that he had to consider such an appeal a regrettable mistake on the part of Russia." Nevertheless, Bohr[70]

Bohr notes,
2 Nov 45
English
Full text on p. [301]

"... was pleased to hear from [Fog], just as Kapitza had also assured him, that there was also confidence in him on the part of Russia. The only way in which B[ohr] could perhaps make a small contribution to the great cause, was to work in full openness for a mutual understanding of the great problems that the latest development of science had raised."

Bohr reported the approach to Danish intelligence, which took steps to safeguard him. The British Embassy was also informed. It had to be expected,

---

[69] H.D. Smyth, *A General Account of the Development of Methods of Using Atomic Energy for Military Purposes Under the Auspices of the United States Government, 1940–1945*, The Government Printing Office, Washington, D.C. 1945. Reprinted by His Majesty's Stationary Office, London 1945.

[70] The name "Fog" was later cut out.

as was later confirmed, that Professor Yakov Terletzkii from Moscow University was seeking information about atomic energy on behalf of Soviet Intelligence. Nevertheless, Bohr and the Danish intelligence authorities judged that it would involve little risk to receive Terletzkii for an open meeting at the institute. When Terletzkii visited on 14 November, he was received like any other visiting scientist, though with extensive security measures. Almost five decades later allegations were made that Bohr gave away secret information at the meeting. Aage Bohr refuted them in a press release, in which he reported that during the conversation, at which he was present, Terletzkii "raised some technical questions concerning atomic energy, to which my father answered that he was not acquainted with details". Instead, Bohr referred Terletzkii to the Smyth Report.[71]

Press release,
28 Apr 94
English
Full text on p. [332]

On being received at the institute, Terletzkii handed Bohr Kapitza's letter. In it Kapitza reported that research was now again in progress at his institute and had led to discoveries concerning the superfluidity of liquid helium; these discoveries were going to stand as landmarks in the field. He went on to express his worries "about the question of international collaboration of science which is absolute[ly] necessary for the healthy progress of culture in the world." He observed that "the famous atomic bomb I think proves once more that science is no more the hobby of university professors but is one of the factors which may influence the world politics." Emphasizing the danger in keeping scientific discoveries secret, since they might "be used for selfish interests of particular political or national groups", Kapitza wondered

Kapitza to Bohr,
22 Oct 45
English
Full text on p. [298]

> "... what must be the right attitude of scientists in these cases. I should very much like at the first opportunity to discuss these problems with you personally and I think that it would be wise as soon as possible to bring them up to a discussion at some international gathering of scientists. ...
>
> I should be glad to hear from you what is the general attitude on these questions of the leading scientists abroad. Any suggestions about means to discuss these questions from you I shall welcome mostly. I can indeed inform you what can be done in this line in Russia."

Kapitza ended his letter by introducing its bearer as "a young and able professor of the Moscow University". Many years later Anna Kapitza, Peter Kapitza's wife, confided to Aage Bohr how embarrassing it had been for her husband to

---

[71] For a detailed description by a historian, see D. Holloway, *Beria, Bohr, and the Question of Atomic Intelligence* in *Reexamining the Soviet Experience: Essays in Honor of Alexander Dallin* (eds. D. Holloway and N. Naimars), Boulder, CO: Westview Press 1996, pp. 237–258.

be used in this manner. On the other hand, it provided Kapitza with an opportunity for expressing his views to Bohr that he otherwise might not have had.

Bohr had time to write a response to Kapitza and forward it to Terletzkii before the latter left Copenhagen. He picked up on Kapitza's suggestion of an international conference:

Bohr to Kapitza,
17 Nov 45
English
Full text on p. [302]

"For some time I have hoped that it should be possible to arrange such a meeting here in Copenhagen, and I have already mentioned it to several of our mutual English friends.

If you and some of your colleagues could come, I am sure that a number of the leading physicists from other countries would join us. Such a meeting which we are prepared to arrange at any time would not only give us all the opportunity to exchange views about those matters which are so much on the heart of everyone, but also to take up the question of collaboration in science and to discuss the many advances which have been made in various fields of physics in recent years."

Bohr expressed a desire to meet Kapitza even if the plans for a conference did not work out: "I need not add that quite apart from such plans, it would be an extreme pleasure to see you here in Copenhagen whenever you might be able to come."

Bohr soon wrote to Anderson about the meeting with Terletzkii. He explained that the formulations in the letter to Kapitza, which he enclosed, were intended "to avoid the question of an invitation to me to come to Moscow, as touched upon by Professor Terletzky." Anderson responded in full agreement:

Bohr to Anderson,
20 Nov 45
English
Full text on p. [304]

Anderson to Bohr,
29 Nov 45
English
Full text on p. [305]

"As regards the calling of an international conference of scientists in Copenhagen, I think that would be an excellent idea provided sufficiently long notice were given. Perhaps the late summer or early autumn of next year would be a suitable time. This need not, of course, necessarily rule out an earlier visit by Professor Kapitza which clearly would be far preferable to a visit by you to Moscow which, in present circumstances, might be misunderstood."

The only response to his letter that Bohr received from Kapitza was a New Year's telegram simply saying: "Happy New Year and my best wishes for a real peace". Then, in April 1946, Bohr confirmed, in a private letter to Kapitza, "the official communication which is being sent to you from our Academy"; the communication announced Kapitza's election as a foreign member of the Danish Academy of Sciences and Letters. Peter and Anna Kapitza were also

Bohr to Kapitza,
12 Apr 46
English
Full text on p. [307]

Bohr was able to visit the Soviet Union again in 1961. From left to right: I.V. Obreimov, A.V. Fock, N. Bohr and P. Kapitza.

invited to Denmark, with all expenses paid. Kapitza's thank-you telegram came through, but Bohr never received the letter promised in it.[72]

Not being allowed to visit Copenhagen, Kapitza was formally given the membership diploma by the Danish ambassador in Moscow. After the ceremony, he wired Bohr that in his lecture of appreciation he "also took the opportunity to stress the necessity of maintaining the international collaboration of scientists which is now in danger" and promised to send Bohr a transcript of his talk. In the talk itself, which Bohr did not receive from Kapitza but in Danish translation from the Danish ambassador, Kapitza stressed Bohr's and his own shared concern with "the necessity of holding on to the internationalism of science." He continued, after pointing out the perils in this regard posed by the existence of the atomic bomb:

Kapitza to Bohr,
telegram
12 Jul 46
English
Full text on p. [314]

Kapitza talk,
8 Jul 46
Danish
Full text on p. [310]
Translation on p. [312]

> "I have recently had the opportunity to exchange thoughts in this direction with Bohr. Our views are in agreement on the point that the voice of the

---

[72] Telegram from Kapitza to Bohr, 11 May 1946, BSC (21.4).

researchers must be raised in protest against keeping the work secret and against all possible attempts to transform one of the most remarkable results of science into a toy for narrow imperialistic viewpoints or aggressive plans of some countries."

Døssing to Bohr,
18 Jul 46
Danish
Full text on p. [315]
Translation on p. [316]

In his covering letter the ambassador, Thomas Døssing, told Bohr that "Academician Kapitza has 2 great wishes, the one to see you as a guest of the Soviet Union, the other that he could come to Denmark himself."

Bohr to Døssing,
10 Aug 46
Danish
Full text on p. [317]
Translation on p. [318]

Bohr used the opportunity of Døssing's letter to state briefly his endeavour to spread "the hope that the development of science ... will in the near future lead to decisive improvements in the relationship between nations", enclosing reprints of his relevant writings. Bohr told the ambassador that he would "make sure that this invitation is sent to Kapitza once more under the most official forms."

During the first year after his return to Denmark, Bohr thus attempted to open opportunities for exchanges between scientists from East and West on the issues raised by the atomic bomb, as well as for a resumption of the exchange within science itself. However, neither an informal visit from Kapitza nor a conference including Russian physicists materialized. Bohr's attempt to communicate his vision of an open world to the Russian public and its physicists had run into the barrier of the closed Soviet system. Intercourse between scientists from the Soviet Union and their colleagues in the West would remain almost totally blocked until well into the 1950s.

WRITING FOR A DANISH AUDIENCE

As soon as the news of the atomic bomb broke, Bohr was beleaguered by Danish journalists in London, who were anxious to bring news about the bomb project, and particularly Bohr's role in it. To quote only a few of the headlines dominating Danish newspapers on 8 August, while "Bohr did not aim at making bombs", he was "today the most famous Dane", "whose work led to the release of atomic energy". There would have been "no atomic bomb without the effort of Bohr". It was evidently difficult for the press to distinguish between Bohr's pioneering contributions to the scientific basis for the project and his more limited involvement in the development of the bomb itself.

The circumstances of Bohr's escape two years earlier, as well as his role in occupied Denmark before his escape, were also extensively reported. Bohr was hailed as Denmark's great son, who had fought against the Nazi threat from beginning to end and had been instrumental in winning the war through his contribution to the atomic bomb project. Bohr withdrew from this onslaught, the brunt of which was taken by Aage.

Nevertheless, Bohr did want to inform the Danish public about the developments in which he had participated and in particular the perspectives, dangers and hopes they held. His article in "The Times" of 11 August was published in Danish in the newspapers "Berlingske Aftenavis" and "Politiken", respectively the same afternoon and the day after. But there was a need to address the Danish public independently, and even while in London Bohr worked out an address for the purpose, which he gave on Danish national radio the very day after his return to Denmark.

Bohr's homecoming attracted substantial attention from the press, even though he tried to make as little of his arrival as possible. His insistence on continuing business as usual did not discourage the journalists. On the contrary, the photo of Bohr returning to the institute on his bicycle only became another part of the much publicized image of the great Danish scientist.

In his talk Bohr began by reflecting upon his constant concern for what those at home might have suffered in the difficult period for the country during his absence. While abroad, Bohr reported, he had encountered general appreciation for the strong stand of the entire population to safeguard Danish cultural values as well as admiration for the resistance against the occupying forces.

Bohr had also found everywhere

> "... that the respect shown for the name of Denmark has its deep roots in the contributions our country, despite its modest size, has managed to make in the course of time to common human culture ..."

Radio talk,
26 Aug 45
Danish
Full text on p. [136]
Translation on p. [140]

To explain the basis for this respect he recurred to the viewpoint contained in his introduction to the monumental "Denmark's Culture in the Year 1940":[73]

> "It is our outlook on the rights and responsibilities devolving upon every single member of the community, and our affinity with all humanity, that not only has led to so much progress in enlightenment and welfare in our own country, but also has given us a prominent position as regards cultural efforts in general."

Bohr continued his radio address with a reference to the sympathy he had encountered for the Danish efforts to help refugees from the Nazi regime, which, he added, had been an important consideration in his decision to remain in Denmark for as long as possible.

He then turned to his main theme:

---

[73] See p. [12].

Bohr arriving at his institute for the first time after the war (August 1945).

"It is strange to think that the development in the field of atomic physics, which should result in providing mankind with such extraordinary means of power, is rooted in efforts with the sole purpose of deepening the understanding of the regularities, which the exploration of natural phenomena, despite the richness with which they occur, continues to bring more clearly to light."

Bohr described how atomic research, through an intense international cooperation, in the years before the war "showed the way to a possible utilization of the enormous amounts of energy tied up in the nuclei of atoms." At the same time, however, it had become clear that extraordinary technical facilities would be required. "At that time", he noted, "taking into account the dangers that threatened world peace, it was reassuring to be certain that no attack with such devastating weapons could suddenly be made." Therefore, Bohr stated, "it was a complete surprise for me on arrival in England to learn how far the work had progressed already then, and what enormous facilities had been made available for it in America".

Bohr vividly described the unprecedented features of the project constituting "an interference with the natural course of things, which is far more profound than anything humans have hitherto accomplished". He emphasized the modest character of his own contribution at a late stage of the project, which had "only consisted of participation in consultations about the elaboration of the scientific basis and its viability".

While administrative and military decisions concerning the project "are outside the scientists' sphere proper", Bohr wrote,

"... everybody ... must feel deep concern about how the great advance in our mastery of the forces of nature can be utilized for the benefit of all humanity. ... extreme dangers ... threaten the survival of civilization itself, unless in time a worldwide agreement is reached for accomplishing an effective control of the terrible means of destruction. ... it is vitally important that understanding of what is at stake be created in the widest circles ...

Let us hope that the right path is found so that science, which throughout the ages has stood as one of the strongest testimonies of the progress that can be achieved by common human endeavours, precisely with its most recent lesson about the necessity of concord, will contribute decisively to harmonious coexistence among the peoples."

Bohr's sixtieth birthday on 7 October 1945 received considerable attention in the media.[74] A tribute that must have been greatly appreciated by Bohr was a torchlight procession organized by the students at the University of Copenhagen. While Bohr and guests were listening to a one-hour radio broadcast in his honour from loudspeakers set up in the Pompeii Hall of his home in the

---

[74] On the Royal Danish Academy's celebration of the occasion, see the introduction to Part II, p. [345].

[63]

Carlsberg Mansion, as many as three thousand students were gathering close by donning their traditional white caps. While singing student songs, they marched off to Bohr's garden, which had been duly prepared for the ceremony, at the end of which the torches were thrown in front of Bohr's house in a pile making up an impressive fire. There was a tradition among students in Denmark for honouring their nation's prominent cultural figures in this way.

After the student spokesman had briefly motivated the tribute, another set of loudspeakers sounded the response given by Bohr from the mansion's large outdoor staircase; the talk was also printed and circulated. "Such an honour", Bohr began, "can no one, who with modest powers and in a humble fashion participates in the common great research effort, feel he deserves, but it must be considered as homage to the science which we all serve." Turning to the developments that he had been part of, he continued:

> "It has been my good fortune to be able to participate together with older and younger generations in the investigation of an area of knowledge which hitherto has been beyond human reach, and which has given us experience that in a new way has illuminated fundamental features in the nature of human knowledge."

The resulting increase in our mastery of nature, Bohr emphasized, had provided a forceful reminder of how "intimately the fate of all humanity is interwoven." The occasion brought back memories of his own student days, and he recalled how much the continued contact with the younger generations had meant to him over the years. At the end of his response Bohr pointed to the great tasks and responsibilities facing students in a time when cultural efforts were the way to improve the world situation. Again drawing on the characterization of Danish culture contained in his article of the same name, as well as in his speech on his return to Denmark, Bohr went on:

> "The traditions we have in this country as regards the rights and duties of the individual in society, as well as our sense of affinity with the rest of the world that marks our outlook as a whole, should give you [the students] the most beneficial conditions for working for this purpose, and we should be both happy and proud that these traditions stood their test in dark times, and not least that the life at our University both as regards freedom for research and the spirit in the student community, despite pressures and threats, was not broken but rather purified."

Bohr had another occasion to address the Danish people when he was asked to give a talk on Danish national radio on New Years Eve 1945; the talk was

Speech to Danish students,
7 Oct 45
Danish
Full text on p. [145]
Translation on p. [147]

printed in "Politiken" the following day. Here Bohr returned to the perspectives that he had presented in his public writings and addresses during the past year. Whereas civilized society, he noted, had long since introduced regulations of the behaviour of individuals, now, "if all civilization is not to be doomed," a similar contract must be reached between nations.[75] Bohr suggested that Denmark, with its "lively sense of ... affinity with the rest of the world" was well prepared for the task; but "the realization of these great hopes will require genuine willingness on the part of everyone to learn from the course of development."

4. CONTINUATION OF THE CONFIDENTIAL APPROACH, 1945–1948

While eager to publish his general views to the international and Danish public, Bohr was reluctant to take part in public discussions about political matters pertaining to nuclear weapons. He abstained from taking up specific questions of security and procedure that might give rise to controversy. He was waiting for the statesmen to take an initiative and wanted options open for a continuation of his confidential approaches, if that should prove desirable. While thus waiting, he kept closely in touch with developments through his contacts in Britain and the United States. A breakthrough occurred in June 1948, when he was given the opportunity to present his views in a personal meeting with the U.S. Secretary of State George C. Marshall.[76]

REACTION TO THE FIRST INITIATIVES AFTER THE WAR

A first initiative in the direction Bohr was looking for was a conference held in Washington in November 1945 between British, American and Canadian leaders producing the "Washington Declaration". This document proposed the establishment of an International Atomic Energy Commission (UNAEC) under the auspices of the recently established United Nations. The UNAEC would lay down rules for the exchange of scientific information for peaceful purposes and would control research and development to ensure that all applications were made toward peaceful ends. As the conference took place, Bohr discussed the issues involved with the U.S. ambassador to Denmark, Monnett B. Davis. He

---

[75] N. Bohr, *Humanity's Choice Between Catastrophe and Happier Circumstances*, Politiken, 1 January 1946. Reproduced and translated on pp. [149] ff.

[76] Apart from the sources explicitly quoted in the margin, this section draws on undated notes in the BPP in Margrethe Bohr's handwriting, which list a number of high points of Bohr's political activities from the end of the war until 1950. The notes are transcribed in full on p. [324].

[65]

Bohr to Anderson,
20 Nov 45
English
Full text on p. [304]

wrote to Anderson that he considered the Washington Declaration as "a most important step towards a favourable development", a view with which Anderson agreed. The UNAEC was formally established by a vote in the United Nations' General Assembly on 24 January 1946. It was going to last for only two years.

A next major step in the efforts to establish international control of nuclear energy was the release at the end of March 1946 of the U.S. State Department's Acheson–Lilienthal report, which was intended as a working paper for the American representative on the UNAEC, who was yet to be appointed. It was written on the initiative of Under Secretary of State Dean Gooderham Acheson in his capacity as chairman of the Secretary of State's Committee on Atomic Energy. The report was drafted by a board established for the purpose under the chairmanship of David E. Lilienthal. Oppenheimer, who was also on the board, was the main author of its recommendation to establish an "Atomic Development Authority" under the United Nations. The new Authority would control all "dangerous" aspects of nuclear energy and have a substantial staff of scientists to take a leading role in research and development in the field. The report foresaw that the detailed planning of the Authority would be the object of extensive deliberations in the UNAEC, before being formally approved by the United Nations and its member countries.[77]

Bohr to Bush,
17 Apr 46
English
Full text on p. [308]

Bohr was kept informed about these developments and received an early copy of the report, which he acknowledged in a letter to Bush, who was a member of Acheson's Committee. Bohr praised the report for giving a "most admirable expression for a spirit which contains the very best hopes for the development in which everybody puts his faith."

Bohr to Oppenheimer,
17 Apr 46
English
Full text on p. [309]

Bohr wrote to Oppenheimer that it had given him deep pleasure to read the report. "In every word", he wrote, "I find just the spirit which I think offers the best hopes" and "from page to page I recognized your broad views and refined power of expression." Bohr used the occasion to explain why he had not taken part in the public debate apart from "quite general utterances": "not being able to follow developments at first hand", he had felt it unwise "to take part in discussions about security problems".

After he had written these lines, but before sending off the letter, Bohr received a package of reports from Oppenheimer himself, together with a personal handwritten letter stating that the report "is not all it should be ... For what is good in it, it should be dedicated to you." Upon reading Oppen-

Oppenheimer to Bohr,
30 Mar 46
English
Full text on p. [307]

---

[77] *A Report on the International Control of Atomic Energy*, Prepared for the Secretary of State's Committee on Atomic Energy by a Board of Consultants, U.S. Government Printing Office, Washington, D.C. 1946.

heimer's note, Bohr added a handwritten PS in his letter, expressing his sincere appreciation of Oppenheimer's words and concluding:

"I knew of course that the task would be very difficult but none could have been better fitted for it than you, and I think that things now look most hopeful. Notwithstanding all such reactions as I have felt proper you know of course that you can always count on me. Good luck."

Once appointed, the American UNAEC representative did not accept the Acheson–Lilienthal Report unreservedly. The appointee was the 75-year old financier Bernard Baruch who had been adviser to several presidents. Baruch wanted to make his own statement of policy on the issue of nuclear arms control.

Again Bohr followed events at close hand. He had developed a relationship of trust with the new American ambassador to Denmark, Josiah Marvel, who had moved to Copenhagen in January 1946. Bohr told Marvel in confidence about his efforts during the war. Marvel grasped Bohr's idea that the situation offered unique opportunities for establishing mutual trust and encouraged him to write to Baruch.

Bohr followed Marvel's suggestion, sharing his letter to Baruch with Anderson. In it, he first referred to his own experience and efforts during the war and presented the general views he had developed. He emphasized the importance and timeliness of the Acheson–Lilienthal Report, while expressing his appreciation of "the restraint exerted [in the report] in separating the discussion on technical and administrative problems from the wider prospects". He continued:

Letter to Baruch,
1 Jun 46
English
Full text on p. [156]

"Notwithstanding the decisive importance of technical matters for the shaping of the eventual control arrangement, it is obvious, however, that the promotion of the cause will above all depend on the general recognition of the fact that the formidable power of destruction which has come within reach of man has irrevocably changed the conditions for the life of nations."

On this basis, he added, it should "be possible to obtain agreement from all sides about the necessity of farreaching renunciations on accustomary prerogatives." To buttress his point, Bohr enclosed his articles from "The Times" and "Science".

Baruch's own proposal for an "International Atomic Development Authority", presented at the UNAEC opening session in the middle of June 1946, did seek to provide a broad vision. Moreover, at Baruch's meeting with U.S.

Secretary of State James F. Byrnes and Under Secretary Acheson in preparation of his opening statement to the UNAEC, the name of Niels Bohr was presented as "one ideal type" in "the make-up of the Board of Directors" of Baruch's proposed Authority.[78] Nevertheless, although his proposal of an Atomic Development Agency was adopted from the Acheson–Lilienthal Report, Baruch's version allowed less room for bargaining. He thus insisted on the introduction of a control regime with strong policing power before the issue of forbidding and destroying nuclear weapons could be taken up. The Soviet Union, represented by its UNAEC representative Andrei Gromyko, demanded the opposite order of priorities. Such a hardening of positions was quite contrary to Bohr's vision for a development of trust between nations.

On his visit to the United States in the autumn of 1946 Bohr met Baruch and had an opportunity to explain further his ideas for obtaining a common basis for confidence. Baruch did not seem to be receptive to Bohr's thinking.

FOCUSING ON OPENNESS

The negotiations within the United Nations on international control of atomic energy ran into sharp controversies from the very start, and soon there appeared to be little hope for a successful outcome. At the same time, the general climate in international relations was beset with distrust and suspicion. New barriers between nations had arisen, as Bohr experienced in his attempt to reestablish relations with Soviet scientists.[79]

In the course of this decline Bohr became more and more convinced that it was necessary to introduce openness as the key to the promotion of confidence. Bohr thought that an initiative in this direction would have to come from the United States.

Bohr had discussed such possibilities with American statesmen already during his visit to the United States in 1946.[80] On his return to Copenhagen, Marvel, who continued to take a constructive interest in the matter, arranged meetings between Bohr and representatives of the U.S. Government visiting Copenhagen. Of particular consequence was the visit by John J. McCloy, who at that time was President of the International Bank for Reconstruction and Development (later to be known as the World Bank). McCloy had served

---

[78] "Department of State Atomic Energy Files: Memorandum of Conversation, by Mr. John M. Hancock of the United States Delegation to the Atomic Energy Commission", FRUS 1946, Vol. 1, pp. 802–806, on p. 805. The document is dated 30 May 1946, and the meeting took place on this and the following day.

[79] See above, pp. [53]–[60].

[80] Bohr, *Open Letter*, ref. 1, this volume on p. [181].

as Under Secretary of War under Henry Stimson and had been instrumental in setting up the Interim Committee. He had been a member of Acheson's Committee on Atomic Energy and was thus well versed in questions of control as well as close to the top echelons of the American political system.

McCloy was invited "home to Carlsberg" where he saw Bohr's "papers and manuscripts". He was impressed by Bohr's views and considered it important that George C. Marshall, whom Truman had appointed Secretary of State in January 1947, should learn about Bohr's views at first hand. Marshall had served as Chief of Staff for the U.S. Army throughout the war and is remembered in particular for the "Recovery Plan for Europe", more popularly known as the Marshall Plan.

Bohr notes,
undated
Danish
Full text on p. [324]
Translation on p. [328]

An opportunity to meet Marshall arose in connection with Bohr's visit to the United States in 1948. A vivid account of the preparations for the meeting and of the conversation between the two is contained in a letter that Bohr wrote to Anderson upon his return to Denmark "for your personal information to tell you a little about my experiences." The letter was never sent, possibly because he expected to see Anderson in the near future.

Bohr to Anderson,
unsent draft
23 Jul 48
English
Full text on p. [319]

When the two met in New York, McCloy advised Bohr as a first step to write a brief introductory statement to Marshall. Bohr prepared a selection of statements from the Memorandum and Addendum. In a one-paragraph covering letter, Bohr referred to his wartime work in the atomic bomb project, adding that he had since followed developments closely, without taking part in public discussion. Yet, he wrote, the views he had encountered in "various communities" had strengthened his belief that

**Covering letter,
22 Mar 48
English
Full text on p. [159]**

"... there are still points of view which need to be more fully explored, and the consideration of which, above all against the background of the growing world tension, may be of significance for the statesmen in their search for a constructive approach to the great problems confronting humanity."

In his unsent letter to Anderson Bohr wrote that "it was hoped that I should see Marshall immediately, but as you know it was suddenly decided that he should go for about six weeks to Bogota where also McCloy went."[81] Upon his return McCloy asked Bohr "to write a somewhat more explicit account" of his views. This suggestion resulted in the "Comments", in which Bohr for the first time presented his case for making the stand for openness the paramount issue.

---

[81] Marshall and McCloy were attending the meeting of the Organization of American States.

"Comments",
17 May 48
English
Full text on p. [160]

The Comments begin by recalling the "deep-rooted divergencies" which had "grown out of social and political developments in the last decades" and which "were bound to present a serious strain on international relations at the conclusion of the second world war." Already on this basis, Bohr argued, it was clear that a whole-hearted cooperation would demand a deep-going readjustment of international relations, a need which "was even further accentuated by the great scientific and technical developments".

Bohr then pointed to "a unique opportunity for seeking continued cooperation on vital problems" offered precisely by the challenge to civilization posed by the revolutionary development of atomic armaments. In this connection he noted that already during the war it was

> "... felt that a favourable foundation for later developments might be created by an early initiative aimed at inviting confidence ... by assuring them [all partners] of willingness to share in the farreaching concessions as to accustomed national prerogatives which would be demanded from every side."

"Since the war", Bohr continued, "the divergencies in outlook have manifested themselves ever more clearly", leading to "distrust and suspicion between nations and even ... within ... nations". Thus, "the hopes embodied in the establishment of the United Nations organization have met with repeated great disappointments and, in particular, it has not been possible to obtain consent as regards control of atomic energy armaments."

At this point Bohr invoked openness:

> "... the turning of the trend of events requires that a great issue be raised, suited to invoke the highest aspirations of mankind. Here it appears that the stand for an open world, with unhampered opportunities for common enlight[en]ment and mutual understanding, must form the background for such an issue. Surely, respect and goodwill between nations cannot endure without free access to information about all aspects of life in every country."

Although it was true, Bohr wrote, that the need for greater openness already underlay the proposals before the UNAEC, "just the difficulty experienced in obtaining agreement under present world conditions would suggest the necessity of centering the issue more directly on the problem of openness." Bohr suggested

> "... that most careful consideration should be given to the consequences which might ensue from an offer, extended at a well-timed occasion, of immediate measures towards openness on a mutual basis."

Granting that the step envisaged "might seem beyond the scope of conventional diplomatic caution", Bohr maintained that if taken it could only improve the situation. Returning to an argument first made in the Addendum, Bohr wrote that

"... irrespective of the immediate response, the very existence of an offer of the kind in question should deeply affect the situation in a most promising direction. In fact, a demonstration would have been given to the world of preparedness to live together with all others under conditions where mutual relationships and common destiny would be shaped only by honest conviction and good example."

In other words, the offer would have a positive effect even if the Soviet Union rejected it:

"Such a stand would, more than anything else, appeal to people all over the world, fighting for fundamental human rights and would greatly strengthen the moral position of all supporters of genuine international cooperation."

Bohr's new document constituted a general statement of principle and did not provide any detailed description of how to implement openness in practice. Bohr was entirely aware of this, writing in the unsent letter to Anderson that he had "studiously avoided to enter on specific problems of security and procedure." Such specifics were left for their personal meeting, which took place "in all privacy in the home of McCloy in Washington." Bohr and Marshall

"... had a long and frank talk on many aspects of the matter ... during which Marshall entered as well on difficulties of preparing the American public and Congress for a policy of openness, as on the anxieties as regards military security in case a proposal of universal openness should be accepted. ...
To the best of my knowledge I tried to answer on all such questions and at the end of the talk Marshall invited me to write him a personal letter, for his closer consideration of the matter."

Bohr wrote the letter "after consultation with McCloy, who throughout has been most sympathetic and helpful, as well as with Frankfurter and Oppenheimer". Povl Bang–Jensen, a war-time acquaintance at the Danish Embassy in Washington, also provided valuable assistance. "As you will see," Bohr continued in his draft letter to Anderson, "I entered in some detail on objections as regards security and I even ventured to comment on questions of procedure". Dated 10 June 1948, the letter to Marshall is the most detailed of all of Bohr's political documents. The following presentation of the letter aims at helping the reader follow the many-faceted arguments involved. For a full appreciation of Bohr's argument, his letter needs to be read in full.

Bohr started out motivating his plea for an "open world" by the need for

"... a world public opinion, which cannot exist or function as long as great countries are practically closed, and it is neither possible to learn what is happening within their territories, nor for their populations to obtain a proper picture of conditions abroad."

The only way out of this barren situation was "to make a stand for free access, on a mutual and worldwide basis, to all information essential for international relationship, and to make this a paramount issue."

"The revolutionary technical accomplishment", Bohr wrote, might already have served "as a temporary deterrent factor in world politics". At the same time he stressed its potential for opening up the world by

"... providing a possibility for renouncing a momentary advantage in return for far-reaching concessions required for common security, concessions which otherwise hardly would be obtainable."

Bohr allowed that "the American proposal to the Atomic Energy Commission of the United Nations" included steps toward openness. However, he repeated his conviction, already stated in the Comments, that its failure indicated

"... that progress can only be achieved by a line of policy, directly aimed at mutual openness on a still broader and more immediate basis."

Such "openness on a broader basis" involved

"... free access to all information about the conditions in every country, including technical development and military preparations."

Only such comprehensive openness "would enable the people of any country to form a well founded judgment on all questions decisive for its relations with other nations."

Bohr next addressed the possible risks involved in the implementation of openness. He envisaged

"... an arrangement covering [an] initiating period, upon which any formal offer of mutual openness must be conditioned."

Such an initiation period should ensure that all information "would be accessible in an appropriately balanced manner", taking into account that dissimilar conditions in different countries might

"... cause inequality in the ease and speed with which the first observers admitted to the various territories would be able to obtain essential information."

In order to reach a well-functioning arrangement, it was essential that the

"... fixation of final details should be postponed until the offer of openness was accepted in principle, whereafter such details would be worked out and settled in common, before the plan of admitting observers was put into operation."

A situation would thereby be created, Bohr argued,

"... in which good faith on the accepting, as well as on the inviting side, can and must be proved by performance."

Assessing the balancing to be achieved during the initiation period, Bohr noted two particular points:

"One is the very great advantage which a closed country now possesses in withholding information essential for strategic and tactical plans of opponents. The second point is the very minor value of technical information, obtained by personal observation and access to written reports, in relation to the actual implementation of the industrial program for military effectiveness."

Bohr invoked another argument for openness by referring to

"... the new and ominous menace to world security, presented by the possibility of employing the results of the latest development of bacteriological and biochemical science as terrible life-destructive means."

Since this menace under present conditions "cannot be eliminated by any practicable control," it will "remain a latent danger until ... cooperation in openness

[73]

and confidence has been achieved". In this regard, the prior development of nuclear weapons might in fact have proved fortunate.

This argument brought Bohr back to his case for the unique opportunity provided by the recent developments:

"Certainly, the advance in science and technology has presented our generation with a great responsibility, but also with a unique opportunity through the fact, that the only measures which can offer guaranties for common security are the same as those required for mutual understanding and genuine cooperation."

He stressed the need for early action:

"... the advantage of the initiative may be lost, and the possibility of security become remote, as soon as competitive efforts in atomic armaments have reached an advanced stage, or already at the moment when the probability of the existence of such competition can no longer be discounted."

Bohr now turned to Marshall's concern for the question of procedure. He realized that

"... the difficulty of conveying the actual situation in all its aspects ... might easily give rise to misunderstandings and anxiety ... if a proposal of openness were to be submitted to public discussion, without proper and responsible authority behind it."

Indeed, the reaction "might be unfavourable for the achievement of the desired aim." What was needed was

"... a statement from outstanding authority proposing an immediate approach to universal openness."

Such a statement "would constitute a challenge to the nation, which would appeal to the noblest traditions inherent in the birth and growth of American democracy", and would thus alleviate the "wide-spread despair and confusion" about the postwar international situation. "On this background", Bohr wrote,

"... the initiative will surely be greeted with eagerness, here and elsewhere, and provide a rallying point for a high and unifying ideal."

Bohr's preferred recommendation was that the Americans present a fully worked-out proposal for achieving openness; yet he recognized that preparatory steps might be needed:

"If the international and domestic situation favoured it, the most direct step would be a concrete proposal of universal openness, aimed at prompt realization. Yet, if such a procedure would not be deemed timely, it might be found suitable, after proper preparation, to use an early occasion, when restating the general lines of American policy, to stress the urgency of fullest mutual openness and express readiness to entertain proposals to this effect.

Even the preliminary approach, consisting of a declaration of aspiration and intent, might perhaps elicit an answer which could serve as a further stepping stone. But irrespective of such response, this initiative should contribute decisively to clarify the situation; it would greatly strengthen the moral position of all advocates of genuine international cooperation and bring adversaries everywhere in rapidly increasing difficulties."

In either case, a realization of the desired goal would require

"… a firm stand and a courageous fight for principles, in which everybody here and all over the world will be called upon to take part and do his share."

Bohr found

"… good reason to hope that a purposeful and imaginative exploitation of the unique opportunity would help to alleviate international tension and bring the world out of the present crisis."

He concluded his argument with yet another appeal to American values:

"This hope is inspired by the conviction that your country, in human and material resources, possesses the strength required to take the lead in accepting the challenge, with which civilization is confronted, and in establishing that cooperation, in openness and confidence, between all nations on which, not merely progress and prosperity of humanity, but its very existence depends."

Bohr was not to see Marshall again during his U.S. stay. However, after sending his letter to Marshall he did have opportunity to continue discussions in Washington with some of Marshall's advisers in the State Department especially connected with atomic energy developments. In preparation for these

"Annotations",
19 Jun 48
English
Full text on p. [169]

discussions Bohr worked out a set of "Annotations" in which he sought to clarify further such terms as "full mutual openness" and "initiating period" and pointed to the advantageous "psychological effects" of the process once the issue had been opened. In the latter regard Bohr observed that "an offer of full mutual openness and the actual admittance of observers may, in fact, be said to be an irreversible step" because of the "immense loss in political and moral position abroad and at home" for any country going back on its original agreement to take part in the process. The day before he returned to Denmark, Bohr had a last "long talk" with R. Gordon Arneson, one of the State Department advisers described by Bohr in his unsent letter to Anderson as "a most attractive and enlightened person". The conversation took place in New York, after Arneson had travelled all the way from Washington just for this purpose.

Bohr and Marshall did agree to meet again in Paris in the autumn, when Bohr could travel there from Brussels after attending the Solvay Conference from 27 September to 2 October 1948. Marshall, however, was too occupied with the still escalating Berlin crisis as a result of which Bohr was only able to speak to some of Marshall's advisers. There are no records of these discussions, which were not pursued.

By this time Bohr's ideas had been the subject of thorough discussion in the State Department, which produced a document starting out with the cutting conclusion that "Dr. Bohr's proposal for openness should not be made by the United States at this time."[82]

However, the State Department document indicates that Bohr's proposal was indeed taken seriously, the consensus being that "Dr. Bohr had hit upon the nub of the difficulties with the Soviet Union". Yet "only a most meticulous study of all ramifications of the proposal could reveal whether it was feasible and desirable" and such a study was deemed inopportune at this time "in view of the domestic political situation", including the forthcoming Presidential election.[83] For this reason, the document deals only superficially with Bohr's views and the consequences of his proposal, notably whether or not it would be accepted by the Soviet Union.

The seventh and last conclusion of the document states that "Dr. Bohr should be informed of the action decided upon by the Department."[84] Marshall noted in the margin of his copy that "I am not yet clear in my own mind about this".

---

[82] This is the first of seven numbered conclusions stated in the extensive *Memorandum for the Secretary of State* of 20 August 1948. FRUS 1948, Volume 1, Part 1, pp. 388–399, on p. 388.
[83] *Ibid.*, p. 391.
[84] *Ibid.*, p. 389.

Bohr never learned about this discussion of his views and had no opportunity to comment on it.

On 23 September 1949 President Truman announced that the Soviet Union had exploded its first atomic bomb on 28 August. The Soviet atomic explosion added life to the U.S. discussion of whether or not to develop the substantially more powerful hydrogen bomb, and on 31 January 1950 Truman decided, against the advice of some his closest scientific advisers including Oppenheimer, to go ahead with this project.

Bohr continued to follow developments closely, always on the alert to see whether his views might help put them in a new perspective. In preparation for his continued discussions with Marvel, for example, he drafted a statement proposing that the nations taking the initiative to the Atlantic Pact (or NATO) should link an offer of openness to the public announcement of NATO's establishment. By such an action, Bohr argued, "a decisive strengthening of the cause of freedom could be achieved." There is no indication that Bohr's suggestion was followed up.

Bohr notes,
1949
Danish
Full text on p. [322]
Translation on p. [323]

5. OPEN LETTER TO THE UNITED NATIONS, 1950

From the end of the war until early 1950 Bohr abstained from taking part in the public debate on specific issues raised by the advent of atomic armament. Instead, he continued to make use of his personal contacts with statesmen in Britain and the United States. In his confidential conversations and writings Bohr commented on developments and advocated the crucial importance of openness as the key to mutual trust and a prerequisite for reaching agreement on vital matters. He attempted to persuade the American government to take steps towards making openness a paramount issue.[85]

During the same period Bohr addressed the public on his general views on the implications of the existence of the atomic bomb.[86] His public writings made no mention of the views and recommendations that he had presented to governments during and after the war. In 1950 he decided to make these confidential reports public.

WIDENING THE AUDIENCE

In February 1950 Bohr travelled to the United States once more. He had originally contemplated spending some of his time in the United States writing

---

[85] See above, Section 4.
[86] See above, Section 3.

a comprehensive account of the general philosophical views he had presented in the Gifford Lectures in Edinburgh the preceding autumn.[87] These plans, however, had to be postponed due to developments in the political arena.

From discussions with members of the State Department, acquaintances in American political life and colleagues, Bohr realized that the political climate in Washington was not conducive to initiating an official offer of openness on the part of the United States government. He drew the conclusion that he must present the case himself in a public appeal. As he explained in the resulting publication:[88]

> "While possibilities still existed of immediate results of the negotiations within the United Nations on an arrangement of the use of atomic energy guaranteeing common security, I have been reluctant in taking part in the public debate on this question. In the present critical situation, however, I have felt that an account of my views and experiences may perhaps contribute to renewed discussion about these matters so deeply influencing international relationship."

Bohr addressed his appeal to the United Nations in an Open Letter which he wrote while a visitor to the Institute for Advanced Study in Princeton. It was composed with the help and advice of many colleagues and political acquaintances. By the time Bohr left the United States in early May, the letter was practically completed. Returning via London, Bohr had the opportunity to show it to Anderson, who fully supported Bohr's new approach. Anderson recommended a postponement of the letter's publication until June when Jan Christian Smuts was scheduled to visit London from South Africa. Both Bohr and Anderson considered it important to secure Smuts's advice and support. Smuts, however, was compelled to cancel his visit for reasons of health, and the letter was published without his advice.

THE OPEN LETTER AND ITS PRESENTATION

The Open Letter describes Bohr's wartime and postwar efforts, quoting extensively from the main documents that he had presented to the British and American governments. This account is followed by a comprehensive presen-

---

[87] On the Gifford Lectures, see Vol. 10, pp. [172]–[181]. Bohr's intention to prepare an article on the basis of these lectures while in the United States can be seen from the letters from Margrethe Bohr to Aage Bohr 28 October 1949, and Aage Bohr to Margrethe Bohr 21 November 1949. I am grateful to Aage Bohr for allowing me to see these letters which are part of his personal papers.

[88] Bohr, *Open Letter*, ref. 1, this volume on p. [175].

tation of the case for openness as the way to meet the challenge posed by the revolutionary development of science and technology. Bohr had not lost his conviction[89]

> "... that quite unique opportunities exist to-day for furthering co-operation between nations on the progress of human culture in all its aspects.
>
> I turn to the United Nations with these considerations in the hope that they may contribute to the search for a realistic approach to the grave and urgent problems confronting humanity. The arguments presented suggest that every initiative from any side towards the removal of obstacles for free mutual information and intercourse would be of the greatest importance in breaking the present deadlock and encouraging others to take steps in the same direction. The efforts of all supporters of international co-operation, individuals as well as nations, will be needed to create in all countries an opinion to voice, with ever increasing clarity and strength, the demand for an open world."

Bohr's "Open Letter to the United Nations" was delivered on 12 June 1950 by Aage Bohr, who was in New York at the time, to the secretary of Trygve Lie, the U.N. Secretary General. On the same day in Copenhagen, Bohr gave a brief address to the press at the Carlsberg Mansion. The letter, he said, was

> "... a contribution to the standing discussion all over the world about ways of meeting the serious crisis in civilization, resulting from the great development which has taken place in our time in the domain of science and technology."

Press release,
12 Jun 50
English
Full text on p. [188]

Referring to the letter itself for particulars, Bohr stressed his main point:

> "... the rapid development of technical resources created by science makes urgently necessary an adjustment of relations between nations in order to avoid mortal dangers to civilization."

Bohr considered two phrases in the press release sufficiently important to underline them. First, he emphasized the need for *"full mutual openness* as regards information about all aspects of the life of the individual nations, including social structure and technical development". Second, he pointed to the

---

[89] *Ibid.*, p. [185].

"unique opportunities for a *co-operation between all nations for the progress and safeguarding of civilization.*"[90]

IMMEDIATE REACTIONS

The reaction of the Danish press was as impressive as its coverage of Bohr's homecoming from the war.[91] It ran the story for several days after the publication of the Open Letter, and the major newspapers published Bohr's long letter in full the day after it was officially released.[92]

The Danish government's reaction as delivered in a statement by the Danish Prime Minister, Hans Hedtoft, also appeared in the press. While confirming the statement in the Open Letter that Bohr was "acting entirely on my own responsibility and without consultation with the government in any country",[93] Hedtoft stated:[94]

Press release,
13 Jun 50
Danish
Full text on p. [331]
Translation on p. [331]

"We in Denmark sincerely hope that the thoughts expressed in the Open Letter may be an impulse to serious deliberations in the minds of everybody who has a share in the responsibility for the future of our world."

The Copenhagen communist daily "Land og Folk" ("Land and People") took a different line, seeing a chance to exploit Bohr's views in favour of its own campaign for the "Stockholm appeal". This appeal for peace, signed by a great number of people, demanded the "prohibition of the atomic weapon" and the "establishment of international control", thus reflecting the Soviet position in the UNAEC negotiations. Two days after the release of the Open Letter, "Land og Folk" carried the banner headline, "Niels Bohr's Open Letter serious inducement for signing the peace appeal".[95] This claim led Bohr to write yet another press release containing the Stockholm appeal itself followed by Bohr's explanation of why he could not support it:

**Press release,
14 Jun 50
English
Full text on p. [189]**

"From the content of the open letter it will be understood that I cannot join any appeal, however well-meant it might be, which does not include the clearly expressed demand of access to information about conditions in

---

[90] The underlined passages in the original are reproduced in italics.

[91] See above, p. [60].

[92] See the references on p. [172].

[93] Bohr, *Open Letter*, ref. 1, this volume p. [175].

[94] The translation, which is not quite exact, is taken from an *aide-mémoire* of 21 June 1950 prepared by Danish ambassador Henrik Kauffmann for the U.S. Department of State, FRUS 1950, Vol. 1, pp. 78–79, on p. 78.

[95] Land og Folk, 14 June 1950.

all countries and of fully free exchange of ideas within every country and across the boundaries."

In contrast to the Danish response to the Open Letter, only limited interest was shown by media abroad; despite requests from Bohr's contacts, no British or American newspaper printed it. In the U.S. it appeared in the weekly "Science" as well as in the "Bulletin of the Atomic Scientists". The lack of interest in the Open Letter in the general press was compounded by the outbreak of the Korean war only days after the release of Bohr's letter. The Korean crisis, in which the United Nations was strongly involved, served to divert the attention not only of the media, but also of the statesmen and the United Nations itself. Bohr's letter would never be put on the United Nation's agenda.

CONTINUATION

After he had presented his case to the public in his Open Letter, Bohr's activities for an "open world" turned into a different kind of campaign. From now on, people and institutions sympathizing with his views could help promote them, either publicly or in confidence within the international political environment. For example, the governments of the Scandinavian countries – Denmark, Norway and Sweden – joined Bohr in moving towards the realization of his idea.

Bohr's main hope and strategy were still directed towards persuading the United States of the benefit that would accrue from a universal offer of an "open world", and it was essentially with this in mind that the Scandinavian countries sought to advance his views. At the same time, Bohr's interaction with people from many nations sharing his concerns expanded considerably, and he found widespread response to his ideas.

Bohr's home at Carlsberg became a meeting place for statesmen and other people in public life from many countries, as well as scientists and artists. The discussions tended to range over many issues, but almost invariably they would turn to the great change in the relationship between nations demanded by the advance of science and technology. Similarly, in his many talks and articles during this period concerning the epistemological lesson of atomic physics, he closely integrated the issue of the "open world" with his general philosophical outlook.[96]

In 1956, at the time of the Hungarian crisis, Bohr sent what he termed a "continuation" of his Open Letter to the U.N. Secretary General Dag Ham-

---

[96] See also Vols. 10 and 12.

marskjöld who had succeeded Trygve Lie. Bohr now proposed that the United Nations itself take an initiative toward openness. He considered that such an action would strengthen the potential of the organization for creating confidence between nations and, consequently, for upholding the peace:[97]

> "Besides lending support to immediate urgent endeavours, it would seem that an essential strengthening of the potentialities of the Organization for creating confidence between nations might result from a general declaration of adherence to the principle of openness.
>
> In particular, all nations might be called upon to declare their readiness to assist the Organization in obtaining the information necessary for impartial judgement of any issue with which the organization may be confronted."

Bohr's new letter received little attention.

The Indian Prime Minister Jawaharlal Nehru visited Bohr at Carlsberg in 1957, and when Bohr travelled to India in 1960, the two continued their discussion about how to reduce tensions in the world and the role that openness might play in this connection. On a visit to the Soviet Union the following year, Bohr sought a meeting with the Russian leader Nikita Khruschev which, however, did not materialize.

Thus, after 1950, just as before, Bohr persevered in promoting understanding for the issues he considered so vital for the future of mankind. Indeed, this stage in Bohr's efforts constitutes an instructive and many-faceted story in its own right. Because the issues were now openly discussed, the archival material is substantially richer than for the preceding period. Since Bohr's efforts, however, were now based to a greater extent on informal personal contacts, and since others helped in advancing his views, they ceased to produce publications or "publishable" writings from Bohr's own hand, the only exception being the second letter to the United Nations just described. Bohr's political activities after 1950, therefore, lie outside the scope of the Niels Bohr Collected Works.

6. CONCLUSION

Bohr's political efforts did not lead to the radical initiatives that he called for. However, his argument for openness as a key element in international relationships, presented with such conviction and perseverance, left a lasting legacy. To Bohr, it was a question of guiding human civilization in a direction that the progress of science and technology makes necessary for its very survival.

---

[97] Open letter from Bohr to Hammarskjöld, ref. 2; this volume, pp. [191] ff.

The point of departure for Bohr's campaign for openness had been his recognition, which went back to his efforts for international scientific cooperation during the prewar years, that the lack of free communication between the Soviet Union and the rest of the world was a decisive obstacle to any kind of real cooperation. It was against this background that Bohr saw the advent of the atomic bomb as a unique opportunity to break the isolation.

Bohr lived to see a thaw in international scientific cooperation connected with President Dwight D. Eisenhower's "Atoms for Peace" programme first presented publicly in Eisenhower's speech to the General Assembly of the United Nations on 8 December 1953.[98] The thaw continued with the establishment of the International Atomic Energy Agency a year later and the Geneva Conference on Peaceful Uses of Atomic Energy in 1955, where Bohr gave an important address.[99] Bohr's institute was among the first to take advantage of the new possibilities. Indeed, the Copenhagen institute's tradition for international cooperation proved particularly useful in the opening of this new phase in the relationship among scientists.

It was only many years after Bohr's death that the end of the Cold War led to far-reaching steps towards freer communication both inside the former Eastern bloc and in the relationship of these nations with the rest of the world. In this historical process the quest for freer communication was itself an important driving factor, and a gradual removal of barriers to communication between East and West occurred at many levels. Thus, with regard to military matters, the need for verification became a crucial element in the efforts to introduce arms control. At a critical stage, the Soviet Union itself attempted to introduce the policy of "glasnost" involving greater openness. As for the role of science, the Soviet government's recognition of the vital need for full participation in international developments made it possible for scientists there to make increasingly strong demands for freer international contact and cooperation.

There were, of course, many different factors and special circumstances responsible for the end of the Cold War. Among these, Bohr's efforts helped in creating an atmosphere conducive to the process. It would have given him special satisfaction to see how the bonds between scientists contributed to promoting confidence between nations, thereby fulfilling promises that, throughout his career as a scientist, he had considered a necessary and integral part of scientific endeavour.

---

[98] See also the introduction to Part II, p. [366].

[99] N. Bohr, *Physical Science and Man's Position* in *Proceedings of the International Conference on the Peaceful Uses of Atomic Energy, Volume 16, Record of the Conference*, United Nations, New York 1956, pp. 57–61. Reproduced (from Ingeniøren **64** (1955) 810–814) in Vol. 10, pp. [102]–[106].

I. DOCUMENTS, WORLD WAR II

.

Doc. no.	Description	Date	Reproduced p.	Quoted p.
1	Letter to Anderson	16 February 1944	86	17
2	"Message" to Anderson	April 1944	89	22
3	Memo	2 April 1944	90	22
4	Letter to Churchill	22 May 1944	96	27
5	Memo	23 May 1944	99	28
6	"Memorandum"	3 July 1944	101	32
7	Letter to Roosevelt	7 September 1944	109	36
8	"Comments"	18 January 1945	111	43
9	"Addendum"	24 March 1945	113	44
10	Memo	8 May 1945	118	47
11	Memo	12 July 1945	119	48

Document 1:
LETTER TO ANDERSON, 16 February 1944
[Typewritten]
[Source: CAB 126/39; typewritten transcript in BPP 2.2]

Washington, D.C.   February 16th 1944.

PERSONAL.   MOST SECRET.

The Rt. Hon. Sir John Anderson M.P.
Chancellor of the Exchequer.
Westminster, London.

Dear Sir John.

Through your introduction I was, as you will know, received by Lord Halifax shortly after my first arrival in Washington. On my return from a three weeks visit to "Y",[1] I called again at the Embassy and have in the last weeks had various conversations with the Ambassador[2] and with Sir Ronald Campbell[3] who, according to the Ambassador's request was brought into the discussion of the matter. I understand that Lord Halifax will himself communicate with you, but I presume that you may be interested in a brief report about my experiences and endeavours since my departure from London.

As regards my own status no change has so far been proposed from the American side, and my stay in "Y" took place according to an invitation from General Groves who, without any special commitments except for security purposes, has agreed that I and my son go out to "Y" on temporary visits for consultations with the American and British scientists working there.

I was most impressed by the magnitude of the effort and the progress of the work already achieved. Thus all problems concerning the mechanism of the fundamental processes have been thoroughly investigated and in particular it has been proved that in accordance with the theoretical expectations no disturbing time delay in the neutron emission occurs. As regards the scientific basis no doubt is therefore left as to the realizability of the project.

[1] Los Alamos National Laboratory.
[2] Lord Halifax (1881–1959).
[3] Halifax's deputy at the British Embassy in Washington.

Also on the preparations for the construction of the actual weapon a great amount of work has been done in "Y" and various ingenious methods by which smaller amounts of material than previously estimated may effect a release of nuclear energy on a large scale have recently been proposed. Much experimental work, however, will still be needed to test the efficiency of such methods.

In view of the decisive importance that all such preparations are completed before the material is at hand I have on my return to Washington in conversations with General Groves stressed the urgency that just this aspect of the work be promoted to the outmost, and he agrees upon the necessity of extending the work at "Y" with new facilities for carrying out large scale experiments on the latest devices for the assembly of the weapon. I know that Dr. Chadwick has proposed that the British collaboration on the project, not least in "Y", if possible be still further extended, and I wish most heartily to support such advice.

In Washington I have been in close connection with British scientists who frequently come for meetings from the various places where the work on the production of the materials take[s] place and to which I have hitherto had no access. Through discussions, especially with Dr. Oliphant[4] who is here on his way to England, I have obtained most encouraging information about the recent great progress of these truly gigantic efforts, which has secured an impending realization of the whole project.

This situation certainly calls for the most thorough attention to all implications of the project, and ever since our last conversation in London during which you showed me the confidence of indicating to me your concern about this aspect of the matter, such problems have continually been on my mind.

I know that the question of control has already been considered within your committee, but the more I have learned and thought about the possible development in this new field of science and technique, the more I am convinced that no kind of customary measures will suffice for this purpose and that no real safety can be achieved without a universal agreement based on mutual confidence.

An effective control would in fact not only involve intricate technical and administrative problems, but would also demand such concessions regarding exchange of information, and openness about industrial efforts including mili-

---

[4] The Australian physicist Mark Oliphant (1901–2000) had worked with Rutherford at the Cavendish Laboratory from 1927 to 1936, the last two years as Assistant Director of Research, whereupon he accepted a professorship at the University of Birmingham. He was a leading figure in the British group of scientists working on the atomic bomb.

[87]

tary preparations, as would hardly be conceivable unless at the same time all partners were assured of a compensating guarantee of common security against dangers of unprecedented acuteness.

In this connection it might even be necessary temporarily to renounce the great promises of the project as regards peaceful industrial development and leave the exploitation of such possibilities until by universal consent it is deemed that an established control will prevent any illegitimate use of the powerful new materials.

All such details will depend on technical developments which cannot be surveyed at the moment, but the main point of the argument is that the impending realization of the project would not only seem to necessitate, but should also, due to the urgency of confidence, facilitate, a new approach to the problem of international relationship.

I am well aware that I here enter upon questions about which only the responsible statesmen who alone have insight in the political possibilities are able to judge, but stressing the uniqueness of the situation which the progress of science has brought about I venture to point to the favourable consequences which might result from an early initiative from the side which by good fortune has obtained a lead in the efforts to master forces of nature hitherto beyond human reach.

Such an initiative, aimed at preventing a future competition about the formidable weapon should in no way impede the importance of the project for immediate military objectives, but might by strengthening the confidence within the United Nations turn this unparalleled enterprise to lasting benefit for the common cause.

Of course, I appreciate the necessity of separating sharply between the actual work on the project and its possible implications of general political character, just as I realize the serious complications for the handling of the whole situation which might arise from discussions of such implications even within the circle of officials and technicians charged with the development of the project.

I have, therefore, been most grateful for the privilege you have given me of bringing my impressions to the notice of Lord Halifax who, as well as Sir Ronald Campbell, has taken great interest in the matter. I understand that the Ambassador shares the view that the situation requires an early attention in order that no opportunity be lost.

Yours sincerely

N. Bohr

Document 2:
"MESSAGE" TO ANDERSON, April 1944
[Carbon copy]
[Source: BPP 2.5]

April 1944.

Most confidential and secret message to the Chancellor   personal.

Due to my participation in international scientific and cultural endeavours I had, as you will know, already many years before the war formed intimate personal relations not only in England, but also in America. These general interests, which as arranged were given as the official object for my journey to America, brought me on my arrival there naturally in connection with various confidential friends who share such interests. Of course, I could never, not even in confidence, have talked about the project with anyone of whom I did not on beforehand know that he was already informed about it through legitimate sources. Some of my American friends, however, told me without my asking that they possessed such legitimate information and that they were deeply concerned with the implications of the project. I felt it therefore not only proper, but almost a duty, to tell them in a general way, without entering upon matters of military secrecy, about the hopes which I thought the situation inspired. On my return to Washington from my recent visit to "X"[5] and "Y", I was, in utmost confidence and under the most urgent pledge not to disclose any personal sources, trusted with the information that the President, quite apart from all connections with his technical advisers, in his own mind is deeply occupied with the immense consequences of the project, and that he realizes he shares the responsibility for the handling of the matter to the greatest possible benefit for the common cause solely with the Prime Minister; further I was assured that the President would not only welcome, but is eager from the Prime Minister to receive any suggestions for dealing with this problem, so fraught with danger, but also with great hope for mankind. I need hardly add that the President is quite unaware of my activities and even of my presence in U.S.A. so that in no sense did I take part in even unofficial political discussions. Only in view of my old American relations I had the fortunate opportunity of being able to learn with assurance the directions of the President's mind.

---

[5] The isotope separation plant at Oak Ridge, Tennessee.

Document 3:
MEMO, 2 April 1944
[Carbon copy]
[Source: BPP 7.1]

April 2nd 1944.

Confidential comments on the project of exploiting the latest
discoveries in atomic physics for industry and warfare.

I. Foundations of the project.

The recent fundamental advances of physical science have opened unexpected ways of releasing the energy bound in matter, and thus not only created quite new perspectives for future technical and industrial development, but have at the same time given birth to the project of procuring a military weapon with a destructive power far beyond all prior possibilities.

The radical difference between this project and all previous technical efforts, including industrial endeavours as well as military preparations, originates from the essentially different character of the atomic processes involved.

As we have learned in course of the last generation, each atom consists of a cluster of electrons held together by the attraction from an electrified nucleus, which, although it contains practically the whole mass of the atom, has a size extremely small compared with atomic dimensions, determined by the extension of the electron configuration around the nucleus.

When materials are exposed to ordinary physical agencies, any change in atomic constitution is confined to distortion or disrupture of the electron structure. Also in chemical processes the atomic nuclei are left entirely unaltered, and the formation and dissociation of the molecules of chemical compounds involve only an exchange of electrons between the atoms of the elements concerned.

A whole new epoc[h] of science has, however, been initiated by the discovery that it is possible, by special agencies like the high-speed particles emitted by Radium, to produce changes in the constitution of the atomic nuclei themselves and to cause nuclear disintegrations accompanied by a release of energy, millions of times larger than the energy liberated in the most violent chemical processes.

These discoveries have not only made it possible for the first time to transform one element into another, but have even disclosed the origin of the hitherto quite unknown energy sources present in the interior of the stars, and thereby afforded the explanation of, how our sun has been able through billions of years to emit the powerful radiation upon which all organic life on the earth is dependent.

The rapid exploration of this novel field of research, promoted by a most energetic and fruitful international collaboration, has led to a number of important discoveries regarding the intrinsic properties of atomic nuclei and has in particular revealed the existence of a non-electrified nuclear constituent, the so-called neutron, which, when set free, is an especially active reagent in producing nuclear transmutations.

The actual basis for the project in question was the discovery made just in the year before the war, that the nuclei of the heaviest elements like Uranium, by neutron bombardment may split in fragments ejected with enormous energies, and that this process is accompanied by the release of further neutrons, which may themselves effect the splitting of other heavy nuclei.

Just the possibility of such nuclear chain reactions, closely resembling the mechanism of ignition of combustible substances, has opened the perspectives of obtaining an energy release from heavy materials of a magnitude which per weight is about 100,000,000 times as large as that obtainable by the chemical processes used in ordinary explosives.

An immediate accomplishment of this prospect was hindered, however, by certain peculiar circumstances preventing the development of chain reactions in the heavy materials directly obtainable from natural ores. All Uranium on Earth consists in fact of a mixture of two modifications, or isotopes, of which only the comparatively rare [one] lends itself to such effects, while the more abundant isotope is not capable of maintaining chain reactions.

It was, therefore, realized that it would be necessary, either from ordinary Uranium samples to extract the rare active isotope in sufficiently large amounts, or through special devices involving the admixture of certain light elements to common Uranium, to secure that the presence of the inactive isotope does not prevent chain reactions between the nuclei of the active isotope to develop. Efforts in both these directions were at once pursued in several countries, but up to the war no decisive progress had been obtained.

Since the outbreak of hostilities no further contributions to this subject has been made public, but it is known that in connection with its possible importance as a war effort much attention has been given to the project by

[91]

various governments. All results of the work are, of course, kept as military secrets, but from leakage regarding the activities of German scientists it is practically certain that no substantial progress has been achieved by the Axis Powers.

Within the United Nations the project has been taken up as a joint enterprise of the United States of America and the British Commonwealth and a most efficient scientific and engineering organization has been formed, at the disposal of which enormous material and technical resources are placed. Even in these confidential comments it is hardly proper to enter upon details, guarded with utmost secrecy, and for the purpose it may be sufficient to state that a realization of the project as originally planned is not only impending, but that the vigorous exploration of this new field of research has steadily revealed further possibilities of even more stupendous implications.

No matter how the latest perspectives ultimately will take shape, it is already evident that we are presented with one of the greatest triumphs of science and technique, destined deeply to influence the future of mankind.

II. General implications of the project.

It surely surpasses the insight and imagination of anyone to survey the consequences which the project will bring about in coming years. The enormous energy sources which will be available may be expected in the long run to revolutionize industry and transportation and the exploration of the multivaried effects which accompany the nuclear transmutations promises results of perhaps equal importance for human welfare.

The fact of immediate preponderance is, however, that a weapon of an unparalleled power is being created which will completely change all future conditions of warfare. Quite apart from the question of the role which this weapon may play in the present war, its impending accomplishment raises a number of problems which call for most urgent attention. Unless, indeed, some agreement about the control of the use of the new active materials can be obtained in due time, any temporary advantage, however great, may be outweighed by a perpetual menace to human security.

Ever since the possibilities of releasing atomic energy on a vast scale came in sight, much thought has naturally been given to the question of control, but the further the exploration of the scientific problems concerned is proceeding, the clearer it becomes that no kind of customary measures will suffice for this purpose and that especially the terrifying prospect of a future competition between nations about a weapon of such formidable character can only be avoided through a universal agreement implying a radical revision of present ideas of national sovereignty.

An effective control will in fact not only involve intricate technical and administrative problems, but will also demand such concessions regarding exchange of information, and openness about industrial efforts including military preparations, as would hardly be conceivable unless at the same time all partners were assured of a compensating guarantee of common security against dangers of unprecedented acuteness.

For this purpose it might even be necessary temporarily to renounce the great promises of the project as regards peaceful industrial development and leave the exploitation of such possibilities until by universal consent it is deemed that an established control will prevent any illegitimate use of the dangerous materials.

The ultimate form of the control arrangement must, of course, largely depend on the future development and at any stage, advice of scientists and technicians will surely be needed. The main point of the argument is, however, that the accomplishment of the project would not only seem to necessitate, but should also, due to the urgency of mutual confidence, facilitate, a new approach to the problem of international relationship.

Far from being a hindrance to the free development of national communities or to the rivalry between nations as regards human progress, a universal agreement of the kind indicated should just offer a solid foundation for such endeavours in eliminating the menace of a suppression by materially superior powers and thus securing a lasting peaceful international collaboration.

III. Possibilities of the momentary situation.

The present moment, where almost all nations are entangled in a deadly struggle for freedom and humanity, might at first sight seem most unsuited for any committing arrangement concerning the project. Not only have the ag[g]ressive powers still great military strength, although their original plans of world domination have been frustrated and it seems certain that they must ultimately surrender, but even when this happens, the nations united against ag[g]ression will face most serious causes of disagreement due to conflicting attitudes towards social and economic problems.

By a closer consideration, however, it would appear that the potentialities of the project as a means of inspiring confidence just under these circumstances acquire most actual importance. Moreover, the momentary situation would in various respects seem to afford quite unique possibilities which might be forfeited by a postponement.

Although according to all available information no substantial progress as regards the project has been made outside the United States of America and the British Commonwealth, this state may change as soon as a relief of the strain of war permits other powers like the Soviet Union, where the necessary resources are at disposal, to concentrate on similar efforts.

Not least in view of such eventualities the situation would, therefore, seem to offer a most favourable opportunity for an early initiative from the side which by good fortune has achieved a lead in the efforts of mastering mighty forces of nature hitherto beyond human reach.

Such an initiative, aiming at forestalling a fateful competition about the formidable weapon, need in no way impede the importance of the project for the immediate military objectives, but should serve to uproot any cause of distrust between the powers on whose harmonious collaboration the fate of coming generations will depend.

An attitude inviting confidence by extending generosity might indeed be the only way of furthering a development of the relations between nations, not detrimental to the hopes which in spite of all sufferings and sacrifices have kept up the courage everywhere in the world where the ideals of freedom and humanity prevail.

Many reasons would also seem to justify the conviction that a proper approach, with the object of establishing common security from ominous menaces without excluding any nation from participating in the promising industrial development which the accomplishment of the project entails, will be welcomed with sincerity and be responded with a loyal cooperation on the effectuation of the necessary far reaching control measures.

The enthusiasm and gratitude with which an agreement ensuring real international cooperation everywhere would be received is certainly a source of greatest encouragement and inspiration. Notwithstanding divergent views about ways and means, every nation will at that great moment find itself united in the deliverance from sombre anxieties and in pride and admiration for the gigantic scientific and engineering effort which as a contribution to the common cause will rank with the most heroic and successful military undertakings.

Of course, the responsible statesmen alone are in a position to judge the actual political possibilities, but in this unprecedented situation, brought about by the advance of science, a helpful support for their endeavours might perhaps be found in the world-wide scientific collaboration which, until it was hampered by the revival of racial prejudices and the fear of impending ag[g]ression, developed so fruitfully in the years between the world wars and embodied such bright promises for common human striving.

This international collaboration was shared not only by the western democracies where the pursuit of science through centuries has been offered so favourable conditions; but not least in the Soviet Union, where purposeful endeavours were made to encourage scientific research, such collaboration was actively promoted by arranging international scientific conferences and facilitating the participation of Soviet citizens in research work in other countries.

It is true that these efforts were gradually abandoned in the critical years before the war, where military preparations against ag[g]ression demanded a rigid control of all information. Still, the ties have not been broken and among eminent Russian scientists, many of whom no doubt have rendered great services in their country's heroic defence, one can reckon to find enthusiastic supporters of universal cooperation.

On this background personal connections between scientists in confidence of the American and British Governments on the one hand and the Soviet Government on the other hand might possibly be of help in establishing preliminary and non-committing contact, in case the two first Governments, after weighing all circumstances, decide that before the surrender of Germany an attempt should be made to achieve, at least in principle an agreement about the project with the other great powers among the United Nations.

Such suggestions imply no underrating of the difficulty and delicacy of the steps to be taken by the Governments in order to obtain an arrangement satisfactory to all concerned, but should only point to some aspects of the situation which may ease such endeavours. Should the efforts be crowned with success, the project will surely have brought about a turning point in history and this wonderful adventure will stand as a symbol of the benefit to mankind, which science can offer when handled in a truly human spirit.

Document 4:
LETTER TO CHURCHILL, 22 May 1944
[Carbon copy]
[Source: BPP 2.1]

22nd May 1944.

The Rt. Hon. Winston S. Churchill, C.H., M.P.

Sir,

In accordance with your kind permission, I have the honour to send you a brief report about my impressions of the great Anglo–American enterprise, in the scientific aspects of which I have been given the opportunity to participate together with my British colleagues.

The principles on which the enormous energy stored in the nuclei of atoms may be released for practical purposes were, as a result of international scientific collaboration, already perceived in outline before the war and are, therefore, common knowledge to physicists all over the world. It was, however, by no means certain whether the task would surpass human resources, and it was therefore a revelation to me, on my arrival in England last October, to learn with what courage and foresight the effort had been undertaken and what an advanced stage the work had already reached.

In fact, what until a few years ago might be considered as a fantastic dream is at present being realized within great laboratories and huge production plants secretly erected in some of the most solitary regions of the United States. There a larger group of physicists than ever before collected for a single purpose, working hand in hand with a whole army of engineers and technicians, are preparing new materials capable of an immense energy release, and are developing ingenious devices for the most effective use of these materials.

To everyone who is given the opportunity to see for himself the refined laboratory equipment and the gigantic production machinery, it is an unforgettable experience, of which words can only give a poor impression. Moreover it was to me a special pleasure to witness the most harmonious and enthusiastic cooperation between the British and American colleagues, and on my departure I was expressly asked by the leaders of the American organization to convey their genuine appreciation of the help they are receiving, on an ever increasing scale, from their British collaborators.

I will not tire you with any technical details, but one cannot help comparing the situation with that of the alchemists of former days, groping in the dark in their vain efforts to make gold. To-day physicists and engineers are, on the basis of firmly established knowledge, controlling and directing violent reactions by which new materials far more precious than gold are built up, atom by atom. These processes are in fact similar to those which took place in the early stages of development of the universe and still go on in the turbulent and flaming interior of the stars.

The whole undertaking constitutes, indeed, a far deeper interference with the natural course of events than anything ever before attempted, and it must be realized that the success of the endeavours has created a quite new situation as regards human resources. The revolution in industrial development which may result in coming years cannot at present be surveyed, but the fact of immediate preponderance is, that a weapon of devastating power far beyond any previous possibilities and imagination will soon become available.

The lead in the efforts to master such mighty forces of nature, hitherto beyond human reach, which by good fortune has been achieved by the two great free nations, entails the greatest promises for the future. The responsibility for handling the situation rests, of course, with the statesmen alone. The scientists who are brought into confidence can only offer the statesmen all such information about technical matters as may be of importance for their decisions.

In this connection it is significant that the enterprise, immense as it is, has still proved to demand a much smaller effort than might have been anticipated, and that the development of the work has continually revealed unsuspected possibilities for facilitating the production of the materials and for intensifying their effects.

These circumstances obviously have an important bearing on the question of an eventual competition about the formidable weapon, and on the problem of establishing an effective control, and might therefore perhaps influence the judgment of the statesmen as to how the present favourable situation can best be turned to lasting advantage for the cause of freedom and world security.

I hope you will permit me to say that I am afraid that, at the personal interview with which you honoured me, I may not have given you the right impression of the confidential conversation in Washington on which I reported. It was, indeed, far from my mind to venture any comment about the way in which the great joint enterprise has been so happily arranged by the statesmen; I wished rather to give expression to the profound conviction I have met everywhere on my journey that the hope for the future lies above all in the most brotherly friendship between the British Commonwealth and the United States.

[97]

It was just in this spirit of co-operation that the President's friend,[6] believing the matter to be of the highest importance for the two countries, and knowing that, at the Chancellor's request, I was coming to England for technical consultations, entrusted me, in strictest confidence, to convey to you, that the President is deeply occupied in his own mind with the stupendous consequences of the project, in which he sees grave dangers, but also unique opportunities, and that he hopes together with you to find ways of handling the situation to the greatest benefit of all mankind.

Most respectfully,
[Niels Bohr]

---

[6] Felix Frankfurter.

Document 5:
MEMO, 23 May 1944
[Carbon copy]
[Source: BPP 7.4]
[Editor's comment: Written in Aage Bohr's handwriting on the top of the document: "Note given to the Chancellor".]

23rd May 1944.

The starting point for the argument is that the success of the project, achieved in wartime as a joint Anglo–American enterprise, would not only seem to have secured a decisive lead in the production of the new formidable weapon, but would also seem to offer a unique opportunity of establishing a foundation for international collaboration, in conformity with the ideals for which the war is being fought.

Even without inside knowledge of political affairs one cannot indeed help feeling that the situation which will arise, when the nations united against aggression shall settle their future relationship, must contain great complications due to different attitudes towards social and economic problems.

If in this situation the promises which the project entails could not be taken into account, the prospects would hardly be comforting. Not least would the huge armaments, which will have been accumulated by the end of the war, represent a threat to the peaceful co-operation so needed for human progress.

The situation created by the project should indeed facilitate a new approach to international relationship through the recognition that the project presents such obvious dangers to everyone, if a competition about it is allowed to start, that most unaccustomed steps will be necessary to restore world security.

I am well aware that the considerations expressed in the notes I showed you on my return to England must appear very superficial, but as I explained, they were only intended, without entering on political problems in which I have no insight, to point to some aspects of the situation which might be of interest for the statesmen to consider. I wish, however, to stress that I in no way underrate the difficulties, which truly are so great that they might be discouraging to anyone who was not convinced of the compelling nature of the task.

The question will surely be raised what kind of arrangement is actually suggested, but I think that it must first be asked what the situation would be if an arrangement about the project, however difficult, cannot be obtained. In particular it seems to me that unless the question of the concessions which the various nations would be prepared to make was brought up in the negotiations among the great powers, it would never be possible for any one of them to know what their own interests really are and with what intentions from the side of the others they will have to reckon.

Only when it is recognized that security cannot be obtained without an arrangement in true confidence, will a foundation be created for negotiations about an effective control, not only of the new weapon but also of all other military preparations, which would offer all partners the possibility of developing their own problems without fear of aggression.

Quite apart from the situation here presented, where it is presupposed not only that America and England have the lead in the project, but that no real progress has been made elsewhere, there are perhaps even immediate dangers which can only be eliminated by an early initiative to establish confidence among the United Nations.

Not only may preparations for a competition be under way in U.S.S.R., and if matters are left to themselves until the war against Germany and Japan is finished the great opportunity may be lost, but it cannot even be excluded that in Germany, especially if information of the success of the Anglo–American enterprise has leaked out, some initial progress may recently have been made which may bring them into a position to offer substantial help to their present enemies if they could persuade them of the insincerity of intentions of others among the United Nations.

I do not wish to suggest that any such developments are likely, but only to point to the fateful consequences which might arise if the problem of handling the project to the benefit of the common cause is considered as a question which can be separated entirely from the conduct of the war.

Document 6:
"MEMORANDUM", 3 July 1944
[Carbon copy]
[Source: BPP 8.2; also in CAB 126/39]
[Editor's comment: Some of the passages are taken, even literally, from Documents 3 and 4. The **boldface** represents the parts of the "Memorandum" that Bohr chose to quote in his *Open Letter to the United Nations* (reproduced on p. [171]) six years later. The BPP contains versions with and without the last paragraph. The entire Memorandum, as received by Roosevelt (including the additional paragraph), has been reproduced in *Roosevelt and Frankfurter: Their Correspondence 1928–1945* (ed. M. Freedman), Little, Brown and Company, Boston 1967, pp. 728–735.]

TOP SECRET.

July 3rd 1944.

M E M O R A N D U M.

The project of releasing, to an unprecedented scale, the energy bound in matter is based on the remarkable development of physical science in our century which has given us the first real insight in the interior structure of the atom.

This development has taught us that each atom consists of a cluster of electrified corpuscles, the so-called electrons, held together by the attraction from a nucleus which, although it contains practically the whole mass of the atom, has a size extremely small compared with the extension of the electron cluster.

By contributions of physicists from nearly every part of the world, the problems of the electron configuration within the atom were in the course of relatively few years most successfully explored and led above all to a clarification of the relationship between the elements as regards their ordinary physical and chemical properties.

In fact, all properties of matter like hardness of materials, electric conductivity and chemical affinities, which through the ages have been exploited for technical developments to an ever increasing extent, are determined only by the electronic configuration and are practically independent of the intrinsic structure of the nucleus.

This simplicity has its root in the circumstance that by exposure of materials to ordinary physical or chemical agencies, any change in the atomic constitution is confined to distortion or disrupture of the electron cluster while the atomic nuclei are left entirely unchanged.

The stability of the nuclei under such conditions is in fact the basis for the doctrine of the immutability of the elements which for so long has been a fundament for physics and chemistry. A whole new epoque of science was therefore initiated by the discovery that it is possible by special agencies, like the high-speed particles emitted by Radium, to produce disintegrations of the atomic nuclei themselves and thereby to transform one element into another.

The closer study of the new phenomena revealed characteristic features which differ most markedly from the properties of matter hitherto known, and above all it was found that nuclear transmutations may be accompanied by an energy release per atom millions of times larger than the energy exchanged in the most violent chemical reactions.

Although at that stage no ways were yet open of releasing for practical purposes the enormous energy stored in the nuclei of atoms, an immediate clue was obtained to the origin of the so far quite unknown energy sources present in the interior of the stars, and in particular it became possible to explain how our sun has been able through billions of years to emit the powerful radiation upon which all organic life on the earth is dependent.

The rapid exploration of this novel field of research in which international co-operation has again been most fruitful led within the last decenniums to a number of important discoveries regarding the intrinsic properties of atomic nuclei and especially revealed the existence of a non-electrified nuclear constituent, the so-called neutron, which when set free is a particularly active reagent in producing nuclear transmutations.

The actual impetus to the present project was the discovery made in the last year before the war, that the nuclei of the heaviest elements like Uranium by neutron bombardment, in the so-called fission process, may split in fragments ejected with enormous energies, and that this process is accompanied by the release of further neutrons which may themselves affect the splitting of other heavy nuclei.

This discovery indicated for the first time the possibility, through propagation of nuclear disintegrations from atom to atom, to obtain a new kind of combustion of matter with immense energy yield. In fact, a complete nuclear combustion of heavy materials would release an energy 100,000,000 times larger than that obtainable by the same amount of chemical explosives.

This prospect not only at once attracted the most wide spread interest among physicists, but of its appeal to the imagination of larger circles I have vivid recollections from my stay in U.S.A. in the spring of 1939 where, as guest of the Institute of Advanced Studies in Princeton, I had the pleasure to participate together with American colleagues in investigations on the mechanism of the fission process.

Such investigations revealed that among the substances present in natural ores, only a certain modification of Uranium fulfils the conditions for nuclear combustion. Since this active substance always occurs mixed with a more abundant, inactive Uranium modification, it was therefore realized that in order to produce devastating explosives, it would be necessary to subject the available materials to a treatment of an extremely refined and elaborate character.

The recognition that the accomplishment of the project would thus require an immense technical effort, which might even prove impracticable, was at that time, not least in view of the imminent threat of military aggression, considered as a great comfort since it would surely prevent any nation from staging a surprise attack with such super weapons.

Any progress on nuclear problems achieved before the war was, of course, common knowledge to physicists all over the world, but after the outbreak of hostilities no further information has been made public, and efforts to exploit nuclear energy sources have been kept as military secrets.

During my stay in Denmark under the German occupation nothing was, therefore, known to me about the great enterprise in America and England. It was, however, possible, due to connections originating from regular visits of German physicists to the Institute for Theoretical Physics in Copenhagen in the years between the wars, rather closely to follow the work on such lines which from the very beginning of the war was organized by the German Government.

Although thorough preparations were made by a most energetic scientific effort, disposing of expert knowledge and considerable material resources, it appeared from all information available to us, that at any rate in the initial for Germany so favourable stages of the war, it was never by the Government deemed worth while to attempt the immense and hazardous technical enterprise which an accomplishment of the project would require.

Immediately after my escape to Sweden in October 1943, I came on an invitation of the British Government to England where I was taken into confidence about the great progress achieved in America and went shortly afterwards together with a number of British colleagues to U.S.A. to take part in the work. In order, however, to conceal my connection with any such enterprise, post war

[103]

planning of international scientific co-operation was given as the object of my journey.

Already in Denmark I had been in secret connection with the British Intelligence Service, and more recently I have had the opportunity with American and British Intelligence Officers to discuss the latest information, pointing to a feverish German activity on nuclear problems. In this connection it must above all be realized that if any knowledge of the progress of the work in America should have reached Germany, it may have caused the Government to reconsider the possibilities and will not least have presented the physicists and technical experts with an extreme challenge.

Definite information of preparations elsewhere is hardly available, but an interest within the Soviet Union for the project may perhaps be indicated by a letter which I have received from a prominent Russian physicist with whom I had formed a personal friendship during his many years stay in England and whom I visited in Moscow a few years before the war, to take part in scientific conferences.

This letter contained an official invitation to come to Moscow to join in scientific work with Russian colleagues who, as I was told, in the initial stages of the war were fully occupied with technical problems of immediate importance for the defence of their country, but now had the opportunity to devote themselves to scientific research of more general character. No reference was made to any special subject, but from pre-war work of Russian physicists it is natural to assume that nuclear problems will be in the centre of interest.

The letter, originally sent to Sweden in October 1943, was on my recent visit to London handed to me by the Counsellor of the Soviet Embassy[7] who in a most encouraging manner stressed the promises for the future understanding between nations entailed in scientific collaboration. Although, of course, the project was not mentioned in this conversation I got nevertheless the impression that the Soviet Officials were very interested in the effort in America about the success of which some rumours may have reached the Soviet Union.

Even if every physicist was prepared that some day the prospects created by modern researches would materialize, it was a revelation to me to learn about the courage and foresight with which the great American and British enterprise had been undertaken and about the advanced stage the work had already reached.

[7] See the appendix to Part I, p. [238], ref. 45.

What until a few years ago might have been considered a fantastic dream is at the moment being realized in great laboratories erected for secrecy in some of the most solitary regions of the States. There a group of physicists larger than ever before assembled for a single purpose, and working hand in hand with a whole army of engineers and technicians are producing new materials capable of enormous energy release and developing ingenious devices for their most effective use.

To everyone who is given the opportunity for himself to see the refined laboratory equipment and the huge production machinery, it is an unforgettable experience of which words can only give a poor impression. Truly no effort has been spared and it is hardly possible for me to describe my admiration for the efficiency with which the great work has been planned and conducted.

Moreover it was a special pleasure to me to witness the complete harmony with which the American and British physicists, with almost every one of whom I was intimately acquainted through previous scientific intercourse, were devoting themselves with the utmost zeal to the joint effort.

I shall not here enter on technical details, but one cannot help comparing with the Alchemists of former days, groping in the dark in their vain efforts to make gold. To-day physicists and engineers are on the basis of well established knowledge directing and controlling processes by which substances far more precious than gold are being collected atom by atom or even built up by individual nuclear transmutations.

Such substances must be assumed to have been abundant in the early stages of our universe where all matter was subject to conditions far more violent than those which still persist in the turbulent and flaming interior of the stars. Due, however, to their inherent instability the active materials now extracted or produced have in the course of time become very rare or even completely disappeared from the household of nature.

The whole enterprise constitutes indeed a far deeper interference with the natural course of events than anything ever before attempted and its impending accomplishment will bring about a whole new situation as regards human resources. Surely, we are being presented with one of the greatest triumphs of science and engineering destined deeply to influence the future of mankind.

It certainly surpasses the imagination of anyone to survey the consequences of the project in years to come, where in the long run the enormous energy sources which will be available may be expected to revolutionize industry and transport. The fact of immediate preponderance is,

however, that a weapon of an unparalleled power is being created which will completely change all future conditions of warfare.

Quite apart from the questions of how soon the weapon will be ready for use and what role it may play in the present war, this situation raises a number of problems which call for most urgent attention. Unless, indeed, some agreement about the control of the use of the new active materials can be obtained in due time, any temporary advantage, however great, may be outweighed by a perpetual menace to human security.

Ever since the possibilities of releasing atomic energy on a vast scale came in sight, much thought has naturally been given to the question of control, but the further the exploration of the scientific problems concerned is proceeding, the clearer it becomes that no kind of customary measures will suffice for this purpose and that especially the terrifying prospect of a future competition between nations about a weapon of such formidable character can only be avoided through a universal agreement in true confidence.

In this connection it is above all significant that the enterprise, immense as it is, has still proved far smaller than might have been anticipated and that the progress of the work has continually revealed new possibilities for facilitating the production of the active materials and of intensifying their effects.

The prevention of a competition prepared in secrecy will therefore demand such concessions regarding exchange of information and openness about industrial efforts including military preparations as would hardly be conceivable unless at the same time all partners were assured of a compensating guarantee of common security against dangers of unprecedented acuteness.

The establishment of effective control measures will, of course, involve intricate technical and administrative problems, but the main point of the argument is that the accomplishment of the project would not only seem to necessitate but should also, due to the urgency of mutual confidence, facilitate a new approach to the problem of international relationship.

The present moment where almost all nations are entangled in a deadly struggle for freedom and humanity might at first sight seem most unsuited for any committing arrangement concerning the project. Not only have the aggressive powers still great military strength, although their original plans of world domination have been frustrated and it seems certain that they must ultimately surrender, but even when this happens, the nations united against aggression may face grave causes of disagreement due to conflicting attitudes towards social and economic problems.

By a closer consideration, however, it would appear that the potentialities of the project as a means of inspiring confidence just under these circumstances acquire most actual importance. Moreover the momentary situation would in various respects seem to afford quite unique possibilities which might be forfeited by a postponement awaiting the further development of the war situation and the final completion of the new weapon.

Although there can hardly be any doubt that the American and British enterprise is at a more advanced stage than any similar undertaking elsewhere, one must be prepared that a competition in the near future may become a serious reality. In fact, as already indicated, it seems likely that preparations possibly urged on by rumours about the progress in America, are being speeded up in Germany and may even be under way in the Soviet Union.

Further it must be realized that the final defeat of Germany will not only release immense resources for a full scale effort within the Soviet Union, but will presumably also place all scientific and technical experience collected in Germany at the disposal for such an effort.

In view of these eventualities the present situation would seem to offer a most favourable opportunity for an early initiative from the side which by good fortune has achieved a lead in the efforts of mastering mighty forces of nature hitherto beyond human reach.

Without impeding the importance of the project for immediate military objectives, an initiative, aiming at forestalling a fateful competition about the formidable weapon, should serve to uproot any cause of distrust between the powers on whose harmonious collaboration the fate of coming generations will depend.

Indeed, it would appear that only when the question is taken up among the United Nations of what concessions the various powers are prepared to make as their contribution to an adequate control arrangement, it will be possible for anyone of the partners to assure themselves of the sincerity of the intentions of the others.

Of course, the responsible statesmen alone can have the insight in the actual political possibilities. It would, however, seem most fortunate that the expectations for a future harmonious international co-operation which have found unanimous expression from all sides within the United Nations, so remarkably correspond to the unique opportunities which, unknown to the public, have been created by the advancement of science.

[107]

Many reasons, indeed, would seem to justify the conviction that an approach with the object of establishing common security from ominous menaces without excluding any nation from participating in the promising industrial development which the accomplishment of the project entails will be welcomed, and be responded with a loyal co-operation on the enforcement of the necessary far reaching control measures.

Just in such respects helpful support may perhaps be afforded by the world wide scientific collaboration which for years has embodied such bright promises for common human striving. On this background personal connections between scientists of different nations might even offer means of establishing preliminary and non-committal contact.

It need hardly be added that any such remark or suggestion implies no underrating of the difficulty and delicacy of the steps to be taken by the statesmen in order to obtain an arrangement satisfactory to all concerned, but aims only at pointing to some aspects of the situation which might facilitate endeavours to turn the project to lasting benefit for the common cause.

Should such endeavours be successful, the project will surely have brought about a turning point in history and this wonderful adventure will stand as a symbol of the benefit to mankind which science can offer when handled in a truly human spirit.

Document 7:
LETTER TO ROOSEVELT, 7 September 1944
[Carbon copy]
[Source: BPP 4.2]
[Editor's comment: A different version of the letter, deposited in the Frankfurter file of the Oppenheimer Papers, Library of Congress, Washington, D.C., seems to be an earlier draft.]

TOP SECRET.

Washington, September 7th 1944.

My dear Mr. President.

I wish to thank you most heartily for the honour and confidence you showed me by receiving me and talking with me of your concern about the great and urgent problems raised by the recent extraordinary development of physical science.

The impending realization of the prospects of releasing atomic energy on a large scale, secured by the great pioneer work in this country, is surely destined deeply to influence the future of mankind. But, as you appreciate so fully, the bright promises which the wonderful adventure entails may be overshadowed by most ominous threats to human security unless in due time international agreement can be obtained as regards an effective control of the new formidable weapon.

As a physicist who has been given the privilege to follow the latest development at close hand, I was, however, most grateful for the opportunity to bring before you some considerations concerning the technical aspects of the unparalleled enterprise with special regard to the problem of control measures and the possibilities of competing efforts in other countries.

In the last respect it should be borne in mind that, as a result of a fruitful international scientific co-operation, the principles on which a large scale release of atomic energy can be effected were, at any rate in outlines, perceived already before the war and it is known that attempts of realizing the prospects have since been made in various countries.

It seems certain that the American and British effort undertaken with so happy a foresight and backed with such great resources has achieved a decisive lead. Still, it must be realized that any information, however scanty, about the success of the work in this country which may have leaked out is likely to have caused the utmost intensification of similar efforts elsewhere.

Fortunately the course of the war has probably removed any danger of a military use of the purposeful endeavours on such lines made in Germany. One must, however, be prepared that with the final defeat of Germany all expert knowledge and technical experience collected there will presumably become available in equal measure for the great victorious nations.

It is from the point of view of proper timing that I venture to suggest that the technical factors of the situation make the present moment most favourable for considerations of the question of control by the friendly governments most concerned, lest opportunities be forfeited of forestalling a fateful competition about the new weapon and of turning the great triumph of science and engineering to lasting benefit for the common cause.

Of course, it is wholly outside my province to suggest the procedures appropriate for such delicate problems of statesmanship. But as a scientist it occurs to me that in this unique situation pre-war scientific connections may prove helpful in conveying, with entire regard to security, an understanding of how much would be at stake should the great prospects of atomic physics materialize, and in preparing an adequate realization of the great benefit which would ensue from a whole-hearted co-operation on effective control measures.

I wish once more to thank you for the great kindness you showed me, and for the wish you expressed to see me again at some later occasion. I need not assure you that I shall be most happy to respond to the honour of a call from you.

Most respectfully yours
[Niels Bohr]

The President.

Document 8:
"COMMENTS", 18 January 1945
[Carbon copy]
[Source: BPP 7.7]

TOP SECRET.

January 18th 1945.

Comments to memorandum of July 3rd 1944.

In the memorandum arguments have been presented pointing to the favourable consequences which might ensue from early negotiations among the Allied Nations with the aim of eliminating the dangers to world security and turning the great scientific and technical advance to lasting benefit for the common cause.

Such argumentation raises naturally the question what prospects consultations at the moment might have of achieving the desired result and whether an initiative of the kind suggested might not, in case of failure, involve the risk of impeding the advantageous position of the powers which, according to all available evidence, have acquired a lead in the great enterprise.

As an opening of consultations with the primary purpose of inspiring confidence and relieving disquietude, it would appear, however, to be necessary only to bring up the problem what the attitude of each partner would be, if the prospects created by the progress of science, which in outlines are common knowledge, should be realized to an extent which will revolutionize all aspects of warfare.

It would thus seem that no information about technical developments of importance for the conduct of the war or the balance of power need be exchanged before agreement is reached about the principles on which a control arrangement, constituting an inherent part of any ultimate world-wide security organization, must be based.

The unprecedented common dangers which can be averted only through a co-operation in mutual confidence should offer a unique opportunity of appealing to an unprejudiced attitude and of creating a new background for international understanding, helpful for the settlement of numerous other urgent issues.

The full mutual access to information about scientific researches and technical enterprises including military efforts which will be necessary for the operation of an effective control should also in itself have far-reaching consequences in furthering such knowledge of material and spiritual aspects of life within the various countries, without which respect and good-will between nations can hardly endure.

Should, against expectations, no satisfactory response be obtainable, it would seem excluded that the initiative could have led to any disadvantage. Rather would the inviting nations be better able to judge what precautions might be necessary to safeguard themselves and their position be strengthened through the proof they have given of their sincere intentions to strive for a solution beneficial to all.

As regards the question of proper timing, it should be added to the arguments for the urgency of the whole matter, discussed in the memorandum, that it would appear most important that an agreement between the Governments is attempted before any public discussion of the matter can arise. It would, therefore, seem essential that consultations on these problems are initiated sufficiently long before there can be question of an actual use of the new potentialities of warfare.

Document 9:
"ADDENDUM", 24 March 1945
[Carbon copy]
[Source: BPP 8.4; also in CAB 126/39]
[Editor's comment: Some of the passages are taken, even literally, from Document 8. The **boldface** represents the parts of the "Addendum" that Bohr chose to quote in his Open Letter to the United Nations (reproduced on p. [171]) five years later.]

TOP SECRET. March 24th 1945.

ADDENDUM TO MEMORANDUM OF JULY 3RD 1944.

In a previous memorandum some outlines of the development of the project of utilizing atomic energy for industrial and military purposes were described, and a brief discussion attempted of the implications of the impending accomplishment of this project which involves so radical a revolution of material resources destined to have most far-reaching consequences for civilization.

In particular, arguments were presented pointing to the unique opportunity, which early consultations among the Allied Nations might offer, for eliminating imminent menaces to world security and turning the great scientific and technical advance to lasting benefit for the common cause.

With regard to the question of what prospects such consultations would have of achieving the desired result, these additional notes aim at recalling certain aspects of the extraordinary situation so fraught both with dangers and with great hopes for mankind.

———————

As explained in the memorandum, the possibilities of practical application of atomic energy sources had already been perceived before the war but, although endeavours were undertaken in various countries, all available evidence indicates that so far only the joint American and British effort has succeeded in developing methods for the production of super-explosive materials in sufficient amounts for the assembly of a weapon of formidable destructive power.

Still, it must be realized that the pioneer effort, immense as it has been, has proved far smaller than might have been anticipated and that any information, however scanty, which may have leaked out about the success of the enterprise will have greatly stimulated efforts elsewhere.

Above all, it should be appreciated that we are faced only with the beginning of a development and that, probably within the very near future, means will be found to simplify the methods of production of the active substances and intensify their effects to an extent which may permit any nation possessing great industrial resources to command powers of destruction surpassing all previous imagination.

Humanity will, therefore, be confronted with dangers of unprecedented character unless, in due time, measures can be taken to forestall a disastrous competition in such formidable armaments and to establish an international control of the manufacture and use of the powerful materials.

—————

Any arrangement which can offer safety against secret preparations for the mastery of the new means of destruction would, as stressed in the memorandum, demand extraordinary measures. In fact, not only would universal access to full information about scientific discoveries be necessary, but every major technical enterprise, industrial as well as military, would have to be open to international control.

In this connection it is significant that the special character of the efforts which, irrespective of technical refinements, are required for the production of the active materials, and the peculiar conditions which govern their use as dangerous explosives, will greatly facilitate such control and should ensure its efficiency, provided only that the right of supervision is guaranteed.

Detailed proposals for the establishment of an effective control would have to be worked out with the assistance of scientists and technologists appointed by the governments concerned, and a standing expert committee, related to an international security organization, might be charged with keeping account of new scientific and technical developments and with recommending appropriate adjustments of the control measures.

On recommendations from the technical committee the organization would be able to judge the conditions under which industrial exploitation of atomic energy sources could be permitted with adequate safeguards to prevent any assembly of active material in an explosive state. All material prepared for armaments, however, might ultimately be entrusted to the security organization to be held in readiness for eventual policing purposes.

—————

As argued in the memorandum, it would seem most fortunate that the measures demanded for coping with the new situation, brought about by the advance of science and confronting mankind at a crucial moment of world affairs, fit in so well with the expectations for a future intimate international co-operation which have found unanimous expression from all sides within the nations united against aggression.

Moreover, the very novelty of the situation should offer a unique opportunity of appealing to an unprejudiced attitude; and it would even appear that an understanding about this vital matter might contribute most favourably towards the settlement of other problems where history and traditions have fostered divergent view-points.

With regard to such wider prospects, it would in particular seem that the free access to information, necessary for common security, should have far-reaching effects in removing obstacles barring mutual knowledge about spiritual and material aspects of life in the various countries, without which respect and goodwill between nations can hardly endure.

Participation in a development, largely initiated by international scientific collaboration and involving immense potentialities as regards human welfare, would also reinforce the intimate bonds which were created in the years before the war between scientists of different nations. In the present situation these bonds may prove especially helpful in connection with the deliberations of the respective governments and the establishment of the control.

In preliminary consultations between the governments with the primary purpose of inspiring confidence and relieving disquietude, it should be necessary only to bring up the problem of what the attitude of each partner would be if the prospects opened up by the progress of physical science, which in outline are common knowledge, should be realized to an extent which would necessitate exceptional international action.

It would in fact seem that no information about technical developments of importance for the possible use of the new weapon in the present war, or for the future balance of power, need be exchanged before understanding is reached and guarantees are obtained for co-operation on a control arrangement to form an integral part of an organization for world security.

In this respect it would appear to be of significance that the very nature of the measures called for should ensure their effectiveness regardless of subsequent developments, and thus guarantee that the control, if once established, would actually mean a decisive change in international affairs.

The openness as regards national efforts, demanded for an effective control, will indeed amount to a revision of the relationship between sovereign nations so radical that it would hardly be feasible unless there were a question of unprecedented common dangers which can be averted only by co-operation in true confidence.

In case the approach should not in the first instance meet with a satisfactory response, the initiative could hardly have led to any disadvantage. Rather it would seem that the position of those nations which, with so much foresight, have acquired the lead in the new development would have been greatly strengthened by this proof of a sincere intention to strive for an arrangement beneficial to all.

It would thus appear not only that a better judgment would be possible as to the procedure to be followed to bring about an ultimate solution corresponding to the requirements of the situation, but also that a moral background would have been created without which a resort to unilateral measures might not be reconcilable with the ideals to which these nations adhere.

Should other ways remain closed, it may, in fact, be of decisive importance that the experience gained through the pioneer enterprise makes it possible, within a very short time, by means of expenditures and resources not exceeding those of other major military efforts, to multiply the production of the active materials to an extent which, at least temporarily, would secure complete supremacy.

In all the circumstances it would seem that an understanding could hardly fail to result, when the partners have had a respite for considering the consequences of a refusal to accept the invitation to co-operate, and for convincing themselves of the advantages of an arrangement guaranteeing common security without excluding anyone from participation in the promising utilization of the new sources of material prosperity.

All such opportunities may, however, be forfeited if an initiative is not taken while the matter can be raised in a spirit of friendly advice. In fact, a postponement to await further developments might, especially if preparations for competitive efforts in the meantime have reached an advanced stage, give the approach the appearance of an attempt at coercion in which no great nation can be expected to acquiesce.

In this connection it must be realized that the development of the war may soon create a situation where, as mentioned in the memorandum, any possible progress achieved by the much advertised German preparations to use atomic energy sources for military purposes may become common knowledge of the great victorious nations.

As an additional argument for the urgency of the matter, it would appear most important that an agreement between the governments be invited before public discussion can arouse sentiments and cause unpredictable complications. It would, therefore, seem essential that consultations be initiated sufficiently long before there can be a question of the actual use of the new means of warfare.

Indeed, it need hardly be stressed how fortunate in every respect it would be if, at the same time as the world will know of the formidable destructive power which has come into human hands, it could be told that the great scientific and technical advance has been helpful in creating a solid foundation for a future peaceful co-operation between nations.

Document 10:
MEMO, 8 May 1945
[Typewritten]
[Source: BPP 8.5]

May 8th 1945.

The background for the considerations indicated in the memorandum of July 3rd 1944 and further explained in the addendum of March 24th 1945 is the grave responsibility, resting upon our generation, that the great pioneer effort, based on the recent advance of physical science, is used to the benefit of all humanity and does not become a menace to civilization.

The essence of the arguments presented is that an early invitation to all nations united against aggression, to co-operate on the enforcement of an effective universal control of the new formidable means of destruction, should afford a unique opportunity for furthering a harmonious relationship between nations.

In fact, more than perhaps in any other matter, it is here indisputably a question of deepest common interest and, due to their very nature, the control measures required will in themselves go far in removing main obstacles for an international understanding.

In preliminary consultations no information as regards important technical developments should, of course, be exchanged; on the contrary, the occasion should be used frankly to explain that all such information must be withheld until common safety against the unprecedented dangers has been guaranteed.

For this purpose far-reaching concessions from all partners will obviously be demanded, but for every nation there should be most pertinent reasons realistically to consider the consequences of a refusal to co-operate on the establishment of control.

In all the circumstances a timely assurance of good–will from the side which has acquired the lead in the new development would seem to form the proper basis for a purposeful policy meeting the requirements of a situation, so fraught with dangers, but at the same time offering the greatest opportunities.

Document 11:
MEMO, 12 July 1945
[Typewritten]
[Source: BPP 7.9]

TOP SECRET   12/7/45.

In the memorandum of July 3rd [1944] and the addendum of March 24th 1945 an attempt has been made to call attention to certain aspects of the extraordinary situation, which have naturally impressed themselves on a physicist, who for years has had the opportunity to participate in the international collaboration by which the scientific foundation for the project was created and to whom has been given the privilege to follow the great pioneer enterprise.

The success of this enterprise which has initiated a veritable revolution of material resources is indeed confronting civilization with an actual crisis involving on the one hand unprecedented menaces to world security, but on the other hand affording a unique opportunity for a new approach to international relationships to meet the requirements of the situation.

The great fortune that the new formidable means of destruction should not have fallen into the hands of the nations seeking world domination through aggression, but that by so great foresight a lead in the new development has been acquired as a part of the efforts to safeguard the ideals of freedom and humanity, surely forms the very best background for turning the great advance of science and engineering to the benefit of all mankind.

Considering that a universal control of the utilisation of the powerful forces of nature which, due to the advance of science, have come into the reach of man is a matter of the deepest common concern, it would seem that an appeal to the nations united against aggression to co-operate in the establishment of such a control would go far in inviting confidence and promoting the efforts to establish a firm foundation for world security.

All decisions as regards procedure to obtain this goal rest, of course, with the responsible governments, but in a matter so different in many respects from ordinary political and military issues, a valuable help for the endeavours of the government may be found in the worldwide ties between scientists, created by international co-operation, which in the last decades has been so fertile, not least in the domain of atomic physics.

If given opportunity, every scientist who has taken part in the basic development will certainly feel it the greatest obligation not only to co-operate loyally in making an eventual control effective, but also to contribute to the utmost of his powers in preparing an understanding of what is at stake and an adequate appreciation of the immense opportunities, which the development presents.

In this connection it may be most essential, however, that the scientists, on whom the governments of every country will depend for advice, from the very beginning feel assured that the unique situation brought about by so fruitful and promising an exploration of a new domain of knowledge is being handled in a spirit conforming with the ideals of common striving for human progress for which science through the ages has stood as a symbol.

II. SCIENCE AND CIVILIZATION

The Times, 11 August 1945

See Introduction to Part I, p. [50].

[121]

SCIENCE AND CIVILIZATION (1945)

Versions published in English and Danish

English: Science and Civilization
 A   *Energy from the Atom – An Opportunity and a Challenge – The Scientist's View*, The Times, 11 August 1945
 B   "One World or None" (eds. D. Masters and K. Way), McGraw-Hill, New York 1946, pp. IX–X

Danish: Videnskab og Civilisation
 C   *Energi fra Atomet – En Mulighed og en Udfordring – Videnskabsmændenes Syn*, Berlingske Aftenavis, 11 August 1945
 D   *Den fysiske Videnskab og den menneskelige Civilisation*, Politiken, 11 August 1945
 E   "Een Verden eller ingen" (eds. D. Masters and K. Way), Haase & Søn, Copenhagen 1946, pp. 11–15

All of these versions agree with each other, with the exception that all of C, D and E are translated differently.

REPRINTED FROM

THE 🦁👑🦄 TIMES

Saturday August 11 1945

SCIENCE AND CIVILIZATION

By Niels Bohr

The possibility of releasing vast amounts of energy through atomic disintegration, which means a veritable revolution of human resources, cannot but raise in the mind of everyone the question whither the advance of physical science is leading civilization. While the increasing mastery of the forces of nature has contributed so prolifically to human welfare, and holds out even greater promises, it is evident that the formidable power of destruction which has come within reach of man may become a mortal menace unless human society can adjust itself to the exigencies of the situation. Civilization is presented with a challenge more serious, perhaps, than ever before, and the fate of humanity will depend on its ability to unite in averting common dangers and jointly to reap the benefit from the immense opportunities which the progress of science offers.

In its origin science is inseparable from the collecting and ordering of experiences, gained in the struggle for existence, which enabled our ancestors to raise mankind to its present position among the other living beings which inhabit our earth. Even in highly organized communities where, within the distribution of labour, scientific study has become an occupation by itself, the progress of science and the advance of civilization have remained most intimately interwoven. Of course, practical needs are still an impetus to scientific research, but it need hardly be stressed how often technical developments of the greatest importance for civilization have originated from studies aimed only at augmenting our knowledge and deepening our understanding. Such endeavours know no national borders and where one scientist has left the trail another has taken it up, often in a distant part of the world. For long scientists have considered themselves as a brotherhood working in the service of common human ideals.

In no domain of science have these lessons received stronger emphasis than in the exploration of the atom, which just now is bearing consequences of such overwhelming practical implications. As is well known, the roots of the idea of atoms as the ultimate constituents of matter go back to ancient thinkers searching for a foundation to explain the regularity which, in spite of all variability, is ever more clearly revealed by the study of natural phenomena. After the Renaissance, when science entered so fertile a period, atomic theory gradually became of the greatest importance for the physical and chemical sciences, although, until half a century ago, it was generally accepted that, due to the coarseness of our senses, any direct proof of the existence of atoms would always remain beyond human scope. Aided, however, by the refined tools of modern technique, the development of the art of experimentation has removed such limitation and even yielded detailed information about the interior structure of atoms.

In particular, the discovery that almost the entire mass of the atom is concentrated in a central nucleus proved to have the most far-reaching consequences. Not only did it become evident that the remarkable stability of the chemical elements is due to the immutability of the atomic nucleus when exposed to ordinary physical agencies but a novel field of research was opened up by the study of the special conditions under which disintegrations of the nuclei themselves may be brought about. Such processes, whereby the very elements are transformed, were found to differ fundamentally in character and violence from chemical reactions, and their investigation led to a rapid succession of important discoveries through which ultimately the possibility of a large-scale release of atomic energy came into sight. This progress was achieved in the course of a few decades, and was due not least to most effective international cooperation. The world community of physicists was, so to say, welded into one team, rendering it more difficult than ever to disentangle the contributions of individual workers.

The grim realities which are being revealed to the world in these days will no doubt, in the minds of many, revive terrifying prospects forecast in fiction. With all admiration for such imagination, it is, however, most essential to appreciate the contrast between these fantasies and the actual situation confronting us. Far from offering any easy means to bring destruction forth, as it were by witchcraft, scientific insight has on the contrary made it evident that use of nuclear disintegration for devastating explosions demands most elaborate preparations, involving a profound change in the atomic composition of the materials found on earth. The astounding achievement of producing an enormous display of power on the basis of experience gained by the study of minute effects, perceptible only by the most delicate instruments, has, in fact, besides a most intensive research effort, required an immense engineering enterprise, strikingly illuminating the potentialities of modern industrial development.

Indeed, not only have we left the time far behind where each man, for self-protection, could pick up the nearest stone, but we have even reached the stage where the degree of security offered to the citizens of a nation by collective defence measures is entirely insufficient. Against the new destructive powers no defence may be possible, and the issue centres on world-wide cooperation to prevent any use of the new sources of energy which does not serve mankind as a whole. The possibility of international regulation for this purpose should be ensured by the very magnitude and the peculiar character of the efforts which will be indispensable for the production of the new formidable weapon. It is obvious, however, that no control can be effective without free access to full scientific information and the granting of the opportunity of international supervision of all undertakings which, unless regulated, might become a source of disaster.

Such measures will, of course, demand the abolition of barriers hitherto considered necessary to safeguard national interests but now standing in the way of common security against unprecedented dangers. Certainly the handling of the precarious situation will demand the good will of all nations, but it must be recognized that we are dealing with what is potentially a deadly challenge to civilization itself. A better background for meeting such a situation could hardly be imagined than the earnest desire to seek a firm foundation for world security, so unanimously expressed from the side of all those nations which only through united efforts have been able to defend elementary human rights. The extent of the contribution which an agreement about this vital matter would make to the removal of obstacles to mutual confidence and to the promotion of a harmonious relationship between nations can hardly be exaggerated.

In the great task lying ahead, which places upon our generation the gravest responsibility towards posterity, scientists all over the world may offer most valuable services. Not only do the bonds created through scientific intercourse form some of the firmest ties between individuals from different nations, but the whole scientific community will surely join in a vigorous effort to induce in wider circles an adequate appreciation of what is at stake and to appeal to humanity at large to heed the warning which has been sounded. It need not be added that every scientist who has taken part in laying the foundation for the new development, or has been called upon to participate in work which might have proved decisive in the struggle to preserve a state of civilization where human culture can freely develop, is prepared to assist in any way open to him in bringing about an outcome of the present crisis of humanity worthy of the ideals for which science through the ages has stood.

Printed by The Times Publishing Company, Limited, Printing House Square, London, E.C.4, England.

III. A CHALLENGE TO CIVILIZATION

Science **102** (1945) 363–364

See Introduction to Part I, p. [51].

A CHALLENGE TO CIVILIZATION (1945)

Versions published in English and Danish

English: A Challenge to Civilization
A   Science **102** (1945) 363–364

Danish: En krise for civilisationen
B   "Atom Alderen: Skandinavisk–Britisk Udstilling, Charlottenborg 1. til 18. september 1949", Een Verden, Copenhagen 1949, pp. 5–6

Both of these versions agree with each other.

Reprinted from SCIENCE, October 12, 1945, Vol. 102,
No. 2650, pages 363–364.

A CHALLENGE TO CIVILIZATION

By Dr. NIELS BOHR

COPENHAGEN

THE advance of physical science which has made
it possible to release vast amounts of energy through
atomic disintegration has initiated a veritable revolution of human resources, presenting civilization
with a most serious challenge. Man's increasing
mastery of the forces of nature, which has provided
ever richer possibilities for the growth of culture,
may indeed threaten to upset the balance vital for
the thriving of organized communities, unless human
society can adjust itself to the exigencies of the
situation. The great technical developments of the
last century already deeply affected the social structure of every country but, evidently, we have now
reached a stage which calls for a new approach to
the whole problem of international relationship.

The formidable means of destruction which have
come within reach of man will obviously constitute
a mortal menace to civilization unless, in due time,
universal agreement can be obtained about appropriate measures to prevent any unwarranted use of
the new energy sources. An agreement to this purpose will surely demand the abolition of barriers
hitherto considered necessary to protect national interests but now standing in the way of common safety
against unprecedented dangers. In fact, only international control of every undertaking which might
constitute a danger to world security will in future
permit any nation to strive for prosperity and cultural development without constant fear of disaster.

Whatever renunciations as regards customary prerogatives such regulation will involve, it should be clear to all that, in contrast to other issues where history and traditions may have fostered divergent viewpoints, we are here dealing with a matter of the deepest interest to all nations. Moreover, the free and open access to information about all scientific and technical progress, which will be a basic condition for the efficiency of the control, will in itself go far towards promoting mutual knowledge and understanding of the cultural aspects of life in the various countries, without which respect and good-will between nations can hardly endure.

In all the circumstances it would appear that the possibility of producing devastating weapons, against which no defence may be feasible, should be regarded not merely as a new danger added to a perilous world, but rather as a forceful reminder of how closely the fate of all mankind is coupled together. Indeed, the crisis with which civilization is at present confronted should afford a unique opportunity to remove obstacles to peaceful collaboration between nations and to create such mutual confidence as will enable them jointly to benefit from the great promises, as regards human welfare, held out by the progress of science.

The attainment of this goal, which places upon our generation the gravest responsibility towards posterity, will of course depend on the attitude of all people. Valuable services, however, may be rendered by scientists all over the world in bringing about a genuine appreciation of what is at stake and in pointing out how the great development of our resources may contribute to progress for humanity. In a matter of such universal scope, help may also be found in the intimate connections between scientists, created by international cooperation which proved so fertile just in a domain of research that was to have such overwhelming consequences.

For meeting the challenge to civilization in the proper spirit it should be a most fortunate omen that we have to do with implications of pure scientific studies pursued with no other aim than to widen the borders of our knowledge and to deepen our understanding of that nature of which we ourselves are a part. Let us hope that science which, through the ages, has stood as a symbol of the progress to be obtained by common human striving, by its latest emphasis on the necessity of concord, may contribute decisively to a harmonious relationship between all nations.

PROCEEDINGS

of the

American Philosophical Society

Contents of Volume 91, Number 2

Price for complete number one dollar

AMERICAN PHILOSOPHICAL SOCIETY
INDEPENDENCE SQUARE
PHILADELPHIA 6, PA.

[130]

IV. ATOMIC PHYSICS AND INTERNATIONAL COOPERATION

Proceedings of the American Philosophical Society **91** (1947) 137–138

Address at Symposium of the National Academy of Sciences
Present Trends and International Implications of Science,
Philadelphia, 21 October 1946

See Introduction to Part I, p. [52].

ATOMIC PHYSICS AND INTERNATIONAL COOPERATION (1946)

Versions published in English and German

English: Atomic Physics and International Cooperation
 A   Proc. Amer. Phil. Soc. **91** (1947) 137–138

German: Atomphysik und internationale Zusammenarbeit
 B   Universitas **6** (1951) 547–550

The two versions agree with each other, with the exception that the first two paragraphs in the English version are omitted in the German one.

ATOMIC PHYSICS AND INTERNATIONAL COOPERATION

NIELS H. D. BOHR

Professor of Theoretical Physics, University of Copenhagen

(Read October 21, 1946, in Philadelphia in the Symposium of the National Academy of Sciences on Present Trends and International Implications of Science)

AFTER expressing on behalf of the Danish delegates gratitude towards the American Philosophical Society and the National Academy of Sciences for the generous invitation to participate in this international gathering devoted to the prospects of cooperation in science, which offers such great opportunites not only for advancement of knowledge but also for promotion of understanding between peoples, the speaker opened the discussions of the special problems created by the recent development of atomic physics with some general remarks the essence of which is given in the following abstract.[1]

As an introduction to the subject, stress was laid on the decisive part international cooperation has played in creating, in the course of comparatively few years, a whole new branch of science which, at the same time as it has to such a high degree deepened our insight in the phenomena of nature, has led to practical consequences of immense implication for the future of civilization, calling for a thorough revision of our attitude to the whole question of international relationship.

The fundamental discoveries of the constituents of the atom and the quantum of action, which were the outset of the development, originated, as is well known, in lines of inquiry pursued in various countries by scientific schools with different traditions and outlooks, the confluence of which gave rise to international cooperation of an intensity and enthusiasm which has indeed only few counterparts in the history of science. This cooperation proved ever more fruitful in the last decades, where the interpretation of the ordinary physical and chemical properties of the elements, based on specific ideas of atomic constitution, reached such a remarkable level of consistency and completion and where, at the same time, studies

of the transformations of the very nucleus of the atom opened the quite new field of research which was to bear such great consequences.

Moreover, this development has established, in increasing measure, close contact between many domains of science in which natural phenomena have been approached from different starting points and attacked by different methods. Not only have physics and chemistry become intimately amalgamated on a common basis, but the tools created by atomic research have proved most powerful in wide fields of technology and biology. It need, thus, hardly be stressed how great are the new possibilities which have been opened for the most purposeful applications of the properties of matter and for the refinement of the studies of the structure of organisms and the metabolic processes connected with the upholding of life. The latest progress, which entails such great promise as regards new energy sources for industry, opens also prospects for the medical sciences which can hardly be overestimated.

Indeed, the whole development should offer immense opportunities for a world-wide collaboration on the advance of science and the promotion of human welfare, but all such perspectives may be overshadowed by deadly menaces to civilization, unless in due time universal agreement be obtained about proper measures to avert misuse of the formidable power of destruction which has come into the hand of man. Such measures will obviously demand sacrifices of accustomed national prerogatives and abolition of barriers hitherto considered necessary to safeguard national interests, but now standing in the way of common security against unprecedented dangers. In removing obstacles to mutual confidence, an agreement on this vital matter should certainly go far in furthering a harmonious relationship between nations.

The achievement of the great goal which places the gravest responsibility upon our generation will, of course, demand the good will of all peoples and, in this respect, important services may be rendered by scientists in every country, not only in their capacity as advisers to the responsible govern-

---

[1] Reference may also be made to some previous articles of the writer: Biology and atomic physics (Reports of the Galvani Congress), *Il Nuovo Cemento* n.s. 15: 429–438, 1938; Natural philosophy and human cultures, *Nature* 143: 268, 1939; Science and civilization, *The Times,* August 11, 1945 (reprinted in *One world or none* ed. by D. Masters and K. Way, New York, Whittlesey, 1946); A challenge to civilization, *Science* 102: 363, 1945.

PROCEEDINGS OF THE AMERICAN PHILOSOPHICAL SOCIETY, VOL. 91, NO. 2, APRIL, 1947
Reprint   *Printed in U. S. A.*

ments but also by bringing about a proper public appreciation of the situation, just as it is to be hoped that the intimate bonds between scientists of various nations, created by years of close intercourse, may prove to be of essential help.

The circumstance that we have to do with implications of scientific studies, undertaken with no other aim than to widen the borders of our knowledge and to deepen our understanding of that nature of which we ourselves are a part, should present the best background imaginable for meeting the exigencies of the situation in the proper spirit and, in this respect, it is perhaps not unimportant that the very philosophical lesson impressed upon us by the development of physics in our generation is especially suited to further common human understanding.

In fact, again and again we have been confronted with experiences which in an unforeseen way have revealed limitations of customary ideas and demanded a revision of the foundation of our whole conceptual frame. Just as the recognition of general relativity, which has lent such unity to our picture of the universe, has strikingly illuminated the extent to which the presentation of any event depends on the standpoint of the observer, so has the elucidation of the apparent paradoxes which were encountered in the exploration of the world of atoms disclosed further presumptions for the interpretation of observations and reminded us of fundamental aspects as regards analysis and synthesis of experience.

The impossibility of making a sharp distinction between an independent behavior of the objects and their interaction with the measuring agencies entails, thus, a limitation as regards the analysis of individual atomic processes incompatible with accustomed demands of explanation. In particular, phenomena observed under different experimental arrangements cannot be directly compared, but must be regarded as complementary in the sense that only together do they exhaust all obtainable knowledge regarding the objects. In offering a frame wide enough to allow a harmonious synthesis of the peculiar regularities of atomic phenomena, the conception of complementarity may be regarded as a rational generalization of the very ideal of causality.

New as this way of description is in physical science, it is nevertheless familiar in other fields of knowledge. Thus, in biology we meet with suggestive analogies in the opposite points of view of mechanistic and vitalistic argumentation, and especially in psychological studies are we presented with problems which, in epistemological respects,

exhibit a close analogy to the intricacies of atomic physics. Thus, in introspection we have always to do with an arbitrary separation between the object of conscious analysis and the subjective background for the judgment, and it is essential to recognize how intimately the choice of this separation depends on those aspects of psychological experience to which attention in the given situation is directed.

In fact, the latitude for all feeling of volition is afforded by the very circumstance that the situations in which we can speak of freedom of will are mutually exclusive to those where causal analysis of mental phenomena can reasonably be attempted. Likewise, the background for all deeper religious attitude is the more or less intuitive recognition of the complementary relation between situations where such words as love or justice can be unambiguously applied. Indeed, we are here reminded of the old wisdom that, in searching for harmony, we must never forget that we are as well actors as spectators in the great drama of existence.

As regards that education in mutual understanding which at present is more necessary than ever, it should also be remembered that the comparison between different human cultures besides obvious relative features discloses typical complementary relationships. Not merely are we here presented with such contrasting pictures as meet observers attempting from different standpoints to describe the behavior of physical objects in terms of well established general concepts, but the very background for the attitude of different peoples to the various sides of life is engendered by traditions which have deep roots in historical developments. A full mutual appreciation of the cultural achievements of humanity is indeed irreconcilable with the complacency inherent in any national culture.

An apology for entering on these general topics in connection with the acute problems facing the world today may perhaps be found in the circumstance that the pursuit of knowledge in a domain of science as remote from human passions as the study of the most elementary physical phenomena has not only forcefully impressed upon us how necessary is openness of mind for the search of truth, but also suggestively reminded us of the common basis of all endeavors for the elevation of human culture. Surely, this circumstance should prove helpful for the strengthening of that spirit which now is incumbent if mankind shall be able jointly to reap the fruits which the progress of science offers.

V. DOCUMENTS, 1945:
STATEMENTS UPON
RETURN TO DENMARK

Niels and Margrethe Bohr return from England to Denmark in 1945 after Bohr's war exile.

Document 1:
RADIO TALK, 26 August 1945
[Stencil]
[Source: NBA]

Udtalelse af Professor Niels Bohr
fremsat i Pressens Radioavis den 26. August 1945.

Jeg er taknemmelig for at faa denne Lejlighed til at sige et Par Ord efter at være kommet hjem fra den lange Fraværelse, hvorunder der, saavel ude i Verden som herhjemme, har fundet saa store Begivenheder Sted.

I disse Aar, hvor Tankerne stadig har kredset om Hjemlandet, og hvor Sindet har været fuldt af Bekymring for, hvad de derhjemme maatte gennemgaa, har det samtidig været en inderlig Glæde at føle, i hvor høj Grad man ude forstod og værdsatte det stærke Sammenhold inden for alle Kredse af Befolkningen og den ubrydelige Vilje, der blev lagt for Dagen til med alle Midler at værne om de Traditioner og Idealer, der kendetegner vor Kultur.

Det var jo med Stolthed og Beundring, at man fulgte den Kamp mod Voldsmagten, der maatte organiseres i det skjulte, men efterhaanden naaede Resultater, hvis Betydning for den fælles Sag rakte ud over Landets Grænser, og som fandt saa oprigtig en Anerkendelse hos de Magter, hvis Sejre førte til den endelige Befrielse.

Man følte ogsaa overalt, at den Respekt, der staar om Danmarks Navn, har sin dybe Baggrund i de Bidrag, som vort Land trods sin Lidenhed i Tidernes Løb har formaaet at give til den fælles menneskelige Kultur, og de Forudsætninger vi har for ogsaa i Fremtiden at kunne yde en saadan Indsats.

Det er jo vort Syn paa de Rettigheder og Pligter, der tilfalder hver enkelt af Samfundets Medlemmer, og vor Følelse af Samhørighed med hele Menneskeheden, der ikke alene har ført til saa megen Fremgang i Oplysning og Velfærd i vort eget Land, men tillige har givet os en fremtrædende Plads inden for almene kulturelle Bestræbelser.

Som Videnskabsmand ligger det særlig nær for mig at tænke paa den Støtte og Forstaaelse, der saavel fra Statsmagten som fra samfundsindstillede Borgeres Side har betinget vor Deltagelse i det internationale videnskabelige Arbejde, der efterhaanden som Opgaverne voksede har faaet stedse større Betydning.

Den Sympati, hvormed danske modtages i videnskabelige Kredse overalt i Verden, er ogsaa blevet styrket ved vore Bestræbelser for at yde et Fristed her i Landet til Videnskabsmænd, der i Aarene før Fjendtlighedernes Udbrud som Følge af Fordomme, man længst troede udryddede, forjoges fra deres Hjemland, og ved at det med heltemodig Hjælp fra Modstandsorganisationernes Side lykkedes at bringe de landflygtige i Sikkerhed, da Forholdene herhjemme blev uholdbare.

Saadanne Forpligtelser ved Siden af mange andre maatte jo ogsaa for mit eget Vedkommende være medbestemmende for min Beslutning om at blive herhjemme saa længe som muligt til Trods for Opfordringer til at deltage med Kolleger i Udlandet i Arbejdet paa at udnytte Videnskabens Resultater i Kampen for Friheden, et Arbejde der paa saa mange Omraader skulde faa afgørende Betydning for Krigens Gang.

———————————

Det er ejendommeligt at tænke paa, at den Udvikling paa Atomfysikkens Omraade, der skulde ende med at give Menneskene saa overordentlige Magtmidler i Hænde, har sin Rod i Bestræbelser med det ene Formaal at uddybe Forstaaelsen af de Lovmæssigheder, som Udforskningen af Naturfænomenerne, trods den Righoldighed, hvormed de udfolder sig, stadig bringer klarere for Dagen.

Selv om Atomforestillingerne, der gaar helt tilbage til Oldtidens Tænkere, fandt en stedse frugtbarere Anvendelse inden for Naturvidenskabens forskellige Grene, lykkedes det dog først de seneste Generationer at bøde paa vore Sansers Grovhed ved, igennem Instrumenternes Forfinelse, ikke alene direkte at paavise Atomernes Eksistens, men at naa til Indsigt i selve Atomernes Bygning.

Især skulde Opdagelsen af Atomernes Kerner og af Muligheden for disses Omdannelser, der er knyttet til en af de største Skikkelser i Videnskabens nyere Historie, Lord Rutherford, faa Konsekvenser af den allerstørste Rækkevidde.

Inden for dette nye Forskningsomraade, der saa at sige hvert Aar bragte overraskende Opdagelser, har internationalt Samarbejde spillet en ganske afgørende Rolle. Fysikere hele Verden over har arbejdet i saa nær Forbindelse med hverandre, at det maaske i højere Grad end paa noget andet af Videnskabens Felter, ofte var vanskeligt at udrede den enkelte Forskers Bidrag.

Ogsaa herhjemme har vi paa Grund af den Understøttelse, der blev givet os baade til at tage Arbejdet op i vor egen Kreds og til at yde Gæstfrihed for værdifuldt Medarbejderskab, været i Stand til at tage Del i denne Udvikling, der netop i Aarene før Krigen viste Vejen til en mulig Udnyttelse af vældige Energimængder bundet i Atomernes Kerner.

Den forøgede Indsigt gjorde det imidlertid samtidigt klart, at de Stoffer, hvoraf vor Klode bestaar, ikke kunde benyttes til at fremstille Sprængstoffer af forfærdende Virkning uden efter en Ændring i deres Sammensætning af en saa indgribende Art, at dette vilde kræve store Forberedelser og overordentlige tekniske Hjælpemidler. Paa det Tidspunkt var det, i Betragtning af de Farer, der truede Verdensfreden, en Beroligelse at kunne være forvisset om, at intet Angreb med saadanne alt-ødelæggende Vaaben pludselig vilde kunne foretages.

Selv om man ikke kunde tvivle om, at der under Krigen fra mange Sider vilde blive gjort Forsøg paa at udnytte alle Muligheder, og man endda i Skræmmeøjemed fra tysk Side kom med aabne Antydninger, der for de indviede dog nærmest røbede, hvor langt man var fra Maalet, var det en fuldstændig Overraskelse for mig ved Ankomsten til England at faa at vide, hvor langt Arbejdet allerede da var fremskredent, og hvilke uhyre Hjælpemidler der var stillet til Raadighed for det i Amerika i Forstaaelse af den Katastrofe, det vilde være, om de Nationer, hos hvilke Tanken om at søge Verdensherredømme gennem andres Undertrykkelse var blevet raadende, skulde blive de første til at faa de nye Ødelæggelsesmidler i Hænde.

[138]

Beretninger om, hvordan det store Foretagende udviklede sig og sluttelig gennemførtes, er offentliggjort af de engelske og amerikanske Myndigheder,[1] og jeg skal her ikke komme ind paa de der beskrevne Enkeltheder. Man kan dog knapt undlade at tænke paa, hvor meget der er hændt i Videnskabens og Teknikkens verden, siden Middelalderens Guldmagere famlede i Mørket i deres forgæves Søgen efter de Vises Sten. Vi er nu Vidne til, hvorledes, paa Grundlag af klare videnskabelige Principper, Fysikere og Ingeniører i Fællesskab leder kæmpemæssige Laboratorier, hvori nye Stoffer langt kosteligere end Guldet saa at sige opsamles Atom for Atom eller endda opbygges gennem Omdannelse af hver enkelt Atoms Kerne.

Saadanne Stoffer maa antages at have været til Stede i rigeligt Maal paa tidligere Stadier af Universets Udvikling, hvor alt Stof var underkastet Forhold og Omskiftninger, hvortil vi nu kun har svage Efterklange i Solens og Stjernernes varmeste Indre. Paa Grund af Ændringer, som disse Stoffer bestandig undergaar, er de imidlertid nu enten yderst sjældne eller allerede forlængst helt forsvundne.

Det hele Foretagende betyder et Indgreb i den naturlige Udviklings Gang, der er langt dybere end noget, Mennesker hidtil har magtet, og Foretagendets Gennemførelse maa betragtes som en af de største videnskabelige og tekniske Bedrifter, der ikke vil kunne undgaa at faa de mest gennemgribende Indvirkninger paa Menneskehedens Fremtid.

Hvad en Videnskabsmand, der paa et sent Tidspunkt bragtes ind i den omfattende Organisation, har kunnet bidrage, er naturligvis yderst beskedent og har for mit Vedkommende kun bestaaet i Deltagelse i Raadslagninger vedrørende det videnskabelige Grundlags Uddybning og dets Bærekraft, der efter Sagens Natur først kunde bestaa sin Prøve, naar hele Værket var fuldført.

Medens Opbygningen og Administrationen af de vældige tekniske Virksomheder og Afgørelser af militær–strategisk Art falder uden for Videnskabsmændenes egentlige Omraade, maatte det jo for alle, og især for dem, der igennem Aarene havde deltaget i det internationale videnskabelige Samarbejde, hvorved de nye Muligheder aabnedes, ligge dybt paa Sinde, hvorledes den store Landvinding i vor Beherskelse af Naturens Kræfter vil kunne udnyttes til hele Menneskehedens Vel.

---

[1] The Smyth Report. See the introduction to Part I, ref. 69.

Jeg tænker jo ikke alene paa de Muligheder, der holdes os for Øje for at udnytte de nye Energikilder i Samfundets Tjeneste, men først og fremmest paa Erkendelsen af de uhyre Farer, der truer selve Civilisationens Bestaaen, dersom der ikke i Tide kan opnaas en verdensomspændende Overenskomst for at gennemføre en virkningsfuld Kontrol af de frygtelige Ødelæggelsesmidler, der ved Videnskabens Fremskridt er kommet inden for Menneskets Rækkevidde.

Ingen bedre Baggrund for Behandlingen af denne livsvigtige Sag, der paalægger hele Menneskeheden det alvorligste Ansvar, kunde imidlertid tænkes end de Bestræbelser for en Sikring af Verdensfreden, der har fundet en saa enstemmig Tilslutning fra alle Sider inden for de Nationer, som det kun ved forenede Kræfter lykkedes at forsvare de mest elementære Menneskerettigheder.

For at forberede den Indstilling, der vil kræves for Gennemførelsen af den nødvendige Overenskomst, er det af afgørende Betydning, at der i videste Kredse skabes Forstaaelse for, hvad der staar paa Spil, og i denne Forbindelse turde de Baand, der sammenknytter Videnskabsmænd hele Verden over som maaske ingen anden Gruppe inden for Samfundene, kunne blive af største Betydning.

Lad os haabe, at de rette Veje findes til, at Videnskaben, der igennem Tiderne har staaet som et af de stærkeste Vidnesbyrd om de Fremskridt, der kan naas ved fælles menneskelige Bestræbelser, netop ved dens seneste Belæring om Sammenholdets Nødvendighed vil bidrage afgørende til et harmonisk Samliv mellem Folkeslagene.

TRANSLATION

Statement by Professor Niels Bohr
Broadcast on the [Danish National] Radio News 26 August 1945.

I am grateful to have this opportunity to say a few words after returning home from my long absence, during which such great events have taken place both in the world at large and here at home.

During these years when my thoughts have constantly dwelt upon the homeland, and my mind has been full of concern for what those at home might have had to endure, it has at the same time been a heartfelt joy to feel how much people abroad understood and appreciated the strong unity within all circles of the population and the unbreakable will that was displayed in order to safeguard by any means the traditions and ideals that characterize our culture.

It was with pride and admiration that I followed the fight against the aggressor, which had to be organized in secret but which, as time passed, achieved results whose significance for the common cause reached out across the borders of the country and which found such sincere appreciation by those powers whose victories led to the final liberation.

One also felt everywhere that the respect shown for the name of Denmark has its deep roots in the contributions our country, despite its modest size, has managed to make in the course of time to common human culture, and the qualifications we have for also being able to make such an effort in the future.

It is our outlook on the rights and responsibilities devolving upon every single member of the community, and our affinity with all humanity, that not only has led to so much progress in enlightenment and welfare in our own country, but also has given us a prominent position as regards cultural efforts in general.

As a scientist I find it most natural to think of the support and understanding from the state authorities as well as from the public-minded citizens that have been the basis for our participation in international scientific work, which has assumed increasingly greater significance as the tasks have grown.

The sympathy with which Danes are received in scientific circles all over the world has also been strengthened by our efforts to provide a haven in this country for scientists who in the years before the outbreak of hostilities, because of prejudices one thought had been eradicated long ago, were driven out of their homeland and by the fact that with heroic help on the part of the resistance organizations the refugees were successfully brought to safety when the conditions here at home became untenable.

Such responsibilities together with many others must necessarily, also in my case, contribute to my decision to remain in this country for as long as possible despite invitations to take part with colleagues abroad in the work to utilize scientific results in the fight for freedom, work which in so many fields was to have decisive significance for the course of the war.

It is strange to think that the development in the field of atomic physics, which should result in providing mankind with such extraordinary means of power, is rooted in efforts with the sole purpose of deepening the understanding of the regularities, which the exploration of natural phenomena, despite the richness with which they occur, continues to bring more clearly to light.

Even though conceptions of the atom, which go back as far as the thinkers of Antiquity, found an ever more fruitful application within the various branches of science, only the most recent generations have succeeded in compensating for the coarseness of our senses by, through the refinement of instruments, not alone demonstrating the existence of atoms directly, but obtaining insight into the structure of the atoms themselves.

Especially the discovery of atomic nuclei and the possibility of transforming them, which is linked with Lord Rutherford, one of the greatest figures in the recent history of science, was to have consequences of the greatest importance.

Within this new field of research, which brought surprising discoveries more or less every year, international cooperation has played a quite decisive role. Physicists all over the world have worked so closely with each other that, perhaps to a greater degree than in any other field of science, it was often difficult to specify the contribution of the individual scientist.

Owing to the support we were given, both to take up work within our own circle and to offer hospitality for valuable cooperation, we here at home have also been able to take part in this development, which precisely during the years before the war showed the way to a possible utilization of the enormous amounts of energy tied up in the nuclei of atoms.

At the same time, however, the increased knowledge made it clear that the substances which make up our globe could not be used to make explosives of terrifying effect without a change in their composition so radical that it would require a great deal of preparation and extraordinary technical facilities. At that time, taking into account the dangers that threatened world peace, it was reassuring to be certain that no attack with such devastating weapons could suddenly be made.

Even though there could be no doubt that during the war efforts would be made in many quarters to utilize all possibilities and, with the intent of spreading fear, the German side even made open indications, which for the informed, however, more or less revealed how far from success they were, it was a complete surprise for me on arrival in England to learn how far the work had progressed already then, and what enormous facilities had been made available for it in America, in recognition of the catastrophe it would be if those nations, in which the idea of seeking world domination through the oppression of others had become prevalent, should be the first to get their hands on the new means of destruction.

Accounts of how the great project developed and was finally realized have been made public by the British and American authorities,[1] and I shall not here go into the details described there. Nevertheless, one can hardly refrain from thinking about how much has happened in the world of science and technology since the alchemists in the Middle Ages groped in the dark in their futile search for the philosophers' stone. We now witness how, on the basis of clear scientific principles, physicists and engineers together head gigantic laboratories in which new substances much more precious than gold are so to speak assembled atom by atom or are even constructed by the transformation of every individual atomic nucleus.

Such substances must be assumed to have been present in rich measure at earlier stages of the formation of the Universe, when all matter was subject to conditions and changes, of which we now only have faint echoes in the hottest core of the sun and the stars. Because of changes that these substances constantly undergo, they are now, however, either extremely rare or have already disappeared long ago.

The whole project signifies an interference with the natural course of things, which is far more profound than anything humans have hitherto accomplished, and the completion of the project must be regarded as one of the greatest scientific and technical achievements, which must necessarily have the most fundamental impact on the future of humanity.

The contribution that a scientist, brought into the large organization at a late stage, has been able to make is naturally extremely modest and has in my case only consisted of participation in consultations about the deepening of the scientific basis and its viability, which under the circumstances could prove its worth only upon the completion of the whole enterprise.

Whereas the construction and administration of the huge technical activities and decisions of a military–strategic kind are outside the scientists' sphere proper, everybody – and especially those who through the years have taken part in the international scientific cooperation whereby the new possibilities were revealed – must feel deep concern about how the great advance in our mastery of the forces of nature can be utilized for the benefit of all humanity.

I do not think only of the possibilities which are borne in mind for utilizing the new energy sources for the benefit of society, but first and foremost of the recognition of the extreme dangers which threaten the survival of civilization itself, unless in time a worldwide agreement is reached for accomplishing an effective control of the terrible means of destruction which, through the progress of science, have come within the reach of man.

No better background could be imagined, however, for treating this vital matter, which places the most serious responsibility on all humanity, than

[143]

the efforts for ensuring world peace that have found such unanimous support from all quarters within those nations who, only by joint forces, succeeded in defending the most elementary human rights.

In order to prepare the attitude that will be required for the realization of the necessary agreement, it is vitally important that understanding of what is at stake be created in the widest circles, and in this connection the ties that unite scientists all over the world, as perhaps no other group in society, could be of the greatest significance.

Let us hope that the right path is found so that science, which throughout the ages has stood as one of the strongest testimonies of the progress that can be achieved by common human endeavours, precisely with its most recent lesson about the necessity of concord will contribute decisively to harmonious coexistence among the peoples.

The students' torchlight procession on Bohr's sixtieth birthday, 7 October 1945.

Document 2:
SPEECH TO DANISH STUDENTS, 7 October 1945
[Stencil]
[Source: NBA]

Tale ved Studenternes Fakkeltog til 60-Aarsdagen den 7. Oktober 1945.

Jeg er mere taknemmelig, end jeg kan sige over, at danske Studenter paa denne Maade har villet bringe mig en Hilsen, og dybt rørt over de Ord, som Deres Talsmand har rettet til mig. En saadan Ære kan dog ingen, der med ringe Kræfter og paa beskeden Maade deltager i det fælles store Forskningsarbejde, føle tilkommer ham, men den maa opfattes som en Hyldest til den Videnskab, som vi alle tjener. At være med til at løfte blot en Flig af det Slør, hvorunder Sandheden skjuler sig og maaske derved komme paa Spor efter dybere Sammenhæng[e] end de, som umiddelbart frembyder sig, er al den Lykke, der kan gives en Forsker.

Det har været min gode Skæbne sammen med ældre og yngre at kunne deltage i Udforskningen af et Kundskabsomraade, der hidtil har ligget udenfor Menneskets Rækkevidde, og som har givet os Erfaringer, der paa ny Maade har belyst dybe Træk i vor Erkendelses Væsen. At denne Udvikling tillige saa stærkt skulde forøge vort Herredømme over Naturens Kræfter er et Vidnesbyrd om, hvilken Indflydelse Stræben efter at udvide Rammerne for vore Kundskaber og uddybe Grundlaget for vor Forstaaelse af den Natur, hvoraf vi selv er en Del, kan faa for de Vilkaar, hvorunder Menneskelivet maa udfolde sig. Den seneste Paamindelse om, hvor uløseligt hele Menneskehedens Skæbne er sammenknyttet, kan næppe overhøres, og vi tør sætte vor Lid til, at Videnskaben, som gennem Tiderne har staaet som et Symbol paa Frugtbarheden af fællesmenneskelige Bestræbelser, paa afgørende Maade vil bidrage til et harmonisk Samliv mellem Folkeslagene.

Naar jeg i Dag staar overfor Ungdommen, gaar mine Tanker tilbage til den Tid, hvor jeg selv med Sindet fyldt af Forventning begyndte et Studium ved vort gamle Universitet, hvor jeg ikke alene fandt Lærere, hvis Vejledning blev mig saa værdifuld, men ogsaa Kammerater, hvis Venskab skulde blive en Kilde til saa stor Glæde og Opmuntring. Samtidig tænker jeg paa, hvor meget det i Aarenes Løb har betydet for mig, at det skulde falde i min Lod at komme i Berøring med unge, der til Arbejdet bragte friske Kræfter og et opladt Sind, der ofte blev afgørende for at bringe Udviklingen ind paa nye Baner.

I den Tid der kommer, hvor Højnelsen og Udbredelsen af Kulturen maa blive Vejen til at bringe lysere Tider i Verden, vil den studerende Ungdom faa Opgaver og Ansvar maaske større end nogensinde før. Saavel de Traditioner, som vi herhjemme besidder vedrørende den enkeltes Rettigheder og Pligter i Samfundet, som det Syn paa vor Samhørighed med den øvrige Verden, der præger vor hele Indstilling, turde give Dem de gunstigste Betingelser for at virke for dette Formaal, og vi tør være baade lykkelige og stolte over, at disse Traditioner stod sin Prøve i mørke Tider, og ikke mindst over, at Livet ved vort Universitet, saavel hvad Forskningens Frihed som Aanden i Studenterverdenen angaar, trods Tryk og Trusler ikke knægtedes, men snarere lutredes.

Jeg vil slutte med at udtale Haabet om, at de unge, der har faaet den Lykke at kunne vie Dem til en Gerning i Kulturens Tjeneste, maa kunne opleve og medvirke til, at de Løfter om Menneskelivets Berigelse, som Videnskaben holder frem for os, maa blive indfriet.

[146]

TRANSLATION

Speech on the Occasion of the Students' Torchlight Procession
for the 60th Birthday on 7 October 1945.

I am more grateful than I can say that you Danish students have in this way wanted to bring me a greeting and profoundly moved by the words that your spokesman has addressed to me. Such an honour can no one, who with modest powers and in a humble fashion participates in the common great research effort, feel he deserves, but it must be considered as homage to the science which we all serve. To take part in lifting only a corner of the veil under which truth is hiding and perhaps thereby getting on the track of deeper connections than those immediately apparent, is all the happiness that a researcher can be given.

It has been my good fortune to be able to participate together with older and younger generations in the investigation of an area of knowledge which hitherto has been beyond human reach, and which has given us experience that in a new way has illuminated fundamental features in the nature of human knowledge. That this development should furthermore so strongly increase our mastery of the forces of nature is evidence of the influence which the striving to enlarge the framework for our knowledge and elaborate the foundation for our understanding of that nature, of which we ourselves are part, can have for the conditions under which human life must thrive. The most recent reminder as to how inextricably the fate of all humanity is interwoven, can hardly be ignored, and we dare to be confident that science, which through the ages has stood as a symbol of the fruitfulness of shared human endeavours, will contribute in a decisive way to harmonious coexistence between the peoples of the world.

When I stand today before the younger generation, my thoughts go back to the time when I myself, full of expectation, started my studies at our old University, where I found not only teachers whose guidance became so valuable to me, but also comrades whose friendship should become a source of such great joy and encouragement. At the same time, I think of how much it has meant to me through the years that it should be my lot to come into contact with young people who to their work brought fresh strength and a dedicated mind, which often proved decisive for giving the development a new turn.

In the time to come, when the heightening and the dissemination of culture must be the way to bring brighter times to the world, students will be given tasks and responsibilities perhaps greater than ever before. The traditions we have in this country as regards the rights and duties of the individual in society, as well as our sense of affinity with the rest of the world that marks our outlook as a whole, should give you the most beneficial conditions for working for this purpose, and we should be both happy and proud that these traditions stood their test in dark times, and not least that the life at our University both as regards freedom for research and the spirit in the student community, despite pressures and threats, was not broken but rather purified.

I will conclude by expressing the hope that you young people, who have the fortune to be able to dedicate yourselves to work in the service of culture, may be able to experience and contribute to the realization of the promises for the enrichment of human life which science holds out for us.

VI. HUMANITY'S CHOICE BETWEEN CATASTROPHE AND HAPPIER CIRCUMSTANCES

MENNESKEHEDENS VALG MELLEM KATASTROFE
OG LYKKELIGERE KAAR
Politiken, 1 January 1946

Talk on Danish national radio, New Year's Eve 1945

TEXT AND TRANSLATION

See Introduction to Part I, p. [65].

Menneskehedens Valg mellem Katastrofe og lykkeligere Kaar

Atomenergien giver de rigeste Løfter for Menneskehedens Velfærd, men rummer samtidig dødelig Fare

Professor Niels Bohrs Tale i Radioen i Aftes

Professor Niels Bohr ved Mikrofonen.

I Radioen holdt *Niels Bohr* i Aftes denne tankevækkende Nytaarstale, som vi bringer efter Talerens Manuskript.

I det Aar, vi forlader, oplevede vi det endelige Sammenbrud af Historiens maaske farligste Forsøg paa ved Vold og umenneskelige Forfølgelser at berøve Folkene Friheden til at indrette deres Liv i Overensstemmelse med egne Traditioner og de Idealer, der danner Grundlaget for selve Civilisationen, til hvis Udvikling gennem Tiderne alle Folkeslagene, hvert paa sin Vis, har givet deres Bidrag.

De frygtelige Rædsler, Verden har været Vidne til, og den forfærdelige Nød, som har fulgt i deres Spor, har skabt en almindelig Forstaaelse af, at der maa findes nye Former for Samlivet mellem Nationerne for at hindre Gentagelse af saadanne Trusler, som det denne Gang kun ved de største Ofre og de yderste fælles Anstrengelser lykkedes at afværge, og for at sikre et fredeligt Samarbejde paa Kulturens Fremskridt.

Som et af de stærkeste Vidnesbyrd om, hvad der i Fællesskab kan naas, staar den videnskabelige Forskning, der søger sin Rod i en almen menneskelig Trang til at forøge vore Kundskaber og udvide Rammerne for vor Erkendelse. Saadanne Bestræbelser kender efter deres Art ingen nationale Grænser, og til Videnskabens stadig mere omfattende og grundfæstede Bygningsværk har Folkeslag fra Jordens mest forskellige Egne baaret Sten.

[150]

Verdensaltets dybe Gaade

I vor Tidsalder, der i mange andre Henseender rummede saa store Modsætningsforhold, oplevede vi paa Videnskabens Omraade et internationalt Samarbejde af stedse større Frugtbarhed, der medførte en Fremgang, som, samtidig med, at vi har vundet Kundskaber, der skulde komme til at faa en gennemgribende Indflydelse paa Menneskelivets Kaar, har givet os en dybere Indsigt i vor Erkendelses Væsen og derved et mere frigjort Syn paa mange af de Problemer, der fra de tidligste Tider har beskæftiget Menneskeaanden.

Dette gælder ligefuldt for Opklaringen af Gaader, vi møder ved Forsøgene paa at udforske Verdensaltet, der i Rum og Tid spænder saa uendelig vidt i Forhold til de Omgivelser, hvori vi er vant til at finde os til Rette, som for Udredelsen af de Love, der behersker Vekselspillet mellem Stoffernes mindste Byggestene, som paa Grund af deres Lidenhed saa længe unddrog sig hvert nærmere Studium.

Det Indblik, vi har faaet i Atomernes Verden, har ikke alene klarlagt de Forhold, der er bestemmende for alle Egenskaber hos Stofferne, vi hidtil har betjent os af for de mest forskellige Formaal, men tillige har vi gennem Opdagelsen af Muligheden for Grundstoffernes Forvandlinger lært helt nye Naturfænomener, der er knyttet til Omdannelser i selve Atomernes Kerner, at kende.

Udløsningen af de vældige Energi-Mængder og Menneskehedens Fremtid

Først gennem disse Opdagelser kom vi paa Spor efter Kilden til den Energi, som Solen har udstraalet i Milliarder af Aar, og som betinger alt Liv paa Jorden, og omsider har vi ogsaa fundet Vej til ved egen Haand at udløse vældige, hidtil utilgængelige Energimængder og derved paa den mest indgribende Maade forøget Menneskets Herredømme over Naturens Kræfter.

Allerede før dette skelsættende Trin i Udviklingen var naaet, har Videnskabens Fremgang i nyere Tid haft en stadig voksende Indflydelse paa de Betingelser, under hvilke Samfundet lever. Hvad betyder ikke Lægekunstens Højnelse og den øgede Evne til at værne om Menneskenes Sundhed for Trygheden og vore Fremtidshaab, og hvor meget har ikke Teknikens stadige Fremskridt forandret hele vor Tilværelse.

Denne Fremgang har grebet dybt ind i Samfundsforholdene i alle Kulturlande, men de Perspektiver, som nu aabner sig, og som indebærer de rigeste Løfter for Menneskenes Velfærd, rummer samtidig dødelige Farer, der kun kan overvindes gennem en helt ny Indstilling til Forholdet mellem Folkeslagene indbyrdes.

Ligesom Samfundets enkelte Medlemmer for den almene Sikkerheds Skyld for længst har maattet opgive at tage sig selv til Rette, maa der jo nu, dersom hele Civilisationen ikke skal gaa sin Undergang i Møde, opnaas Enighed mellem Folkeslagene om Forholdsregler til at hindre enhver egenmægtig Brug af de vældige Kræfter, der er kommet inden for Menneskenes Rækkevidde, og sikre, at Udnyttelsen af de nye Energikilder tjener Menneskehedens fælles Tarv.

En Overenskomst til dette Formaal vil naturligvis forlange en Nedbrydning af mange af de Skranker, der staar Nationerne imellem, men det skulde være klart, at det her drejer sig om et Livsspørgsmaal for alle. Gunstigere Tidspunkt for Fremkomsten af en saadan uopsættelig Sag kan vel næppe tænkes end netop nu, hvor Begivenhederne i Verden har forenet Nationerne i et brændende Ønske om at finde frem til en varig indbyrdes Forstaaelse.

Her i Danmark, hvor der i lange Tider har været en vaagen Følelse af vor Samhørighed med den øvrige Verden, turde vi vel være forberedt paa en saadan Styrkelse af Baandene mellem Folkeslagene, som nu stilles i Udsigt, men som alle andre maa vi indstille os paa, at Virkeliggørelsen af de store Forhaabninger vil kræve god Vilje af enhver af os til at drage Lære af Udviklingens Gang.

Ved dette Aarsskifte befinder hele Menneskeheden sig paa en Skillevej. Vi har faaet den alvorligste Belæring om Farer, der kan føre os alle i Afgrunden, om vi hver for sig fortsætter blindt, men samtidig er der stillet os Udsigter for Øje, der rummer lysere Haab end nogen Sinde før for i Fællesskab at skabe lykkeligere Kaar for Menneskelivet.

[151]

TRANSLATION

Humanity's Choice Between Catastrophe and Happier Circumstances

———

Atomic energy holds the richest promises for human welfare but contains at the same time fatal dangers

———

Professor Niels Bohr's speech on the radio yesterday evening

> On the radio yesterday evening *Niels Bohr* gave this thought-provoking New Year's speech, which we bring according to the speaker's manuscript.

In the year we are leaving we experienced the final collapse of perhaps the most dangerous attempt in history to rob, by violence and inhuman persecution, the peoples of the freedom to arrange their life in agreement with their own traditions and the ideals which form the basis for civilization itself, to whose development through the ages all peoples, each in their own way, have given their contributions.

The dreadful horrors witnessed by the world and the terrible hardship that has followed in their wake have given rise to a general understanding that new forms of coexistence between nations must be found to prevent a reoccurrence of such threats, which this time were successfully averted only by the greatest sacrifices and the most extreme shared efforts, and to ensure peaceful cooperation based on the progress of culture.

As one of the strongest testimonies of what can be jointly achieved stands scientific research, which seeks its roots in a general human need to increase our knowledge and extend the framework of our understanding. By their nature such endeavours know no national boundaries, and peoples from the most diverse regions of the Earth have carried building blocks to the ever more comprehensive and well-founded edifice of science.

The profound riddle of the universe

———

In our time, which in many respects has held such great antagonisms, we have experienced international cooperation in the field of science of ever greater fruitfulness, bringing advances which, while we have gained knowledge that was going to have a fundamental influence on the conditions of human life, have given us deeper insight into the nature of our knowledge and thereby a more liberal attitude regarding many of the problems which have occupied the human spirit since the earliest times.

This is equally true for the solution of riddles we meet when attempting to explore the universe, which in space and time stretches so infinitely wide in relation to the surroundings in which we are used to feeling at home, as it is for the elucidation of the laws governing the interaction between the smallest constituents of matter, which because of their littleness so long escaped any closer inspection.

The insight we have gained into the world of atoms has not only clarified the relations determining all properties of the substances we hitherto have made use of for the most diverse ends, but in addition we have through the discovery of the possibility of the transmutation of the elements, learned about completely new natural phenomena, which are related to transformations inside the very nuclei of the atoms.

The release of the enormous amounts of energy and the future of humanity

———

It was only through these discoveries that we got on the track of the source of the energy that the sun has radiated for billions of years and which determines all life on Earth, and we have at last also found a way to release on our own enormous hitherto inaccessible amounts of energy and thus in the most radical way increased mankind's mastery over the forces of nature.

Already before this epoch-making step in the development was achieved, the advance of science in recent time has had a steadily growing influence on the conditions under which society lives. What do the improvement of the art of medicine and the increased ability to safeguard people's health not mean for our sense of security and our hopes for the future, and how much have the constant advances of technology not changed our whole existence.

These advances have deeply affected the social conditions in all civilized countries, but the perspectives now opening and holding the richest promises for human welfare contain at the same time fatal dangers which can only be

[153]

overcome by an entirely new attitude towards the mutual relations between nations.

Just as the individual members of society, for the sake of general safety, have had to give up taking the law into their own hands long ago, so must now, if all civilization is not to be doomed, agreement be reached between peoples about measures to prevent any arbitrary use of the enormous forces that have come within the reach of humans, and ensure that the use of the new energy sources serves the common good of humanity.

An agreement to this end will naturally require the breaking down of many of the barriers standing between the nations but it should be obvious that this is a vital matter for everyone. A better time for the appearance of such an urgent matter can hardly be imagined than just now, when the events in the world have united the nations in a burning desire to reach a lasting mutual understanding.

Here in Denmark, where there has long been a lively sense of our affinity with the rest of the world, we ought to be well prepared for such a strengthening of the bonds between nations which is now held in prospect, but like all others we must be prepared that the realization of these great hopes will require genuine willingness on the part of everyone to learn from the course of development.

This New Year the whole of humanity is at a crossroads. We have had the most serious lesson about dangers that can lead us all into the abyss if we each on our own continue blindly, but at the same time we have prospects which hold brighter hope than ever before for together to create happier circumstances for human life.

VII. DOCUMENTS, 1946 AND 1948:
CONTACT WITH BARUCH AND MARSHALL

Document 1:
LETTER TO BARUCH, 1 June 1946
[Typewritten]
[Source: CAB 126/40]

Dear Mr. Baruch,

I am very grateful for the kind invitation from the American Minister in Denmark, Mr Joseph Marvel, to write to you concerning the urgent problems which place so grave a responsibility on our whole generation and the consultations about which among the United Nations will be followed with the greatest expectations all over the world.

As a scientist who through many years has participated in the international collaboration in atomic physics which was to lead to a veritable revolution of human resources I have naturally been very much occupied with the implications of the development and during my stay in England and U.S.A. the last two years of the war I had much occasion to discuss technical aspects of the matter with American and British colleagues.

As you may know, I had also the great privilege to present to the late President Roosevelt some considerations about the opportunities the situation offered, and from a personal conversation with the President I received a deep impression of his firm belief that the immense enterprise which was then nearing its completion would inaugurate a new era in the history of mankind.

Since my departure from the U.S.A., where I was given the opportunity to submit an account of my views to the consideration of the Advisory Committee appointed by President Truman, I have been able to follow developments only through public statements. Like other scientists I have, however, tried to assist in bringing about a general understanding of what is at stake and to this purpose I have written two small articles of which I enclose reprints.[1]

The principal aim of these articles was to stress that, notwithstanding obvious dangers, the present situation should afford quite unique opportunities for furthering international relationships. The very fact that we are presented with a serious challenge to our civilization in the meeting of which the interests of all nations clearly coincide should indeed offer a realistic basis for an agreement about measures which are necessary for common safety and which would in themselves constitute a decisive advance as regards co-operation in mutual confidence.

---

[1] N. Bohr, *Science and Civilization*, The Times, 11 August 1945; this volume, pp. [121] ff. N. Bohr, *A Challenge to Civilization*, Science **102** (1945) 363–364; this volume, pp. [125] ff.

For reasons which I am sure you will appreciate I have, in the articles, confined myself to such arguments as should immediately appeal especially to scientists of all nations and have abstained from entering on any specific questions of security and procedure which might give rise to controversies. In this respect the situation has, of course, been greatly relieved by the recent publication by the American State Department of the report on International Control of Atomic Energy,[2] whereby the discussions on such questions have been initiated in an appropriate manner.

From every point of view it is surely to be greeted most heartily that a constructive proposal which bears witness to so much thought and expert knowledge has now been brought before the public and, as regards the reaction of scientists, I am convinced that especially the idea of a most intimate connection between control and international cooperation on the realization of the great prospects which the progress of science holds out for human welfare will meet with general approval. In this connection I am thinking not least of the role which the ties between scientists throughout the world, established by many years of fruitful collaboration, may play for the creation of an atmosphere of good-will so essential for a favourable development.

Personally I have studied the report with the deepest interest and have not only admired the soundness and ingenuity of the proposals it contains but also appreciated the restraint exerted in separating the discussion on technical and administrative problems from the wider prospects of a whole new approach to the question of international relationship, which the unique situation has opened and which inspire everyone with so great hopes.

Notwithstanding the decisive importance of technical matters for the shaping of the eventual control arrangement, it is obvious, however, that the promotion of the cause will above all depend on the general recognition of the fact that the formidable power of destruction which has come within reach of man has irrevocably changed the conditions for the life of nations. Even the bright promise of industrial applications of atomic energy sources may be regarded as a secondary aspect of the situation in the sense that the fulfillment of such promises will rather fall, in due time, as ripe fruits of a development, the primary step of which must be a genuine common effort aimed directly at averting a deadly menace to civilization.

There should be every reason to believe that it will be possible to obtain agreement from all sides about the necessity of farreaching renunciations on accustomary prerogatives, and that in particular the situation calls for the

---

[2] The Acheson–Lilienthal Report. See the introduction to Part I, p. [66].

[157]

removal of many barriers which present obstacles for mutual confidence and for the upholding of which no arguments are now maintainable. Truly, the United Nations are here presented with a paramount task in the success of which all peoples must put their faith.

Of course, I am well aware of the many difficulties with which the statesmen are confronted in dealing with these stupendous problems, but I have been grateful for this opportunity to give expression for views and sentiments which I know are shared by many fellow scientists. I need not add that I am completely at your disposal if you should think that, in any way, I can be of assistance for your endeavours.

Most respectfully yours,

[Niels Bohr]

Mr. Bernhard M. Baruch.
State Department.
Washington, D.C.

Document 2:
COVERING LETTER, 22 March 1948
[Typewritten]
[Source: BPP 117]
[Editor's comment: Added on top of document (in Danish) in the handwriting of Bohr's secretary: "U.S.A. State Department, given to McCloy for forwarding to Marshall." With the letter, Bohr enclosed selected passages from the "Memorandum" (p. [101]) and the "Addendum" (p. [113]).]

March 22, 1948

In connection with my association with the atomic energy project during the war, I had the opportunity to present to the governments of the United States and Great Britain certain views, a brief account of which is contained in the enclosed memoranda, regarding the opportunities which this revolutionary technical development offered for a new approach to the question of international relationships. Aware of the responsibilities of the statesmen who will have to make momentous decisions in order to advance the cause, and due to the position of confidence I was brought into during the war, I have closely followed the developments, without participating in public discussions. Yet the occasions I have had to learn of the views that are held in these matters in various communities have strengthened my belief that, despite the great efforts which have been made, there are still points of view which need to be more fully explored, and the consideration of which, above all against the background of the growing world tension, may be of significance for the statesmen in their search for a constructive approach to the great problems confronting humanity. With this in mind, I, of course, hold myself ready to discuss these questions within the measure of my knowledge and experience.

Niels Bohr, Director
Institute for Theoretical Physics
Copenhagen, Denmark

Visiting Professor
Institute for Advanced Study
Princeton, N.J.

Document 3:
"COMMENTS", 17 May 1948
[Typewritten]
[Source: BPP 117]
[Editor's comment: Bohr reprinted this document in full as part of his "Open Letter to the United Nations" from 1950 (see p. [171]). Since it is such an essential part of his contact with Marshall, it is also reproduced here. The footnote is in the original.]

Confidential–

COMMENTS[*]

The deep-rooted divergencies in attitudes to many aspects of human relationship which have grown out of social and political developments in the last decades, were bound to present a serious strain on international relations at the conclusion of the second world war. While, during the war, the efforts in common defense largely distracted attention from such divergencies, it was clear that the realization of the hopes acclaimed from all the nations united against aggression of a wholehearted co-operation in true confidence would demand a radically new approach to internation[al] relations.

The necessity of a readjustment of such relations was even further accentuated by the great scientific and technical developments which hold out bright prospects for the promotion of human welfare, but at the same time have placed formidable means of destruction in the hands of man. Indeed, just as previous technical progress has led to the recognition of [the] need for adjustments within civilized societies, many barriers between nations which hitherto were thought necessary for the defense of national interests would now obviously stand in the way of common security.

The fact that this challenge to civilization presents the nations with a matter of the deepest common concern should offer a unique opportunity for seeking

[*] These comments are presented as a supplement to the memoranda of the war years, enclosed with note of March 22, 1948, to suggest how the issues may appear in the present situation.

continued co-operation on vital problems. Already during the war, it was, there-fore, felt that a favourable foundation for later developments might be created by an early initiative aimed at inviting confidence by making all partners aware of the actual situation which would have to be faced, and by assuring them of willingness to share in the farreaching concessions as to accustomed national prerogatives which would be demanded from every side.

In the years which have passed since the war, the divergencies in outlook have manifested themselves ever more clearly and a most desperate feature of the present situation is the extent to which the barring of intercourse has led to distortion of facts and motives, resulting in increasing distrust and suspicion between nations and even between groups within many nations. Under these circumstances the hopes embodied in the establishment of the United Nations organization have met with repeated great disappointments and, in particular, it has not been possible to obtain consent as regards control of atomic energy armaments.

In this situation with deepening cleavage between nations and with spreading anxiety for the future, it would seem that the turning of the trend of events requires that a great issue be raised, suited to invoke the highest aspirations of mankind. Here it appears that the stand for an open world, with unhampered opportunities for common enlight[en]ment and mutual understanding, must form the background for such an issue. Surely, respect and goodwill between nations cannot endure without free access to information about all aspects of life in every country.

Moreover, the promises and dangers involved in the technical advances have now most forcibly stressed the need for decisive steps towards openness as a primary condition for the progress and protection of civilization. The appreci-ation of this point, it is true, underlies the proposals to regulate co-operation on the development of the new resources, brought before the United Nations Atomic Energy Commission, but just the difficulty experienced in obtaining agreement under present world conditions would suggest the necessity of cen-tering the issue more directly on the problem of openness.

Under the circumstances it would appear that most careful consideration should be given to the consequences which might ensue from an offer, extended at a well-timed occasion, of immediate measures towards openness on a mutual basis. Such measures should in some suitable manner grant access to infor-mation, of any kind desired, about conditions and developments in the various countries and would thereby allow the partners to form proper judgment of the actual situation confronting them.

An initiative along such lines might seem beyond the scope of conven-tional diplomatic caution; yet it must be viewed against the background that,

[161]

if the proposals should meet with consent, a radical improvement of world affairs would have been brought about, with entirely new opportunities for co-operation in confidence and for reaching agreement on effective measures to eliminate common dangers.

Nor should the difficulties in obtaining consent be an argument against taking the initiative since, irrespective of the immediate response, the very existence of an offer of the kind in question should deeply affect the situation in a most promising direction. In fact, a demonstration would have been given to the world of preparedness to live together with all others under conditions where mutual relationships and common destiny would be shaped only by honest conviction and good example.

Such a stand would, more than anything else, appeal to people all over the world, fighting for fundamental human rights and would greatly strengthen the moral position of all supporters of genuine international cooperation. At the same time, those reluctant to enter on the course proposed would have been brought into a position difficult to maintain since such opposition would amount to a confession of lack of confidence in the strength of their own cause when laid open to the world.

Altogether, it would appear that, by making the demand for openness a para-mount issue, quite new possibilities would be created, which, if purposefully followed up, might bring humanity a long way forward towards the realization of that co-operation on the progress of civilization which is more urgent and, notwithstanding present obstacles, may still be within nearer reach than ever before.

May 17, 1948

Document 4:
LETTER TO MARSHALL, 10 June 1948
[Typewritten transcript]
[Source: BPP 117]

C o p y.

Washington, D.C.
June 10th, 1948.

PERSONAL AND CONFIDENTIAL

Dear General Marshall,

It was a great privilege and a great source of encouragement to me to be able to speak with you of the present situation, as well of the difficulties as of the hopeful aspects, and I am happy to follow your kind suggestion that I supplement in a personal letter the brief account of the views, presented in my Comments of May 17th, of which I for your convenience enclose an extra copy.

As you yourself so fully realize, genuine international cooperation is hampered and almost paralysed as long as diplomatic exchange, instead of serving as a realistic and constructive approach to practical problems, is largely a tool of political propaganda. Certainly, this present barren situation cannot be essentially improved without a world public opinion, which cannot exist or function as long as great countries are practically closed, and it is neither possible to learn what is happening within their territories, nor for their populations to obtain a proper picture of conditions abroad.

It will appear, therefore, that the only way out of the crisis is to make a stand for free access, on a mutual and worldwide basis, to all information essential for international relationship, and to make this a paramount issue. It is my belief that the pioneering accomplishment in releasing atomic energy offers a unique opportunity for an initiative to this purpose, which may lead directly

The Honourable
George C. Marshall
Secretary of State

[163]

to the desired goal or, at any rate, will strengthen the position of supporters of genuine international cooperation so much that the world situation will be greatly changed.

I shall not here comment upon what the revolutionary technical accomplishment may already have meant as a temporary deterrent factor, in world politics, but only upon the background it presents for an initiative to open up the world, in providing a possibility for renouncing a momentary advantage in return for far-reaching concessions required for common security, concessions which otherwise hardly would be obtainable.

Of course, I appreciate that the promotion of a development along such lines was also the purpose of the American proposal to the Atomic Energy Commission of the United Nations, but in my conviction, the failure of obtaining consent to this proposal demonstrates, that progress can only be achieved by a line of policy, directly aimed at mutual openness on a still broader and more immediate basis than contemplated in the American proposal. The purpose of this policy should be to create a radically changed situation, combining the removal of present obstacles for understanding with establishment of common security.

By openness on a broader basis I mean free access to all information about the conditions in every country, including technical development and military preparations, and limited only by the constitutional rights of the individual and indispensable requirements of sound administrative practices. Such openness which could, of course, only be granted on a mutual and universal basis, would enable the people of any country to form a well founded judgment on all questions decisive for its relations with other nations.

I am well aware that it may be asserted as an argument against the suggested course that it renounces many of the precautionary measures incorporated in the American proposal to the United Nations. Still, consideration of all aspects of the situation would seem to indicate that the immediate full mutual openness would not involve additional risks, but is on the contrary required in order not to lose the opportunity to forestall a fateful competition in atomic armaments and to obtain guaranties for lasting peaceful cooperation.

With regard to the problems which a universal acceptance of an offer of free mutual access to information will at once present, it might be argued that the dissimilar conditions now prevailing in the different countries would cause inequality in the ease and speed with which the first observers admitted to the various territories would be able to obtain essential information. Still, through an arrangement covering this initiating period, upon which any formal offer of mutual openness must be conditioned, it should be possible to secure that all information as to atomic development and other military preparations would be accessible in an appropriately balanced manner.

I have given much thought to, and I am ready to discuss the functioning of such an arrangement, which of course will demand careful consideration in advance of many practical and technical problems. However, the essential point is that fixation of final details should be postponed until the offer of openness was accepted in principle, whereafter such details would be worked out and settled in common, before the plan of admitting observers was put into operation.

Just this procedure will already at this early stage offer a most valuable test as to genuine cooperation through the need of contribution from either side. Unlike the frequently encountered situation, where engagement in negotiations about matters in dispute does not imply readiness to work for constructive solution in mutual interest, we have here to do with a practical task, the agreed object of which is to regulate an exchange of specified information, a situation in which good faith on the accepting, as well as on the inviting side, can and must be proved by performance.

Surely, the acceptance of the offer and the realization of the plan for the initiating period would accelerate a change in international relations, more sweeping than can readily be imagined and described. What it would mean, if the whole picture of social conditions in every country were open for judgment and comparison, need hardly be enlarged upon. Above all, if such openness was first achieved, it would be difficult to think of a reason anybody could have, or maintain, against accepting an arrangement abolishing competition in atomic armaments, along the lines of the American proposal, or against even more far reaching arrangements for common security.

Still, before the initiative is decided upon, punctilious examination must of course be given to the balance of gain and loss during the transitory period, including the most searching inquiry into relative military position. Two points would seem to be relevant: One is the very great advantage which a closed country now possesses in withholding information essential for strategic and tactical plans of opponents. The second point is the very minor value of technical information, obtained by personal observation and access to written reports, in relation to the actual implementation of the industrial program for military effectiveness.

In the evaluation of the effect of release of information about technical details, it must be kept in mind that the exploitation of scientific progress, largely achieved by open international collaboration, owed its importance in the wartime effort to the rapidity with which industrial developments could be undertaken, due to the advanced stage of technology and enterprise in this country. It must also be recognised, how difficult it is to transplant a technological development, attuned to a specific industrial potential, into another community, lacking the corresponding background.

Indeed, the enterprise is of such character and size that if information was collected in full openness – instead of obtained by secretive means or granted stepwise – the witnessing of its magnitude and diversity will produce a so overwhelming impression that, rather than facilitate competitive efforts, it might discourage such attempts and serve as a powerful inducement to accept the offer, held out in the American proposal to the United Nations, of participating in the venture of turning the great enterprise to common benefit for humanity.

With regard to the points here mentioned, the large group of experts in this country should be able to assist in reaching a sound evaluation of the actualities. It has not been my intention to prejudge the outcome of such studies but only to indicate, how necessary they are in order to put arguments of security in their proper perspective. I know, of course, that in some respects similar studies have been made in connection with the safeguards embodied in the United States proposal for international control of atomic energy, the substance of which will surely remain in the ultimate arrangements for effective longrange security.

Incidentally, the new and ominous menace to world security, presented by the possibility of employing the results of the latest development of bacteriological and biochemical science as terrible life-destructive means, cannot be eliminated by any practicable control and will, therefore, remain a latent danger until such cooperation in openness and confidence has been achieved that a common higher moral atmosphere is gradually created. Indeed, it may be regarded as a good fortune of mankind that, before such alarming prospects have become imminent, the development of atomic science has presented the world with an immediate challenge which, due to the specific nature of the technical problems involved, can be met by practical measures of such a kind that they, when realized, will constitute a decisive step in the direction humanity must proceed if civilization shall survive.

Certainly, the advance in science and technology has presented our generation with a great responsibility, but also with a unique opportunity through the fact, that the only measures which can offer guaranties for common security are the same as those required for mutual understanding and genuine cooperation. However, the advantage of initiative may be lost, and the possibility of security become remote, as soon as competitive efforts in atomic armaments have reached an advanced stage, or already at the moment when the probability of the existence of such competition can no longer be discounted.

In the foregoing I have tried to the best of my knowledge to comment upon the question raised during our conversation, whether a policy aimed at immediate world-wide openness would be in conformity with reasonable demands for security. I feel, of course, more reluctant in offering suggestions as to the question, you also touched upon, of procedure suitable to carry out this policy.

In relation to this point, it must be taken into consideration that the difficulty of conveying the actual situation in all its aspects to the American public might easily give rise to misunderstandings and anxiety about the purpose and implications of the policy. Indeed, the reaction which could be expected, if a proposal of openness were to be submitted to public discussion, without proper and responsible authority behind it, might be unfavourable for the achievement of the desired aim. If exploited by unfriendly propaganda, such reaction could even be harmful to the position and endeavors of this country.

The situation, however, should be quite different in case the Government of the United States, after careful consideration and with full cognizance of the problems of security, should come to the conclusion that the initiative would serve the true interests of this country as well as of humanity at large. Surely, in that case, a statement from outstanding authority proposing an immediate approach to universal openness would constitute a challenge to the nation, which would appeal to the noblest traditions inherent in the birth and growth of American democracy.

On my many visits to this country I have, as I told you, received the deepest impression of enthusiasm for human progress and of that feeling of brotherhood with all humanity, which just now has manifested itself in the magnanimous efforts to help the nations staggering under the calamities brought about [by] the war. I have, however, also here, as in many other countries, found a wide-spread despair and confusion, due to the disappointment of the great hopes which united so many nations during the war. On this background, the initiative will surely be greeted with eagerness, here and elsewhere, and provide a rallying point for a high and unifying ideal.

If the international and domestic situation favoured it, the most direct step would be a concrete proposal of universal openness, aimed at prompt realization. Yet, if such a procedure would not be deemed timely, it might be found suitable, after proper preparation, to use an early occasion, when re-stating the general lines of American policy, to stress the urgency of fullest mutual openness and express readiness to entertain proposals to this effect.

Even the preliminary approach, consisting of a declaration of aspiration and intent, might perhaps elicit an answer which could serve as a further stepping stone. But irrespective of such response, this initiative should contribute decisively to clarify the situation; it would greatly strengthen the moral position of all advocates of genuine international cooperation and bring adversaries everywhere in rapidly increasing difficulties.

A refusal to accept an invitation to mutual openness would, as indicated in my previous Comments, mean an admission, which no propaganda could effectively conceal, of fear of laying actual conditions open to the world. Certainly, any anxiety which has to be overcome within the inviting country must be measured against the effect which a bold and candid action would have in compelling others to reconsider their situation and inducing them to a change of attitude.

Surely, the momentous goal of turning the trend of events and inaugurating a new era for humanity cannot be achieved without a firm stand and a courageous fight for principles, in which everybody here and all over the world will be called upon to take part and do his share. Still, there should, it seems to me, be good reason to hope that a purposeful and imaginative exploitation of the unique opportunity would help to alleviate international tension and bring the world out of the present crisis.

This hope is inspired by the conviction that your country, in human and material resources, possesses the strength required to take the lead in accepting the challenge, with which civilization is confronted, and in establishing that cooperation, in openness and confidence, between all nations on which, not merely progress and prosperity of humanity, but its very existence depends.

In concluding this letter, I wish once more to thank you for the kindness and trust you have shown me, and I need hardly repeat that I am always at your disposal, if in any way I can be of assistance in connection with the purposes which are so foremost in your mind and for which you carry so heavy a burden.

In gratitude and deep respect
(signed) Niels Bohr

ENCLOSURE

Document 5:
"ANNOTATIONS", 19 June 1948
[Typewritten]
[Source: BPP 117]
[Editor's comment: Added on top of document (in Danish) in the handwriting of Bohr's secretary: "Remarks added to letter to Marshall 10/6 1948 and developed for negotiations in the State Department during the following weeks."]

A n n o t a t i o n s.

1. The policy of full openness would, apart from the access to information, not entail apriori commitment as to disarmament of any kind, but should create a situation which would remove objections for consent to security measures in mutual interest as well as obstacles for effective control to this purpose.

2. By "full mutual openness" is meant not only free access to collect information desired, but also obligation for every country to assist in the collection of such information within its territory. Thus the acceptance of an offer of free [sic] mutual openness would involve a declaration of readiness to furnish all information requested and to facilitate direct access to such information, including the admittance of observers with full freedom of movement and communication.

3. An "arrangement covering the initiating period to insure appropriately balanced access to information" should not imply any limitation in the freedom to collect information during this period, but should only serve to provide proper assistance from all sides before observers are admitted to the various territories.

4. It is contemplated that after the offer has been accepted in principle by everyone, but before full openness is granted, a conference should be held of authorized representatives for the various Governments with the purpose of exchanging preliminary information, in order that no observers are handicapped in finding and reaching the locations, where the most urgently desired knowledge can be gathered. After agreement has been reached about the number of observers initially to be admitted and about regulations for their functions, the conference should fix the date from which full openness should enter into effect.

5. As regards to functions of observers, a practical solution with all reasonable guaranties should be easy to find, particularly considering modern means of short wave communication. At first sight, however, it might perhaps seem questionable whether the preliminary information exchanged during the conference would be accurate and complete. Still, it must here be taken into consideration that not only will it be possible already during the conference with expert advice to form a general opinion about the appropriateness of the information, but also that its essence will be tested by the observers as soon as they begin to function.

6. In connection with the foregoing point, it should again be remembered that no information about scientific progress and industrial development will have immediate military value and will in this respect be greatly exceeded by the many improvements in short range security, which an admittance of observers to previously closed countries would entail. The importance of reasonable initial access to information is indeed far more to be found in the impression on the public mind in either nation, that each country retains something valuable to offer until, simultaneously or at any rate in close proximity in time, equally important information can be gathered. Incidentally, such psychological effects may be an inducement for the parties at the conference not to minimize own accomplishments.

7. Altogether, the eventual acceptance of an offer of full mutual openness and the actual admittance of observers may, in fact, be said to be an irreversible step. It can hardly be imagined, how any country, by repudiating at a later stage such commitments under any pretext, should obtain a gain comparable with the immense loss in political and moral position abroad and at home, considering that its original acceptance no doubt would have been greeted with jubilant enthusiasm all over the world as the inauguration of a new era for humanity.

VIII. OPEN LETTER TO THE UNITED NATIONS, JUNE 9TH, 1950

J.H. Schultz,
Copenhagen 1950

See Introduction to Part I, pp. [49], [78].

[171]

OPEN LETTER TO THE UNITED NATIONS, JUNE 9TH, 1950

Versions published in English, Danish and German

English: Open Letter to the United Nations, June 9th, 1950

A   Private print, J.H. Schultz, Copenhagen 1950

B   Science **112** (1950) 1–6

C   *The Ideal of an Open World*, Impact of Science on Society **1** (No. 2, 1950) 68–76

D   *For an Open World*, Bulletin of the Atomic Scientists **6** (No. 7, July 1950), 213–217, 219.

Danish: Åbent brev til De Forenede Nationer, 9. juni 1950

E   Private print, J.H. Schultz, Copenhagen 1950

F   Berlingske Tidende, 13 June 1950

G   Information, 13 June 1950

H   Politiken, 13 June 1950

German: Für eine "offene Welt"

I   Die neue Gesellschaft, 2 Jg., 12. Heft, März/April (1955) 3–9

A and *E* were printed and distributed simultaneously. *D* contains minor stylistic changes in relation to *A*, *B* and *C* and omits the paragraph from the Memorandum starting with "In preliminary consultations ...". *D*, *F*, *H* and *I* include subheadings. *I* is an abbreviated and considerably revised version published without Bohr's knowledge or consent.

Niels Bohr

OPEN LETTER

TO

THE UNITED NATIONS

June 9th, 1950

J. H. SCHULTZ FORLAG

COPENHAGEN · DENMARK

I address myself to the organization, founded for the purpose to further co-operation between nations on all problems of common concern, with some considerations regarding the adjustment of international relations required by modern development of science and technology. At the same time as this development holds out such great promises for the improvement of human welfare it has, in placing formidable means of destruction in the hands of man, presented our whole civilization with a most serious challenge.

My association with the American-British atomic energy project during the war gave me the opportunity of submitting to the governments concerned views regarding the hopes and the dangers which the accomplishment of the project might imply as to the mutual relations between nations. While possibilities still existed of immediate results of the negotiations within the United Nations on an arrangement of the use of atomic energy guaranteeing common security, I have been reluctant in taking part in the public debate on this question. In the present critical situation, however, I have felt that an account of my views and experiences may perhaps contribute to renewed discussion about these matters so deeply influencing international relationship.

In presenting here views which on an early stage impressed themselves on a scientist who had the opportunity to follow developments on close hand I am acting entirely on my own responsibility and without consultation with the government of any country. The aim of the present account and considerations is to point to the unique opportunities for furthering understanding and co-operation between nations which have been created by the revolution of human resources brought about by the advance of science, and to stress that despite previous disappointments these opportunities still remain and that all hopes and all efforts must be centered on their realization.

For the modern rapid development of science and in particular for the adventurous exploration of the properties and structure of the atom, international co-operation of an unprecedented extension and intensity has been of decisive importance. The fruitfulness of the exchange of experiences and ideas between scientists from

3

[175]

all parts of the world was a great source of encouragement to every participant and strengthened the hope that an ever closer contact between nations would enable them to work together on the progress of civilization in all its aspects.

Yet, no one confronted with the divergent cultural traditions and social organization of the various countries could fail to be deeply impressed by the difficulties in finding a common approach to many human problems. The growing tension preceding the second world war accentuated these difficulties and created many barriers to free intercourse between nations. Nevertheless, international scientific co-operation continued as a decisive factor in the development which, shortly before the outbreak of the war, raised the prospect of releasing atomic energy on a vast scale.

The fear of being left behind was a strong incentive in various countries to explore, in secrecy, the possibilities of using such energy sources for military purposes. The joint American-British project remained unknown to me until, after my escape from occupied Denmark in the autumn of 1943, I came to England at the invitation of the British government. At that time I was taken into confidence about the great enterprise which had already then reached an advanced stage.

Everyone associated with the atomic energy project was, of course, conscious of the serious problems which would confront humanity once the enterprise was accomplished. Quite apart from the role atomic weapons might come to play in the war, it was clear that permanent grave dangers to world security would ensue unless measures to prevent abuse of the new formidable means of destruction could be universally agreed upon and carried out.

As regards this crucial problem, it appeared to me that the very necessity of a concerted effort to forestall such ominous threats to civilization would offer quite unique opportunities to bridge international divergencies. Above all, early consultations between the nations allied in the war about the best ways jointly to obtain future security might contribute decisively to that atmosphere of mutual confidence which would be essential for co-operation on the many other matters of common concern.

In the beginning of 1944, I was given the opportunity to bring such views to the attention of the American and British governments. It may be in the interest of international understanding to record some of the ideas which at that time were the object of serious deliberation. For this purpose, I may quote from a memorandum which I submitted to President Roosevelt as a basis for a long conversation which he granted me in August 1944. Besides a survey of the scientific background for the atomic energy project, which is now public knowledge, this memorandum, dated July 3rd, 1944, contained the following passages regarding the political consequences which the accomplishment of the project might imply:

4

„It certainly surpasses the imagination of anyone to survey the consequences of the project in years to come, where in the long run the enormous energy sources which will be available may be expected to revolutionize industry and transport. The fact of immediate preponderance is, however, that a weapon of an unparalleled power is being created which will completely change all future conditions of warfare.

Quite apart from the question of how soon the weapon will be ready for use and what role it may play in the present war, this situation raises a number of problems which call for most urgent attention. Unless, indeed, some agreement about the control of the use of the new active materials can be obtained in due time, any temporary advantage, however great, may be outweighed by a perpetual menace to human security.

Ever since the possibilities of releasing atomic energy on a vast scale came in sight, much thought has naturally been given to the question of control, but the further the exploration of the scientific problems concerned is proceeding, the clearer it becomes that no kind of customary measures will suffice for this purpose and that especially the terrifying prospect of a future competition between nations about a weapon of such formidable character can only be avoided through a universal agreement in true confidence.

In this connection it is above all significant that the enterprise, immense as it is, has still proved far smaller than might have been anticipated and that the progress of the work has continually revealed new possibilities for facilitating the production of the active materials and of intensifying their effects.

The prevention of a competition prepared in secrecy will therefore demand such concessions regarding exchange of information and openness about industrial efforts including military preparations as would hardly be conceivable unless at the same time all partners were assured of a compensating guarantee of common security against dangers of unprecedented acuteness.

The establishment of effective control measures will of course involve intricate technical and administrative problems, but the main point of the argument is that the accomplishment of the project would not only seem to necessitate but should also, due to the urgency of mutual confidence, facilitate a new approach to the problems of international relationship.

The present moment where almost all nations are entangled in a deadly struggle for freedom and humanity might at first sight seem most unsuited for any committing arrangement concerning the project. Not only have the aggressive powers still great military strength, although their original plans of world domination have been frustrated and it seems certain that they must ultimately surrender, but even when this happens, the nations united against aggression may face grave causes of disagreement due to conflicting attitudes towards social and economic problems.

By a closer consideration, however, it would appear that the potentialities of the project as a means of inspiring confidence just under these circumstances acquire most actual importance. Moreover the momentary situation would in various respects seem to afford quite unique possibilities which might be forfeited by a postponement awaiting the further development of the war situation and the final completion of the new weapon".

„In view of these eventualities the present situation would seem to offer a most favourable opportunity for an early initiative from the side which by good fortune has achieved a lead in the efforts of mastering mighty forces of nature hitherto beyond human reach.

Without impeding the importance of the project for immediate military

5

objectives, an initiative, aiming at forestalling a fateful competition about the formidable weapon, should serve to uproot any cause of distrust between the powers on whose harmonious collaboration the fate of coming generations will depend.

Indeed, it would appear that only when the question is taken up among the united nations of what concessions the various powers are prepared to make as their contribution to an adequate control arrangement, it will be possible for anyone of the partners to assure themselves of the sincerity of the intensions of the others.

Of course, the responsible statesmen alone can have the insight in the actual political possibilities. It would, however, seem most fortunate that the expectations for a future harmonious international co-operation which have found unanimous expression from all sides within the united nations, so remarkably correspond to the unique opportunities which, unknown to the public, have been created by the advancement of science.

Many reasons, indeed, would seem to justify the conviction that an approach with the object of establishing common security from ominous menaces without excluding any nation from participating in the promising industrial development which the accomplishment of the project entails will be welcomed, and be responded with a loyal co-operation on the enforcement of the necessary far reaching control measures.

Just in such respects helpful support may perhaps be afforded by the world-wide scientific collaboration which for years has embodied such bright promises for common human striving. On this background personal connections between scientists of different nations might even offer means of establishing preliminary and non-committal contact.

It need hardly be added that any such remark or suggestion implies no underrating of the difficulty and delicacy of the steps to be taken by the statesmen in order to obtain an arrangement satisfactory to all concerned, but aim only at pointing to some aspects of the situation which might facilitate endeavours to turn the project to lasting benefit for the common cause".

The secrecy regarding the project which prevented public knowledge and open discussion of a matter so profoundly affecting international affairs added, of course, to the complexity of the task of the statesmen. With full appreciation of the extraordinary character of the decisions which the proposed initiative involved, it still appeared to me that great opportunities would be lost unless the problems raised by the atomic development were incorporated into the plans of the allied nations for the post-war world.

This viewpoint was elaborated in a supplementary memorandum in which also the technical problem of control measures was further discussed. In particular, I attempted to stress that just the mutual openness, which now was obviously necessary for common security, would in itself promote international understanding and pave the way for enduring co-operation. This memorandum, dated March 24th 1945, contains, besides remarks which have no interest to-day, the following passages:

„Above all, it should be appreciated that we are faced only with the beginning of a development and that, probably within the very near future, means will be found to simplify the methods of production of the active substances and intensify their effects to an extent which may

6

permit any nation possessing great industrial resources to command powers of destruction surpassing all previous imagination.

Humanity will, therefore, be confronted with dangers of unprecedented character unless, in due time, measures can be taken to forestall a disastrous competition in such formidable armaments and to establish an international control of the manufacture and use of the powerful materials.

Any arrangement which can offer safety against secret preparations for the mastery of the new means of destruction would, as stressed in the memorandum, demand extraordinary measures. In fact, not only would universal access to full information about scientific discoveries be necessary, but every major technical enterprise, industrial as well as military, would have to be open to international control.

In this connection it is significant that the special character of the efforts which, irrespective of technical refinements, are required for the production of the active materials, and the peculiar conditions which govern their use as dangerous explosives, will greatly facilitate such control and should ensure its efficiency, provided only that the right of supervision is guaranteed.

Detailed proposals for the establishment of an effective control would have to be worked out with the assistance of scientists and technologists appointed by the governments concerned, and a standing expert committee, related to an international security organization, might be charged with keeping account of new scientific and technical developments and with recommending appropriate adjustments of the control measures.

On recommendations from the technical committee the organization would be able to judge the conditions under which industrial exploitation of atomic energy sources could be permitted with adequate safeguards to prevent any assembly of active material in an explosive state".

„As argued in the memorandum, it would seem most fortunate that the measures demanded for coping with the new situation, brought about by the advance of science and confronting mankind at a crucial moment of world affairs, fit in so well with the expectations for a future intimate international co-operation which have found unanimous expression from all sides within the nations united against aggression.

Moreover, the very novelty of the situation should offer a unique opportunity of appealing to an unprejudiced attitude, and it would even appear that an understanding about this vital matter might contribute most favourably towards the settlement of other problems where history and traditions have fostered divergent viewpoints.

With regard to such wider prospects, it would in particular seem that the free access to information, necessary for common security, should have far-reaching effects in removing obstacles barring mutual knowledge about spiritual and material aspects of life in the various countries, without which respect and goodwill between nations can hardly endure.

Participation in a development, largely initiated by international scientific collaboration and involving immense potentialities as regards human welfare, would also reinforce the intimate bonds which were created in the years before the war between scientists of different nations. In the present situation these bonds may prove especially helpful in connection with the deliberations of the respective governments and the establishment of the control.

In preliminary consultations between the governments with the primary purpose of inspiring confidence and relieving disquietude, it should be necessary only to bring up the problem of what the attitude of each partner would be if the prospects opened up by the progress of physical science, which in outline are common knowledge, should be realized to an extent which would necessitate exceptional action".

7

„In all the circumstances it would seem that an understanding could hardly fail to result, when the partners have had a respite for considering the consequences of a refusal to accept the invitation to co-operate, and convincing themselves of the advantages of an arrangement guaranteeing common security without excluding anyone from participation in the promising utilization of the new sources of material prosperity.

All such opportunities may, however, be forfeited if an initiative is not taken while the matter can be raised in a spirit of friendly advice. In fact, a postponement to await further developments might, especially if preparations for competitive efforts in the meantime have reached an advanced stage, give the approach the appearance of an attempt at coercion in which no great nation can be expected to acquiesce«.

„Indeed, it need hardly be stressed how fortunate in every respect it would be if, at the same time as the world will know of the formidable destructive power which has come into human hands, it could be told that the great scientific and technical advance has been helpful in creating a solid foundation for a future peaceful co-operation between nations".

Looking back on those days, I find it difficult to convey with sufficient vividness the fervent hopes that the progress of science might initiate a new era of harmonious co-operation between nations, and the anxieties lest any opportunity to promote such a development be forfeited.

Until the end of the war I endeavoured by every way open to a scientist to stress the importance of appreciating the full political implications of the project and to advocate that, before there could be any question of use of atomic weapons, international co-operation be initiated on the elimination of the new menaces to world security.

I left America in June 1945, before the final test of the atomic bomb, and remained in England, until the official announcement in August 1945 that the weapon had been used. Soon thereafter I returned to Denmark and have since had no connection with any secret, military or industrial, project in the field of atomic energy.

When the war ended and the great menaces of oppression to so many peoples had disappeared, an immense relief was felt all over the world. Nevertheless, the political situation was fraught with ominous forebodings. Divergencies in outlook between the victorious nations inevitably aggravated controversial matters arising in connection with peace settlements. Contrary to the hopes for future fruitful co-operation, expressed from all sides and embodied in the Charter of the United Nations, the lack of mutual confidence soon became evident.

The creation of new barriers, restricting the free flow of information between countries, further increased distrust and anxiety. In the field of science, especially in the domain of atomic physics, the continued secrecy and restrictions deemed necessary for security

8

reasons hampered international co-operation to an extent which split the world community of scientists into separate camps.

Despite all attempts, the negotiations within the United Nations have so far failed in securing agreement regarding measures to eliminate the dangers of atomic armament. The sterility of these negotiations, perhaps more than anything else, made it evident that a constructive approach to such vital matters of common concern would require an atmosphere of greater confidence.

Without free access to all information of importance for the inter-relations between nations, a real improvement of world affairs seemed hardly imaginable. It is true that some degree of mutual openness was envisaged as an integral part of any international arrangement regarding atomic energy, but it grew ever more apparent that, in order to pave the way for agreement about such arrangements, a decisive initial step towards openness had to be made.

The ideal of an open world, with common knowledge about social conditions and technical enterprises, including military preparations, in every country, might seem a far remote possibility in the prevailing world situation. Still, not only will such relationship between nations obviously be required for genuine co-operation on progress of civilization, but even a common declaration of adherence to such a course would create a most favourable background for concerted efforts to promote universal security. Moreover, it appeared to me that the countries which had pioneered in the new technical development might, due to their possibilities of offering valuable information, be in a special position to take the initiative by a direct proposal of full mutual openness.

I thought it appropriate to bring these views to the attention of the American government without raising the delicate matter publicly. On visits to the United States in 1946 and in 1948 to take part in scientific conferences, I therefore availed myself of the opportunity to suggest such an initiative to American statesmen. Even if it involves repetition of arguments already presented, it may serve to give a clearer impression of the ideas under discussion on these occasions to quote a memorandum, dated May 17th, 1948, submitted to the Secretary of State as a basis for conversations in Washington in June 1948:

„The deep-rooted divergencies in attitudes to many aspects of human relationship which have grown out of social and political developments in the last decades, were bound to present a serious strain on international relations at the conclusion of the second world war. While, during the war, the efforts in common defense largely distracted attention from such divergencies, it was clear that the realization of the hopes acclaimed from all the nations united against aggression of a whole-hearted co-operation in true confidence would demand a radically new approach to international relations.

The necessity of a readjustment of such relations was even further accentuated by the great scientific and technical developments which hold out bright prospects for the promotion of human welfare, but at

9

the same time have placed formidable means of destruction in the hands of man. Indeed, just as previous technical progress has led to the recognition of need for adjustments within civilized societies, many barriers between nations which hitherto were thought necessary for the defense of national interests would now obviously stand in the way of common security.

The fact that this challenge to civilization presents the nations with a matter of the deepest common concern should offer a unique opportunity for seeking continued co-operation on vital problems. Already during the war, it was, therefore, felt that a favourable foundation for later developments might be created by an early initiative aimed at inviting confidence by making all partners aware of the actual situation which would have to be faced, and by assuring them of willingness to share in the far-reaching concessions as to accustomed national prerogatives which would be demanded from every side.

In the years which have passed since the war, the divergencies in outlook have manifested themselves ever more clearly and a most desperate feature of the present situation is the extent to which the barring of intercourse has led to distortion of facts and motives, resulting in increasing distrust and suspicion between nations and even between groups within many nations. Under these circumstances the hopes embodied in the establishment of the United Nations Organization have met with repeated great disappointments and, in particular, it has not been possible to obtain consent as regards control of atomic energy armaments.

In this situation with deepening cleavage between nations and with spreading anxiety for the future, it would seem that the turning of the trend of events requires that a great issue be raised, suited to invoke the highest aspirations of mankind. Here it appears that the stand for an open world, with unhampered opportunities for common enlightenment and mutual understanding, must form the background for such an issue. Surely, respect and goodwill between nations cannot endure without free access to information about all aspects of life in every country.

Moreover, the promises and dangers involved in the technical advances have now most forcibly stressed the need for decisive steps towards openness as a primary condition for the progress and protection of civilization. The appreciation of this point, it is true, underlies the proposals to regulate co-operation on the development of the new resources, brought before the United Nations Atomic Energy Commission, but just the difficulty experienced in obtaining agreement under present world conditions would suggest the necessity of centering the issue more directly on the problem of openness.

Under the circumstances it would appear that most careful consideration should be given to the consequences which might ensue from an offer, extended at a well-timed occasion, of immediate measures towards openness on a mutual basis. Such measures should in some suitable manner grant access to information, of any kind desired, about conditions and developments in the various countries and would thereby allow the partners to form proper judgment of the actual situation confronting them.

An initiative along such lines might seem beyond the scope of conventional diplomatic caution; yet it must be viewed against the background that, if the proposals should meet with consent, a radical improvement of world affairs would have been brought about, with entirely new opportunities for co-operation in confidence and for reaching agreement on effective measures to eliminate common dangers.

Nor should the difficulties in obtaining consent be an argument against taking the initiative since, irrespective of the immediate response, the very existence of an offer of the kind in question should deeply affect

10

the situation in a most promising direction. In fact, a demonstration would have been given to the world of preparedness to live together with all others under conditions where mutual relationships and common destiny would be shaped only by honest conviction and good example.

Such a stand would, more than anything else, appeal to people all over the world, fighting for fundamental human rights, and would greatly strengthen the moral position of all supporters of genuine international co-operation. At the same time, those reluctant to enter on the course proposed would have been brought into a position difficult to maintain since such opposition would amount to a confession of lack of confidence in the strength of their own cause when laid open to the world.

Altogether, it would appear that, by making the demand for openness a paramount issue, quite new possibilities would be created, which, if purposefully followed up, might bring humanity a long way forward towards the realization of that co-operation on the progress of civilization which is more urgent and, notwithstanding present obstacles, may still be within nearer reach than ever before".

The consideration in this memorandum may appear utopian, and the difficulties of surveying complications of non-conventional procedures may explain the hesitations of governments in demonstrating adherence to the course of full mutual openness. Nevertheless, such a course should be in the deepest interest of all nations, irrespective of differences in social and economic organization, and the hopes and aspirations for which it was attempted to give expression in the memorandum are no doubt shared by people all over the world.

While the present account may perhaps add to the general recognition of the difficulties with which every nation was confronted by the coincidence of a great upheaval in world affairs with a veritable revolution as regards technical resources, it is in no way meant to imply that the situation does not still offer unique opportunities. On the contrary, the aim is to point to the necessity of reconsidering, from every side, the ways and means of co-operation for avoiding mortal menaces to civilization and for turning the progress of science to lasting benefit of all humanity.

Within the last years, world-wide political developments have increased the tension between nations and at the same time the perspectives that great countries may compete about the possession of means of annihilating populations of large areas and even making parts of the earth temporarily uninhabitable have caused widespread confusion and alarm.

As there can hardly be question for humanity of renouncing the prospects of improving the material conditions for civilization by atomic energy sources, a radical adjustment of international relationship is evidently indispensable if civilization shall survive. Here, the crucial point is that any guarantee that the progress of science is used only to the benefit of mankind presupposes the same

11

attitude as is required for co-operation between nations in all domains of culture.

Also in other fields of science recent progress has confronted us with a situation similar to that created by the development of atomic physics. Even medical science, which holds out such bright promises for the health of people all over the world, has created means of extinguishing life on a terrifying scale which imply grave menaces to civilization, unless universal confidence and responsibility can be firmly established.

The situation calls for the most unprejudiced attitude towards all questions of international relations. Indeed, proper appreciation of the duties and responsibilities implied in world citizenship is in our time more necessary than ever before. On the one hand, the progress of science and technology has tied the fate of all nations inseparably together, on the other hand, it is on a most different cultural background that vigorous endeavours for national self-assertion and social development are being made in the various parts of our globe.

An open world where each nation can assert itself solely by the extent to which it can contribute to the common culture and is able to help others with experience and resources must be the goal to be put above everything else. Still, example in such respects can be effective only if isolation is abandoned and free discussion of cultural and social developments permitted across all boundaries.

Within any community it is only possible for the citizens to strive together for common welfare on a basis of public knowledge of the general conditions in the country. Likewise, real co-operation between nations on problems of common concern presupposes free access to all information of importance for their relations. Any argument for upholding barriers for information and intercourse, based on concern for national ideals or interests, must be weighed against the beneficial effects of common enlightenment and the relieved tension resulting from openness.

In the search for a harmonious relationship between the life of the individual and the organization of the community, there have always been and will ever remain many problems to ponder and principles for which to strive. However, to make it possible for nations to benefit from the experience of others and to avoid mutual misunderstanding of intentions, free access to information and unhampered opportunity for exchange of ideas must be granted everywhere.

In this connection it has to be recognized that abolition of barriers would imply greater modifications in administrative practices in countries where new social structures are being built up in temporary seclusion than in countries with long traditions in governmental organization and international contacts. Common readiness

12

to assist all peoples in overcoming difficulties of such kind is, therefore, most urgently required.

The development of technology has now reached a stage where the facilities for communication have provided the means for making all mankind a co-operating unit, and where at the same time fatal consequences to civilization may ensue unless international divergencies are considered as issues to be settled by consultation based on free access to all relevant information.

The very fact that knowledge is in itself the basis for civilization points directly to openness as the way to overcome the present crisis. Whatever judicial and administrative international authorities may eventually have to be created in order to stabilize world affairs, it must be realized that full mutual openness, only, can effectively promote confidence and guarantee common security.

Any widening of the borders of our knowledge imposes an increased responsibility on individuals and nations through the possibilities it gives for shaping the conditions of human life. The forceful admonition in this respect which we have received in our time cannot be left unheeded and should hardly fail in resulting in common understanding of the seriousness of the challenge with which our whole civilization is faced. It is just on this background that quite unique opportunities exist to-day for furthering co-operation between nations on the progress of human culture in all its aspects.

I turn to the United Nations with these considerations in the hope that they may contribute to the search for a realistic approach to the grave and urgent problems confronting humanity. The arguments presented suggest that every initiative from any side towards the removal of obstacles for free mutual information and intercourse would be of the greatest importance in breaking the present deadlock and encouraging others to take steps in the same direction. The efforts of all supporters of international co-operation, individuals as well as nations, will be needed to create in all countries an opinion to voice, with ever increasing clarity and strength, the demand for an open world.

Copenhagen, June 9th, 1950.

Niels Bohr.

[185]

Bohr presents the Open Letter to journalists at Carlsberg, 12 June 1950.

IX. DOCUMENTS, 1950:
PRESENTATION OF AND REACTION TO
THE OPEN LETTER

Document 1:
PRESS RELEASE, 12 June 1950
[Stencil]
[Source: NBA]

June 12th, 1950

Professor Bohr's Address to the Press on the Release of
his Open Letter to the United Nations.

I thank the representatives of the Danish and foreign press for coming to-day to give me an opportunity of informing you of the contents of an open letter to the United Nations, which just to-day has been handed to the Secretary General of the United Nations.

From the text of the letter, of which you will all receive a copy, it will be seen that it is a contribution to the standing discussion all over the world about ways of meeting the serious crisis in civilization, resulting from the great development which has taken place in our time in the domain of science and technology.

To this end I have attempted to state some views which during the years have impressed themselves on me and to point to possibilities which in my opinion still remain despite previous disappointments.

I shall not enter into details of the letter but merely state that the main point is that the rapid development of technical resources created by science makes urgently necessary an adjustment of relations between nations in order to avoid mortal dangers to civilization.

In my opinion this goal can only be reached on the basis of *full mutual openness* as regards information about all aspects of the life of the individual nations, including social structure and technical development.

However utopian such views may appear there still seem to exist quite unique opportunities for a *co-operation between all nations for the progress and safeguarding of civilization*. In this respect it appears to me that every initiative, from any side, towards the removal of obstacles for information and intercourse would be of decisive importance.

It is the principal task of the press to inform and guide the public, and so I hope that you will regard this letter as a serious effort to contribute to the best of my ability to a realistic view on the great and vital problems facing civilization.

As in the open letter I have tried to express my opinion in as much detail as possible, I have nothing more to add here, and no further commentaries to make on the letter.

Document 2:
PRESS RELEASE, 14 June 1950
[Stencil]
[Source: NBA]

"The Stockholm appeal".

We demand prohibition of the atomic weapon, this terrible weapon for mass destruction of men.

We demand establishment of a strict international control in order to secure the accomplishment of this prohibition.

We consider that the government which first uses the atomic weapon against whichever country not only commits a war crime, but also a crime against humanity, and we hold that it should be treated as a war criminal.

We ask all honest men all over the world to sign this appeal.

Professor Niels Bohr's answer
through Ritzau's Bureau:

At the publication of an open letter to the United Nations on June 12th I told the representatives of the press that the letter contained some ideas presented for the consideration of everybody and that I had no further comments to add. Since the Copenhagen newspaper "Land og Folk" has publicly addressed the question to me, whether I can join in the "Stockholm appeal", I should like, however, to give the following answer:

From the content of the open letter it will be understood that I cannot join any appeal, however well-meant it might be, which does not include the clearly expressed demand of access to information about conditions in all countries and of fully free exchange of ideas within every country and across the boundaries. As repeatedly expressed in the open letter this demand is to my conviction a necessary basis for that mutual respect and confidence which is indispensable for fruitful international co-operation on the development of civilization. If an open world can be realized the main obstacle to agreement about measures to guarantee that the progress of science is used only to the benefit of humanity would be removed, while without openness no measures can be expected to lead to the desired result.

Copenhagen,
June 14th 1950.   Niels Bohr.

[189]

Niels Bohr and Dag Hammarskjöld at the 200th anniversary of Columbia University in 1954, where they received honorary degrees. See the introduction to Part II, p. [370].

X. OPEN LETTER TO THE SECRETARY GENERAL OF THE UNITED NATIONS, NOVEMBER 9TH, 1956

Private print in English and Danish

See Introduction to Part I, p. [81].

My dear Mr. Secretary General.

In view of the widespread anxiety called forth by serious divergencies among member states of the United Nations Organization, as regards the interpretation of their rights as sovereign nations and of their obligations towards the Organization, I take the liberty of addressing myself to you with some considerations, in continuation of my letter to the United Nations of June 9th, 1950.

As stressed in the letter, full access to information regarding the conditions in every country for life in all its aspects and free intercourse and exchange of opinion across all boundaries must form the foundation for that co-operation in confidence between nations, which in our time is so vital for the future of mankind. Indeed, it is only by such co-operation that the promises for improving the welfare of peoples all over the globe, held out by the development of science, can be fulfilled and the menace to civilization arising from the new powerful means of destruction can be eliminated.

An open world where nations, in mutual security, can assert themselves solely by the extent to which they can contribute to the common culture and are able to help others with experience and resources may at the moment appear as a remote ideal. Still, this goal is in conformity with the avowed aspirations of all nations, and it appears that an initiative from the United Nations Organization in this direction might contribute to a development in the interest of all.

Not least in times of crises, the hopes of people all over the world are centered on the United Nations Organization and on its efforts to uphold the peace and safeguard human rights. Besides lending support to immediate urgent endeavours, it would seem that an essential strengthening of the potentialities of the Organization for creating confidence between nations might result from a general declaration of adherence to the principle of openness.

In particular, all nations might be called upon to declare their readiness to assist the Organization in obtaining the information necessary for impartial judgement of any issue with which the organization may be confronted. Such assurances might not only serve to relieve present tensions but might also, by helping to harmonize the interests and obligations of the individual nations, be a step towards the realization of world-wide co-operation.

In the hope that these considerations may in some modest way be of aid to your untiring endeavours in the service of the United Nations Organization,

respectfully yours,

NIELS BOHR

HIS EXCELLENCY,

THE SECRETARY GENERAL OF THE UNITED NATIONS,

DAG HAMMARSKJÖLD

APPENDIX. VARIOUS LETTERS AND NOTES

TEXTS AND TRANSLATIONS

This appendix contains the complete documents which have been partially quoted in the introduction to Part I. It starts with a list of the documents providing the page numbers of where they (and, when relevant, their translations) are reproduced in the appendix. A third column indicates where the documents are quoted in the introduction. As the documents constitute a continuing story, they are presented in chronological order.

Because the documents stem from a variety of repositories, the source is indicated before each document, followed, when required, by an editor's note. In addition, persons, institutions and events not described in the introduction are introduced in footnotes. With the exception of documents reproduced as facsimiles, the footnotes are provided in the original text, with the footnote numbers repeated in the translations, as well as elsewhere, as necessary.

LIST OF DOCUMENTS

Description	Date	Reproduced p.	Translated p.	Quoted p.
Before the war				
Izvestia interview	12 May 1934	196	199	8
Politiken interview	28 May 1934	203	204	9
Letters to the Editor	30 & 31 May 1934	206	208	9
Rutherford to Bohr	6 December 1934	212	–	9
Bohr to Bukharin	15 December 1934	214	215	9
Bohr to Rutherford	15 December 1934	216	–	10
Rutherford to Bohr	25 December 1934	218	–	10
Bohr to Rutherford	7 January 1935	221	–	10
Bohr to Langevin	7 January 1935	222	–	10
Bohr to Stalin	23 September 1938	223	225	10
War and exile				
Bohr–Chadwick letters	early 1943	227	–	13
Kapitza to Bohr	28 October 1943	229	–	24
Bohr to Halifax	28 December 1943	231	–	16
Halifax to Anderson	18 February 1944	232	–	19
Anderson to Churchill	21 March 1944	233	–	21
Anderson to Churchill	27 April 1944	236	–	24
Churchill to Roosevelt	27 April 1944	237	–	25
Bohr to Kapitza	29 April 1944	238	–	24
Bohr to Anderson	9 May 1944	239	–	25
Dale to Churchill	11 May 1944	241	–	26
Bohr to Møller	May 1944	243	244	29
Anderson to Campbell	2 June 1944	244	–	30
Smuts to Churchill	15 June 1944	245	–	34
Bohr to Frankfurter	5 July 1944	247	–	31
Bohr to Frankfurter	6 July 1944	248	–	31
Frankfurter to Bohr	10 July 1944	248	–	31
Frankfurter to Roosevelt	10 July 1944	249	–	32
Bohr to Anderson	13 July 1944	250	–	30, 33
Bohr to Frankfurter	14 July 1944	251	–	33
Anderson to Bohr	21 July 1944	252	–	33
Campbell to Anderson	25 August 1944	253	–	34
Halifax to Anderson	27 August 1944	254	–	34
Anderson to Halifax	1 September 1944	256	–	36
Frankfurter to Roosevelt	8 September 1944	257	–	37
Bohr to Anderson	12 September 1944	258	–	35, 36
Aide-Mémoire	18 September 1944	259	–	37
Churchill to Cherwell	20 September 1944	260	–	37

Description	Date	Reproduced p.	Translated p.	Quoted p.
Cherwell to Churchill	23 September 1944	261	–	38
Anderson to Cherwell	29 September 1944	262	–	38
Møller to Bohr	29 September 1944	264	264	29
Bohr memo	6 October 1944	265	–	40
Frankfurter to Bohr	14 October 1944	266	–	40
Bohr to Anderson	17 October 1944	267	–	42
Bohr report	30 December 1944	268	–	39, 41
Bohr to Anderson	18 January 1945	269	–	42
Halifax to Anderson	18 January 1945	271	–	42
Bohr notes	7 March 1945	272	–	41, 42, 43
Bohr to Anderson	27 March 1945	278	–	45
Anderson to Halifax	28 March 1945	279	–	46
Anderson to Chadwick	3 April 1945	280	–	43
Frankfurter report on Bohr	18 April 1945	281	–	21
Halifax to Anderson	18 April 1945	285	–	46
Anderson to Halifax	undated	286	–	47
Bohr notes	19 April 1945	287	–	47
Bohr notes	undated	288	–	34
Bohr notes	3 May 1945	290	–	34
Bohr report on Frankfurter	6 May 1945	291	–	20
Bohr notes	22 June 1945	293	–	48
After the exile				
Press release	4 October 1945	294	295	56
Kapitza to Bohr	20 October 1945	295	–	53
Bohr to Kapitza	21 October 1945	296	–	54
Kapitza to Bohr	22 October 1945	298	–	57
Bohr to Anderson	22 October 1945	299	–	54
Bohr notes	2 November 1945	301	301	56
Bohr to Kapitza	17 November 1945	302	–	58
Bohr to Anderson	20 November 1945	304	–	58, 66
Anderson to Bohr	29 November 1945	305	–	58
Oppenheimer to Bohr	14 December 1945	306	–	51
Bohr to Oppenheimer	14 December 1945	306	–	51
Oppenheimer to Bohr	30 March 1946	307	–	66
Bohr to Kapitza	12 April 1946	307	–	58
Bohr to Bush	17 April 1946	308	–	66
Bohr to Oppenheimer	17 April 1946	309	–	66
Kapitza talk	8 July 1946	310	312	59
Kapitza to Bohr	12 July 1946	314	–	59
Døssing to Bohr	18 July 1946	315	316	60
Bohr to Døssing	10 August 1946	317	318	60
Bohr to Anderson	23 July 1948	319	–	69
Bohr notes	1949	322	323	77
Bohr notes	undated	324	328	56, 69 ff.
Press release	13 June 1950	331	331	80
Aage Bohr Press Release	28 April 1994	332	–	57

BEFORE THE WAR

IZVESTIA INTERVIEW, 12 May 1934
[Typed manuscript, with handwritten corrections]
[Source: NBA]
[Editor's comment: Transcribed and translated from German manuscript, with handwritten corrections by Bohr. The interview was published in Izvestia on 15 May 1934.]

INTERVIEW von PROFESSOR BOHR
vom 12. Mai 1934.

Es ist schwer zu sagen, was in der modernen Physik besonders aktuell ist, und es ist auch schwer festzustellen, was aktuell sein wird. Erscheinungen, von denen man gestern noch kaum etwas wusste, gliedern sich heute in die Wissenschaft hinein. Es ist auch überhaupt unmöglich, der Wissenschaft etwas vorzuschreiben. Denn es gibt z.B. Leute, die sich für die Physik des festen Zustandes interessieren, und sie tun recht, denn man kann sich nicht vorstellen, welche grosse Möglichkeiten sich auf diesem Gebiete eröffnen. Andere beschäftigen sich mit dem wunderbaren Gebiete der Kern-Physik – einem wirklichen Abenteuerland. Es ist nur sehr wesentlich, dass die Physiker sich nicht auf ein einziges Gebiet beschränken. Es ist von grösster Wichtigkeit, dass sämtliche Gebiete sich möglichst weit entwickeln; denn je mehr wir erfahren, desto mehr vergrössern wir unser Bestreben nach neuem Wissen; aber man muss immer den Umstand vor Augen halten, dass die Praxis von der Theorie nicht getrennt werden kann. In allen Ländern versucht man selbstverständlich die Wissenschaft anzuwenden, aber eben hier in der Sowjet-Union dürften die Verhältnisse besonders geeignet sein für eine fruchtbare Verbindung der Theorie mit der Praxis. Ein jeder fühlt hier, dass er mit dem ganzen Aufbau eng verbunden ist. Einen besonders erfreulichen Eindruck macht hier die Begeisterung, die ein jeder in seine Arbeit hineinlegt.

Wenn man die Frage aufwirft, welche philosophischen Folgerungen aus der modernen Physik entspringen, so darf man darunter nicht die Frage verstehen, welche alte philosophischen Schulen der modernen Physik Genüge tun. Jede neue Generation der Philosophen lernt von den neuen Entdeckungen der übrigen Wissenschaften seiner Zeit. Obgleich einige Folgerungen der modernen Physik etwas Gemeinsames haben mit den Gesichtspunkten vieler grossen Philosophen, scheint es mir jedoch, dass falls Männer wie Spinoza oder Marx augenblicklich lebten, sie wahrscheinlich mit uns anderen sich freuen würden aus der modernen Physik Neues für die allgemeine Philosophie zu lernen.

Sehr oft spricht man von der Mystik in der modernen Physik. In Wirklichkeit ist die Rede meistens von den noch nicht klar formulierten Vorstellungen. Wir versuchen eben auf Grund der neuen Ergebnisse der Physik nachzuprüfen, ob nicht irgend welche Elemente der Mystik in den alten Vorstellungen stecken, und wir versuchen die Mystik von überall zu vertreiben, um die Wissenschaft rationell zu machen. Ich glaube, dass die Physik sehr stark auf alles um sie her wirkt. Wir erfahren aus ihr, welche sonderbare Dinge in den einfachsten Dingen stecken.

Es ist durchaus notwendig, das Zusammenwirken und die gegenseitige Beihilfe der Wissenschaftler verschiedener Länder zu fördern. Es ist wesentlich, dass man im Auslande sich darüber klar wird, welche grosse ausserordentliche Bestrebungen zur Entwicklung der Wissenschaft in Ihrem Lande herrschen, und wie man hier die sogar besonders weit von der unmittelbaren Praxis entfernten Gebiete der Wissenschaft unterstützt. Sehr wesentlich ist dazu ein unmittelbarer Kontakt und gegenseitige persönliche Verbindungen zwischen den ausländischen und den Sowjet-Physikern. Es ist durchaus notwendig, dass sie sich möglichst oft und in grösserer Zahl gegenseitig besuchen. Und besonders gilt dies der jüngeren Generation der Wissenschaftler, die besonders fähig sind, Neues aufzunehmen. Ich glaube, dass eben in den Wissenschaften die Gemeinschaft, die man nicht überall aufstellen mag, existieren wird und muss. Die Wissenschaft der ganzen Welt muss auf das Wohl der Menschheit gerichtet sein. In dieser Hinsicht darf man nicht die sozialen Probleme von den wissenschaftlichen Problemen fernhalten.

Ich habe nur beschränkte Zeit gehabt und habe nicht alles sehen können, was ich hier in Leningrad besichtigen möchte. Aber von dem, was ich hier gesehen habe, hat auf mich den grössten Eindruck die Sachlage gemacht, mit wieviel Verständnis Ihre Staatsleute die Wissenschaft fördern. Jedes neue Unternehmen, jedes neue Gebiet der Wissenschaft wird sofort vom Staat unterstützt. Ich habe gesehen die bemerkenswerten Institute, die von Professor Joffe[1] geschaffen sind, und ich fühle mich sehr erfreut, dass so grosse Mittel im Sowjetlande für die Wissenschaft gegeben werden. Grosse Mittel sind darunter für die Kernphysik gegeben worden, und eben in den letzten Tagen sind im physikalisch–technischen Institut[2] auf diesem Gebiet sehr interessante Entdeckungen gemacht worden, deren Perspektiven unbegrenzt sind. Man vergisst niemals die engste Verbindung zwischen der reinen Wissenschaft und

[1] Abram Feoderovich Ioffe (1880–1960), prominent Russian physicist and friend of Bohr.
[2] Physico–Technical Institute of the Academy of Sciences of the U.S.S.R., headed by Ioffe.

ihren möglichen Anwendungen. In dieser Hinsicht bildet das Optische Institut[3] ein besonders charakteristishes Beispiel. Ich besuchte auch das Phisiologische Institut von Professor Pawlow,[4] und auch dort konnte ich sehen, wie die Leute, die auf einem theoretischen Gebiet arbeiten, nach den Wegen suchen, um der Menschheit zu helfen, und die im Laboratorium gemachten Entdeckungen anzuwenden. Ich meine, die neuesten Entdeckungen von Professor Pawlow auf dem Gebiet der höheren Nerventätigkeit.

In Leningrad habe ich das Turbinenwerk von Stalin besucht und nun habe ich gesehen, wie die höchste Technik sich mit den sozialen Problemen vereinigt. Ganz bemerkenswert ist diese Verbindung einer Fabrik mit einer Lehranstalt. Ich habe hier nur einen Tag verbracht und trotzdem konnte ich bemerken, welche grosse zielbewusste Einigung bei diesem tausendköpfigen Kollektiv herrscht. Jeder Arbeiter für sich und alle sie zusammen fühlen, dass sie etwas sehr Wichtiges tun, was die Lebensverhältnisse der ganzen Gesellschaft verbessern wird. Ein jeder weiss, wozu er es tut und wohin das Produkt seiner Arbeit gehen soll. Ich bin auch im Hause der Kultur des Wassili Ostrow gewesen.[5] Einen besonders grossen Eindruck hat dort auf mich die Bibliothek gemacht. Ich habe gesehen, welche grosse Mühe man sich hier gibt, um die Arbeiter dazu zu veranlassen ihre Kenntnisse zu vergrössern, und wie man sich bemüht, den Bedürfnissen der Arbeiter Genüge zu tun. Besonders wichtig ist dabei der Umstand, dass alles dieses sich auf die ganze Bevölkerung ausdehnt. Das, was man im Auslande nur versuchen kann zu machen, wird hier in einem kolossalen Masstab durchgeführt. Wenn alle Ihre geplanten, soziellen und kulturellen Massnahmen vollständig durchgeführt werden, so wird es etwas ganz Wunderbares ergeben.

Unter den Eindrücken möchte ich den wichtigsten hervorheben: die Sowjet-Union gibt einem jeden zu denken. Es ist schwer zu sagen, inwiefern alles was hier geschieht, sich auf andere Länder übertragen lässt, aber es ist nicht daran zu zweifeln, dass hier in manchen Hinsichten etwas sehr Wunderbares geschaffen wird.

Ich bin niemals fern von der Wissenschaft der Union gewesen und habe viele Freunde unter den russischen Physikern, aber ich hoffe jetzt, diese Verbin-

---

[3] State Optical Institute, founded in 1918 by Dmitrii Sergeevich Rozhdestvenskii (1876–1940).

[4] Ivan Petrovich Pavlov (1849–1936), prominent Russian physiologist.

[5] The House of Culture in Leningrad was the first of several such institutions set up in major Russian cities after the 1917 revolution. Established in part with the help and insistence of the playwright Maxim Gorky (1868–1936), they were designed as meeting places for scientists, academics, writers and other leading intellectuals. The Leningrad House of Culture is not, as Bohr indicates, located on the island Vasilevskii Ostrov but across the river Neva.

dungen zu verstärken und die Union wieder zu besuchen. So geht es allen, die hierher kommen. Das weiss ich von vielen Wissenschaftlern und besonders von den jungen Physikern. Professor Dirak[6] hat mehrmals die Sowjet-Union besucht und mit welcher Liebe spricht er davon.

Ich möchte wieder sagen: die Wissenschaft ist international. Ich möchte hier noch gerne meinen Standpunkt zu der nationalen Frage äussern, besondern im Zusammenhang mit den unwissenschaftlichen Rassentheorien, die jetzt in einigen Ländern kultiviert werden. Ich glaube, dass die geistigen Einstellungen der Nationen ausschliesslich durch Traditionen und durch die äusseren Bedingungen erklärt werden können, unter denen die einzelnen Menschen sich entwickelten und erzogen wurden. Alle sind sie letzten Endes Menschen, und die sogenannte Rassenfrage dürfte in dieser Hinsicht nicht existieren. Die Sowjet-Union bildet doch gerade ein Beispiel, wo zahlreiche Völker trotz ihrer Sprachunterschiede eine sozusagen gemeinsame Sprache gefunden haben, um in ihrer Weise auf das Wohl der Menschheit zusammen zu arbeiten. –
Durchgesehen und korrigiert
  gez.: N. Bohr

Translation

INTERVIEW with PROFESSOR BOHR
on 12 May 1934.

It is difficult to say what is especially topical in modern physics, and it is also difficult to establish what will become topical in the future. Phenomena about which hardly anything was previously known, have today become part of science. It is also completely impossible to lay down rules for science. For there are, for example, people who are interested in solid state physics, and they are doing the right thing, for one cannot imagine what great possibilities will open up in this field. Others work in the wonderful field of nuclear physics – truly a world of adventure. What is very essential is only that the physicists do not confine themselves to a single field. It is most important that all fields develop as far as possible; for the more we learn, the more we extend our striving for new knowledge; but the circumstance that practice cannot be separated

---

[6] Paul Adrien Maurice Dirac (1902–1984), British physicist who played a major role in the development of quantum mechanics and who visited Bohr's institute several times between the world wars.

[199]

from theory must always be kept in mind. In all countries one naturally tries to apply science, but precisely here in the Soviet Union the conditions might be particularly suited for a fruitful connection between theory and practice. Everyone feels that here he is in close touch with the entire construction process. A particularly pleasing impression here is made by the enthusiasm that everyone puts into his work.

When one raises the question of which philosophical consequences arise from modern physics, one may not thereby understand the question to mean which old philosophical schools comply with modern physics. Every new generation of philosophers learns from the new discoveries of other sciences of its time. Although some consequences of modern physics have something in common with the viewpoints of many great philosophers, yet it seems to me, that if men such as Spinoza or Marx were alive today they would probably, together with the rest of us, enjoy learning new things from modern physics of relevance for general philosophy.

One often speaks of the mysticism of modern physics. In reality it is mostly a question of ideas that are not yet clearly formulated. We try precisely to check on the basis of the new results in physics whether any elements of the mysticism are contained in the old ideas, and we try to drive out the mysticism everywhere, in order to make science rational. I believe that physics affects everything around it very strongly. We learn from it what strange things are contained in the most simple things.

It is extremely necessary to promote in every way the scientific cooperation between, and the mutual support to, scientists of the various countries. It is important that it is recognized abroad what large exceptional efforts are being made in your country towards developing science and how one here supports even the fields of science that are particularly far from direct practical application. Very important in this regard are the direct contact and mutual personal connections between foreign and Soviet scientists. It is extremely necessary that they pay mutual visits to one another as often as possible and in greater numbers. And this applies in particular to the younger generation of scientists, which is particularly able to pick up what is new. I believe that in science there will and must exist a sense of community and cooperation that one may not be able to create in all fields. Science all over the world must be directed towards the welfare of humanity. In this regard one should not distinguish the social problems from the scientific problems.

I have only had limited time and have not been able to see everything I would have liked to see here in Leningrad. But from what I have seen here, the great understanding with which your statesmen have supported science has made the greatest impression on me. Every new project, every new field of science is instantly funded by the State. I have seen the remarkable institutes founded by Professor Ioffe,[1] and I am very pleased that such great funds are allocated to science in the Soviet Union. Among these, extensive funds have been allocated to nuclear physics, and just in the last few days very interesting discoveries, with unlimited perspectives, have been made in this field in the Physico–Technical Institute.[2] The close connection between pure science and its possible applications is never forgotten. In this regard, the Institute for Optics[3] represents a particularly characteristic example. I also visited Professor Pavlov's[4] Institute of Physiology, and I could also there see how people who work in a theoretical field are searching for ways to help humanity and to apply the discoveries made in the laboratory. I here refer to the most recent discoveries by professor Pavlov in the field of higher nervous activity.

In Leningrad I visited the Stalin turbine factory and saw then how the most advanced technology is combined with social concerns. This connection between a factory and an institute of learning is quite remarkable. I spent only one day there and I was able nonetheless to notice what great purposeful agreement prevails in this collective of a thousand people. Every worker for himself and all of them together feel that they are doing something very important, something that will improve the living conditions of the whole society. Everyone knows why he does it and for what the product of his work will be applied. I have also been to the Vasilevskii Ostrov House of Culture.[5] The library there made an especially great impression on me. I saw to what great lengths one goes here to enable the workers to increase their knowledge, and how one strives to satisfy the workers' needs. Particularly important in this regard is the circumstance that all this extends to the entire population. What one in foreign countries can only try to achieve is carried out here on a colossal scale. When all your planned social and cultural measures are completely implemented, then it will result in something quite wonderful.

I would like to emphasize the most important among my impressions: the Soviet Union gives everyone something to think about. It is difficult to say to what an extent everything taking place here can be transferred to other countries, but there is no doubt that in many respects something wonderful is being created here.

I have never been far away from the science of the Union and have many friends among the Russian physicists, but I now hope to strengthen these ties and visit the Union again. This happens to everyone who comes here. I know this from many scientists and especially from the young physicists. Professor Dirac[6] has visited the Soviet Union several times and speaks of it with great affection.

I will say this again: science is international. I would here only like to express my position as regards the question of nations, especially in connection with the unscientific racial theories that are currently being cultivated in some countries. I believe that the intellectual attitudes of nations can be explained solely by traditions and by the external conditions under which each individual developed and was brought up. In the end, all are human beings and the so-called racial question ought not to exist in this regard. The Soviet Union represents of course precisely an example where numerous peoples despite their linguistic differences have found, so to speak, a common language, in order to work together in their way for the benefit of humanity. –

Seen and corrected
   sign.: N. Bohr

POLITIKEN INTERVIEW, 28 May 1934
[Newspaper article]

Niels Bohr kom i Gaar hjem fra Rusland

Prof. Bohr fortæller om sine Indtryk fra Rejsen.

Niels Bohr kom i Aftes hjem fra Leningrad, og da vi traf ham, bad vi ham fortælle om sin Rejse og dens Formaal.

— Jeg rejste til Rusland for efter Indbydelse af det russiske Akademi og Universiteterne Kharkov og Moskva at holde en Række Forelæsninger og deltage i Diskussioner om aktuelle fysiske Spørgsmaal. Det var en stor Glæde at se med hvilken Begejstring videnskabelige Undersøgelser paa disse Felter drives i Rusland i de mange nye Institutter, som man i de sidste Aar har indrettet saavel i Leningrad som andre Steder i Unionen. Selv om det er karakteristisk for disse Institutioner, at man alle Vegne søger at forene det rent videnskabelige Arbejde med Anvendelse af Videnskabens Resultater som Led i den tekniske Opbygning, har Myndighederne alligevel i høj Grad Forstaaelsen af, at man ved videnskabelige Undersøgelser i stort Omfang maa se bort fra Spørgsmaalet om enhver øjeblikkelig Anvendelighed. Det er derfor ikke saa meget Formen for selve den videnskabelige og tekniske Forskning, som de for Videnskab og Teknik fælles Rammer, der betinger den ydre Forskel mellem det naturvidenskabelige Arbejde i Sovjet og mange andre Steder i Verden.

For øvrigt er min personlige Forbindelse med russisk videnskabeligt Arbejde paa Fysikens Omraade ikke begyndt med denne Rejse, men jeg har haft Lejlighed til ved videnskabelige Møder uden for Rusland tidligere at lære adskillige fremragende russiske Fysikere at kende, ligesom jeg har haft stor Glæde af blandt mine Medarbejdere paa Universitetets Institut for teoretisk Fysik herhjemme at kunne tælle flere unge lovende russiske Videnskabsmænd, der har givet selvstændige Bidrag af afgørende Betydning for Atomteoriens seneste Udvikling.

— Hvad var Deres almindelige Indtryk af Forholdene i Rusland, spørger vi Prof. Bohr.

— Det er svært at give noget enkelt og kort Svar paa. Man viste mig med stor Elskværdighed foruden de videnskabelige Institutter adskillige Fabriksvirksomheder og nye sociale Institutioner; men selv om jeg derigennem har faaet mange tankevækkende Indtryk, er det naturligvis umuligt i saa kort Tid og ved en saadan Lejlighed at naa til nogen Bedømmelse af, hvorledes Forholdene i Almindelighed former sig. Hvis jeg ikke ved min Hjemkomst til min store Forbavselse havde erfaret, at man paa Grundlag af avtentiske Referater om min Rejse har ment at kunne tillade sig en offentlig Diskussion om mit Syn paa de russiske Forhold, behøvede jeg heller næppe at understrege, at jeg naturligvis ikke har indladt mig paa nogen Art af Stillingtagen til almindelige politiske Spørgsmaal.

Specielt ser jeg, at man har misforstaaet en enkelt Sætning i et Interview, der hovedsagelig drejede sig om rent videnskabelige Forhold, som en Udtalelse om Arbejdernes almindelige Stilling i Sovjet-Unionen.

Den paagældende Udtalelse fremkom imidlertid som Led i mit Svar paa Spørgsmaalet om mit Indtryk af et Besøg i et stort Turbineværk i Leningrad, som man havd vist mig, fordi man her har gjort et interessant Forsøg paa at forbinde en rent industriel Virksomhed med en videnskabelig teknisk Læreanstalt, hvorved det paa ejendommelig Maade var lykkedes at interessere alle paa Værket beskæftigede Ingeniører og Arbejdere i Virksomhedens forskelligste Afskygninger og faa dem til at føle sig som Led i et samfundsnyttigt Hele.

I denne Sammenhæng behøver jeg vist heller ikke at betone, at jeg altid er meget tilbageholdende med at udtale mig om nogen Sag, som jeg ikke har studeret nærmere, og at mine Samtaler med Regeringsmyndighederne i Moskva udelukkende drejede sig om Spørgsmaal af Betydning for det internationale videnskabelige Samarbejde. *M—e.*

[203]

Translation

Niels Bohr returned home from Russia yesterday

Professor Bohr tells about his impressions from the journey.

Niels Bohr returned home from Leningrad yesterday evening, and when we met him we asked him to tell about his journey and its purpose.

– I travelled to Russia on invitation from the Russian Academy and the Universities of Kharkov and Moscow to give a series of lectures and to take part in discussions on topical questions in physics. It was a great pleasure to see the enthusiasm with which scientific research is carried out in these fields in Russia in the many new institutes that in recent years have been established in Leningrad as well as other places in the Union. Although it is characteristic for these institutions that one tries everywhere to combine the purely scientific work with the application of the results of science as a part of building up technology, the authorities have nevertheless to a great degree the understanding that with scientific research one must to a large extent ignore the question of any immediate application. This is why it is not so much the form of the scientific and technical research itself as it is the common framework for science and technology that determines the external difference between scientific work in the Soviet Union and many other places in the world.

Incidentally, my personal connection with Russian scientific work in the domain of physics did not begin with this journey, but I have had the opportunity, through scientific meetings outside Russia, to get to know several excellent Russian physicists earlier, just as it has been very rewarding for me to count among my collaborators at the University Institute for Theoretical Physics here at home several young promising Russian scientists who have made independent contributions of decisive importance for the recent development of atomic theory.

– What was your general impression of the conditions in Russia, we ask Prof. Bohr.

– It is difficult to give any simple and short answer to this. Apart from the scientific institutes, I was shown with great kindness several factories and

[204]

new social institutions; but although I thereby have received many thought-provoking impressions, it is of course impossible in so short a time and on such an occasion to reach any judgement of how conditions are in general. If on my return I had not found out, to my great astonishment, that it was thought to be possible, on the basis of authentic[7] accounts of my journey, to have a public discussion of my views on the conditions in Russia, I would hardly either need to emphasize that I have of course not assumed any kind of attitude to general political questions.

In particular, I see that there has been a misunderstanding about one sentence in an interview mainly concerning purely scientific matters, as being a statement about the general position of the workers in the Soviet Union.

The statement in question appeared, however, as part of my answer to the question of my impression of a visit to a big turbine works in Leningrad, which was shown to me because an interesting experiment had been made here to combine a purely industrial enterprise with a scientific technical institute of higher education, whereby in a special way one had succeeded in getting all the engineers and workers employed at the works interested in the most diverse aspects of the enterprise and to get them to feel as parts of a whole to the benefit of society.

In this connection I need probably not emphasize that I am always very cautious about pronouncing on any matter that I have not studied more closely, and that my talks with the government authorities in Moscow concerned only questions of significance for international scientific cooperation.   M–e.[8]

---

[7] Bohr must here have meant "unauthentic". See his reply to Victor Pürschel below.
[8] Merete Bonnesen (1901–1980), prominent Danish journalist.

LETTERS TO THE EDITOR, 30 and 31 May 1934
[Newspaper articles]

[Editor's comment: The comments by Holger Iacobæus (which have been abbreviated) and Bohr's response appeared in the same issue of Politiken on 30 May 1934. The comments by Victor Pürschel were printed on 30 May in Berlingske Aftenavis, with Bohr's response appearing the day after.]

Niels Bohr og Sovjetrusland

Overlæge Dr. med. *H. Jacobæus* har anmodet os om Optagelse af nedenstaaende Indlæg:

Ifølge et officielt Referat af et Interview med Hr. Prof. *Niels Bohr* udtalte han sig ikke blot med stor Begejstring om de Summer, Sovjetrusland ofrede i videnskabeligt Øjemed, men ogsaa om det gode Samarbejde mellem Høje og Lave og den smukke Samfølelse.

Da Undertegnede — siden jeg i Vinteren 1917—18 søgtes som Læge af en Del russiske Emigranter — har forsøgt at holde mig à jour med Forholdene i Sovjetrusland, gjorde jeg i Berlingske Aftenblad 18. Maj opmærksom paa, at talrige Kilder gav et ganske andet Billede af Forholdene end der fremgik af Professorens Udtalelser.

Jeg tror — ja undskyld mig, Hr. Professor — at det er lidt naivt overhovedet at tale om Misforstaaelse, hvad enten Sovjetisterne skulde have misforstaaet Dem eller Publikum det officielle Referat — det fremgaar nemlig ikke helt klart af Interviewet i *Politiken* den 28. Maj. Med gode Grunde tror jeg, at Professorens velkomne Udtalelser med velberaad Hu er omformede, saa de passer i Bolsjevikernes Kram.

I al Beskedenhed forekommer det mig, at de Herrer Verdensberømtheder — som *Bernard Shaw* og *Niels Bohr* — inden de fremsætter Udtalelser om Tilstanden i Sovjetrusland bør erindre, at de er mellem Mennesker, som for en Del lider af „Moral insanity", samt at deres Udsagn kan volde den største Skade.
Med Tak for Optagelsen.
H. Jacobæus,

Prof. Niels Bohrs Svar.

Vi forelagde i Aftes Professor *Niels Bohr* dette Indlæg fra Overlæge Jacobæus, og Professoren udtalte som Svar herpaa følgende til os:

— Det forundrer mig unægteligt, at man efter mine Udtalelser til *Politiken* forleden har villet fortsætte med Bestræbelserne paa at inddrage min Person i den politiske Diskussion om de almindelige Leveforhold i Sovjetunionen. Som jeg sagde, var det mig naturligvis som Udlænding under mit korte Ophold i Rusland umuligt at danne mig en tilstrækkelig begrundet Mening om saadanne Forhold og de Ændringer, de stadig undergaar, til at jeg paa Rejsen eller efter min Hjemkomst vilde indlade mig paa at give nogen offentlig Udtalelse derom.

Interviewet i *Isvestia*, hvoraf nogle ved Sammentrækning forvanskede Gengivelser har givet Anledning til Misforstaaelse, fremtraadte som Svar paa bestemt formulerede Spørgsmaal fra Redaktionens Side vedrørende mit Syn paa Betingelserne og Formerne for det videnskabelige Arbejde i Rusland og mit Indtryk af visse nye kulturelle og industrielle Institutioner, som jeg i Leningrad havde haft Lejlighed til at bese. Med det i Interviewet gengivne Forbehold, at jeg ikke kunde overse i hvilket Omfang de nye Forsøg lod sig overføre til andre Lande, kunde jeg ikke undlade at udtrykke min store Beundring og Interesse for meget af det nye, som jeg havde set og lært; og jeg tror, at mine Indtryk i denne Henseende vil deles af enhver, der faar Lejlighed til ved Selvsyn at lære de paagældende Institutioner nærmere at kende.

PROFESSOR NIELS BOHRS UDTALELSE

EFTER sin Hjemkomst fra Rusland har Hr. Professor Niels Bohr til i „Politiken" udtalt sin Forbavselse over, at man herhjemme har diskuteret hans Udtalelse om Forholdene i Sovjetrusland. Han har ikke villet udtale sig om Forholdene i Almindelighed, men kun om Forholdene ved en bestemt Virksomhed.

I denne Anledning maa det være tilladt at gentage den paagældende Passus i den udsendte Udtalelse. Den lød i Ritzaus Bureaus Gengivelse:

„Sovjetunionen tvinger enhver til Overvejelse; thi enhver føler, *at man der skaber noget overordentligt,* noget *prægtigt.* Der anvises uhyre Pengesummer til Videnskaben. Mænd, der arbejder paa Teoriens Omraade, søger at finde Veje til at være Menneskeheden nyttige og føre de Resultater, de er naaede til i deres Laboratorier, ud i det praktiske Liv. *Enhver Arbejder for sig,* saavel som alle i Fællesskab føler, at de udretter noget, der kan forbedre Eksistensvilkaarene for Samfundet som Helhed. *I Sovjetrusland ved enhver, hvorfor han arbejder,* hvortil Udbyttet af hans Arbejde vil tjene, eller hvor det vil blive anvendt. Sovjetunionen .... tjener som Eksempel paa, hvorledes et stort Antal Folk, til Trods for Forskel i Sprog har fundet et fælles Sprog til gensidigt Samarbejde i hele Menneskehedens Interesse."*

Professoren har ikke beklaget sig over, at Udtalelsen er urigtigt gengivet. Er den det, burde Professoren have sagt det klart og utvetydigt. Er den rigtigt gengivet, tør det siges, at de her understregede Sætninger saavel som Slutningen er et ualmindelig uheldigt valgt Udtryk for en Omtale af noget specielt En jævnt begavet Læser kan sprogligt logisk ikke faa andet ud af det, end at der er Tale om en almindelig Vurdering. Professoren har altsaa udtrykt sig uheldigt. Dekan ske for os alle.

Men nu ved vi altsaa, at Professoren erkender, at han ikke kender tilstrækkeligt til Forholdene derovre til at udtale sig om dem i al Almindelighed. Hans store Autoritet som Videnskabsmand kan altsaa ikke mere med Rette bruges som Reklame for det kommunistiske „Lykkeland".

Med dette Resultat har vi alle Grund til at være tilfredse, undtagen de kommunistiske Agitatorer. Men for Professorens Forbavselse er der oprigtigt talt ikke saa megen Plads, som et Atom normalt udfylder.

Victor Pürschel.

En Udtalelse af Professor Niels Bohr.

I Anledning af et i „Berlingske Tidende" i Gaar Aftes fremkommet Indlæg i den Diskussion, som et Interview i „Isvestia" under min Ruslandsrejse har fremkaldt i forskellige Blade herhjemme, vil jeg gerne have Plads for følgende korte Bemærkning:

I de i „Politiken" af 28. og 30. Maj gengivne Udtalelser har jeg efter min Hjemkomst formentlig tilstrækkelig tydeligt gjort Rede for Arten af de Spørgsmaal, man stillede mig i det omhandlede Interview og mine Svar derpaa, samt for de Misforstaaelser, som et af Ritzaus Bureau udsendt kortfattet uautentisk Referat af Interviewet, hvori enkelte Sætninger i mere eller mindre korrekt Gengivelse er sammenstillet paa misvisende Maade, har givet Anledning til. I det nævnte Indlæg i „Berlingske Tidende" fortolkes mine Udtalelser i „Politiken" imidlertid paa en Maade, som jeg ikke kan vedkende mig. Med Udtalelsen om, at jeg ikke have indladt mig paa en Stillingtagen til almindelige politiske Spørgsmaal, mente jeg at oplyse om, at jeg i Interviewet udtrykkelig havde taget Afstand fra enhver Sammenligning mellem Forholdene i Rusland og i andre Lande. Som det klart maatte fremgaa af Baggrunden for denne Udtalelse og den hele Sammenhæng, hvori den fremkom, mente jeg derimod ingenlunde, at jeg havde ønsket at afholde mig fra at udtale mig om de tankevækkende Indtryk, som de Forsøg, der foretages i Rusland paa at skaffe nye Rammer for almindelige videnskabelige og kulturelle Bestræbelser, maa gøre paa enhver, der faar Lejlighed til ved Selvsyn at stifte nærmere Bekendtskab dermed.

Niels Bohr.

Translation

Niels Bohr and Soviet Russia

Senior physician H. Iacobæus, MD,[9] has requested the publication of the following contribution:

According to an official account of an interview with Prof. Bohr, he commented not only with great enthusiasm about the sums of money Soviet Russia gave for scientific purposes, but also about the good cooperation between high and low and about the beautiful fellow feeling.

As – since the winter of 1917–1918, when I was consulted as a doctor by many Russian immigrants – this writer has tried to keep up to date with the conditions in Soviet Russia, I pointed out in the Berlingske Evening News of 18 May that numerous sources gave a quite different picture of the conditions than it appeared from the Professor's statements. ...

...

Excuse me, Professor, I think it is a little naïve to talk at all about a misunderstanding, whether the Sovietists should have misunderstood you or whether the public should have misunderstood the official account – this is not quite clear from the interview in Politiken on 28 May. With good reason I think that the Professor's welcome statements have deliberately been reshaped so they fit the Bolsheviks' ends.

...

In all modesty it appears to me that the world famous gentlemen – such as *Bernard Shaw*[10] and *Niels Bohr* – should recall, before pronouncing opinions about the situation in Soviet Russia, that they are among people who to some extent suffer from "moral insanity" as well as that their statements can cause the greatest harm.

With thanks for publication.

H. Iacobæus

[9] Holger Iacobæus (1865–1934).
[10] George Bernard Shaw (1856–1950).

Prof. Niels Bohr's reply.

We showed this contribution by Doctor Iacobæus to Professor *Niels Bohr* yesterday evening, and in reply to this the professor stated the following to us:

– It certainly surprises me that after my statements to *Politiken* recently one has wished to continue the endeavours to involve my person in the political discussion about general living conditions in the Soviet Union. As I said, as a foreigner it was of course impossible for me during my short stay in Russia to form a sufficiently well-founded opinion about such conditions, and the changes they are still undergoing, to allow myself, either on the journey or after my return home, to make any public statement on this.

The interview in *Izvestia*, from which some quotations – distorted by abbreviation – have given rise to misunderstanding, appeared as answers to precisely formulated questions on the part of the editors regarding my view of the conditions and forms for scientific work in Russia and my impression of certain new cultural and industrial institutions which I had had the opportunity to visit in Leningrad. With the proviso stated in the interview that it was not possible for me to assess to what extent the new experiments could be transferred to other countries, I could not fail to express my great admiration for and interest in many of the new things I had seen and learnt; and I think that my impressions in this respect will be shared by everyone who gets the opportunity through personal experience to become more closely acquainted with the institutions in question.

PROFESSOR
NIELS BOHR'S
STATEMENT

After his return from Russia, Professor Niels Bohr has in "Politiken" expressed his amazement that his statement about conditions in Soviet Russia has been subject to discussion here at home. It was not his intention to express himself about the conditions in general, but only about the conditions at a certain enterprise.

In this connection it must be permissible to repeat the relevant passage in the published statement. It read as reported by Ritzau's Bureau:[11]

"*The Soviet Union forces everyone to reflect*; for everyone feels *that there something extraordinary*, something splendid, is being created. Enormous sums

---

[11] Danish news agency, established in 1866 by Erik Ritzau (1839–1903).

of money are being allocated to science. People who work in the theoretical field try to find ways to be of use to humanity and to bring the results they have reached in their laboratories out into everyday life. *Everyone works on his own*, just as all together feel that they are achieving something which can improve the living conditions of society as a whole. *In Soviet Russia everyone knows why he is working* and what purpose the results of his labour will serve or where they will be used.

The Soviet Union ... serves as an example of how a great number of people, in spite of language differences, have found a common language for mutual cooperation in the interest of humanity as a whole."[12]

The Professor has not complained that the statement is incorrectly reproduced. If it is, the Professor should have said this clearly and plainly. If it is correctly reproduced, then it must be said that the sentences here emphasized, as well as the conclusion, are an unusually badly chosen expression of a description of something specific. As regards language a normally gifted reader cannot logically get anything else out of it than that a general assessment is involved. The Professor has thus expressed himself badly. This can happen for us all.

But now we know, then, that the Professor admits that he does not know enough about the conditions over there to express an opinion about them in general. His great authority as a scientist can thus no longer rightly be used to advertise the communist "paradise".

All of us, except the communist agitators, have reason to be satisfied with this result. But for the professor's amazement, quite honestly, there is not as much space as an atom normally fills.

Victor Pürschel.[13]

---

[12] This is an exact quotation (except for the italics and the ellipses, which are Pürschel's) of Ritzau Bureau's account of the interview which was brought in Danish newspapers on 16 May 1934 and which Bohr in his response (below) described as "unauthentic". The ellipses replace the words "sluttede Bohr" ("Bohr concluded") in the original Ritzau report.

[13] (1877–1963). Conservative member of the Danish parliament.

A Statement by Professor Niels Bohr.

In connection with a contribution, appearing in "Berlingske Tidende" yesterday evening, to the discussion in various newspapers here at home, arising from an interview in "Izvestia" during my journey to Russia, I would like to request space for the following brief comment:

In the statements reproduced in "Politiken" of 28 and 30 May, I have, I think, explained clearly enough, after my return home, the nature of the questions put to me in the interview under discussion and my answers to these, as well as the misunderstandings arising from a brief unauthentic account of the interview published by Ritzau's Bureau, where single sentences, more or less correctly quoted, are put together in a misleading way. In the contribution to "Berlingske Tidende" in question, my statements in "Politiken" are, however, interpreted in a way that I cannot recognize. With the statement that I had not assumed an attitude to general political questions, I meant to relate that in the interview I had expressly taken exception to any comparison between conditions in Russia and in other countries. As it should clearly appear from the background for this statement and the whole connection in which it appeared, I did not at all mean, however, that I had wished to refrain from commenting on the thought-provoking impressions which the attempts being made in Russia to create new frameworks for general scientific and cultural endeavours must make on everyone who gets the opportunity through personal experience to become more closely acquainted with them.

Niels Bohr.

[211]

RUTHERFORD TO BOHR, 6 December 1934
[Typewritten]
[Source: BSC-Supp.]

Cavendish Laboratory,[14]
Cambridge.
6th December, 1934.

My dear Bohr,

You may have heard the news about the retention of P. Kapitza in Russia, but I had better give you the information we have.

Kapitza and his wife went to Russia in the summer with return visas, and he gave some lectures in the University of Kharkov and attended the Mendeléef Conference. A few days before his return, he was peremptorily summoned to Moscow and told they wished him at once to work on Physics problems in Russia. I believe Kapitza refused to do so unless he was given his liberty and allowed to return to England. He was refused his passport, but Mrs Kapitza was allowed to return to England to look after his affairs.

On hearing of this from Mrs K, I wrote an unofficial letter to the Russian Ambassador here to explore the situation, and I enclose herewith copies of my letter and of his reply.[15] From this you will see that he definitely states that K. as a Soviet citizen is required to stay in Russia. Various informal representations have been made by interested parties to see what can be done about the matter, but no indication has been received of any change of view of the Soviet authorities.

We held a meeting of the Mond Laboratory[16] Committee a few days ago and reported the situation to the University and to the Royal Society, suggesting that Kapitza should be given leave of absence for one year in order to give time for further negotiations. I have little doubt that both the University and the Royal Society will move in the matter and make representations both here and in Russia to the appropriate authorities.

While it is quite clear that as a Soviet citizen, the U.S.S.R. has a complete right of using K's services as it thinks fit, yet the present situation is very

---

[14] Renowned laboratory in Cambridge established in 1874.

[15] The Soviet ambassador in London from 1934 to 1943 was Ivan Mikhailovich Maisky (1884–1975). Copies of the letters from Rutherford to Maisky, 12 October 1934, and from Maisky to Rutherford, 30 October 1934, are retained in the NBA.

[16] Laboratory built in Cambridge for Kapitza's own research.

unfortunate for the relations between English and Russian Science. They have been well aware throughout of Kapitza's work, and allowed him to be made a Royal Society Professor and to be responsible for the building of a special Mond Laboratory without any indication that his services would be required in Russia. As soon as the new laboratory is in working order and Kapitza is about to reap the fruits of the organization and services, he is summarily required to drop his work in England and work in Russia.

I imagine the best form of protest is to indicate that this summary treatment of Science in this country is very unfortunate all round, and will leave an unhappy impression in this country which will inevitably interfere with good scientific relations between ourselves and Russia.

Langevin in Paris has been informed of the situation by Mrs K and I think has expressed his view through the Soviet Ambassador in Paris. I was wondering whether you might be in a position either directly or indirectly to express your views on the matter to the proper authorities in Russia. Probably it would be best to get the ear of the Foreign Secretary, Litvinoff,[17] who I understand was absent when the Kapitza question was officially dealt with.

Of course, I have always had at the back of my mind the probability that K. would return to Russia eventually, but it is exceedingly awkward for him and for us to leave his laboratory and his research students in the air in this way. I think that K. feels he has been badly treated, and I believe has definitely refused to undertake any work in Physics until he is allowed to return.

In the meantime, I understand he has started work with Pavlov on Physiology, and therefore possibly an approach through the Academy, of which Pavlov I believe is President, might be appropriate; but I leave the matter to your judgment. In the meantime I have taken charge of the laboratory to keep things running for a year to give time for discussion and consideration.

I am glad to say we are all very well. I have been kept very busy all this term: I go to Manchester on Sunday where I give a lecture on Monday, and hope to get a few days' golf afterwards. After Christmas we shall go down to our cottage for a breathing spell.

You will be sorry to hear that Sir Horace Lamb[18] died quietly and peacefully – a heart attack – on Tuesday. The funeral service is on Friday in Trinity Chapel. I think you knew Lamb well in the old days, and recognised what a

---

[17] Maxim Litvinov (1876–1951) had lived in London for several years before the Russian revolution of 1917 whereupon he played a major role in improving the relations between the Soviet Union and Great Britain as well as other European countries as "roaming ambassador". He was appointed Commissar of Foreign Affairs by Stalin in 1930.

[18] (1849–1934), British mathematical physicist.

fine character he had. He is one of the few men I have seen that grow old gracefully without grasping for power at every stage. His death removes one of the outstanding figures of the age of Kelvin[19] and Rayleigh[20] who contributed so much to the theoretical foundations of the older Physics.

With best wishes to Mrs Bohr and your family.

Yours ever,
Rutherford

BOHR TO BUKHARIN, 15 December 1934
[Carbon copy]
[Source: BSC-Supp.]

15. Dezember [193]4.

Lieber verehrter Herr Bucharin,

Nach meinem Besuch in Russland diesen Sommer, der für mich ein so schönes und interessantes Erlebnis war, habe ich oft die Absicht gehabt, an Sie zu schreiben, um Ihnen zu danken für die grosse Freundlichkeit, mit der Sie mich in Moskau empfangen haben. Ich habe oft an unser lebhaftes Gespräch gedacht über die allgemeinen philosophischen Probleme, die uns beiden so sehr am Herzen liegen. Ich habe nur bedauert, dass die geplante Fortsetzung unserer Unterhaltung nicht möglich war wegen Ihrer Abwesenheit von Moskau bei meiner Rückkehr von der Konferenz in Kharkow, wo mein grosser Eindruck von allem, was in Ihrem Lande für die Förderung der Wissenschaft gemacht wird, noch verstärkt wurde.

Ich ergreife die Gelegenheit heute zu schreiben, weil ich eben von meinem Freund, Lord Rutherford in Cambridge eine vertrauliche Mitteilung bekommen habe über die Schwierigkeiten, die für die Fortsetzung der erfolgreichen Tätigkeit in Cambridge seines geschätzten Mitarbeiters, Prof. Kapitza, entstanden sind. Die Nachricht, dass Prof. Kapitza, als er in diesem Sommer auf Besuch nach Russland kam, um unter anderen an der Mendelejew-Feier teilzunehmen,[21] die Erlaubnis nicht bekam, nach Cambridge zurückzukehren,

---

[19] (1824–1907), also known as William Thomson.
[20] (1842–1919), also known as John William Strutt.
[21] This was the centennial of the birth of the Russian chemist Dmitryi Ivanovich Mendeléev (1834–1907), famous for the periodic table of elements (1869).

um die Leitung des für ihn von der Royal Society gestifteten speziell für seine Arbeiten geplanten, neuen grossen Laboratoriums fortzusetzen, hat Lord Rutherford nicht nur sehr überrascht, sondern auch sehr leid getan.

Lord Rutherford, der, wie Sie wissen, sich immer für die internationale Zusammenarbeit der Gelehrten eingesetzt hat, und den besonders freundliche Beziehungen mit den russischen Kollegen verbinden, denkt dabei nicht nur an die beklagenswerten Konsequenzen, die von einem eventuellen Abbruch der Tätigkeit von Prof. Kapitza in Cambridge für die lokalen Interessen sich ergeben würden, sondern er fürchtet eben so sehr, dass das allgemeine Bekanntwerden einer solchen Situation eine sehr bedauerliche Einwirkung haben könnte auf die Zusammenarbeit zwischen den ausländischen und den russischen Wissenschaftlern, die sich jetzt so vielversprechend entwickelt, besonders dank der verständnisvollen und grossartigen Unterstützung der Wissenschaft von Seiten der Sowjetregierung.

Wenn ich mich mit dieser Angelegenheit an Sie zu wenden wage, so ist es, weil ich weiss, wie grosses Gewicht Sie sowie alle für die Leitung der Union verantwortlichen Persönlichkeiten, auf eine grösstmögliche Harmonie der internationalen wissenschaftlichen Bestrebungen legen, und ich dachte daher, dass Sie vielleicht dazu beitragen möchten, dass die betreffende Angelegenheit von allen Seiten betrachtet werden könnte, bevor eine endgültige Ordnung getroffen wird.

Mit herzlichen Grüssen, und in der Hoffnung, dass ich in nicht zu ferner Zukunft die Gelegenheit haben werde, Sie wieder zu treffen,

Ihr sehr ergebener
[Niels Bohr]

Translation

15 December [193]4.

Dear Mr. Bukharin,

After my visit to Russia this summer, which was such a wonderful and interesting experience for me, I have often intended to write to you to thank you for the great kindness with which you received me in Moscow. I have often thought of our lively conversation about general philosophical issues which are so close to our hearts. I only regret that the planned continuation of our conversation was not possible because of your absence from Moscow on my return from the conference in Kharkov, where my great impression of

[215]

everything that is done in your country for the advancement of science was strengthened even further.

I seize the opportunity to write today, because I have just now received a confidential message from my friend Lord Rutherford in Cambridge concerning the difficulties that have arisen as regards the continuation of the successful work in Cambridge of his highly valued collaborator, Prof. Kapitza. The news that Prof. Kapitza, when he visited Russia this summer, among other things in order to take part in the Mendeléev celebration,[21] did not receive permission to return to Cambridge to continue the directorship of the new large laboratory established by the Royal Society and planned especially for his work, has not only greatly surprised but also greatly saddened Lord Rutherford.

Lord Rutherford who, as you know, has always been committed to the international cooperation among scholars, and to cultivating the particularly friendly relations with the Russian colleagues, does not only consider the unfortunate consequences for the local interests which would result from a possible interruption of Prof. Kapitza's work in Cambridge, but he just as much fears that if such a situation became publicly known, it could have a very regrettable effect on the cooperation between foreign and Russian scientists, which now develops so very promisingly, particularly thanks to the astute and magnificent support of science on the part of the Soviet government.

When I dare to turn to you in this matter, it is because I know how great significance you, like all public figures responsible for the leadership of the Union, attach to a greatest possible harmony in international scientific endeavours, and hence I thought that you might perhaps contribute to having the matter in question considered from all sides before a final arrangement is made.

With cordial regards and with the hope that I will have the opportunity to meet you again in the not too distant future,

Very humbly yours
[Niels Bohr]

BOHR TO RUTHERFORD, 15 December 1934
[Carbon copy]
[Source: BSC-Supp.]

December 15th [193]4.

Dear Rutherford,

Although through Gamow I had heard rumours about Kapitza's difficulties I was quite shocked from your kind letter to learn about the actual situation and your correspondence with the Russian ambassador in London. I shall of course be glad if I can offer any assistance in the matter, but I know that I agree with you in the greatest caution to avoid any premature publicity in a delicate case like this. If the ambassador in London feels himself bound to take the attitude explained in his letter, I think that it is quite impossible for any official professional body like the Russian academy, for which even the President Prof. Karpinski[22] is a very feeble old man, to move in this case. As you suggest, the best way might be to get into touch with Litvinoff, whom I have never met and whom you will of course yourself be the nearest to approach in an official manner. It happens, however, that on my journey to Russia I became through Joffe acquainted with Mr. Bucharin, who was a close friend of Lenin and is a most attractive and intelligent man. I am not sure that he has still the great influence which he exerted for a long time after the revolution, but as the chief editor of the government paper "Iswestia" he is certainly one of the leading personalities in Moskau. When I got your letter, I thought therefore that I would write quite informally to him and I hope that my letter, of which I enclose a copy, even if it may perhaps have no effect can hardly do any harm.

The political state in the world is surely most uncomfortable, and as I wrote to Fowler[23] the other day, the anxiety and trouble, which the refugee problem is causing us all, overshadows almost the pleasure of the truly marvellous development in atomic physics. Of course the stay of Franck[24] and Hevesy[25] here is a great help and encouragement to us all, and at the moment they are effectively reorganising the laboratory, so that we here may take some share in the work on the nuclear problems. I hope also that in this spring the question of the travelling stipend for Jacobsen[26] will be satisfactorily settled, so that he will be able to follow your kind invitation to work for a time in the Cavendish Laboratory. I need not say that my own thoughts are often there, and that I am longing to be able again to talk with you about all the new prospects.

---

[22] Alexander Petrovich Karpinsky (1846–1936), Russian geologist who was President of the Academy of Sciences of the U.S.S.R from 1917 until his death.

[23] Ralph Howard Fowler (1889–1944), mathematical physicist at the Cavendish Laboratory.

[24] James Franck (1882–1964), German experimental physicist who worked closely with Bohr and played a crucial role in verifying his theories. Franck stayed in Copenhagen as a Jewish refugee from 1934 until moving permanently to the United States a year later.

[25] George de Hevesy (1885–1966), Hungarian-born physical chemist and another close friend of Bohr after the two met in Manchester in 1912. He had worked at Bohr's institute from 1920 to 1926 and returned in 1934. He stayed until escaping to Stockholm in 1943, where he spent the rest of his life.

[26] Jacob Christian Jacobsen (1895–1965), Danish experimental physicist who worked at Bohr's institute from its establishment.

I was very sorry indeed to hear about the death of Lamb, who certainly was one of the most capable scientists and noblest characters I ever met. I believe also that in spite of the misfortunes and illness in his family he was a happy man, and I know, how much your faithfull friendship meant to him. We ourselves have socially had a quiet life this year and are going away to Sweden with the children for the Christmas week, which I am afraid will be a very sad time for Margrethe, who after all is wonderfull in keeping up strength and courage.[27]

With our kindest greetings and heartiest wishes for you all,

yours ever,
[N. Bohr]

RUTHERFORD TO BOHR, 25 December 1934
[Typewritten, with handwritten corrections]
[Source: BSC-Supp.]

Cavendish Laboratory,
Cambridge.
25th December, 1934.

PERSONAL

My dear Bohr,

I received your letter and thank you for writing to Bukharin. I thought your letter excellent and it may prove helpful if the matter is further considered.

I would like to ask for your good offices in another direction. We have had a discussion of representatives of the University and the Royal Society on the Kapitza question. We feel that at the moment little would be gained by any direct representations from this country through the Foreign Office, for, as you may guess, there is not much sympathy between our Foreign Office and that

[27] The Bohrs' eldest son, Christian (1916–1934), had died in a sailing accident earlier in the year.

of Russia. In addition the Russian Government would be likely to assume that the University and the Royal Society are annexes of our Government – which we know is not true – and this would weaken the force of our representations.

For the time being, we think it would be better to base any representations on the broad question of international scientific cooperation and the danger of an episode like Kapitza's prejudicing severely that close understanding which exists among scientists of different countries. We think that this point of view could be best expressed by representative scientific men in Europe who are not, like ourselves, primarily connected with Kapitza and his Laboratory. It is on this matter that I would like your good offices, and I hope it will not be throwing too much of a burden on you.

I would suggest that a memorial be drawn up, say by yourself, and circulated among a limited number of scientific men in the various European nations, say not more than a dozen in all, of high reputation in the scientific world, and who know in general about Kapitza's scientific work. In order to help you in this matter, we have drawn up a draft statement which may be of use to you in formulating your memorial.[28]

It is of great importance that such a representation should appear to be a spontaneous opinion of scientists outside this country, and not initiated by the members of the Royal Society or this University, and I hope that you would be willing to take the responsibility of drawing up the type of memorial you think would meet with the approval of yourself and your colleagues, and circulate it to those whose signature you think would be valuable. As I have said, the enclosed draft is merely to save you time, but I hope it embodies the points which we believe to be most likely to produce an effect.

I may mention that Mrs Kapitza[29] has been consulted and strongly supports such a mode of action. I leave the question of the choice of signatories to yourself, for you know better than I do the most suitable people to ask. I think you are aware that both Langevin[30] and Debye[31] know Kapitza well personally and have been in close touch with his work.

It is believed that the pivotal man in the situation is Litvinoff. It is known that he was absent from Russia when the decision to retain Kapitza was taken by the Soviet Government, and he is also the member of the Government most likely to realise the danger of international repercussions. In view of the

---

[28] The draft statement is held together with Rutherford's letter in BSC-Supp.
[29] Kapitza had married Anna Krylova (1903–1996) in 1927.
[30] Paul Langevin (1872–1946), prominent French physicist with close ties to the Soviet Union.
[31] Peter Josephus Wilhelmus Debye (1884–1966), Dutch physicist who emigrated to the United States in 1940. In 1934 he was Director of the Physical Institute of the University of Leipzig.

newly established Franco–Russian entente, it is probable that the best method of approach to Litvinoff would be through some distinguished Frenchman, for instance, a French politician of importance. It is important that the memorial should be sent to Litvinoff direct, and not to a Soviet Ambassador who might not pass it on. As I have mentioned, *it is essential that the English source of this memorial should not be mentioned to anyone*, because Cambridge and the Royal Society are interested parties, and therefore no English scientists should be asked to sign it.

For your guidance, I should mention that Langevin has been visited by Mrs Kapitza and has himself made strong representations against Kapitza's detention. He might be able to advise you as to the most suitable Frenchman to mention the matter to Litvinoff and to forward the memorial.

It is probably preferable for these reasons that the text of the memorial to be circulated should be in *French*, and accordingly we enclose the French text of our suggestions.

I am sorry to give you all this trouble, but I feel that something should be done to try and make the Russian Government aware of the dangerous situation to scientific relations they have brought about by their precipitate action. You will no doubt let me know before long what is your opinion in the matter, and whether you feel that the suggested type of memorial is likely to be the most effective under the difficult conditions that surround this affair.

I am glad to hear that you and Margrethe are going to Sweden for a little holiday at Christmas time. My wife and I will spend Christmas with the grandchildren, and will leave for our country cottage on December 26th for a fortnight's rest. I am glad to say we are all very well, including Fowler, and the grandchildren are very lively.

I note what you say about Jacobsen. We will do what we can to make his visit to Cambridge a profitable one.

We have made good but not sensational progress in the work on transmutation. One point of interest appears, namely that probably H^2 and H^3 may be liberated as units in certain nuclear reactions. The proof of this has involved a determination of the velocity and e/m of the particles.

I am sure it will be a great satisfaction to you to have such fine people as Frank [sic] and Hevesy with you. Please give them my best wishes. Give my love to Margrethe, and best wishes to you all for a happy New Year.

Yours sincerely
Rutherford

P.S. I think it undesirable that my name should appear in any prominent fashion in the proposed memorial. If it occurs at all it should only appear in an explanation of the facts of the case. They will naturally assume that it is my duty to support Kapitza's claims for reconsideration.

R

BOHR TO RUTHERFORD, 7 January 1935
[Carbon copy]
[Source: BSC-Supp.]

January 7th [193]5.

Dear Rutherford,

A few days ago just before our return to Copenhagen I received your kind letter and your proposal for a memorial concerning the retainment of Kapitza in Russia and to be handed over to Litvinoff with the signature of a number of scientists in various countries. Of course I am prepared to offer any assistance I can in this most unhappy case, to which I have given much thought; and if there is a general belief that the step proposed by you is likely to have a good effect, I shall be very glad to take part in this action and help to bring it into effect. Before proceeding with the matter, however, I have thought it best to ask Langevin's opinion about the adviceability of such a step, as I am somewhat afraid that any semi-public international appeal might harden the case prematurely and make it a matter of prestige of the Russian government to keep to its original decision. Another difficulty which may perhaps only carry small weight is that the deplorable political development in several other European countries may make it difficult for many of those, whom it otherwise would be natural to invite to sign such a memorial, to take a public stand against the Russian methods. I therefore still think that the procedure suggested in your former letter of a more private communication with Litvinoff from some independent individual, and if not yourself, perhaps me or perhaps better Langevin, should not be given up before further discussion, and also about this I think it would be good to have Langevin's opinion. I am writing by the same post quite confidentially to Langevin, asking him for an early reply on account of the urgency of the matter. As soon as I get his answer, I shall of course write to you again.

[Niels Bohr]

BOHR TO LANGEVIN, 7 January 1935.
[Carbon copy]
[Source: BSC-Supp.]

January 7th [193]5.

Dear Langevin,

I am writing to you very confidentially about the most unhappy retainment of Kapitza in Russia, of which I understand that you know already. In order, however, to put the whole matter before you I am enclosing all the correspondence I have had with Rutherford concerning it. As explained in my last letter to Rutherford, my intention is before all to ask your opinion regarding the procedure proposed by Rutherford, that a memorial undersigned by a number of scientists from various countries is handed over to Litvinoff. Personally I am not sure that a more private communication with Litvinoff might not be more effective, and at the same time less dangerous for Kapitza, about whose present condition in Russia I have no knowledge. As I have suggested to Rutherford, you might yourself be the best fitted person for a direct communication with Litvinoff, whom I have never met, but I am of course prepared to consider any other step you may advise either together with you or alone. As I have promised Rutherford to write soon again on account of the urgency of the matter, I shall be very thankful to hear about your views as early as possible.

With the kindest regards and best wishes for the new year for you and your family and all common friends in Paris from my wife and

yours,
[Niels Bohr]

[222]

BOHR TO STALIN, 23 September 1938
[Typewritten transcript]
[Source: NBA]

Abschrift.

Den 23. September 1938.

An J. Stalin
Sekretär der Kommunistischen Partei
der Sowjet-Union
Moskau.

Nur meine Dankbarkeit für die rege und fruchtbare Zusammenarbeit mit Wissenschaft[l]ern der Sowjet-Union, in der zu stehen es mir seit vielen Jahren vergönnt ist, und der tiefe Eindruck, den ich bei wiederholten Besuchen in der USSR von der Begeisterung und dem Erfolg, womit wissenschaftliche Forschung dort unterstützt und gefördert wird, erhalten habe, bewegen mich, Ihre Aufmerksamkeit auf eine Sache betreffen einen der bedeutendsten Physiker der jungen Generation, nämlich Professor L.D. Landau[32] vom Institut für physikalische Probleme bei der Sowjet-Akademie der Wissenschaften zu lenken.

Prof. Landau hat in der Tat nicht nur durch eine Reihe sehr bedeutsamer Beiträge zur Atomphysik die Anerkennung der wissenschaftlichen Welt gewonnen, sondern durch seinen begeisternden Einfluss auf jüngere Wissenschaft[l]er auch in entscheidender Weise an der Gründung einer Schule theoretischer Physiker in der USSR mitgewirkt, aus der unentbehrliche Mitarbeiter hervorgegangen sind für die grossartigen experimentellen Forschungen, die in den neuen grosszügig ausgestatteten Laboratorien in allen Teilen der USSR betrieben werden.

Während vieler Jahre habe ich die grosse Freude gehabt, mit Prof. Landau in sehr enger Verbindung zu stehen und regelmässig mit ihm über wissenschaftliche Probleme, die uns beide aufs tiefste interessieren, zu korrespondieren. Auf meine letzten Briefe habe ich jedoch zu meiner grossen Besorgnis keine Antwort empfangen, und soviel ich weiss hat auch keiner der vielen anderen ausländischen Physiker, die seine Arbeit mit grösstem Interesse

---

[32] Lev Davidovich Landau (1908–1968), Russian theoretical physicist who visited Bohr's institute in 1930 and remained close to Bohr.

[223]

verfolgen, Nachrichten von ihm erhalten. Durch eine Anfrage an die Sowjet-Akademie der Wissenschaften, deren Mitglied zu sein ich die Ehre habe, habe ich auch versucht, mit Prof. Landau in Verbindung zu kommen; aber die Antwort des Präsidenten der Akademie, die ich soeben erhielt, enthält keinerlei Auskunft über Aufenthalt oder Schicksal von Prof. Landau.[33]

Hierüber bin ich tief bekümmert, besonders deshalb weil neuerdings Gerüchte von einer Verhaftung Prof. Landaus mich erreicht haben. Ich hoffe noch immer, dass diese Gerüchte jeder Grundlage entbehren; sollte aber Prof. Landau wirklich verhaftet worden sein, so bin ich davon überzeugt, dass es sich um ein unseliges Missverständnis handeln muss; denn ich kann mir nicht vorstellen, dass Prof. Landau, der sich immer ganz der wissenschaftlichen Forschung widmete und dessen aufrichtige Persönlichkeit ich aufs höchste schätze, irgendetwas getan haben könnte, was eine Verhaftung rechtfertigen würde.

In Anbetracht der grossen Bedeutung dieser Angelegenheit sowohl für die Wissenschaft in der USSR wie für die internationale wissenschaftliche Zusammenarbeit wende ich mich an Sie mit der dringenden Bitte, eine Untersuchung über das Schicksal von Prof. Landau zu veranlassen, sodass, wenn wirklich ein Missverständnis vorliegen sollte, dieser so ausserordentlich begabte und erfolgreiche Wissenschaft[l]er wieder die Gelegenheit bekommt, an dem für den Fortschritt der Menschheit so wichtigen Forschungswerk teilzunehmen.

(N. Bohr)
Professor der Universität
Kopenhagen.

---

[33] Academy President A.P. Karpinsky was succeeded upon his death in 1936 by the botanist Vladimir Leontievich Komarov (1869–1945). The letters from Bohr to Komarov, 27 August 1938, and from Komarov to Bohr, 15 September 1938, are deposited in the NBA.

Translation

To J. Stalin
Secretary of the Communist Party
of the Soviet Union

23 September 1938.

Moscow.

Only my gratitude for the active and productive cooperation with scientists in the Soviet Union, in which for many years I have been privileged to take part, and the deep impression which through repeated visits to the USSR I have received of the enthusiasm and the success with which scientific research is funded and promoted, move me to direct your attention to a matter concerning one of the most important physicists of the young generation, namely Professor L.D. Landau[32] from the Institute for Physical Problems at the Soviet Academy of Science.

Prof. Landau has actually not only gained the recognition of the scientific world through a number of very significant contributions to atomic physics, but has through his encouraging influence on younger scientists also helped decisively in establishing a school of theoretical physicists, from which have emerged indispensable collaborators in the splendid experimental studies pursued in the new magnificently equipped laboratories everywhere in the USSR.

I have for many years had the great pleasure of being in very close contact with Prof. Landau and of corresponding with him regularly about scientific problems that interest both of us most deeply. I have, however, to my great dismay not received a reply to my most recent letters, and as far as I know none of the many other foreign physicists who follow his work with the greatest interest, have received word from him either. I have also tried to get into contact with Prof. Landau through an inquiry to the Soviet Academy of Science, of which I have the honour of being a member, but the answer from the President of the Academy, which I have just received, contains no information at all about Prof. Landau's whereabouts or fate.[33]

I am deeply concerned about this, especially because rumours of Prof. Landau's arrest have recently reached me. I still hope that these rumours are completely unfounded, but should Prof. Landau really have been arrested, I am convinced that it is a matter of an extremely unfortunate misunderstanding, for

I cannot imagine that Prof. Landau, who always devoted himself completely to scientific research and whose sincere personality I hold in the highest esteem, should have done anything that would justify arrest.

In view of the great importance of this matter for science in the USSR, as well as for international scientific cooperation, I turn to you with the urgent plea that an investigation into the fate of Prof. Landau is implemented, so that if there really is a misunderstanding this so exceptionally intelligent and successful scientist again gets the opportunity to take part in the research enterprise that is so important for the progress of humanity.

(N. Bohr)
Professor at the University
of Copenhagen.

WAR AND EXILE

BOHR–CHADWICK LETTERS, early 1943
[Typewritten transcript]
[Source: BPP 1.5]
[Editor's comment: The introduction is part of the transcript. The individual letters are indicated as "items" in the margin.]

Messages exchanged during the German occupation of Denmark between Professor Niels Bohr and Sir James Chadwick, acting on behalf of Sir John Anderson, and committed through the British Secret Service and the Danish Underground Intelligence Service,

1) Message from Chadwick 25.1.1943
2) Answer from Bohr, February 1943
3) Supplementary remarks from Bohr later in the spring of 1943.

Afskrift af papirer om forbindelsen med England, som har været nedgravet i haven på Carlsberg[34]

The University of Liverpool   Item 1)
Physics Laboratories
25th January 1943.

I have heard in roundabout way, that you have considered coming to this country if the opportunity should offer. I need not tell you how delighted I myself should be to see you again; and I can say you, there is no scientist in the world, who would be more acceptable both to our university people and to the general public. I think you would be very pleased by the warmth of welcome you would receive. A factor which may influence you in your decision is that you would work freely in scientific matters. Indeed I have in my mind a particular problem in which your assistance would be of the greatest help. Darwin[35] and Appleton[36] are also interested in this problem and I know

---

[34] In English: "Transcript of papers about the connection with England, which have been buried in the garden at Carlsberg".

[35] Charles Galton Darwin (1887–1962), British physicist and friend of Bohr's after the two met in Manchester in 1912.

[36] Edward Victor Appleton (1892–1965), prominent British physicist, was technical head of the Department of Scientific and Industrial Research, which was responsible for Tube Alloys, the British side of the atomic bomb project.

they too would be very glad to have your help and advice. You will, I hope, appreciate that I cannot be specific in my reference to this work, but I am sure it will interest you. I trust you will not misunderstand my purpose in writing this letter. I have no desire to influence your decision, for you alone can weigh all the different circumstances, and I have implicit faith in your judgment, whatever it should be. All I want to do, is to assure you that, if you decide to come, you will have a very warm welcome and an opportunity of service in the common cause. With my best wishes for the future and my deepest regards to Mrs. Bohr,

Yours sincerely,
(sign.) J. Chadwick

Item 2)

I can hardly express how deeply I appreciate your kind letter for which I thank you most heartily. As you know, I am with my whole heart in the struggle for freedom and humanity and it is a great encouragement to feel, that my friends have not forgotten me and are endeavouring to support my ardent wish to participate in the great common cause. However tempting it would be to me to follow your invitation, I find it after much deliberation to my great regret impossible for the present to leave this country. Not only I feel it to be my duty in our desperate situation to help to resist the threat against the freedom of our institutions and to assist in the protection of the exiled scientists, who have sought refuge here. Still neither such duties nor even the dangers of retaliation to my collaborators and relatives might perhaps not carry sufficient weight to detain me here, if I felt, that I could be of real help in other ways, but I do not think that this is probable. Above all I have to the best of my judgment convinced myself, that in spite of all future prospects any immediate use of the latest marvellous discoveries of atomic physics is impracticable. However there may, and perhaps in a near future, come a moment where things look different and where I, if not in other ways, might be able modestly to assist in the restoration, which is bound to come of international collaboration on human progress. At that moment, whether it will come before or after the cessation of hostilities, I shall make an effort to join my friends and I shall be most thankful for any support they might be able to give me for this purpose. I need not add, that I leave it to you to judge to whom you may convey the content of this letter. With my heartiest greetings and most deepfelt wishes.

Item 3)

In view of the rumours going round the world, that large scale preparations are being made for the production of metallic Uranium and heavy water to be used in atom bombs, I wish to modify my statement as regards the impracticability of an immediate use of the discoveries in nuclear physics. Taking for granted that it is impossible to separate the U-isotopes in sufficient amount, any use of the natural isotopic mixture would as well-known depend on the possibility to retard the fission neutrons to such a degree that their effect on the rare U, 235 isotope excedes neutron capture in the isotope U, 238. Although it might not be excluded to obtain this result in a mixture of Uranium and Deuterium, there will surely be a limit to the degree to which atomic energy could be released in such mixture, due to the decrease of the retarding effect with rising temperature. This limit would seem to be set by a temperature of the D corresponding to about 1 Volt pro atom, and this would, therefore, be the limit of the explosion power of a thorough mixture. If, however, as often suggested, solid pieces of U are placed in a large tank of heavy water, it might be possible to obtain a far higher temperature within the U before the critical D-temperature is reached. Since, however, at least 1 pct. of the average energy set free will independently of the amount of the U used obviously never be greater than about 100 times that required to heat the water to the critical temperature. Even if this is very great compared with that obtainable with ordinary chemical explosives, it would in view of the large scale bombing already achieved hardly be responsible to rely on the effect of a single bomb of this type procurable only with an enormous effort. The situation, however, is of course quite different, if it is true that enough heavy water can be made to manufacture a large number of eventual atom-bombs, and although I am convinced that the arguments here outlined are familiar to experts, I hasten therefore to modify my statement.

KAPITZA TO BOHR, 28 October 1943
[Typewritten]
[Source: BPP 9.4–5; BSC (21.4); CAB 126/39]
[Editor's comment: The PRO version has two "X"'s in the margin (as indicated below).]

MOSCOW.

Oct; 28/1943.

Kaloujskoe Shosse 32
Institute for Physical Problems.

Dear Bohr.

We learnt here that you left Denmark and are now in Sweden. Indeed we do not know all the circumstances of your departure but, considering the mess which the whole of Europe is now in, all we Russian scientists feel very

[229]

anxious about your fate. Of course you are the best judge of what path you must take through all this tempest but I want to let you know that, you will be welcome in the Soviet Union where everything will be done to give you and your family a shelter and where we now have all the necessary conditions for carrying on scientific work. You only have to let me know your wishes and what practical means are open to you and I have all reasonable hope that we could help you anytime you find it convenient for yourself and family.

As you already know, we had a pretty hard time of it at the beginning of the war but now the worst is well over. I think that it is in no way an exaggeration to say that all the people of our country are so closely united in their effort to free themselves from the barbaric invasion that throughout history it would be almost impossible to find a similar case. Now our complete victory is evidently only a question of time. We scientists have done everything in our power to put our knowledge at the disposal of this war cause. Our living conditions are now much easier. We are all back in Moscow and have time to spare for scientific work. At our Institute we hold a scientific gathering every week where you will find a number of your friends. The Academy of Sciences has also started its activities in Moscow and has just concluded its session at which a number of new members were elected. If you come to Moscow you will find yourself joined with us in our scientific work. Even the vague hope that you might possibly come to live with us is most heartily applauded by all our physicists: Joffe, Mendelshtam,[37] Landau, Vavilov[38] Tamm,[39] Alihanov,[40] Semenov[41] and many others, who all ask me to send you their kindest regards and best wishes.

Mrs. Kapitza and the boys are all well and the latter have grown very big. Peter has already joined a technical school. We are most anxious to know how Mrs. Bohr and your boys are.

We have very little information about English physicists. Occasional exchange telegram is all that we know. They, like us, are hard at work fighting for our common cause against Nazism.

Accept my best wishes for the future. Most kind regards from myself and

---

[37] Leonid I. Mandelshtam (1879–1944), Russian theoretical physicist.

[38] Sergei Ivanovich Vavilov (1891–1951) was at this time director of the P.N. Lebedev Institute of Physics of the Soviet Academy. In 1945 he succeeded V.L. Komarov as President of the Academy itself.

[39] Igor Yevgenyevich Tamm (1895–1971), prominent physicist and director of the theoretical division of the Lebedev Institute from 1934.

[40] Abraham Isaak Alikhanov (1904–1970), Russian physicist.

[41] Nikolai Nikolaevich Semyonov (1896–1986), physical chemist. From 1931 he was director of the Institute for Chemical Physics of the Soviet Academy.

Mrs. Kapitza to you and your family. Let me assure you once more that we consider you not only as a great scientist but also as a friend of our country and we shall count it a great privilege to do our utmost for you and your family. And from my personal point of view I always couple your name with that of Rutherford and the great affection we both feel for him is a strong bond between us. It will be the greatest pleasure to me to help you in any respect.

With my very best wishes,

I am most sincerely yours

Peter Kapitza   P.K.

P.S. You may answer this letter through the channel by which you receive it.

P.K.

BOHR TO HALIFAX, 28 December 1943
[Typewritten transcript]
[Source: BPP 2.3]

COPY

Danish Legation   28–12–1943
Mass. Ave. 2343
Washington D.C.
p.t. New York.

Dear Lord Halifax.

I wish to thank you most heartily for the kindness with which you received me and offered me the privilege of further talks about the great project concerning which Sir John Anderson has taken me into his confidence.

The unique hopes and peculiar dangers involved in this project are, of course, fully appreciated by the responsible statesmen, but nevertheless, it must be the duty of a scientist, to whom opportunity is given to follow the gradual development of the technical aspects of the project, to call to their attention any detail of this development which may have a bearing on the realization of such hopes and on the elimination of such dangers. All suggestions in this direction which can be given at the present stage of the development are admittedly of a tentative and preliminary character, and I need not say that this refers in particular to any thoughts to which I tried to give expression during our conversation.

Since I saw you, arrangements have been made permitting me to make a short visit to one of the places where British and American physicists collaborate on the project, and I hope within a few weeks to be able to return to Washington and to communicate to you such opinions as I may be able to form on the implications of the progress of the work.

Respectfully yours
[Niels Bohr]

HALIFAX TO ANDERSON, 18 February 1944
[Typewritten]
[Source: CAB 126/39]

BRITISH EMBASSY
WASHINGTON, D.C.

<u>MOST SECRET.</u>   18th February, 1944.
<u>PERSONAL.</u>

Dr. B has written to you to tell you of certain ideas which have arisen in his mind as the result of his visit here.[42]

He has discussed them several times with me and I myself think that there may be some hope in what he has in mind. These scientists find it very difficult to make their thought precise on political problems, and Ronnie Campbell[43] and I have had to do a lot of work with B to get any clear idea of how his thought worked.

But I think we succeeded in doing it fairly well in the end.

His thought is, I think, that, as long as no competition has begun (and the situation in this respect may change very rapidly after the surrender of Germany), Great Britain and the United States would seem to have in their

---

[42] Bohr to Anderson, 16 February 1944, reproduced on p. [86].
[43] Ronald Campbell, Halifax's deputy at the British Washington Embassy.

hands a card which they could use in their negotiations with others of the United Nations for the improvement of the world situation with a view to the assurance of future peace. This idea of his about the *political* treatment of the project seems sensible, if it can be translated into a practical course of action.

If you thought it would be useful, I could ask Ronnie Campbell to come over to tell you what we have got out of B. as regards his thought on this political side, in greater detail than it is probably either good or possible to put in a letter. For I do believe that B's ideas call for very urgent and deep consideration by the Prime Minister and yourself. I shall be glad to hear what are your reactions.

Y[ours] Sin[cerely]

Edward

P.S. You won't, I hope, ever show B. this letter!

ANDERSON TO CHURCHILL, 21 March 1944
[Typewritten]
[Source: PREM, Series 3, folder 139/2]
[Editor's comment: When reading the document, Churchill circled the passages here reproduced within squares and added the margin notes and last comment (reproduced here in italics) by hand.]

MOST SECRET.

PRIME MINISTER
Tube Alloys

It is now virtually certain that sub-atomic energy will be available within a comparatively short time for use both as a military weapon and as a source of power.

The military weapon, on which the major effort has been concentrated, will probably come first. It is estimated that the Americans will have the first bomb ready at some date between the end of 1944 and the middle of 1945. The autumn of 1944 has even been mentioned as a possibility.

A brief note on the military and industrial potentialities of Tube Alloys is attached.

2.   The fact that serious work is being done on Tube Alloys is, inevitably, pretty widely known. Knowledge of the prospects and implications is, however, at present confined to a very narrow circle.

The military implications of the project are so vast that this no longer seems right. In my view the time has come when the War Cabinet, the Service Ministers and the Chiefs of Staff should be given ⟨full⟩ information.

What can they do about it?

3.   Nor can we afford to ignore the political implications.

It seems certain that the Americans, thanks partly to our help but mainly to their own vast effort and expenditure, will get there first. Fortunately, all the evidence we have goes to show that the Germans are not working seriously on the project. But it would be foolish to suppose that the Russians, even though they may be unable at present to put forth an effort on the gigantic American scale, will not do so as soon as they are free of the German invaders.

Moreover, the likelihood – certainty almost – is that, as experience and knowledge increases, the scale of the effort required will so decrease as to become within the capacity of a large number of countries.

4.   This being so, there are, as it seems to me, two alternatives: –

(a)   Either there will be a particularly vicious form of armaments race, in which at best the United States, or the United States and the United Kingdom working together as a team, will for a time enjoy a precarious and uneasy advantage, or

(b)   A form of international control must be devised which will ensure that sub-atomic energy, if used at all, is used for the common benefit of mankind and is not irresponsibly employed as a weapon of military or economic warfare.

5.   To devise any effective scheme of international control will be a matter of the greatest difficulty. And control will have to cover industrial as well as military exploitation; for it appears that a power-producing plant will inevitably produce simultaneously material for the military weapon.

But no plans for world organisation, which ignore the potentialities of Tube Alloys, can be worth the paper on which they are written. Indeed it may well be that our thinking on these matters must now be on an entirely new plane.

I am myself convinced that we must work for effective international control.

6.   Our relations with the Americans on this subject are such that we cannot

possibly make any move in this direction without them. Indeed it would be difficult for us to take the initiative even with them alone; and I would not recommend our doing so at this stage.

But Sir Ronald Campbell, who is home for a few days' consultation with me, tells me that one or two Americans close to the White House, who have some knowledge of Tube Alloys, are becoming increasingly concerned about the future and are thinking of urging the President to give the problem of international control his serious and urgent attention. It may be, therefore, that he will soon be wishing to discuss it with you.

In such an event it is essential that we should know what we want. It is, therefore, urgently necessary that we should know what are the practical possibilities as regards international control and should then consider, in the light of those possibilities, what our policy should be.

7.   As soon as we get into discussion with the Americans, one point which we shall have to settle quickly with them is whether, and if so when, we should jointly say anything to the Russians. If we jointly decide to work for international control, there is much to be said for communicating to the Russians in the near future the bare fact that we expect, by a given date, to have this devastating weapon; and for inviting them to ⎡collaborate⎤ with us   *On no* in preparing a scheme for international control. If we tell them nothing, they   *account.* will learn sooner or later what is afoot and may then be less disposed to co-operate. At the same time, there would seem to be little risk of the Russians, if they chose to be unco-operative, being assisted in the development of their   *No* own plans by a communication of the kind suggested.

8.   My immediate recommendations are: –

(a) Lord Cherwell and I should take an early opportunity of explaining the present position of Tube Alloys and its implications to the members of the War Cabinet, the three Service Ministers and ⎡the Chiefs of Staff.⎤   *Already know* It is for consideration whether the Minister of Supply and the Minister of Aircraft Production, who already have some vague knowledge of the matter, should be included.

(b) Thereafter regular reports on the position should be given to the Chiefs of Staff, whose duty it would be to bring to the notice of yourself and/or the War Cabinet any current military implications of importance.

(c) The Foreign Secretary should be asked, as a matter of urgency, to set on foot an immediate study of the practical and political problems of international control.

9.   This minute has been prepared in consultation with Lord Cherwell, who agrees.

I do not agree.

Great George Street, S.W.1.
21st March, 1944.

ANDERSON TO CHURCHILL, 27 April 1944
[Typewritten]
[Source: PREM, Series 3, Folder 139/2]
[Editor's comment: The last italicized comment is in Churchill's handwriting.]

TOP SECRET

PRIME MINISTER
    I have been looking at the series of Memoranda on future world organisation circulated with W.P.(44) 220.[44]
    I cannot help feeling that plans for world security which do not take account of Tube Alloys, must be quite unreal. When the work on Tube Alloys comes to fruition – as we now know it will in a comparatively short time – the future of the world will in fact depend on whether it is used for the benefit or the destruction of mankind. We shall be entering an entirely new era.
    I realise that, for the time being, discussions on future world organisation must go on on the present basis. But I submit that it is urgently necessary that a restricted circle in this country should be asked to examine the question whether arrangements for the control of Tube Alloys can be devised and, if so, how they would affect plans for future world organisation.

---

[44] The reference is to document no. 220 circulated to the War Cabinet in 1944. This substantial set of documents, classified Secret, which were circulated on 24 April 1944, lay out detailed plans for the establishment of the United Nations. Starting out with "Future World Organisation: Memorandum by the Lord President of the Council", 22 April 1944, they are deposited in CAB 66/49.

I have reason to believe that President Roosevelt is also giving serious thought to this matter and would not be averse from discussing it with you. It would clearly be a delicate matter for you to take the initiative, seeing that the United States have incurred virtually all the expenditure on the project and that almost all the plant and the materials are in the United States. But, given the political situation in the United States and the fact that the United States Army Officers concerned almost certainly want to keep Tube Alloys for the exclusive use of the U.S. Army, it is probably even more difficult for the President to do so.

I feel sure, therefore, that the best course would be for you to send the President a message which would break the ice but would not be open to misinterpretation. If you decide to do so, you will no doubt wish to draft the message in your own language. The attached draft is merely designed to show you the sort of thing I have in mind.

If you have any doubts about the above suggestions I should be very grateful for an early opportunity to discuss them with you.

I do not think any such telegram is necessary; nor do I wish to widen the circle who are informed. WC 27-4

DRAFT TELEGRAM, CHURCHILL TO ROOSEVELT, 27 April 1944
[Typewritten]
[Source: PREM, Series 3, Folder 139/2]
[Editor's comment: This document was enclosed with the preceding letter.]

DRAFT MESSAGE FROM THE PRIME MINISTER TO PRESIDENT ROOSEVELT
TOP SECRET AND STRICTLY PERSONAL.

I have been following with admiration extraordinary progress you are making on Tube Alloys.

Object of these efforts is to produce a weapon which may shorten this struggle. But their very success makes me wonder whether any plans for future world security which do not take account of the implications and possibilities of this tremendous development must not be quite unreal. I do not know whether you have any ideas on this. It seems to me to require deep thought.

BOHR TO KAPITZA, 29 April 1944
[Carbon copy]
[Source: BSC (21.4); BPP 9.4–5; CAB 126/39]
[Editor's comment: The versions are identical except that the PRO version has the handwritten comment "Draft".]

London, April 29th 1944.

Dear Kapitza.

I do not know how to thank you for your letter of October 28th which, through the Counsellor of the Soviet Embassy in London, Mr. Zinchenko,[45] I received a few days ago on my return from a visit to America. I am deeply touched by your faithful friendship and most grateful for the extreme generosity and hospitality expressed in the invitation to my family and me to come to Moscow. You know the deep interest with which I have always followed the cultural endeavours within the Soviet Union, and I need not say what pleasure it would be to me for a time to participate with you and my other Russian friends in the work on our common scientific interests.

At the moment, however, my plans are quite unsettled. My wife and three of our boys are still in Sweden while, together with our fourth boy who in later years has been a scientific assistant to me, I came over to England already in October, but I hope that it will soon be possible for her to join me here and that it shall not be long before we both can come to Moscow and see you and your family again. Ever since our last visit to Russia we have constantly had you all in our thoughts and treasured the memory of our stay at your beautiful institute. It has also been a very great pleasure to me to learn, what certainly was no surprise, how wonderfully you have succeeded in making the facilities, which you with so liberal support have created there, fruitful to science and your country and thereby to the benefit for mankind.

On my travels in England and America it has been most encouraging to meet a greater enthusiasm than ever for international scientific cooperation in which you know I have always seen one of the brightest hopes for a true universal understanding. Just about this subject I had a very pleasant and interesting talk with Mr. Zinchenko and we spoke especially of the new promises inspired by the mutual sympathy and respect among the United Nations arisen from the

---

[45] Most likely Constantine E. Zinchenko, who went on to become Under-Secretary of the Security Council of the United Nations from 1949 to 1953.

comradeship in the fight for the ideals of freedom and humanity. Indeed, it is hardly possible to describe the admiration and thankfulness which the almost unbelievable achievements of the Soviet Union in these years have evoked everywhere and not least in the countries which have experienced the cruel German suppression.

Notwithstanding my urgent desire in some modest way to try to help in the war efforts of the United Nations, I felt it my duty to stay in Denmark as long as I had any possibility to support the spiritual resistance against the invaders and to assist in the protection of the many refugee scientists which after 1933 had escaped to Denmark and found working conditions there. When, however, last September I learned that they all and besides them a large number of Danes like my brother and myself were to be arrested and taken to Germany, my family and I had the great luck of escaping in the last moment to Sweden among the many others who, due to the unition of the whole Danish population against the Nazis, succeeded in counterfoiling the most elaborate measures of the Gestapo.

For many reasons indeed I am hoping that I shall soon be able to accept your most kind invitation and come to Russia for a longer or shorter visit, and as soon as I know a little more about my plans I shall write to you again. To-day it is above all upon my heart to express my deepfelt thanks to you and my warmest wishes to you and your family and our common friends in Moscow.

Yours ever

[Niels Bohr]

BOHR TO ANDERSON, 9 May 1944
[Typewritten]
[Source: CAB 126/39 and BPP 2.2]

TOP SECRET.   London,

May 9th 1944.

The Right Hon. Sir John Anderson,
Chancellor of the Exchequer.

Dear Sir John,

Thinking over my conversations with you I am a little worried lest I may not have given you an entirely correct idea of the nature of the mission with which I returned to this country. In fact, I fear that I may have given you the impression that my purpose was to urge you to advocate a certain line of policy to the Prime Minister.

The aim of my delicate mission, however, was to convey to the Prime Minister the information that the President does not only regard future policy concerning the project as a matter for Mr. Churchill and himself but that he, without discussing the greater issues with his technical advisers, in his own mind is deeply concerned with the unique hope the situation inspires, and eager to receive any suggestion to this effect from Mr. Churchill.

This very fact was considered by the President's friend,[46] as well as by Lord Halifax and myself, as a cause of great encouragement, and my only duty is to make sure that Mr. Churchill receives a full impression of the aim of the message and realizes the extraordinary circumstances under which it was obtained so that he, in his own mind, can consider the best way of turning to the benefit of England and of all mankind the opportunity which thus offers itself.

I cannot help feeling that the only way in which I can carry out my mission is if I can be given the opportunity to see Mr. Churchill. I fully understand how preoccupied he must be at the present time, and I would not venture to make this suggestion if, in view of the confidence shown me, I did not consider it my duty to give the Prime Minister a first hand account of the confidential conversations which have taken place in Washington, so that he can have a clear idea of what lies behind the message with which I have been entrusted.

As you know, I must according to my arrangements with the American organization be back in U.S.A. by May 26th. Time is therefore very short and I should greatly welcome an early opportunity of discussing with you the aspects of the matter which I have tried to explain in this letter.

Yours sincerely,
Niels Bohr

[46] Felix Frankfurter (1882–1965).

DALE TO CHURCHILL, 11 May 1944
[Typewritten]
[Source: CAB 126/39]

The Royal Society
Burlington House, W.1.
11 May, 1944.

Dear Prime Minister,

I beg you to pardon my intrusion of this letter upon your notice, at a time of your known preoccupation with tremendous responsibilities. Nothing could excuse my doing so but the deepest sense of obligation to science and to the world. From knowledge in which I am one of the very few to share, I cannot avoid the conviction that science is even now approaching the realization of a project which may bring either disaster or benefit, on a scale hitherto unimaginable, to the future of mankind.

Since the beginning of the war I have consistently used any influence which my position has given me among men of science, in advocacy of the duty to give scientific service as soldiers give military service, and to leave high policy to those properly carrying the responsibility for it. It is therefore with a strong reluctance, overpowered by a stronger sense of duty, that I venture to appeal to you on behalf of the scientific community, even under present conditions, to give to Professor Niels Bohr the opportunity of imparting to you a message which he has been charged to deliver to you in person, by a most intimate personal adviser of President Roosevelt. I should add, in that connexion, that before he made contact with the President's friend, Professor Bohr appears to have been fully in the confidence of Lord Halifax. In support of my appeal, I ask you to give attention to the following considerations.

1) Professor Niels Bohr is second to no authority in the world in theoretical physics. I think it probable, indeed, that a vote of the world's scientists would place him first among all the men of all countries who are now active in any department of science. The theoretical basis of the work which has brought to its present stage the "Tube Alloys" project, concerned with the methods of producing an explosive disintegration of atomic nuclei and with its consequences, was largely provided by Professor Bohr's own investigations shortly before the outbreak of this war. Bohr was earlier a pupil and collaborator of J.J. Thomson,[47] and Rutherford, and has a deep and affectionate loyalty to this country, ranking second only to his devotion to his own country of Denmark.

---

[47] Joseph John Thomson (1856–1940), Rutherford's predecessor as head of the Cavendish Laboratory, with whom Bohr spent his first months in England in 1911.

[241]

2) I have been in touch from the beginning with the development of the 'Tube-Alloys' project, as the only independent representative of unofficial science on Sir John Anderson's small and highly confidential Committee. Till recently, as you are aware, its successful outcome, in time for use even at the latest stage of this war, seemed to be doubtful. Progress in recent months has been rapid beyond all expectations. Bohr brings the latest news and his own expert appraisement of its meaning, after the fullest opportunity for inspection and discussion in the U.S.A. with the American and British scientists and engineers directly engaged with the project. His new information further increases the probability of a practical realization early in next year, with a rapid increase towards certainty with every month thereafter.

3) The devastating weapon, of which the early realization is now almost in view, must apparently put the power of world mastery into hands having a one-sided control of its use. This position will have been created by the work of leading men of science of the U.S.A. and Britain, largely on the basis of Bohr's theoretical work and recently with his active participation; but these men of science cannot concern themselves with its tremendous political implications. It is impossible, nevertheless, for a man of science who has been allowed to see what is happening and what may be involved, to neglect any opportunity of furthering the timely consideration of these implications by the only two men in whose power it may yet lie to take effective action – yourself and President Roosevelt. It is my serious belief that it may be in your power, even in the next six months, to take decisions which will determine the future course of human history. It is in that belief that I dare to ask you, even now, to give Professor Bohr the opportunity of brief access to you.

Lord Cherwell, with whom I have discussed the position, has kindly offered to bring this letter to your hands, and Sir John Anderson knows of my intention to send it.

With deep apology, but a clear conviction of my duty,

I am,
Yours faithfully,

(Sgd.) H.H. DALE.

The Rt. Hon. Winston S. Churchill,
C.H. F.H.S., M.F.,

10 Downing Street,
S.W.1

BOHR TO MØLLER, May 1944
[Typewritten]
[Source: BPP 3.3]
[Editor's comment: The date, which has been added to the document in Aage Bohr's handwriting, is uncertain.]

Maj 1944

Dette brev, der straks maa tilintetgøres, sendes paa lignende Maade som de Meddelelser, jeg selv modtog og besvarede, medens jeg endnu var i Danmark. Det gælder ligesom tidligere Spørgsmaalet om, hvorvidt der fra tysk Side er Bestræbelser i Gang for at udnytte Atomkerneenergien som Krigsvaaben. Som De ved, har jeg været bekymret derfor lige siden Heisenberg's Besøg i København i Efteraaret 1941. I lang Tid følte vi os jo beroligede, især gennem Jensen's[48] Meddelelser, men i den sidste Tid har der atter været Rygter fremme om tysk Virksomhed paa dette Omraade og enhver nærmere Oplysning vil derfor kunne være betydningsfuld. Man er herovre især interesseret i at vide, om man i København skulde have nogen som helst Oplysninger om tyske Fysikere, der eventuelt kunde give Fingerpeg om hvor de arbejder og hvad de foretager sig. Dette gælder alle tyske Fysikere som f.Ex. Heisenberg, Weizsäcker,[49] Bothe,[50] Süss,[51] Harteck,[52] Houtermanns,[53] Jensen, Mattauch,[54] Flügge,[55] Fiertz[56] og Clusius;[57] ikke mindst er man naturligvis interesseret i, om man kan være helt sikker paa Jensen's Indstilling. Det vil ogsaa være af Betydning om noget kendes om Opstilling af og Arbejde med Cyclotroner i Tyskland samt om Arbejde med Paris Cyclotronen og Joliot's[58] Skæbne. Det er jo meget at spørge om, men man vil være taknemmelig for selv den mindste Oplysning, hvor ubetydelig den end kunde synes. De vil bedst selv forstaa med hvor stor Forsigtighed og Fortrolighed en Brevveksling som denne maa behandles, men i alle saadanne Henseender vil De kunde regne med Overbringerens

---

[48] Johannes Hans Daniel Jensen (1907–1973), German nuclear physicist.

[49] Carl Friedrich von Weizsäcker (1912–), Heisenberg's younger colleague who had lived in Copenhagen in his youth.

[50] Walter Bothe (1891–1959), German physicist.

[51] Hans Suess (1909–), Austrian–American geochemist.

[52] Paul Harteck (1902–1985), German chemist.

[53] Friedrich Georg (Fritz) Houtermans (1903–1966), German physicist.

[54] Josef Mattauch (1895–1976), German physicist.

[55] Siegfried Flügge (1912–1997), German physicist.

[56] Markus E. Fierz (1912–), Austrian physicist.

[57] Klaus Clusius (1903–1963), German physicist.

[58] Frédéric Joliot (1900–1958).

Raad. Jeg behøver jo ikke at tilføje, hvormeget jeg tænker paa, hvordan det gaar paa Instituttet og hvor inderligt jeg haaber, at alle har det saa godt, som Omstændighederne tillader.

Translation

May 1944

This letter, which must be destroyed immediately, is being sent in a way similar to the messages that I personally received and answered while I was still in Denmark. As before, it concerns the question of whether there are efforts underway on the part of Germany to utilize nuclear energy as a weapon of war. As you know, I have been worried about this ever since Heisenberg's visit to Copenhagen in the autumn of 1941. For a long time we felt reassured, especially through Jensen's[48] communications, but lately rumours about German activity in this area have again come to the fore and any further information might therefore be significant. Over here, it is of particular interest to know whether there is any information at all in Copenhagen about German physicists, which could possibly give us a hint about where they are working and what they are doing. This holds for all German physicists such as, for example, Heisenberg, Weizsäcker,[49] Bothe,[50] Suess,[51] Harteck,[52] Houtermanns,[53] Jensen, Mattauch,[54] Flügge,[55] Fierz[56] and Clusius;[57] naturally, it is not of least interest to know whether there is complete certainty about Jensen's attitude. It would also be of significance if anything is known about the construction of and work with cyclotrons in Germany as well as about work with the Paris cyclotron and the fate of Joliot.[58] This is of course much to ask, but we would be grateful for even the smallest piece of information, however insignificant it might seem. You will best understand yourself with how great care and confidentiality a correspondence such as this must be treated, but in all such respects you will be able to trust the advice of the bearer. Of course, I do not need to add how much I think about how things are going at the Institute, and how sincerely I hope that all are as well as circumstances allow.

ANDERSON TO CAMPBELL, 2 June 1944
[Typewritten]
[Source: CAB 126/39]

2 June 1944

I have had some discussion with Bohr, who will be returning to the United States by direct fast boat leaving on the 9th of June, about his position. In my view the most satisfactory arrangement would be for Bohr to continue in the somewhat anomalous and undefined position which he has occupied during the last few months – that is to say that he should continue to be given access to the various sections of the work in the U.S.A., and to be allowed to talk with both the American and British scientists engaged in the work; and that his personal advice should be available to the U.S. authorities and to myself alike.

I have considered how best to achieve this object. I thought at one time that the best course might be to put to General Groves the direct proposal that the present arrangement should be continued. On reflection, however, I am inclined to think that it would be wiser to write the General a letter which is likely to lead him to this conclusion himself.[58]

I enclose a letter to General Groves which I have accordingly written. I have asked Bohr to get in touch with you on his arrival and would like to leave it to you, when you have seen him, to decide whether or not he should deliver the letter.

Bohr will give you an account of his other activities while he has been in London and I have told him that I hope that he will continue to keep in touch with you and with the Ambassador on this aspect of the matter.

SMUTS TO CHURCHILL, 15 June 1944
[Handwritten]
[Source: PREM, Series 3, folder 139/11A]

TOP SECRET

SOUTH AFRICA HOUSE,

TRAFALGAR SQUARE,

LONDON, W.C.2

15.6.1944.

My dear Prime Minister,

You mentioned to me the agreement between you and the President about the scientific secret and also that you were not willing to disclose it to a third party, and did not hope to obtain more from the President than the existing

---

[58] CAB 126/39 contains two different drafts of a letter from Anderson to Groves, both dated 2 June 1944.

agreement. Anderson and the Prof[59] have also raised the matter with me at your request.

I think you are quite right in not raising the matter with Stalin, at any rate at present. Both the discovery and the physical means of its production are in the possession of the U.S.A. and any move to consult others and to make a disclosure should therefore come from the President. He may have the same feeling as yourself against making any disclosure. In any case you could not do anything without full consultation and agreement between you two.

Of course the discovery is both for war and peace, for destruction and beneficent use, the most important ever made by science. I have discussed its possibilities with the scientific expert who saw you also,[60] and it is clear that it opens a new chapter in human history. Something will have to be done about its control, but exactly what is at present far from clear. While it may be wise to keep the secret to ourselves for the moment, it will not long remain a secret, and its disclosure after the war may start the most destructive competition in the world. It would therefore be advisable for you and the President once more to consider this matter, and especially the question whether Stalin should be taken into the secret. There must of course be the fullest trust and confidence between you as a condition precedent of any such disclosure. But the matter cannot be allowed to drift indefinitely.

If ever there was a matter for international control this is one. And immediately after the war steps should be taken by international agreement and control of the strictest character to regularise the production and use of subatomic energy and the like sources of power. This may be even more important than the control of international aviation. A small committee of foremost physicists should advise the Powers responsible for maintaining world Peace in the carrying out of the policy in the interests of mankind.

You may decide to discuss the matter once more with the President and this note is to remind you of the matter.

Yours sincerely,\
J.C. Smuts[61]

---

[59] Lord Cherwell (1886–1957), also known as Frederick Lindemann.\
[60] Niels Bohr.\
[61] Jan Christian Smuts (1870–1950). See the introduction to Part I, p. [26].

BOHR TO FRANKFURTER, 5 July 1944
[Typewritten transcript]
[Source: BPP 4.1]

COPY.

Washington, July 5th 1944.

Dear Dr. Frankfurter.

Unfortunately my papers have not yet arrived, and due to this as well as to the Washington climate, the completion of the memorandum[62] has been somewhat delayed, what I hope has not caused you inconvenience. In accordance with your advice I have written it for yourself to make what use of you find fit. As you will see, the points are the same as in my previous notes, but I have tried throughout to stress international scientific co-operation as the most natural apology for my interest in the matter. I have omitted some of the more sentimental passages, but left a sentence of such character at the very end. For the case you should think that also this passage should be omitted, I am sending a version of the last page without it. I am very anxious to know whether you think the memorandum is quite suited for its purpose and shall be thankful for any suggestion of alterations in which case I shall at once send it back to you in an improved form. I shall be staying here for another week before I have to go away for some weeks to take part in the work elsewhere, but I have arranged that any message sent to the Legation will reach me rapidly.

With my most heartfelt thanks for all your kindness and sympathy, and with the best wishes for a good summer for your wife and yourself.

Yours ever

(Sign.)   NIELS BOHR

PS: I should like to add that my correspondence with Kapitza has so far not been mentioned to any American authorities.

---

[62] The "Memorandum" is reproduced on p. [101].

BOHR TO FRANKFURTER, 6 July 1944
[Carbon copy]
[Source: BPP 4.1]

Washington, July 6th 1944.

Dear Dr. Frankfurter.

Since I sent the memorandum to you I have had serious anxieties that it may not correspond to your expectations and perhaps not at all be well suited for the purpose. When writing it I felt most strongly that a statement on a matter like this which presents so many sentimental and practical aspects is very difficult unless one knows exactly the background on which it will be received. Of course, I leave everything to your judgment, but I am only eager to emphasize that I am not only ready to change the form and content of the memorandum entirely, in accordance with any criticism and suggestion, but even prepared that you may find it better to wait with any written statement until the situation has further developed and I perhaps through a personal interview have learned on what points an opinion from me might be requested.

Counting on your kind forbearance

Yours ever

NIELS BOHR

FRANKFURTER TO BOHR, 10 July 1944
[Handwritten]
[Source: BPP 4.1]

New Milford, Conn.
July 10th [1944].

Dear Professor Bohr:

The memorandum, you must let me say, is just right – precisely what it should be for its purpose. It has gone on its way, with an appropriate covering note, including information of your continuing stay in Washington.

I cannot forego saying that you are conducting matters of the deepest concern for mankind with a delicacy and wisdom worthy of the enterprise.

I do hope the heat will reluct.

Very sincerely

Felix Frankfurter

P.S. Please send me copy of the memo.

FRANKFURTER TO ROOSEVELT, 10 July 1944
[Handwritten]
[Source: Roosevelt Papers, Roosevelt Presidential Library]
[Editor's comment: Also reproduced in *Roosevelt and Frankfurter: Their Correspondence 1928–1945* (ed. M. Freedman), Little, Brown and Company, Boston 1967, p. 728.]

New Milford, Conn
July 10, 1944

Dear Frank:

After my recent talk with you it occurred to me that, if by chance it should not be possible for you to see Professor Niels Bohr, you might like to have on paper the direction of his thoughts, in so far as these could be put on paper. Here is the memorandum which he wrote exclusively for this use for me. Indeed, as I told you, he has not spoken to a soul about these aspects of the matter to anyone except me on this side of ocean, not even to the Danish Minister, whose guest he is.[63] The memorandum is in his able but quaint English and of course couched in the most abstract language for security reasons. Not even I have made a copy of this memorandum.

Since leaving Washington I have learned that Professor Bohr's stay here has been extended. He will be in Washington through Saturday, July 15.

But I hope that you won't have to stay in that hell's hole. Holmes[64] used to speak of the "peculiar" heat of Washington. It sure is "peculiar". From all that

---

[63] Henrik Kauffmann (1888–1963) was appointed Danish ambassador to Washington in 1939 and was accepted as such throughout the war by the U.S. even though the Danish government felt compelled to announce his dismissal in 1941. During the war he became a close friend of Bohr, whose views he supported for the rest of his life.

[64] Oliver Wendell Holmes, Jr. (1841–1934), who served at the United States Supreme Court from 1902 to 1932, was an example for Frankfurter, who was appointed to the court in 1939.

one can gather from the press, the visit of Jean D'Arc to Clemenceau came off well.[65] And what grand news from the fronts!

Take care of yourself! Marilyn and I send on affectionate greetings

Sincerely

Felix Frankfurter

BOHR TO ANDERSON, 13 July 1944
[Typewritten]
[Source: CAB 126/39]

<u>TOP SECRET.</u>   <u>PERSONAL.</u>   Washington, July 13th 1944.

The Rt. Hon. Sir John Anderson M.P.
Chancellor of the Exchequer
Westminster, London.

Dear Sir John.

I wish to thank you most heartily for all your kindness to me during my visit to London, and I need hardly say how much it is upon my mind to prove worthy of the confidence you have showed me.

As you will know, I have had to stay in Washington somewhat longer than anticipated in order to await the arrival here of Sir Ronald Campbell and Dr. Chadwick. As soon, however, as I came to Washington, I had a most pleasant and encouraging talk with General Groves about recent developments of the project. Yesterday I again saw the General in order to make final arrangements for my visit to "Y"[66] to where I am proceeding to-morrow. On this occasion I delivered your letter which in the mean time I had received from Sir Ronald and for which the General was very thankful.

A few days after my arrival in Washington I also had the opportunity to

---

[65] The reference is to General Charles de Gaulle's visit to Washington from 6 July to gain U.S. support for his Free French Forces fighting the German occupation of France.
[66] Los Alamos National Laboratory.

[250]

reestablish my connection with the President's friend[46] who in a subsequent conversation asked me confidentially to inform you that he, learning about the reception of the message in London, had given the President a full account of his own endeavours in the matter, which the President not only heartily approved of, but also welcomed as a fortunate development.

On the request of the President's friend I have attempted, for his personal orientation and for what use he may himself find appropriate, in a memorandum to present the various aspects of the matter. This memorandum, of which I enclose a copy, contains the same points which I have had the privilege to discuss with you. The repeated allusions to international scientific co-operation are, as you will understand, primarily meant as an apology for the interest of a scientist in such matters. In this connection I have also made a reference to the invitation from Dr. Kapitza, but in an accompanying letter I have called attention to the fact that the correspondence with Kapitza has so far not been mentioned to any other American authority.

Before Lord Halifax left Washington I was given the opportunity to report about my journey to London to the Ambassador who was deeply interested in the development, and later I have had the privilege to discuss the whole situation thoroughly with Sir Ronald Campbell who is kindly forwarding this letter.

Yours sincerely
Niels Bohr

BOHR TO FRANKFURTER, 14 July 1944
[Carbon copy]
[Source: BPP 4.1]

Washington, July 14th 1944.

Dear Dr. Frankfurter.

I was most thankful for your kind letter and am happy that you think the memorandum may be useful. I have just sent a copy of it to Anderson to whom I have confidentially conveyed such information as you advised me. I will have to leave Washington already to-day, but as soon as I, after a few weeks absence, return from my journey I shall get in touch with you and send you a copy of the papers which unfortunately have not yet arrived.

With the kindest regards and best wishes

Yours ever
[Niels Bohr]

P.S. I enclose a copy of the memorandum

ANDERSON TO BOHR, 21 July 1944
[Typewritten]
[Source: BPP 2.2 and CAB 126/39]

TOP SECRET

21st July, 1944.

Dear Dr. Bohr,

I was very pleased to receive your letter of 13th July forwarded to me by Sir Ronald Campbell. I think it very satisfactory indeed that the President has been informed of the general course of your activities and has reacted favourably. On this side, too, there is a satisfactory development to report, for the Prime Minister now fully recognises that it would be appropriate that he should raise the long-term problem personally with the President when they next meet, which I personally hope will be before very long. I shall naturally keep in the closest possible touch myself with any developments.

I was much interested in reading the notes enclosed with your letter, which seem to me, if I may say so, to present the whole matter in a very clear and compact form suitable for the enlightenment of a non-scientific reader.

I have already had a good talk with Lord Halifax and hope to see him again before he returns to Washington.

With very kind regards, I am,

Yours sincerely,

John Anderson

Dr. Niels Bohr.

CAMPBELL TO ANDERSON, 25 August 1944
[Typewritten]
[Source: CAB 126/139]

TOP SECRET CYPHER TELEGRAM
RECEIVED BY TELEKRYPTON.

IZ 6268
TOO 252345Z
TOR 260030Z

IMPORTANT

FROM:– J.S.M. Washington
TO:– A.M.S.S.O.

ANCAM 99   25th August 1944.

Following Private and Personal for Sir John Anderson from R.I. Campbell.

The Ambassador has seen Bohr. B. was very shortly to see his contact[46] to whom he was going to suggest that he should present following considerations to his highly-placed friend:[67] –

(1) We must assume that Russians knew that great efforts were being made in the United States to reach practical solutions in this field.

(2) Russians themselves were studying matter and would be free to develop full effort at end of German war.

(3) Moreover Russians would be likely either to obtain from the Germans or be offered all their secrets by the Germans in the hope of making mischief between Russians and Anglo-Americans.

(4) It was therefore all the more urgent to decide line to pursue in respect of Russians.

---

[67] President Roosevelt.

(5) If we said nothing we would (A) arouse their suspicions and therefore create greater risk of competition, (B) lose prospect of using an approach to them for purpose of establishing the confidence on their side on which could be based action for beneficial use of our knowledge and achievement in order to create confidence among all nations. (You are familiar with B's reasoning).

(6) It was not necessary to start off by giving Russians detailed information of our activities. We should say to them that as they know everybody was working along these lines and ask them what they thought should be done if anybody reached solutions. If they responded in co-operative spirit way would be open for frank discussion. If not, we should know where we were and have to consider other means of dealing with the thing.

2. B's contact is away and Ambassador will not as you will remember he hoped have any natural opportunity of seeing him but will have immediate report from Bohr of his conversation and any sequel it may have.

3. Ambassador thought it would be useful for you to know the foregoing. He will report after seeing B. again.

T.O.O. 252345Z

HALIFAX TO ANDERSON, 27 August 1944
[Typewritten]
[Source: CAB 126/39 and BPP 2.7]

TOP SECRET CYPHER TELEGRAM.
RECEIVED BY TELEKRYPTON.

IZ6316
TOO 271745Z
TOR 271852Z

IMMEDIATE

FROM:– J.S.M.
TO:– A.M.S.S.O.

ANCAM 100   August 27th 1944.

Private and Personal for Sir J. Anderson from Lord Halifax.

1. Campbell and I saw B. on August 26 and he asked us to send you the

following:

2. BEGINS:

"B's contact has forwarded B's memorandum to his highly placed friend, who expressed wish to see B. The interview took place on August 26 and had a most satisfactory course. In particular B had opportunity thoroughly to discuss the views he had explained to Ambassador at earlier interview (see ANCAM 99) and as far as B understood these views were fully appreciated. Above all, the highly placed friend agreed that nothing could be lost, but very much gained by an early approach to the S[oviet] U[nion] and said that he would raise the question when meeting his British colleague in the near future. (H.P.F.[68] had told B. of place and approximate date). The H.P.F. asked whether you would attend the meeting yourself. If not so, perhaps Gorell Barnes[69] could come. Ambassador discussed with B. whether B. could go to London immediately and report to you about the conversation as he was authorized by H.P.F. to do, but it will hardly be possible to arrange the necessary formalities with the American organization in due time. If you are coming it would, of course, be quite natural if you asked B. to come up and meet you, but if Gorell Barnes came he could perhaps come here on the way since in that case it might be more doubtful for B. to ask permission to leave this country. In the mean time B. intends to visit his contact who is on vacation and again discuss the whole matter in order to find out whether any further step on this side before the meeting may be useful."

ENDS.

3. Campbell and I got impression from B's report that H.P.F. would be glad to see you with P.M. I cannot however make sure judgement on how far this impression, if correct, would represent H.P.F.'s considered mind, having regard to stage II discussions procedure (see my tel No. 4478).[70]

---

[68] Highly Placed Friend, i.e., President Roosevelt.

[69] John Anderson's assistant.

[70] "Tel No. 4478" of 20 August 1944 is in PREM 4/75/2, part 1, folio 341. The "stage II discussions procedure" seems to refer to Roosevelt's disinclination to discuss financial matters at the Quebec Meeting and, as a consequence of this, the decision not to include Henry Morgenthau (1891–1967) (who served as Treasury Secretary from 1934 to 1945) in the US delegation. For this reason, the British thought it inappropriate to include Anderson in his capacity as Chancellor of the Exchequer in the British delegation. As noted in the introduction to Part I (p. [35]), Anderson was subsequently replaced by Lord Cherwell.

4. Unless P.M. has already asked direct question, or feels inclined to do so, and if you and he feel it desirable for you to come, I should advise that P.M. should tell H.P.F. so, and leave it to him to state his difficulties on stage II grounds if he feels them.

T.O.O. 271745Z

ANDERSON TO HALIFAX, 1 September 1944
[Typewritten]
[Source: CAB 126/39]

TOP SECRET CYPHER TELEGRAM
DESPATCHED BY TELEKRYPTON

OZ 4919
TOO 011550Z
TOD 011710Z

IMMEDIATE
FROM:– A.M.S.S.O.
TO:– J.S.M. WASHINGTON

GUARD   4919   1st September, 1944
   Following private and personal for Lord Halifax and Sir R.I. Campbell from Sir John Anderson.
   I am very glad to have the information in your telegrams numbers 99 and 100.
   2 As you know it has now been decided that I should not go and that the Prime Minister should be accompanied by Cherwell.
   3 I have put Cherwell fully into the picture. He and I both think that it is essential to leave it to H.P.F.[68] to take the initiative on this aspect of T[ube] A[lloys]; and it seems from your telegrams that he is intending to do so. It would we think be dangerous from all points of view for B to make any attempt to come to place of meeting. And, as you have so clearly explained in your

telegrams the position at your end, we do not think any useful purpose would be served by Gorell Barnes or anyone else going to Washington.

4 No one here except Cherwell, Gorell Barnes and myself knows the contents of your telegrams.

T.O.O. 011550Z

FRANKFURTER TO ROOSEVELT, 8 September 1944
[Handwritten]
[Source: *Roosevelt and Frankfurter: Their Correspondence 1928–1945* (ed. M. Freedman), Little, Brown and Company, Boston 1967, p. 735]

Washington, D.C., September 8, 1944

Dear Frank:

Here is a letter from my Danish friend.[71]

From many long talks with him I gather that there are three solid reasons for believing that knowledge of the pursuit of our project can hardly be kept from Russia:

(1) they have very eminent scientists, particularly Peter Kapitza, entirely familiar through past experience with these problems.

(2) some leakage, even if not of results and methods, must inevitably have trickled to Russia.

(3) Germans have been similarly busy, and knowledge of their endeavors will soon be open to the Russians. Therefore, to open the subject with Russia, without of course making essential disclosures before effective safeguards and sanctions have been secured and assured, would not be giving them anything they do not already – or soon will – substantially have.

In a word, the argument is that appropriate candor would risk very little. Withholding, on the other hand, might have grave consequences. There may be answers to these considerations. I venture to believe, having thought a good deal about it, that in any event these questions are very serious.

My very best wishes for successful days in the talks immediately ahead.

Affectionately yours,
F.F.

---

[71] Bohr to Roosevelt, 7 September 1944, reproduced on p. [109].

BOHR TO ANDERSON, 12 September 1944
[Typewritten]
[Source: CAB 126/39 and BPP 2.2]

TOP SECRET.   Washington, September 12th 1944.

The Rt. Hon. Sir John Anderson, M.P.
Chancellor of the Exchequer
Westminster, London.

Dear Sir John.

I wish to thank you most heartily for your kind letter of July 21st which I received through Sir Ronald Campbell during my last stay in "Y", and which was a very great encouragement to me. I was also glad to learn that you found that the notes which I had written on the request of the President's friend gave a reasonable representation of the matter.

As you know from Sir Ronald Campbell's telegram, the friend thought it fit to forward the notes to the President on whose wish to see me an interview was arranged two weeks ago in a manner avoiding all publicity. The President who received me alone talked quite frankly with me about the whole situation and most kindly gave me opportunity to stress the arguments stated in the notes.

As far as I understood the President fully appreciated these arguments and entirely agreed as to the urgency of the matter. As you know, the President told me that at the meeting with the Prime Minister he would take up the question of an early initiative. At the end of the interview which was, I think, in every respect most satisfactory the President said that he would like to see me again when he returned from the meeting with the Prime Minister.

Since that time I have had various talks with Lord Halifax and Sir Ronald Campbell as well as with the President's friend on whose advice I wrote a letter to the President which was delivered through the friend on the very eve of the President's departure for Quebec.

The purpose of the letter, of which I enclose a copy, was to remind the President of the main points of the conversation, and the wording of various passages corresponds closely to the way in which the President expressed himself about the political situation.

As regards my plans for the immediate future, Sir Ronald Campbell has agreed that while I await a call from the President it would hardly be advisable to go up to "Y". In the mean time I hope to complete a report on my impressions of the technical aspects of the project which General Groves has requested me to write.

From Dr. Chadwick you will have had a full account of the present state of the work, and I can only add that from my personal experiences I have received the strongest impression of the vigour and success with which all aspects of the work at "Y" are being pursued. From numerous discussions with American colleagues I can also reaffirm how greatly the valuable contributions of the British collaborators are appreciated.

Yours sincerely
Niels Bohr

AIDE-MÉMOIRE, 18 September 1944
[Typewritten]
[Source: CAB 127/201]
[Editor's comment: Also reproduced in M. Gowing, *Britain and Atomic Energy 1939–1945*, Macmillan, London 1964, p. 447.]

TOP SECRET

TUBE ALLOYS

Aide-memoire of conversation between the President and the Prime Minister at Hyde Park, September 18, 1944.

1. The suggestion that the world should be informed regarding Tube Alloys, with a view to an international agreement regarding its control and use, is not accepted. The matter should continue to be regarded as of the utmost secrecy; but when a "bomb" is finally available, it might perhaps, after mature consideration, be used against the Japanese, who should be warned that this bombardment will be repeated until they surrender.

2. Full collaboration between the United States and the British Government in developing Tube Alloys for military and commercial purposes should continue after the defeat of Japan unless and until terminated by joint agreement.

3. Enquiries should be made regarding the activities of Professor Bohr and steps taken to ensure that he is responsible for no leakage of information, particularly to the Russians.

CHURCHILL TO CHERWELL, 20 September 1944
[Typewritten with handwritten corrections and initials]
[Source: CAB 126/39]

10, Downing Street,
Whitehall.

PRIME MINISTER TO LORD HALIFAX,
BRITISH EMBASSY,
WASHINGTON.

Following for Lord Cherwell.
Personal, Most Secret, Eyes Only.

Begins.   The President and I are much worried about Professor Bohr. How
did he come into this business? He is a great advocate of publicity. He made an
unauthorized disclosure to Chief Justice Frankfurter who startled the President
by telling him he knew all the details. He says he is in close correspondence
with a Russian professor, an old friend of his in Russia, to whom he has written
about the matter and may be writing still. The Russian professor has urged him
to go to Russia in order to discuss matters. What is all this about? It seems to
me Bohr ought to be confined or at any rate made to see that he is very near
the edge of mortal crimes. I had not visualized any of this before, though I did
not like the man when you showed him to me, with his hair all over his head,
at Downing Street. Let me have by return your views about this man.
I do not like it at all.

WC

20 September, 1944.

CHERWELL TO CHURCHILL, 23 September 1944
[Typewritten]
[Source: CAB 126/39]
[Editor's comment: Handwritten annotation by Anderson on top of document: "I entirely agree".]

TOP SECRET

BRITISH EMBASSY,
WASHINGTON, D.C.

PRIME MINISTER   23 September 1944

T.A. Your minute of September 20, 1944

Professor Bohr, a Danish subject with I believe a Jewish mother, is probably the world's greatest authority on the theoretical scientific side. When the Germans took over the government at Copenhagen we assisted in smuggling him out through Sweden to England. At the request of the American Government he was brought over here to survey the work being done and advise both governments. I understand his help on scientific matters has been appreciated.

Some months ago he was asked to call at the Russian Embassy in London for a letter sent to him by an old fellow pupil under Rutherford, Professor Kapitza, who as you may remember visited Russia some years ago and has been held there ever since. This message reached him here and on his next visit to England, with Anderson's agreement, he went to the Russian Embassy to collect it. The letter merely contained a warm invitation to Russia with the offer of all facilities to continue physical research, but Bohr formed the impression that the Russians were not entirely disinterested in their offer to promote pure physics. He therefore concerted a reply with our people politely declining the offer. I should be very much surprised if he had any other communication of any sort with Kapitza.

I think there is good contemporary evidence at the Embassy here that Frankfurter broached the matter to Bohr (and not Bohr to Frankfurter) displaying knowledge that work was going on, that a weapon of immense power might one day be produced and that this was the reason for Bohr's presence in this country. Bohr assured Halifax and Campbell that he did not ever reveal to Frankfurter any scientific or technical details of the processes or of the work going on here – nor do I think Frankfurter would be capable of understanding them or interested in this aspect.

[261]

As you know Bohr like many other people had some rather woolly ideas about using the existence of a super weapon to induce the nations to live in confidence and at peace. It is this aspect which Frankfurter raised with him and which they discussed and it is this which I believe Frankfurter may have raised with the President.

I have always found Bohr most discreet and conscious of his obligations to England to which he owes a great deal, and only the very strongest evidence would induce me to believe that he had done anything improper in this matter.

I do not know whether you realize that the possibilities of a super weapon on T.A. lines have been publicly discussed for at least six or seven years and indeed I should be surprised if all of the earlier patents, dating from before the war, were secret ones. The things that matter are which processes are proving successful, what the main snags are and what stage has been reached. Most of the rest is published every silly season in most newspapers.

Since drafting this I have had an hour's interview with the President, Leahy[72] and Bush when this question was thoroughly discussed. Nothing emerged which could cause me to alter any part of it. Indeed it was confirmed by Bush in all particulars of which he could be expected to have knowledge. Bush undertook however to "check up" on Bohr and I will let you know if any further developments ensue.

Cherwell.

ANDERSON TO CHERWELL, 29 September 1944
[Typewritten]
[Source: CAB 126/39]

TOP SECRET

From London, 29 September 1944

Following for Lord Cherwell from the Chancellor of the Exchequer. Personal and Private. Top Secret. EYES ONLY.

Since replying to your message about Baker,[73] I have seen aide memoire

---

[72] William Daniel Leahy (1875–1959), Roosevelt's personal Chief of Staff during World War II.
[73] Nicholas Baker, Niels Bohr's codename in America during the war.

initialled at Hyde Park on 18th September and Prime Minister's message of 20th September.

2. You will remember that correspondence between Baker and Kapitza last April was started by latter, that Baker's reply was vetted by C's[74] people and by myself and that T.A. is not so much as mentioned in it.

3. Halifax and Campbell have full knowledge of what, so far as we are aware, took place between Baker, his contact and his contact's highly placed friend. Accounts I have received do not bear out at all Prime Minister's impression as conveyed in his message to you.

4. Beginning of paragraph 1 of aide memoire of 18th September and relevant remarks in Prime Minister's message to you are entirely at variance with what we have always understood to be Baker's ideas, and there is no doubt about the reality of the danger which preoccupies him.

5. I am sure you will agree that, in fairness to Baker, Prime Minister should receive, without delay, a correct report of his activities and views so far as they are known to us. I think it would be much better if such a report were to come from you and I hope you will send one.

6. Baker was brought out of Denmark and introduced into T.A. with my approval and in agreement with the Americans. I am prepared to accept full responsibility for this vis-a-vis the Prime Minister. I am sure it was right and, during his present visit, Chadwick has confirmed the great value of his advice and judgment on the technical side.

7. Paragraph 2 of aide memoire of 18th September is most satisfactory.

8. Please let me know what you propose to do about Baker.

---

[74] "C" was the head of the British Secret Intelligence Service. See the introduction to Part I, p. [25].

MØLLER TO BOHR, 29 September 1944
[Typewritten]
[Source: BPP 3.2]

29th September, 1944.

Dear Sir,

Herewith reply to your letter which I sent to Professor Moeller in July:[75]
"Angaaende de Personer, De omtalte i Deres Brev, er det stadig min Overbevisning, at Je. og S. er helt paalidelige og jeg vil tro, at deres Fremstilling
af Udsigterne for en Anvendelse af de omdiskuterede Metoder i det Store
og Hele er korrekt. H. og W. synes i det sidste Aar ikke at have beskaeftiget
sig med Spørgsmaalet, idet jeg positivt ved, at de har vaeret optaget af helt
andre Ting. Dog maa jeg nu tilføje, at et Brev, som jeg sendte til H. i
Midten af Juli indtil Dato ikke er blevet besvaret, til Trods for at Brevet
indeholdt en Del Spørgsmaal, som jeg havde ventet vilde blive besvaret
hurtigt. Det forlød i Foraaret at B. stadig var i H., hvor han ogsaa havde
saadanne Hjaelpemidler til sin Raadighed, som De omtalte i Deres Brev.
Om de andre Personer har jeg ingen Oplysninger, hvis jeg senere skulde
modtage saadanne, skal jeg straks sende dem. Vi glaeder os meget til at se
Dem igen og venter at det nu ikke kan vare saa laenge."

I believe Je. is Jensen,[48] S. Süss,[51] H. Heisenberg, W. Weiszäcker,[49]
B. Bothe[50] and the place H. to be Heidelberg.

Please send me a telegram on receipt of this letter confirming that you agree
with the above interpretation of the names.

I hope you are well. Give my kind regards to Jim.[76]

Yours respectfully,
Eric Welsh

Translation

29th September, 1944.

Dear Sir,

Herewith reply to your letter which I sent to Professor Moeller in July:[75]

---

[75] Bohr to Møller, May 1944, reproduced on p. [243].
[76] James Baker, Aage Bohr's codename.

"Concerning the persons whom you mentioned in your letter, it is still my conviction, that Je. and S. are completely reliable and I would think that their description of the prospects for an application of the methods in question on the whole is correct. H. and W. do not seem to have worked with the question during the last year, as I know for certain that they have been preoccupied with completely different matters. I must now add, however, that a letter, which I sent to H. in the middle of July, has so far not been answered, although the letter contained some questions that I had expected would be answered quickly. It was reported in the spring that B. was still in H., where he also had such facilities available to him, as you mentioned in your letter. I do not have any information about the other persons; if I am later to receive such information, I will send it to you immediately. We much look forward to seeing you again and expect that now it will not be so long."

I believe Je. is Jensen,[48] S. Suess,[51] H. Heisenberg, W. Weizsäcker,[49] B. Bothe[50] and the place H. to be Heidelberg.

Please send me a telegram on receipt of this letter confirming that you agree with the above interpretation of the names.

I hope you are well. Give my kind regards to Jim.[76]

Yours respectfully,
Eric Welsh

BOHR MEMO, 6 October 1944
[Typewritten]
[Source: BPP 6.5]

October 6th 1944.

The whole question of a lack of discretion on my part must rest on a misunderstanding from the side of the American organization, which has arisen from my obligation to keep my activities in the delicate matters, reserved for the British and American Governments, completely apart from my participation in the technical work on the project.

In fact, I have taken utmost care to be discrete about my confidential connections, and when the Officials of the American organization learned about my activities, in ways I do not know, they may have got the wrong impression that

I had violated Security regulations by talking about the project with persons whom they thought had no legitimate access to such information.

FRANKFURTER TO BOHR, 14 October 1944
[Handwritten]
[Source: BPP 4.1]

9018 Dunbarlou Ave,
October 14/[19]44

Supreme Court of the United States
Washington, D.C.

Dear Professor Bohr:

Your visit the other day left me with real joy. I was much relieved that needless anxiety was lifted from you. Of course there could not have been any real occasion for concern. Everything you did and said has been done and said with the most fastidious regard for truth and duty. And yet, a disagreeable episode had arisen, and I am very happy all the clouds have cleared away.

You must let me say that you symbolize for me the finest traditions and aspirations of science – science as the pursuit of truth and the true promoter of humanity.

With all good wishes.

Truly sincerely yours,
Felix Frankfurter.

BOHR TO ANDERSON, 17 October 1944
[Typewritten]
[Source: CAB 126/39 and BPP 2.2]

TOP SECRET.   PERSONAL.   Washington, October 17th 1944.

The Rt. Hon. Sir John Anderson, M.P.
Chancellor of the Exchequer.
Westminster, London.

Dear Sir John.

As you know I have in the last weeks been staying here in Washington awaiting a call from the President. A few days ago, however, I got a message that the President, due to a journey, would be unable to receive me, but that he suggested that I saw Dr. Bush with whom he had talked about the matter.

The conversation with Dr. Bush was most pleasant and enlightening, and we discussed at some length the various aspects of the long-term problems. He expressed the wish to see me again on a later occasion when he had considered the arguments more closely.

During the conversation I also had opportunity to explain my relations with the President's friend, about which Dr. Bush had learned from the President himself. Yesterday I had a conversation about this point with General Groves and Dr. Chadwick, and I believe the misunderstandings have now been cleared up.

After arrangement with Dr. Chadwick I am to-day proceeding to "Y" for consultations about recent developments. I was thankful before leaving to have the privilege of a conversation with Lord Halifax about the whole situation, and the Ambassador has kindly offered to forward this letter.

Yours sincerely
Niels Bohr

BOHR REPORT, 30 December 1944
[Handwritten by Aage Bohr]
[Source: BPP 6.7]

30th Dec. [19]44

Report on misunderstandings arising at meeting between
Pr[esident] and P[rime] M[inister]

While waiting in Washington for a message from Pr. – who had expressed the wish to see B[ohr] again when he returned from Q[uebec] – B learned from Chadw[ick] on his return from London that the Chanc[ellor] immediately before his departure through G[orell] B[arnes] had let him know that complaints about B's talking to unauthorized persons had been expressed during the meeting and that the Chanc. believed it to be a misunderstanding which he was anxious to have completely cleared up.

Chadw. was under the impression that the complaints referred to B's conv[ersation] with the Pres's friend whereupon B assured him that the friend had not only told him without his asking, but even in the message to the Chanc. which he had himself composed definitely stated that he was informed about the project from legitimate sources.

The next day Chadw. told B that G[eneral] G[roves] in a conversation with him, which he had just had, had told that the complaints about B's connection with Dr. F[rankfurter] originated from the Pres. himself.

B felt it was his duty at once to inform F [of] the incident and in the conversation the same evening F not only confirmed his previous statements, but assured B that it must have been a misunderstanding that the Pres. should have uttered such complaints since all their connections had been most scrupulously correct and discrete.

A few days later B received a message from the Pres. deploring that he, due to a journey, could not receive him, but suggested that he saw Dr. Bush with whom he had talked about the matter discussed during the interview.

In the subsequent conversation with Dr. Bush of which the main topic was the general implications of the project also B's connection with Dr. F was discussed, and Dr. Bush expressed great relief in learning that B and F were closely acquainted due to common interest in international cultural co-operation already before the war and that although they had met several times after B['s] arrival to America, the project had, of course, never been mentioned before F without being asked told B that he was informed about it through legitimate sources and was much concerned about its implications.

B further stressed that he had felt it an obligation most strictly to separate between the actual technical work and its political implications which he understood was a question for the statesmen alone. This was the reason that he had not mentioned any such matters to anyone connected with the organizations, like Dr. Bush himself, but when the Pres. showed him the confidence to receive him he felt it, of course, his duty to talk about the ideas he had formed as to how the great enterprise might be turned to lasting benefit for the common cause.

After Dr. Chadw.'[s] return from a journey to Y together with Lord Cherwell, B and Chadw. had a conversation with Gen. Groves about B's connection with Dr. F, in which B told about his conversations with Dr. Bush as regards this point and further added that due to the obligation he had undertaken to avoid all publicity as regards his presence in America he had seriously considered to inform the Security Org[anization] about the interview with the Pres. in case it should wish to take precautions to avoid publicity, but feeling that it was a matter for the Pres. alone to decide who should know about the interview he had instead informed the Pres.'s Secretary about his obligations with the result that the Pres. had arranged that the interview should be off the record. Gen. Groves stressed that his only interest in the matter was the question of security and as far as B could understand the Gen. felt himself satisfied as regards this point.

BOHR TO ANDERSON, 18 January 1945
[Typewritten]
[Source: CAB 126/40 and BPP 2.2]

TOP SECRET   PERSONAL

Washington, January 18th 1945

The Rt. Hon. Sir John Anderson, M.P.
Chancellor of the Exchequer.
Westminster, London.

Dear Sir John.

I wish to thank you most heartily for your kind letter of November 10th which was a great encouragement to me.

On my return to Washington after a two months stay in "Y", I had some conversations with the Ambassador and later with Sir Ronald Campbell about recent developments and I suggested that it might perhaps be useful that I returned to London for a visit to report to you and at the same time to discuss technical and personal problems with your advisers.

A week ago I learned from Sir Ronald who in the mean time had communicated with you that you do not think the moment propitious for such a visit. I am glad, however, to have had the opportunity of some further talks with Sir Ronald who, as I understand, will return to London in the very near future. I am also grateful that it has been possible to arrange that my son goes to London to discuss on my behalf various urgent problems about which I have corresponded with Cdr. Welsh.

In a few days I am returning to "Y" where the work, due to the rapidly increasing amounts of active materials available, has entered into a most interesting phase. It is in fact now possible to perform various kinds of so-called integral tests which already have gone so far in verifying the conclusions drawn from theoretical arguments based on measurements of differential type.

The study of the most efficient methods of assembly are also at the moment being pursued with great vigour. On my last visit I was not least interested in discussions of devices in which the specific nuclear properties of materials are used in a manner analogous to autocatalysis in ordinary chemistry. Especially in case the methods relying on a high initial speed of assembly should not have been fully developed when sufficient materials are at hand, devices of this kind may perhaps be helpful to avoid delay in the completion of the weapon.

I feel it a privilege to be given the opportunity to participate in the unique enterprise at the moment where it is acquiring such actuality, but I need not say that I hold myself ready whenever you think that I can in any way be of assistance in the endeavours concerning which you have shown me a confidence which I so deeply appreciate.

Yours sincerely
Niels Bohr

HALIFAX TO ANDERSON, 18 January 1945
[Typewritten]
[Source: BPP 2.7 and CAB 126/40]
[Editor's comment: The version in the PRO, in which item IV has apparently been cut out, carries the following handwritten comments, respectively by Gorell Barnes and John Anderson: "This arrived without covering note or explanation in an envelope addressed to me by Sir R. Campbell." "Better ask R. Campbell when he arrives."]

TOP SECRET.

Washington, January 18th 1945.

I)   In the memorandum of July 3rd 1944[62] which was prepared for the orientation of the President and of which a copy was simultaneously sent to the Chancellor, certain general arguments regarding the implications of the project have been presented from a scientist's point of view. Whence the repeated references to international scientific co-operation and the very general terms in which practical political matters have been alluded to. The short comments,[77] however, attempt to answer somewhat more specific questions raised in conversations at the Embassy in Washington.

II)   Since B's conversation with Dr. Bush, reported in letter to the Chancellor of October 17th,[78] he has not found it advisable, without directives from the Chancellor or further initiative from the American side, to continue discussions on the matter with members of the American organisation.

III) B has been concerned that the misunderstandings which arose at the meeting between the Prime Minister and the President should have caused inconveniences to the Chancellor. He is not entirely clear as to the character and origin of these unfortunate misunderstandings, but during the conversations at the Embassy he has in detail reported about the development of the whole matter as far as it is known to him.

---

[77] Bohr's "Comments" are reproduced on p. [111].
[78] Bohr to Anderson, 17 October 1944, reproduced on p. [267].

IV) Confidentially B has informed the Ambassador that he has been approached by various scientists deeply concerned about developments, but has strived to prevent any actions which might complicate the situation.

BOHR NOTES, 7 March 1945
[Handwritten by Aage Bohr]
[Source: BPP 2.6]

March 7th [1945].

Points for conversation with Chancellor on March 8th 3p.m.

(1. Problem of Chanc[ellor]'s time.)

2. Thankful for invitation to report about experiences and impressions.

3. Favourable development of technical work.

     a) Solid implosion.

     b) Test of compression

     c) Development of timed neutron sources

     d) Conversation with Gen. Groves before departure.

4. Developments in general problems since last visit:

     a) Report to Lord H[alifax] and P[resident's] F[riend][46] on reception of message in London.

     b) PF's report to the Pres about his activities and his entrusting of message to B[ohr].

c) Pres.'s welcome of information as favourable development.

d) B's preparation of memorandum on PF's request.

Journey to Y.

After return:

e) Conversations with Lord H (and Sir R[onald] C[ampbell]) after Lord H's return from Europe where meeting the P[rime] M[inister]. B's knowledge through Sir RC of Ambassadors reactions and report to London.

f) Pres.'s wish to see B; interview arranged after consultations with Sir R.C.

g) <u>Interview with the Pres.</u>: Detailed account.

Pres.'s utterances about P.M. and Chanc.

Pres.'s views on international co-operation on project; in part. with the USSR.

Pres.'s hope of understanding and his appreciation that internat[ional] col[laboration] in science might be helpful.

B's showing of Kapitza's letter.

Pres.'s declaration of his intentions to propose the desired procedure to P.M. at coming meeting.

Pres.'s wish to see B again after the meeting.

h) Immediate report of interview to Lord H and Sir RC; suggested message to Chanc.

i) B's letter to Pres. after consultations with PF and Lord H.

[j) Pres.'s expression about interview to common, but unknowing friend.[46]][79]

---

[79] Brackets in the original.

5. Account of <u>complications</u> as far as known to B.

a) First information through Chanc. via Gor[ell] Barnes and Chadwick on latters return to Wash.

b) B's astonishment of complaints regarding connection with PF, who himself in written message had declared his legitimate access to information, before the subject was mentioned by B.

PF's renewed assurance about this point after B had informed him about complaints.

c) Chadwick's and B's <u>conversation with Gen. Groves.</u>

B's insistence on separation between work and politics; explained duty to inform Pres about general views when favoured with interview.

No mentioning, of course, in this or other conversations with American officials of B's connection with the Brit. Gov., and all steps explained by interest of scientist, taken into confidence.

Gen. Groves declaration that his only interest in the matter was concerned with security; his anxiety for the fate of any document reaching the White House.

d) B's subsequent conversation about this point with Chadw., who had been able to declare that he had no knowledge of interview.

Chadw.'s assurance that memorandum from Groves' point (military secrets) was most adequate, his agreement that it nevertheless was correct of B not to show it to Groves and that B had been right in declaring that it had to be left to the Pres. himself whom he would inform about matters discussed in his interviews.

e) B's subsequent talk with Lord H to whom he reported about the matter and discussed relations with PF.

f) Just before B left Wash. for Y, he received a personal letter from PF, in which he expressed his desire to give utterance for his feelings about incident.

g) After return to "Y", B learned from Chadw. who had been travelling with Lord Cherwell that the question was not a complaint about B's connections with PF, but that the main question (issue) had been a blame from the side of the P.M. of B's connections with S[oviet] U[nion]

6. Conversation with Bush

A few days after Chadw.'s return from Engl. with the information about the complications at the Quebec-meeting, B received a message from the Pres. that, due to a journey he would not be able to receive B as he had desired, but suggested that B instead saw Dr. Bush with whom he had talked about the matter.

The interview was arranged immediately, (before convers. with Gen. Groves).

a) As a first point B explained about connection with PF and Bush expressed satisfaction to learn that B had known PF before the war (cult[ural] co-op[eration]) and had seen PF several times in Wash before PF, without B's asking told that he from legitimate sources was informed, and deeply concerned.

b) B understood that Bush was informed about interview with Pres., memorandum and corresp. with Kapitza.

c) Bush assured he quite shared hopes that project, in spite of dangers, would essentially facilitate peaceful collab. between nations. Explained promising experiences with Sov. scient., as well from Bush's journeys to S.U. before war as from visits of Rus. scient. to U.S.A. during war.

He said, however, that his views differed somewhat as reg[ards] question of timing. Thought time not ripe for consultations with Rus.

d) B answered that even if time not come for final arrang[ement] or exchange of information, he thought, that when act[ing] the Governments had views as Bush expl[ained], there could be no disadvantage or danger, but much gained, by assurance of intentions, what would invite confidence and relieve disquietude.

e) As far as B understood, Bush was impressed by argument, to which he offered no objections.

On the contrary, Bush expressed wish to see B and for written statement for deliberations of Bush and committee.

f) B immediately prepared letter to Bush.[80]

expressed pleasure for conversation and from learning Bush's attitude and happy exper[ience] with scient[ific] col[laboration].

Next briefly outlined views expressed and conviction of fav[ourable] consequ[ences] of approach.

g) On account of tense sit[uation] B showed letter to Lord Cherw[ell]. who expressed doubt as to consequences.

B decided to wait for directives from Chanc.

h) As the matter stands, it will be quite natural to take up again the matter with Dr. Bush.

7) <u>Anxiety in scient. circles.</u>

a) From many American scientists received impression of extreme disquietude.

b) keeping strictly the line of separating between work and implications; has, however, strived to discourage steps which might complicate matters for responsible statesmen. Without disclosing information and by expressing general hope, such endeavours have hitherto succeeded.

c) In strictest conf[idence], B may use the opport[unity] to inform Chanc. of letter from eminent scientist, no more direct in work, but intimately connected with matter from earliest stages.[81]

Letter at first great embarrassment to B.

Advice of PF, who is himself close friend of X.[82]

---

[80] Draft letter, Bohr to Bush, 16 October 1944, BPP 5.1.
[81] Einstein to Bohr, 12 December 1944, BPP 10.1.
[82] "X" is Albert Einstein.

Took steps to achieve that no complications from such side will for the time being appear.

Question of further efforts

I.) Advisability of removing misunderstandings as regards Pres.'s attitude.

B believes PF will gladly offer his services, since he has often assured B that he considers it a personal duty in every proper way to strive for a beneficial solution.

II.) Advisability of accepting Bush's invitation.

III.) Problem of new approach to P.M.

Perhaps, if Chanc consents, through For[eign] Secr[etary][83] who, as B understands, is in confidence and with whom the Chanc. has already suggested that B might have an interview about Danish matters.

Although B feels it an obligation not to show public interest in such matters, he has, of course, a deep concern about them and would gratefully welcome such an interview which might be a natural background for discussing question concerning B's present work.

IV.) Who else is in position to give P.M. a proper account of the matter? Sir Henry Tizzard?[84]

IV.) Smut's[61] whereabouts.

---

[83] Anthony Eden (1897–1977), British Foreign Secretary throughout the war.
[84] Henry Thomas Tizard (1885–1959), British science adviser.

BOHR TO ANDERSON, 27 March 1945
[Carbon copy]
[Source: BPP 2.2]

London, March 27th 1945.

The Rt. Hon. Sir John Anderson, M.P.,
Chancellor of the Exchequer.

Dear Sir John,

Since our last conversation and my subsequent meeting with Lord Cherwell, I have been giving further thought to the question of the best procedure for bringing to the notice of the President the arguments for the urgency of the matter which I have had the privilege of discussing with you.

The problem is whether the forwarding through the President's friend[46] of an addendum[85] to the memorandum[62] I prepared last summer might give rise to misunderstandings; or whether it might be preferable this time to approach the President through Dr. Bush.

As you know, the President himself wished me to discuss the matter with Dr. Bush, whose attitude was sympathetic and who actually asked me to let him have my views in writing so that he could give them closer consideration. It seems clear, therefore, that an approach through him could not be regarded as improper either by the President or by the American organization for the project.

At the same time Dr. Bush might welcome an opportunity to associate himself with views for which in principle he has already expressed sympathy and which by now he will presumably know have been independently advocated by some of the most trusted American scientists.

Needless to say much will depend on the situation in Washington, where developments may have taken place since my departure; and I would propose, immediately on my arrival, to discuss the whole matter thoroughly with Lord Halifax and the President's friend before decisions are reached.

Whatever procedure may be thought advisable, I will, of course, take the greatest care that no doubt can be left that there is a question only of the views

---

[85] Bohr's "Addendum" is reproduced on p. [113].

of a scientist who from the British as well as from the American side has been given the opportunity to follow the development of the great enterprise.

In case you should not be able to see me again before my departure, which, owing to an unexpected alteration in the transport schedule has been put forward to this evening, may I close this letter by saying how grateful I am to you for all your kindness during my visit.

Yours sincerely,
[Niels Bohr]

ANDERSON TO HALIFAX, 28 March 1945
[Typewritten]
[Source: CAB 126/40]

<u>TOP SECRET</u>

28th March 1945.

My dear Edward

I have just said goodbye to B. who is now on his way back by boat. I have had several very interesting talks with him during his visit. Some of them were concerned with the technical aspects of the work and with these I need not trouble you; but I must let you know at once the position in regard to the problem of high policy which, as you know, has been exercising B's mind for some time.

You no doubt have in mind the memorandum of 3rd July last which B. wrote primarily, I think, for the President's friend. He has now prepared an addendum of which I enclose copies in the form it has assumed after several revisions made in the light of discussions with various people here.

B's first idea was that some initiative might be taken on this side and it was with that idea that he had prepared the addendum. I had no difficulty, however, in convincing him that, having regard to the stage now reached and the enormously greater American contribution, any suggestion from here for widening the area of collaboration would be open to misinterpretation. B. then had the idea that he might send the memorandum[86] to the President's friend[46]

---

[86] I.e., the Addendum.

[279]

with a view to its being passed on to the President. I would not myself have raised any objection to this course, but Cherwell, whom I consulted and who subsequently had a long talk with B., felt very strongly that there might be a real risk, if this course were taken, of B. becoming again involved in the sort of misunderstanding which arose, you will remember, after the Quebec Conference and which Cherwell was at considerable pains to clear up. The suggestion then was made I think by Ronnie Campbell, that B. might take up the whole question with Dr. Bush. This seemed to me dangerous because, apart from the fact that such policy matters seemed to be above Bush's level, he could hardly be expected to keep the thing to himself and if others who are even more security minded got to know of this further activity on B's part, there might be real trouble.

We left it therefore that B. would do nothing without the fullest consultation with you and if you then advised, after hearing his account of his talks here, that he should take action either through the President's friend or through Dr. Bush, he would be free to act accordingly. I rather think, however, from a remark that he made just before leaving that he would in any case wish to communicate with me again before taking any positive action.

B. knows that I am writing to you on these lines and that I am asking you to try to spare time for a full talk with him as soon as you find convenient after his arrival.

J.A.

The Right Hon. the Earl of Halifax, K.G., G.C.S.I., G.C.I.E., T.D.

ANDERSON TO CHADWICK, 3 April 1945
[Typewritten]
[Source: CHAD 4, Box 2, Folder 11]

My dear Chadwick,

I am very grateful for the memorandum on Future T.A. Policy and Programme which you sent me with your letter of the 23rd March, and also for the very interesting, and on the whole encouraging, information contained in your letter itself.[87]

---

[87] Chadwick to Anderson, 23 March 1945, CHAD 4, Box 2, Folder 11.

I am proposing to circulate your memorandum to my Consultative Council and hold a first meeting of the Council towards the end of next week with a view to mapping out a programme of work for the fortnight or so during which Cockcroft[88] will be available. It may be that, after that meeting, I shall wish to put some questions to you by telegram. Meanwhile, I would prefer to withhold comment on your memorandum except to say that there is much in it with which I entirely agree.

I quite understand why you have not found it possible to come and am very distressed indeed to know that your health has been so poor. I am afraid that the burden upon you has been very heavy, particularly during the last few months; and I need hardly say how very greatly I appreciate the skill and devotion with which you have carried it. If there is any means by which you think that it can be lightened I hope you will not hesitate to let me know. In the interests of the project as well as in your own interest, I am very anxious that the strain upon you should at any rate be sufficiently relaxed fully to restore you health.

Bohr was in fact able to give me much interesting information about the situation in Washington and about your views and his. Discussion with him demands time and patience but is always worth while.

FRANKFURTER REPORT ON BOHR, 18 April 1945
[Typewritten]
[Source: BPP 4.3]

Private

During my year at Oxford (1933–34), as George Eastman Visiting Professor, I made the acquaintance of Professor Niels Bohr, the renowned Danish physicist and Nobel prize winner. We met again when Professor Bohr visited Washington in the spring of 1939. Professor Bohr, who has lived most of his life in the laboratory and in the fellowship of other men of scientific distinction, is a shy man, but we happened to hit it off at this second meeting, largely, I suspect, because he found in me a sympathetic response to his anxieties about the gathering clouds of the European War and Hitler's brutal purposes. As a result, our acquaintance developed into a warm friendly relation. But we had no communication after he left this country, and I knew nothing of his fate

---

[88] John Douglas Cockcroft (1897–1967), British experimental nuclear physicist intimately involved in the war effort. He was a close friend of Bohr.

after the war broke out until I learned from the press of his successful escape to Sweden when the Nazis began their deportations and massacres in Denmark.

Some time thereafter, the Danish Minister, Mr. de Kauffmann,[63] asked my wife and me to a small tea to meet Professor Bohr. (The exact date is easily ascertainable but at the moment I have not time for running it down.) A handful of people was present, including the Crown Princess Martha.[89] I had no chance of any private word with Professor Bohr, but on leaving I expressed the hope that I would have an opportunity to talk with him and invited him to lunch at my Chambers. A time was fixed and he duly turned up. We talked about the recent events in Denmark, the probable course of the war, the state of England (for I knew of his affectionate devotion to England – one of the things that brought us together was his deep regard for Lord Rutherford whom I had the good fortune to know), our certainty of German defeat and what lay ahead. Professor Bohr never remotely hinted the purpose of his visit to this country.

Some time before Professor Bohr's arrival in this country I had been approached by some distinguished American scientists, because of past academic association, to advise them on a matter that seemed to them of the greatest importance to our national interest and presented to them difficulties with which they, as scientists, were unable to cope since they were problems not within their technical competence. I had thus become aware of X[90] – aware, that is, that there was such a thing as X and of its significance, but I have never been told and do not at this moment know anything that could convey to anyone any valuable information about it. Naturally enough, however, knowing the general range of Professor Bohr's work in the field of physics, I rejoiced to infer that his gifts were to be availed of for the common cause.

And so, in the course of the lunch to which I have just referred, I made a very oblique reference to X so that if I was right in my assumption that Professor Bohr was sharing in it, he would know that I knew something about it, and, if not, I could easily turn my question into other channels. He likewise replied in an innocent remote way, but it soon became clear to both of us that two such persons, who had been so long and so deeply preoccupied with the menace of Hitlerism and who were so deeply engaged in the common cause, could talk about the implications of X without either making any disclosure to the other. The conversation proceeded on this basis and Professor Bohr then expressed to me his conviction that X might be one of the greatest boons to

---

[89] Märtha, Swedish-born Crown Princess of Norway.
[90] The atomic bomb project.

mankind or might become the greatest disaster. This has been the core of his talk with me in the many conversations that I have since had with him. These conversations have all been variations on the same theme. He never disclosed anything that anybody could call a secret. He never told me anything that might be deemed a piece of information regarding X. Throughout he has been concerned entirely with the political problem of so controlling X as to make it a beneficial instead of a disastrous contribution by science. And he made it clear to me that there was not a soul in this country with whom he could or did talk about these things except Lord Halifax and Sir Ronald Campbell. He was a man weighed down with a conscience and with an almost overwhelming solicitude for the dangers to our people.

He infected me with his solicitude, and finally I deemed it my duty to communicate to President Roosevelt what had come to me both through the American scientists and Professor Bohr. Accordingly I saw the President and told him in full detail what I have just summarized and put to him the central worry of Professor Bohr, that it might be disastrous if Russia should learn on her own about X rather than that the existence of X should be utilized by this country and Great Britain as a means of exploring the possibility of an effective international arrangement with Russia for dealing with the problems raised by X. And Professor Bohr made me feel that on the basis of the scientific situation antedating the war, it would not be too difficult for Russia to gain the necessary information, to say nothing of other factors that militate against assuring non-disclosure.

The President was plainly impressed by my account of the matter. On this particular occasion I was with the President for about an hour and a half and practically all of it was consumed by this subject. He told me the whole thing "worried him to death" (I remember the phrase vividly), and he was very eager for all the help he could have in dealing with the problem. He said he would like to see Professor Bohr and asked me whether I would arrange it. When I suggested to him that the solution of this problem might be more important than all the schemes for world organization, he agreed and authorized me to tell Professor Bohr that he, Bohr, might tell our friends in London that the President was most eager to explore the proper safeguards in relation to X. I wrote out such a formula for Bohr to take to London – a communication to Sir John Anderson, who was apparently Bohr's connecting link with the British Government.[91]

---

[91] Bohr's "Message" to Anderson of April 1944, reproduced on p. [89]. See the introduction to Part I, p. [22].

It is unnecessary to go into the details by which the desire of the President to see Professor Bohr was carried out. Suffice it to say that the President did see Professor Bohr for about an hour and a quarter shortly before the Quebec Conference. Bohr promptly gave me a full account of his conference with the President, who plainly, could not have been more friendly or more open in his discussion of the political problems raised by X. The President asked Professor Bohr to feel entirely free to communicate with him at any time on the subject and to do so either through "Mr. Justice Frankfurter or, since the latter was going to leave Washington soon for the summer months, through the Danish Minister." After Professor Bohr's interview with the President, General Watson[92] told me of it and said that if I had any communication from Professor Bohr to the President, I was to give it to him, the General, who would see to it that it got directly into the President's hands. Professor Bohr did, as a matter of fact, write a letter to the President which he asked me to get into the latter's hands. I saw General Watson in his apartment the day before he left for the Quebec Conference and gave him Bohr's letter. General Watson said, "I will carry that letter on my person and will personally give it to the President."

Then followed Quebec. About happenings there and thereafter I only know what Lord Halifax and Professor Bohr have told me.

(Sgd.) Felix Frankfurter.

April 18, 1945

---

[92] Edwin M. Watson (1883–1945), Aide to the President, 1933–1945.

HALIFAX TO ANDERSON, 18 April 1945
[Typewritten]
[Source: CAB 126/183]

TOP SECRET CYPHER TELEGRAM
RECEIVED BY TELEKRYPTON

IZ   3885
IMMEDIATE   TOO 182035Z
                                             TOR 182155Z

FROM: J.S.M. Washington.
TO: A.M.S.S.O.

ANCAM 251   18 April, 1945.

Personal from Lord Halifax for Sir John Anderson.

Your letter of the 28th March.[93]

I had a full discussion with B. soon after his arrival. He had the day before talked the matter over with his friend[46] and the latter was himself quite convinced that it would not be wise for him (the friend) to put B's memorandum forward.[85] Although the matters discussed in it were of a kind which could only be decided by the President himself, he would certainly not take such a decision without consulting his advisers, who would be likely to feel resentment that the President had been approached over their heads. B's friend was therefore strongly in favour of the suggestion that B. should discuss the matter with Bush.

2. B. has reminded me that in doing this he would only be acting on the request made to him by the President last October after the Quebec discussions. B. saw Bush at that time who said that the matter had recently been considered by the U.S. Military Policy Committee of which he was Chairman and asked B. to let him have a memorandum. B. did not think it right to put forward such a memorandum without further discussion with you. If,

---

[93] Anderson to Halifax, 28 March 1945, reproduced on p. [279].

[285]

therefore, he were to take the matter up again with Bush, there would be nothing really new in such a step. He would merely be resuming the discussions where they were broken off. In doing this he would be talking to Bush merely as one scientist to another and entirely on his own responsibility. No reference would be made by him to the views held in U.K. quarters on this subject or to any conversations which he has had with you and others. This opportunity should however be taken to clear up misunderstandings which may still remain, about his correspondence with Kapitza, and I think he should show the letters to Bush. I think Groves should see them also though I am not suggesting that B. should himself raise any other question with Groves.

3. I have discussed the matter with B's friend who confirmed to me that he was strongly in favour of the proposal that B. should use his card of re-entry and talk to Bush. I do not myself see any reason why, if you see no objection, you should not concur in B's talking to Bush, thus taking up again the discussions which they had together a few months ago at the President's request. As he has to leave before very long for Y it would be helpful if I could let him know your decision soon.

TOO 182035Z

ANDERSON TO HALIFAX, undated
[Typewritten]
[Source: CAB 126/183]
[Editor's note: Date most likely 19 April 1945.]

CANAM NO. 299
Following personal and private for Lord Halifax from Sir John Anderson.

Your ANCAM 251.
In the light of your telegram I do not feel that I can justifiably raise objection to B speaking to Bush in the manner proposed. It must, however, be clearly understood that B will make no reference whatsoever to views held in the U.K. or to discussions which he has had here.

BOHR NOTES, 19 April 1945
[Typewritten]
[Source: BPP 2.7]
[Editor's note: The back of the sheet of paper contains an account, in Aage Bohr's handwriting, of the meeting after it took place. According to the report, Halifax promised to take up the issue of Denmark's interest in joining the United Nations with the British Foreign Minister, Anthony Eden.]

April 19th 1945.

Points for conversation with H[alifax] April 20th 12n.

1) N[94] was very thankful for Amb[assador]'s[95] letter[96] with the information about the Chanc[ellor]'s[97] attitude. He wants to assure the Amb. that he, in form as well as in spirit will act according to the Amb.'s and the Chanc.'s instructions which entirely conform with N's own wishes.

2) N has at once tried to arrange meeting with Dr. Bu[sh], but learned that he is out of town and will not be back before Monday. N will himself have to go to "Y" some time next week and will, therefore, not be able to see the Amb., but he will write a report for the Chanc. and leave a copy at the Embassy for the Amb. N hopes that the conversation will be fruitful and will, of course, only advocate that the whole matter is looked thoroughly into, to find the best way of making use of the lead for the interest of America and England as well as to all mankind.

3) At the end N will say that he would be glad to talk to the Amb. for a moment about a personal matter.

When alone, N will say that he has never to the Amb. spoken of any problems concerning Denmark, which are, of course, very much upon his mind. N would like to tell that during his stay in Denmark as well as later he has had opportunity closely to follow the resistance movement and owe the possibility of his escape from Denmark to this connection. In Denmark there has been great gratitude for the way the P[rime] M[inister] and the late Pres[ident][98] have acknowledged the contribution of the resistance in Denmark to the common cause and, as the Amb. will know, there has also been some expectations that

[94] Niels Bohr.
[95] Lord Halifax.
[96] This letter has not been found.
[97] John Anderson.
[98] Roosevelt had died on 12 April.

Denmark in some or other way could be represented at the great conference for world security at San Francisco.[99] That these expectations have so far not been fulfilled has caused some disappointment and anxiety in Denmark, but it is hoped that ways will be found to ensure that the circumstances cannot be interpreted as a lack of appreciation of the great bravery which has been shown by the united resistance in Denmark where conditions at the moment are more dangerous than ever.

(Perhaps the Amb. will say something; if not N will continue by saying)

N hopes that the Amb. will understand that he is not attempting to raise any political matters, but in view of the great personal kindness which the Amb. has shown him he felt it his duty as a Dane just to touch upon these problems with a few words.

BOHR NOTES, undated
[Handwritten by Aage Bohr]
[Source: BPP 6.13]
[Editor's comment: Undated, but written before 3 May 1945.]

Interview with the Pres[ident]

According to an invitation sent through the Danish Legation B was received by Pres[ident] Roos[evelt] on August 26th 1944 at 5.p.m. at the White House. Acc[ording] to an arrangement between the Pres.'s secretary and the Counsellor of the Legation precautions had been taken to avoid all publicity and the reception was of a completely private character with nobody present.

The Pres. started by saying that he had hoped to see B, but had had to postpone the reception, due to his journey to the Pacific. The Pres. next talked for some minutes with B about the situation in Denmark and he especially expressed his high esteem and appreciation of Min[ister] Kauffmann and of the services he had rendered to the common cause, in particular his management under great difficulties of the important arrangement about Greenland.[100] He used the expression that Kauffmann, no matter what the King of Denmark[101] had thought at the moment, had done a thoroughly good job. The Pres. also

---

[99] Founding conference of the United Nations, 25 April to 26 June 1945.

[100] During the war Kauffmann reached an agreement with the United States on behalf of Denmark regarding establishing military bases in Greenland without the approval of the Danish government.

[101] Christian X.

accounted in a most kind and humorous mode about his recollections about members of the Danish Royal family and esp[ecially] of his adventures together with Prince Axel.[102] In this connection the Pres. said that even if Denmark like England had a king, they were still democracies.

B told about the unision of the whole population in Denmark against the German oppression.

The Pres., however, soon turned to the decisive contributions of the scientists to the conduct of the war and expressed his deep appreciation and concern about the formidable destructive weapons which the development of science has placed into human hands and which would make future wars so terrible that their prevention, if not for other reasons, was incumbent.

Punkter [points]

The Pres. kindly gave B op[portunity] to explain the views he had stated in the memo[randum][62] and esp[ecially] how urgent it was that steps be taken to forestall a competition. The Pres. said that he hoped that an understanding about this vital matter could be obtained with the Russians and said that he had great trust in the wisdom of Stalin personally. In this connection the Pres. told most humorously about the meeting in Teheran with Churchill and Stalin and how he by bringing in jokes at the appropriate moment had often succeeded in stopping Churchill and Stalin flaring violently up. (He told how Stalin during a meeting had said that he thought it necessary to quench German militarism by eliminating 50,000 German officers, to which Churchill had retorted heatingly that it was not the custom of the English to kill prisoners of war. Roosevelt had then said that he would like to introduce an amendment to Stalin's proposal and, asked what it was, he had said he would only propose to eliminate 49,000 German officers, whereafter all burst in laughter and that Stalin had left his chair and gone over to shake Roos[evelt]'s hand and said that he was a good fellow.)

---

[102] Axel Christian Georg (1888–1964), Prince of Denmark married to Princess Margaretha of Sweden. A close friend of Bohr, he was a fervent opponent of Nazism.

BOHR NOTES, 3 May 1945
[Handwritten by Aage Bohr]
[Source: BPP 6.13]

May 3rd 1945

Ang[ående][103] Pres[ident]

The Pres. stressed that he realized himself that a competition was unavoidable within a few years and that the matter might even develop in unpredictable ways.

After the discussion about the necessity of an understanding and the immense improvement in international relations this would entail, the Pres. said that he had the best hopes that an understanding with the Russians could be reached.

According to his experience, Stalin was a realistic thinking and reasonable man with whom he had great hopes to collaborate.

In this connection the Pres. told the humorous story from Teheran already written down.[104]

The Pres. next said that he would propose taking the matter up with Mr. Churchill at the forthcoming meeting in Quebec, which he told B was going to take place in two weeks' time and would propose that an approach to the Russians be made as reg[ards] the matter.

B said that he realized, of course, that everything which had been said during the interview was most confidential, but asked whether he could report to Sir John And[erson] in the Brit[ish] Gov[ernment] about it. The Pres. answered, that he had no secrets at all for Mr. Churchill, and he next asked (or rather expressed the hope?) that Anderson would attend the meeting. He said that he would be very glad to see him soon over on this side.

At the end the Pres. said that he wished to see B again after the meeting in Quebec, and before B left he asked whether he could send the Pres. further communications through Dr. Frankf[urter]. The President answered that B could always write to him either through the intermediary of the Danish Minister[63] or through Frankf[urter].

---

[103] Danish for "re[garding]".
[104] Bohr's undated notes, reproduced on p. [288].

BOHR REPORT ON FRANKFURTER, 6 May 1945
[Typewritten]
[Source: BPP 4.3]

TOP SECRET.

May 6th 1945.

B[ohr] met F[rankfurter] for the first time on a visit to Washington in April 1939 in order to take part in meetings of the National Academy of Sciences and the American Physical Society. On an invitation from F, B came an afternoon to his home to discuss common interests in international cultural relations. In July of the same year B met F again in London at a meeting of the Society for the Protection of Science and Learning, principally organized to support German scientists and scholars persecuted after the revolution of 1933.

When B, after his escape from Denmark in 1943 came to U.S.A. with the ostensible object of taking part in the planning of post-war international scientific co-operation, he met F again at a reception at the Danish Legation in Washington in December, and a few days later B was invited to F's home where problems of cultural co-operation were discussed on the basis of common experiences.

On B's return to Washington after a longer absence he came in the middle of February 1944 to lunch with F at the Supreme Court. In the course of the conversation F indicated that he was aware of the project and that B had relations to it, and that, while he knew nothing about technical details, he was much concerned with the implications of the project. Without entering upon matters of military secrecy B, therefore, told F that in his opinion the accomplishment of the project should offer a unique opportunity of furthering a future harmonious co-operation between nations. On hearing this F said that, knowing President Roosevelt, he was confident that the President would be very responsive to such ideas as B outlined.

After another absence from Washington B met F again one of the last days of March and learned that in the mean time F had had occasion to speak with the President and that the President shared the hope that the project might bring about a turning point in history. F also informed B that as soon as the question had been brought up, the President had said it was a matter for Prime Minister Churchill and himself to find the best ways of handling the project to the benefit of all mankind, and that he should heartily welcome any suggestion to this purpose from the Prime Minister. The President had authorized F to convey this view with proper discretion and accordingly F wrote out a formula for B to convey to the Prime Minister in strictest confidence.

On B's return to Washington after a visit to England he reported to F on June 18th about the receipt of the message and F felt it important to give the President full details of B's mission to England.

About a week later F told B that this information had been heartily welcomed by the President who had said that he regarded the steps taken as a favourable development. During the talk the President had expressed the wish to see B, and as a preliminary step F advised B to give an account of his views in a brief memorandum, for F to use as he thought fit. This memorandum[62] was sent on July 5th to F, and in an answer a few days later F informed B that he had found the memorandum suitable to be handed to the President and had immediately done so.

On August 26th at 5 p.m., B was received by the President in the White House in a completely private manner, and during the interview the President most kindly gave B opportunity to explain his views and spoke in a very frank and encouraging manner about the hopes he himself entertained. At the end of the interview, which lasted till about 6:30 the President said that he would discuss the matter with the Prime Minister at their forthcoming meeting and expressed the wish after this meeting to see B again.

A few days later B met F in Boston and reported about the conversation. F advised that B, for the President's convenience, add to a letter thanking the President for the kindness of his reception, a restatement of the essence of the arguments discussed. This letter was, with a covering note from F, a week later delivered to the President on the eve of his departure from Washington to the meeting in Quebec.[105]

A few weeks after the President's meeting with the Prime Minister B had word from the President that, due to a journey, he could not himself see B, but suggested that B instead saw Dr. Bush with whom in the meantime the President had talked about the matter.

---

[105] Bohr to Roosevelt, 7 September 1944, reproduced on p. [109].

BOHR NOTES, 22 June 1945
[Handwritten by Aage Bohr]
[Source: BPP 4.4]

June 22nd–[19]45.

F[rankfurter]'s relations to S[timson].[106]

F has through many years been an intimate friend of St. under whom he has served as ?[107] in 7 years (see who's who in America.)

When the Pres. (Roosevelt) died, F felt it his duty to go to Sti and tell him everything about his connection with R about the matter and explain B[ohr]'s views. A few days later F told B that St had been most seriously occupied with the matter and was very sympathetic to such views (F used the expression that the talk with St had been as to knock on an open door.)

A few weeks later F told B that he had had another conversation with St who said that he had had the papers of B and was "deeply impressed" by them.

Since in course of this conversation S had asked about B's connection with Kap[itza], F asked B for a complete account of the correspondence which F handed to St in the course of a further conversation in which St said that he wished to see B and talk with him about the whole matter.

A few days before B left for London, F asked B to come and said that he had again talked with St just before St left Wash. and St had said that he was very sorry that due to lack of time he had not been able to see B, whom he hoped, however, to see immediately B returned from England.

---

[106] Serving for as many as five U.S. presidents during his long career, Henry Lewis Stimson (1867–1950) was Secretary of War under Roosevelt from 1940 to 1945.
[107] The question mark is Bohr's.

AFTER THE EXILE

PRESS RELEASE, 4 October 1945
[Stencil]
[Source: NBA]

Meddelelse fra Professor Niels Bohr udsendt igennem Ritzaus Bureau
den 4. Oktober 1945.

I Anledning af Pressekommentarer[108] til et Foredrag, som Professor Niels
Bohr i Aftes holdt i Ingeniørforeningen om de seneste Fremskridt paa Atomforskningens Omraade, ønsker Professoren at meddele, at naar der henføres til
Atombombens Hemmelighed, er dette en Talemaade, der kan give Anledning
til Misforstaaelse, idet de videnskabelige Principper, der kommer i Betragtning
ved Atomenergiens Frigørelse, i Hovedtræk var offentlig bekendt før Krigen,
og deres senere Udvikling er udførligt beskrevet i de af den amerikanske
og engelske Regering udgivne officielle Beretninger. Foredraget omhandlede
alene saadanne rent videnskabelige Fremskridt, og, som Professoren allerede
i en ved sin Hjemkomst fremsat Udtalelse i Radioen[109] meddelte, har Videnskabsmænd som han selv intet nærmere Kendskab til tekniske Enkeltheder
angaaende Produktionen af de aktive Materialer og til strategiske Problemer
vedrørende de nye Vaabens Anvendelsesmuligheder. Hans Bemærkninger om
den Situation, der er skabt ved de videnskabelige og tekniske Fremskridt, tog
først og fremmest Sigte paa at understrege de Farer, som Civilisationen vil
være udsat for, dersom der ikke i rette Tid opnaas Enighed om en international
Kontrol af de frygtelige Ødelæggelsesmuligheder, der nu er kommet inden
for Menneskenes Rækkevidde. Som han allerede andetsteds har fremhævet,
skulde netop den Omstændighed, at det drejer sig om en Sag af dybeste fælles
Interesse for alle Nationer, frembyde en enestaaende Lejlighed til at fjerne
Hindringer for det fredelige Samarbejde mellem Folkeslagene og til at sikre
Betingelserne for, at de i Fællesskab skal kunne nyde godt af de rige Muligheder for Forbedring af Menneskenes Levevilkaar og for Kulturens Fremme,
som de store Fremskridt paa saa mange Omraader af Videnskaben har skabt.

---

[108] See the introduction to Part I, p. [55].
[109] Talk on Danish national radio, 26 August 1945, reproduced on p. [136] and translated on
p. [140].

Translation

Statement from Professor Niels Bohr issued by Ritzau's Bureau
on 4 October 1945.

On account of the comments in the press[108] concerning a talk held by Professor Niels Bohr yesterday evening in the [Danish] Association of Engineers on the most recent developments in the field of atomic research, the professor would like to state that when reference is made to the secret of the atomic bomb, this is a figure of speech which may give rise to misunderstanding, as the scientific principles involved in the release of atomic energy were, in the main outlines, public knowledge before the war, and their later development is described in detail in the official documents published by the American and British Governments. The talk dealt only with such purely scientific advances and, as the professor on his return to this country has already announced in a statement on the radio,[109] scientists such as himself do not have any precise knowledge of the technical details regarding the production of the active materials or of the strategic problems concerning the potential uses of the new weapons. His comments about the situation that has been created by the scientific and technical progress were primarily aimed at underlining the dangers that civilization will face if agreement is not reached in due time concerning international control of the terrible possibilities for destruction which have now come within the reach of man. As he has emphasized already elsewhere, precisely the circumstance that this is a case of the most profound common interest for all nations should present a unique opportunity for removing obstacles to peaceful cooperation between peoples and for securing the preconditions enabling them to benefit jointly from the rich possibilities for the improvement of people's living conditions and for the advancement of culture created by the great progress in so many branches of science.

TELEGRAM, KAPITZA TO BOHR, 20 October 1945
[Typewritten]
[Source: BSC (21.4)]

THIS IS TO WISH YOU WELLCOME ON YOUR HAPPY AND SAFE REUNION WITH YOUR FAMILY IN FREE DENMARK STOP WITH BEST WISHES TO YOU AND MRS BOHR FROM US BOTH

PETER AND ANNA KAPITZA

BOHR TO KAPITZA, 21 October 1945
[Carbon copy]
[Source: BSC (21.4)]

October 21, [19]45.

Dear Kapitza,

I thank you and your wife most heartily for your kind greeting which gave me great pleasure. In these years I have often thought of you and have been very grateful for the kind invitation which you sent me after my escape from Denmark, but which reached me so late that I found no opportunity to follow it. As you kindly expressed, I am happy indeed after the long separation to be back in my home and united with my whole family. It is almost hard to believe that also most of our friends have come through the hard times with life and health.

Now that the war is over with victory for freedom, international cooperation in science which has meant so much to both of us will certainly not only be revived and extended, but I am convinced that it will contribute more than ever to that understanding between nations in which all peoples put faith and for which so promising a background has been created through the comradeship in the defence of elementary human rights.

I need not say that in connection with the immense implications of the development of nuclear physics the memories of Rutherford have constantly been in my thoughts and that I, like his other friends and pupils, have deplored that he should not himself be able to see the fruits of his great discoveries. His wisdom and authority will also be greatly missed in the endeavours to avert new dangers for civilization and to turn the great advance to the lasting benefit of all humanity.

You may have seen a little article which I sent to the "London Times"[110] a few days after the events and in which I tried to give expression for an attitude widely shared among scientists. A few weeks ago, I have, on an invitation of American friends, who take deep interest in the matter, written another short

---

[110] N. Bohr, *Science and Civilization*, The Times, 11 August 1945; this volume, pp. [121] ff.

article to appear in "Science".[111] I am enclosing copies of both articles which I should be glad if you would show to common friends.

I leave it of course to your judgement but, if you think it desirable, I shall greatly welcome if one or both of the articles could be translated and published in the Soviet Union. I need not add that I shall be most interested to learn what you think yourself about this all-important matter which places so great a responsibility on our whole generation.

Another question in which we surely both will be equally interested is the endeavour to organize scientific research and co-operation on the most effective lines. In this connection I was glad on my return to learn that a group of Danish scientists,[112] who have been active in the resistance to the Germans during the occupation, is planning to send a representative to the Soviet Union to study the organization of scientific work there. Dr. Holter,[113] an eminent biochemist, who has been selected for this task, is a close friend of mine and I hope very much that it will be possible to arrange his journey in which the Soviet Legation here is taking a kind interest.

I hope that you and your family are all well and my wife joins me in sending our kindest regards and best wishes to you all.

Ever Yours
[Niels Bohr]

[111] N. Bohr, *A Challenge to Civilization*, Science **102** (1945) 363–364; this volume, pp. [125] ff.

[112] This is most likely the "Science and Society" study group established in the spring of 1944 by six Danish scientists "whose purpose it should be to take stock of Danish science in order thereby to establish a rational basis for its incorporation in the new society that we hoped would arise after the war"; undated letter (written in the second half of the 1940s) from biochemist Kai Linderstrøm–Lang (a founding member of the group) to Bohr, deposited in the NBA. After the war, the group was expanded.

[113] Heinz Holter (1904–1993), Austrian-born biochemist who moved to Copenhagen in 1930 to work in the Division of Chemistry of the Carlsberg Laboratory, where he stayed until he retired. Holter was another founding member of the "Science and Society" group. There is no indication in the records, nor in Danish newspapers, that Holter actually visited the Soviet Union at this stage.

KAPITZA TO BOHR, 22 October 1945
[Typewritten with handwritten date and signature]
[Source: BSC-Supp.]

My dear Bohr,

It is a great relief to feel that the ordeal of the war is over and we may resume our peaceful life. We all are very happy to know that you and your family went safely through all your adventures and are now united in Copenhagen. I was always happy to have news from you and your family but every time they came with great delay.

To the 220th anniversary of the Academy of Science the only physicist who came was Max Born[114] and practically nobody came from USA where you have been at this time. It was Auger[115] who was in USSR during the war and told us about you.

We are all back to Moscow. It is already two years since the Institute resumed normal scientific work. As before the war we have two times a week liquid helium and have found some curious things at low temperatures. I hope you have seen the theoretical work of Landau and the superfluidity of helium which probably you remember we have discovered just before the war. It is now proved that superfluid helium is a mixture of normal helium with helium with zero-entropie. This I have proved experimentally and this originated the theory of Landau. This particular property of helium gives in principle the possibilities to approach indefinitely closely to the absolute zero and this approach is only limited by technical difficulties. Landau also proved that two kinds of elastical waves must propagate simultaneously in superfluid helium; therefore in helium II must exist two kinds of sound velocities, one at 250 m/sec. (already known one), and another (a new one) at 17–20 m/sec. Peshkov[116] has discovered experimentally the second velocity of sound in helium II.

Besides this peaceful work we helped the country in its war efforts and I am proud to tell you that the Institute received the Order of the Red Banner. It is the only Institute in the Academy of Science which received this honour.

At the moment I am much worried about the question of the international collaboration of science which is absolute necessary for the healthy progress of culture in the world. The recent discoveries in the nuclear physics, the famous

---

[114] (1882–1970), German refugee physicist who worked at the University of Edinburgh from 1936 until his retirement in 1953.
[115] Pierre Auger (1899–1993), French experimental physicist.
[116] Vasily Petrovich Peshkov (1913–1980), Russian low-temperature experimental physicist.

atomic bomb I think proves once more that science is no more the hobby of university professors but is one of the factors which may influence the world politics. Nowadays it is dangerous that scientific discoveries, if kept secret, will serve not broadly humanities but be used for selfish interests of particular political or national groups. Sometimes I wonder what must be the right attitude of scientists in these cases. I should very much like at the first opportunity to discuss these problems with you personally and I think it would be wise as soon as possible to bring them up to a discussion at some international gathering of scientists. May be it will be worth while to think over that measures should be included into the status of the "United Nations" which will guarantee a free and fruitful progress of science.

I should be glad to hear from you what is the general attitude on these questions of the leading scientists abroad. Any suggestions about means to discuss these questions from you I shall welcome mostly. I can indeed inform you what can be done in this line in Russia.

This letter will be handed to you by a young Russian physicist Terletzki. He is a young and able professor of the Moscow university and will explain you the aims of his visits abroad. With him you may send me the answer.

I include herewith a letter from Anna and small presents which she sends to your family.

Meanwhile once more best wishes to you and your boys, kindly remember me to Mrs. Bohr.

Hoping to meet you soon, I am

Yours most sincerely
P. Kapitza

22/X/45
Moscow.

BOHR TO ANDERSON, 22 October 1945
[Carbon copy]
[Source: BPP 116]

GL. CARLSBERG   COPENHAGEN, October 22, 1945.

The Rt. Hon. Sir John Anderson, M.P.

Dear Sir John,
    I wish to thank you most heartily for the kind sentiments for which you gave

expression on my birthday and which moved me deeply.[117] I can hardly say how grateful I have been for the sympathy and confidence you have shown me ever since I was given the privilege to cooperate with your great organization. You know how much, from my early youth, I have benefitted from my connection with British scientists and I wish to say how greatly I valued in these last eventful years to come into contact with new circles in your country to which all mankind owes so much.

The support you gave in your greeting to the hope that the bonds created by international co-operation in science may prove helpful in the present situation has also been a source of great pleasure and encouragement to me. Since I wrote to you last, I have, on an invitation from some American friends, written a short article to be published in SCIENCE[111] and I am enclosing a copy of the manuscript. As you will see, I have, as in the TIMES[110] article in which you took so kind an interest, confined myself to the general human aspects of the situation.

I trust that the whole matter will soon develop in a favourable manner but, as I have written to Dr. Bush in a letter of which I enclose a copy,[118] I have been somewhat worried by the present confusion of public opinion. In this letter I have mentioned certain points of which I have thought in connection with the desirability of inviting the confidence of scientists in other countries. I would also like to tell you that I have just received a most friendly telegraphic greeting from Kapitza[119] and that, as an answer, I sent him a letter of which I enclose a copy.[120]

I need not add how deeply I shall always appreciate any advice from you if you think that in some modest way I can be of assistance for your endeavours.

With kindest regards and best wishes,

Yours sincerely
[Niels Bohr]

---

[117] See the introduction to Part I, p. [54], ref. 66.

[118] This letter has not been found.

[119] Telegram from Kapitza to Bohr, 20 October 1945, reproduced on p. [295].

[120] Bohr to Kapitza, 21 October 1945, reproduced on p. [296].

BOHR NOTES, 2 November 1945
[Typewritten]
[Source: NBA]
[Editor's note: The name "Fog"[121] has been cut out and is reproduced below as "[Fog]".]

2 Nov. [1945]

[Fog] ringede om Formiddagen og spurgte om han kunde faa en kort Samtale i Løbet af Dagen, og det aftaltes at han skulde komme Kl. 13³⁰.

Da han kom, sagde han at han havde et meget diplomatisk Ærinde og forklarede at der var kommet en russisk Videnskabsmand til København med et Brev fra Kapitza og ønskede at aflevere dette til B[ohr], og have en fortrolig Samtale med ham, der maatte arrangeres saa hemmeligt at secret service ikke paa nogen Maade kunde faa Underretning derom.

B svarede at han naturligvis ikke kunde indlade sig paa hemmelige Ordninger af nogen Art og at han maatte betragte en saadan Henvendelse som et beklageligt Fejlgreb fra russisk Side. Efter at have fortalt om sit Forhold til Kapitza og vist [Fog] et Brev til K[apitza] af 21/10–1945,[120] forklarede B at K, der var en nær Ven af ham maatte skrive til ham ganske aabent, og hvis nogen russisk Videnskabsmand ønskede at tale med B maatte dette ske i fuld Aabenhed.

B understregede i Tilknytning til sin Ritzaubureau-meddelelse af 5 Okt.[122] at han ingen Hemmeligheder af militær Betydning besad, men at han havde Forpligtelser paa Grund af den Tillid som var vist ham paa engelsk og amerikansk Side. Han var glad for af [Fog] at høre ligesom ogsaa Kapitza havde forsikret ham om, at man ogsaa havde Tillid til ham fra russisk Side. Den eneste Maade hvorpaa B eventuelt kunde give et beskedent Bidrag til den store Sag, var i fuld Aabenhed at virke for en gensidig Forstaaelse af de store Problemer som Videnskabens seneste Udvikling havde rejst.

Translation

2 Nov. [1945]

[Fog] called in the morning and asked whether he could have a brief talk in the course of the day, and it was agreed that he should come at 1:30pm.

---

[121] Mogens Fog (1904–1990). See the introduction to Part I, p. [56].
[122] Bohr must here be referring to his Press release of 4 October, reproduced on p. [294] and translated on p. [295].

When he came, he said that he had a very diplomatic errand and explained that a Russian scientist had come to Copenhagen with a letter from Kapitza and wished to deliver this to B[ohr] and have a confidential talk with him, which must be arranged so secretly that the secret service could not get to know about it in any way.

B replied that naturally he could not engage in secret arrangements of any kind and that he had to consider such an approach a regrettable mistake on the part of Russia. Upon having told about his relations with Kapitza and shown [Fog] a letter to K[apitza] of 10/21/1945,[120] B explained that K, who was a close friend of his, should write to him quite openly, and if any Russian scientist wished to speak to B, this should take place in complete openness.

B stressed in connection with his statement through the Ritzau Press Agency of 5 Oct.[122] that he held no secrets of military significance, but that he had obligations because of the confidence that had been shown him on the part of England and America. He was pleased to hear from [Fog], just as Kapitza had also assured him, that there was also confidence in him on the part of Russia. The only way in which B could perhaps make a small contribution to the great cause, was to work in full openness for a mutual understanding of the great problems that the latest development of science had raised.

BOHR TO KAPITZA, 17 November 1945
[Carbon copy]
[Source: BSC-Supp.]

November 17th [19]45.

Dear Kapitza,

It was a very great pleasure through Professor Terletzky to receive your most interesting and kind letter of October 23rd[123] together with the beautiful presents which you and your wife, so kindly sent our family and which we treasure very highly.

We were very happy in the Institute to welcome a colleague from Moscow and to get news about our friends whom we have not seen for so long. I was also most interested in the articles enclosed in your letter about the progress of your important researches on superfluidity and Landau's ingenious theoretical analysis. So far we had no opportunity to learn about this progress and our

---

[123] Bohr must here be referring to the letter dated 22 October 1945, reproduced on p. [298].

whole group looks forward to study and discuss the articles and, of course, also the many other publications of your Academy from the last five years, which Professor Terletzky brought us and which we were very grateful to receive.

Professor Terletzky has kindly offered to bring you copies of papers from our Institute, published in recent years, and we are just preparing, in collaboration with the Danish Academy, to reestablish the regular exchange of publications and to send out the issues from the last years. In spite of the difficulties, efforts were made to continue scientific work here during the war, and we hope now to be able, according to the best of our facilities, again to participate in international research work. In this connection we also hope in a near future to resume our annual conferences about atomic problems at which, before the war, we had the pleasure here to see so many colleagues from other countries.

As you will have seen from my letter of October 21st,[120] which I hope you have received in the meantime, my own thoughts have, as yours, been very much occupied with the implications of the recent advance in nuclear physics. I agree most heartily that a discussion of these problems at some international gathering of scientists may be most helpful. For some time I have hoped that it should be possible to arrange such a meeting here in Copenhagen, and I have already mentioned it to several of our mutual English friends.

If you and some of your colleagues could come, I am sure that a number of the leading physicists from other countries would join us. Such a meeting which we are prepared to arrange at any time would not only give us all the opportunity to exchange views about those matters which are so much on the heart of everyone, but also to take up the question of collaboration in science and to discuss the many advances which have been made in various fields of physics in recent years.

I need not add, that quite apart from such plans, it would be an extreme pleasure to see you here in Copenhagen whenever you might be able to come. It will, certainly, be a great honour to our Academy and University if you, as their guest, would lecture for us on your wonderful researches or any topics which you may prefer, and Margrethe and I would be happy, indeed, if you and your wife would stay with us in our home, where we would try to return some of your hospitality which made our visit to Moscow such an unforgetful experience to us.

With kindest regards and best wishes to you and your wife and the boys and our other friends in Moscow,

Yours ever,
[Niels Bohr]

BOHR TO ANDERSON, 20 November 1945
[Carbon copy]
[Source: BPP 116]

Gl. Carlsberg, Copenhagen.
November 20th 1945.

The Rt. Hon. Sir John Anderson, M.P.

Dear Sir John.

It may interest you to learn that, a few days ago, we had at our institute a visit of a Russian physicist, Professor Terletzky from Moscow, who brought me a letter from Professor Kapitza. This letter has crossed the letter which I sent him a few weeks ago and about which I told you in my letter of October 22nd which I hope you have received.

In his letter, which is held in most friendly terms, Kapitza expresses great concern about the problems confronting scientists at this moment. He is very eager to discuss these problems with me personally at the earliest opportunity and also suggests that it would be wise as soon as possible to bring the matter up to a discussion at some international gathering of scientists.

I have, therefore, answered Kapitza that we shall be glad at any time to arrange an international scientific meeting here in Copenhagen and that, if he and perhaps some of his colleagues could attend, I have the hope that some of the leading scientists from other countries would join us. I have also assured him that he is, of course, most welcome here whenever he might be able to pay us a visit where we should greatly enjoy a lecture from him on the progress of his important scientific researches.

It appears to me that such a visit might in several respects be very useful, and you will understand that I hope by this course to avoid the question of an invitation to me to come to Moscow, as touched upon by Professor Terletzky. I shall deeply appreciate your advice and to learn what you would think about calling an international conference in Copenhagen if Russian scientists wish to meet their colleagues here.

I need hardly add that it is with deepest interest that I have read the statement issued after the consultations in Washington.[124] I trust that a most important step towards a favourable development has been taken.

---

[124] The "Washington Declaration". See the introduction to Part I, p. [65].

[304]

With kindest regards and best wishes

Yours sincerely
[Niels Bohr]

ANDERSON TO BOHR, 29 November 1945
[Typewritten]
[Source: BPP 116]

OFFICES OF THE WAR-CABINET,
GREAT GEORGE STREET,
S.W.I

29th November, 1945.

Dear Dr. Bohr,

I have to thank you for your two letters,[125] the second of which reached me yesterday.

I can assure you it was a very great pleasure to me to communicate the little note about you which was published on the occasion of your sixtieth birthday.[117]

As regards the calling of an international conference of scientists in Copenhagen, I think that would be an excellent idea provided sufficiently long notice were given. Perhaps the late summer or early autumn of next year would be a suitable time. This need not, of course, necessarily rule out an earlier visit by Professor Kapitza which clearly would be far preferable to a visit by you to Moscow which, in present circumstances, might be misunderstood.

I fully share your hope that the Washington Declaration will prove to have been a most important step towards a better understanding all round.

With very kind regards, in which my wife joins,

Yours sincerely,
John Anderson

Professor Niels Bohr,
    Gl. Carlsberg,
        Copenhagen.

---

[125] Bohr to Anderson, 22 October and 20 November 1945, reproduced, respectively, on pp. [299] and [304].

[305]

TELEGRAM, OPPENHEIMER TO BOHR, 14 December 1945
[Typewritten]
[Source: BSC (24.1)]

WOULD VERY MUCH LIKE TO QUOTE YOUR STATEMENT IN LON-
DON TIMES[110] AS FOREWORD TO BOOK[126] ABOUT WHICH CABLE[127]
WAS SENT TO YOU FEW WEEKS AGO

SINCE THIS CONTINUES TO BE THE CLASSIC STATEMENT OF THE
FEELINGS WITH WHICH SCIENTISTS APPROACH THE NEW SITUA-
TION IF YOU DO NOT OBJECT TO SUCH QUOTATION WILL YOU
CABLE REPLY TO MCGRAWHILL CHAPMAN NEW YORK

IF YOU FEEL THAT ANY NEW OR DIFFERENT WORDING IS NEC-
ESSARY WILL YOU SEND REVISED COPY OF YOUR STATEMENT BY
MAIL TO MCGRAW HILL ALDWYCH HOUSE LONDON WE COUNT
STRONGLY ON YOUR BEING IN THE BOOK

JR OPPENHEIMER

TELEGRAM, BOHR TO OPPENHEIMER, 14 December 1945
[Typewritten]
[Source: BSC (24.1)]
[Editor's comment: Date added by hand]

GRATEFULLY ACCEPT PROPOSAL QUOTATION TIMES ARTICLE
STOP PREFER UNALTERED TEXT WITH DATE ORIGINAL PUBLICA-
TION AND TITLE SCIENCE AND CIVILIZATION

BOHR

14 Dec. 1945

---

[126] *One World Or None: A report to the public on the full meaning of the atomic bomb* (eds.
D. Masters and K. Way), McGraw-Hill, New York 1946.
[127] This cable has not been found.

OPPENHEIMER TO BOHR, 30 March 1946
[Handwritten]
[Source: BSC (30.2)]

DEPARTMENT OF STATE

———

THE UNDER SECRETARY

Dear Uncle Nick,[73]

Even in our gloomy moments we did not succeed quite in thinking how difficult it would get to be. This report[128] that I am sending you is months later than it should be, and is not all it should be; but I think it may still be of interest to you. For what is good in it, it should be dedicated to you.

Before too long I hope we may see one another again. Very slowly here physics is starting again.

With warm greetings to Jim,[76] and my love to you

March 30 –   Robert Oppenheimer
Niels Bohr

BOHR TO KAPITZA, 12 April 1946
[Carbon copy]
[Source: BSC (29.2)]

April 12, 1946.

Dear Kapitza,

As you will know from the official communication which is being sent to you from our Academy, you have just been elected foreign member. It is the first time since war broke out that the Danish Academy has elected foreign members, and it is a great pleasure to us that we have had this occasion to join in the appreciation, shared by all physicists, of your outstanding contributions

---

[128] The Acheson–Lilienthal Report. See the introduction to Part I, p. [66].

to our science. Personally, I wish to take this opportunity to express the hope that we shall be able to meet soon again and discuss the progress of physics and the prospects which it opens for the future of mankind.

As I wrote last, it will give all Danish scientists extreme pleasure if you could visit us in the near future and talk to us about your researches. On behalf of the Danish Academy of Sciences as well as of the University of Copenhagen I wish herewith most cordially to renew this invitation. Also "Danmarks Naturvidenskabelige Samfund",[129] a society for the promotion of co-operation between scientists and leading technicians, wish to join this invitation and ask for the permission to cover all the expenses of you and your wife during the journey and stay here. I need also not repeat that it would give my wife and me the greatest pleasure if you will be guests in our home during your stay in Denmark. I realize, of course, how great your obligations are, but I still venture to hope that you will be able to visit us soon.

My wife joins me in sending you and your family our kindest greetings and best wishes.

Yours
[Niels Bohr]

P.S. As regards the arrangement of an international conference of scientists which you wrote to me about, I am confident that, if you and some of your colleagues could attend, many of our English and American colleagues would most heartily welcome a meeting with us here in Copenhagen, and I am ready, as soon as I hear your reactions, to proceed with the preparations for such a meeting which could, I think, be arranged almost any time you would find suitable.

BOHR TO BUSH, 17 April 1946
[Carbon copy]
[Source: NBA]

GL. CARLSBERG   VALBY–COPENHAGEN, April 17, 1946.

Dear Dr. Bush,

It was a very great pleasure to me a few days ago to receive a copy of the report[128] on the international control of atomic energy, published by the State Department. I think that it gives most admirable expression for a spirit which

---

[129] Science Society of Denmark. See p. [483], ref. 4.

contains the very best hopes for the development in which everybody puts his faith.

I had hoped for some time to come for a visit to America and on that occasion to have an opportunity also to see you, but so far my duties here have prevented such a journey. It may, however, be possible for me to come to U.S.A. in the autumn and I shall then be most interested to learn how things will have developed.

With kindest regards and best wishes,

Yours sincerely
[Niels Bohr]

BOHR TO OPPENHEIMER, 17 April 1946
[Carbon copy, except last handwritten paragraph in postscript]
[Source: BSC (30.2)]

GL. CARLSBERG   VALBY–COPENHAGEN, April 17, 1946.

Dear Robert,

I have just received a copy of the report on international control of atomic energy published by the State Department,[128] and it is very much on my mind to give expression to you for the deep pleasure it was to me to read this report. In every word of it I find just the spirit which I think offers the best hopes for the development in which we all put our whole faith. I was also deeply impressed by the amount of thought and work which lies behind the preparation of the report, and from page to page I recognized your broad views and refined power of expression. Since my return the great matter has, of course, been constantly in my mind, but, apart from quite general utterances, I have been very reluctant to enter on any definite indications of procedure and, above all, I have felt that it would be unwise to me, not being able to follow developments at first hand, to take part in discussion about security problems. It is for this reason that I have hesitated to accept invitations to attend public meetings about such topics and have mainly confined myself to endeavours to promote understanding among scientists of the various nations. Also in this respect the publication of the report was a great encouragement to me and I need not add that I should greatly wellcome hearing from time to time from you how things are developing.

With heartiest greetings and best wishes to you and your whole family,

Yours ever,
[Niels Bohr]

P.S. I received some time ago a most beautiful and touching letter from Miss Warner,[130] which recalled the spell of the natural and historical surroundings of the place where we worked with you and from which Aage and I will always cherish the memory.

Just as this letter was to be posted I received from the American Legation here a package from you containing a number of much useful copies of the report together with your letter[131] which gave me more pleasure than I can say. I knew of course that the task would be very difficult but none could have been better fitted for it than you, and I think that things now look most hopeful. Notwithstanding all such reactions as I have felt proper you know of course that you can always count on me. Good luck

from

Jim[76] and

Uncle Nick[73]

KAPITZA TALK, 8 July 1946
[Typewritten]
[Source: NBA]
[Editor's comment: Danish translation of Kapitza's talk in Russian enclosed with Døssing to Bohr, 18 July 1946 (p. [315]).]

Tale, holdt af Akademiker P.L. Kapitza ved den højtidelige Modtagelse i V.O.K.S.[132] Mandag den 8' Juli 1946 i Anledning af Overrækkelsen af det danske Videnskabernes Selskabs Medlemsdiplom.

Den Største Ære, som kan blive vist en Videnskabsmand af et andet Lands Folk, er den, at han bliver valgt til Medlem af dette Lands førende Videnskabelige Organisation, af Videnskabernes Akademi.

---

[130] Edith Warner (1893–1951), neighbour to the Los Alamos National Laboratory. Oppenheimer took Niels and Aage Bohr, as well as other Los Alamos scientists, to the tea room in her "house at Otowi crossing", where she shared her considerable knowledge of American Indian culture.
[131] Oppenheimer to Bohr, 30 March 1946, reproduced on p. [307].
[132] All-Union Society for Cultural Relations with Foreign Countries, established in 1925.

Den store Ære, som det danske kongelige Videnskabernes Selskab har vist mig ved at vælge mig, sætter jeg af mange Grunde meget stor Pris paa.

Fremfor alt ser jeg i den Anerkendelse af vor sovjetiske Videnskabs Betydning og dens Deltagelse i Verdensvidenskabens Udvikling.

Dette Valg er mig ogsaa kært af den Grund, at det er et Venskabsbevis fra det danske Folk, som jeg har lært at ære, i min Barndom ved at læse Andersens[133] Eventyr, i de modne Aar ved at lære Brandes[134] at kende, og til sidst, da jeg blev Videnskabsmand og begyndte at studere Magnetismen, ved at møde Ørsteds[135] Navn. I direkte Sammenhæng hermed er jeg i mit videnskabelige Arbejde stødt paa mange danske Fysikere, til sidst paa Arbejder af Niels Bohr, hvis Navn tæller blandt Videnskabens Erobrere, og som paa Højden af sin Forskning uden Modsigelse staar i allerførste Række.

Jeg har den store Lykke og Ære at regnes som Bohrs Ven. Vort Bekendtskab har allerede varet 20 Aar, og fra vort Venskabs Begyndelse har Bindeleddet mellem os ikke alene været Interessen for den moderne Fysiks almindelige Spørgsmaal, men ogsaa den Omstændighed, at vi begge er Elever af en af de mest fremtrædende Fysikforskere i dette Aarhundrede, den geniale Rutherford.

For os begge var Rutherford ikke alene den lærde Videnskabsmand, men ogsaa en Livets Lærer, og vi har begge i ham lært at ære den fremtrædende Videnskabsmand og elske det store Menneske.

Ogsaa et andet Spørgsmaal forbandt mig med Bohr. Det var Spørgsmaalet om Nødvendigheden af at holde fast paa Internationalismen i Videnskaben. Den moderne Videnskabs Erobringer er Resultatet af Samarbejde mellem Videnskabsmænd i en Række Lande. Det videnskabelige Samarbejde paa bestemt afgrænsede Omraader forklarer Videnskabens store Fremskridt i det sidste Aarhundrede. Valget af Videnskabsmænd til Verdens forskellige Akademier kendetegner Viljen til Bevarelse og Udbyggelse af disse internationale Forbindelser i Videnskaben. For Tiden bliver i Forbindelse med de kolossale nye Muligheder, som Atomenergien har aabnet for Menneskeheden, dette internationale Samarbejde udsat for store Farer. Saafremt Løsningen af de grundlæggende Problemer indenfor Atomfysiken ikke straks bliver gjort til Genstand for internationalt Samarbejde, vil Kræfterne blive spildt ved Forsøg paa at løse disse Problemer i hvert Land for sig, og Studiet af disse vigtige Naturfænomener vil uden Tvivl blive forsinket, hvad der vil være saa meget desto sørgeligere, som disse Problemers afgørende Betydning uden Tvivl ikke ligger i de nye

---

[133] Hans Christian Andersen (1805–1875).
[134] Georg Brandes (1842–1927), Danish literary and cultural figure with considerable European influence.
[135] Hans Christian Ørsted (1777–1851).

Opdagelsers Anvendelse til Krigsformaal, men i de nye Energikilders ene-staaende Kraft, som Atomenergien lover at stille til Menneskehedens Raadighed, og som maa forventes med Tiden fuldstændigt at forandre vor Kulturs Ud-seende. De Farer, som truer Videnskabens Udvikling paa dette Omraade, vil uden Tvivl paa det mest kritiske vise sig i Videnskabens og Fremskridtets Udvikling. I den sidste Tid har jeg haft Lejlighed til at udveksle Tanker i denne Retning med Bohr. Vore Synspunkter er overensstemmende i den Henseende, at Forskernes Røst skal løfte sig i Protest mod Arbejdets Hemmeligholdelse og mod alle mulige Forsøg paa at forvandle et af Videnskabens mest fremragende Resultater til et Legetøj for snævre imperialistiske Synspunkter eller enkelte Landes agressive Planer. Udfra dette Synspunkt ser jeg i mit Valg til Medlem af det danske Videnskabernes Selskab en Understregning af Videnskabernes internationale Betydning og Samarbejde mellem Forskerne, som vi paa enhver Maade skal forsvare og opretholde.

Jeg er saaledes personlig dybt rørt ved den Opmærksomhed, som er blevet vist mig af de danske Videnskabsmænd, og samtidig er det mig en Tilfredsstil-lelse at notere dette Skridt som et Tegn paa den gode Vilje til Udviklingen af Forskernes internationale Samarbejde, som er Grundlaget for Videnskabens Udvikling.

Jeg takker endnu en Gang Danmarks højtærede Gesandt i Rusland, Hr. Døs-sing, som har overgivet mig Videnskabernes Selskabs Diplom, og anmoder ham om at overbringe det danske kongelige Videnskabernes Selskabs Præsident[136] min Taknemmelighed for den store Ære, som Selskabet har vist mig ved sit Valg.

Translation

Speech, given by Academician P.L. Kapitza at the formal reception at V.O.K.S.[132] *on Monday 8 July 1946 on the occasion of the presentation of the membership diploma of the Danish Academy of Sciences and Letters.*

The greatest honour that can be shown to a scientist by the people of another country, is the election to membership of the leading scientific organization of that country, of the Academy of Science.

---

[136] Niels Bohr was President of the Royal Danish Academy of Sciences and Letters from 1939 until his death in 1962. See Part II, pp. [378] ff.

The great honour that the Royal Danish Academy of Sciences and Letters has shown to me by electing me, I appreciate deeply for many reasons.

Most of all I see in it the recognition of the significance of our Soviet science and its participation in the development of international science.

This election also gives me a great pleasure because it is a proof of friendship from the Danish people, which I learnt to honour in my childhood by reading Andersen's[133] fairy tales, in my mature years by acquainting myself with Brandes,[134] and finally, when I became a scientist and started studying magnetism, by encountering the name of Ørsted.[135] In direct connection hereto, I have in my scientific work encountered many Danish physicists, finally works by Niels Bohr, whose name counts among the pioneers of science, and who at the height of his research unquestionably stands in the front line.

I have the great happiness and honour of being considered Bohr's friend. Our acquaintance has already lasted for 20 years, and from the beginning of our friendship the link between us has not only been the interest in the general questions of modern physics, but also the circumstance that we are both pupils of one of the most renowned physics researchers of this century, the brilliant Rutherford.

For both of us, Rutherford was not only the learned scientist, but also a teacher of life, and in him we both have learnt to honour the renowned scientist and to love the great man.

Also another question tied me to Bohr. That was the question of the necessity of holding on to the internationalism of science. The conquests of modern science are the result of cooperation between scientists in a number of countries. The scientific cooperation in certain limited areas explains the great advances of science in the last century. The election of scientists into the various academies of the world characterizes the will to preserve and expand these international ties in science. At present this international cooperation is exposed to great perils in connection with the colossal new possibilities that atomic energy has opened for humanity. If the solution to the fundamental problems in atomic physics is not immediately made the object of international cooperation, efforts will be wasted in attempts to solve these problems in each country on its own, and the study of these important natural phenomena will doubtlessly be delayed, which will be so much the worse as the decisive significance of these problems without doubt does not lie in the application of the new discoveries for military purposes, but in the unique power of the new energy sources, which atomic energy promises to place at the disposal of humanity, and which must be expected with time to completely change the appearance of our culture. The perils threatening the development of science in this field will without doubt appear most critically in the development of science and progress. I have recently had the opportunity to exchange thoughts in this direction with Bohr. Our views are in agreement on the point that the voice of the researchers must be raised in protest against keeping the work secret, and against all possible attempts to transform one of the most remarkable results of science into a toy for narrow imperialistic viewpoints or aggressive plans of some countries. On the basis of this viewpoint I see in my election to membership of the Danish Academy of Sciences and Letters an underlining of the international significance of the sciences and of the cooperation between the researchers, which we must defend and maintain in every way.

I am thus personally deeply touched by the token of esteem shown to me by the Danish scientists, and at the same time it gives me pleasure to note this step as a sign of the willingness towards the development of international cooperation between researchers, which is the basis of the development of science.

I once more thank the honourable Danish ambassador in Russia, Mr. Døssing, who has presented me with the diploma of the Academy of Sciences and Letters, and ask him to convey to the President[136] of the Royal Danish Academy of Sciences and Letters my gratitude for the great honour, which the Academy has shown me with its choice.

TELEGRAM, KAPITZA TO BOHR, 12 July 1946
[Typewritten]
[Source: BSC (29.2)]

Telegram from Moscou arrived in Copenhagen July 12, 1946.

Niels Bohr Copenhagen.

Dear Bohr: Yesterday the Danish ambassador Minister Døssing handed me over the diploma of membership of the Royal Danish Academy of Sciences stop Mister Døssing said many kind words and made a very clever speech stop I answered and asked him to convey to you as the President of the Academy the feelings of my gratitude for the great honour which I receive becoming the member of your Academy stop I also took the opportunity to stress the

[314]

necessity of maintaining the international collaboration of scientists which is now in danger stop I am sending you the text of my speech by post stop Accept my best wishes and regards

your very sincerely

= Peter Kapitza =

DØSSING TO BOHR, 18 July 1946
[Typewritten, with handwritten footnote]
[Source: NBA]

J.No. 41.USSR.17.

1 Bilag.

Moskva, den 18' Juli 1946.

Kære Professor Bohr,

Akademiker Kapitza har bedt mig sende Dem hoslagte Oversættelse[137] af hans Tale ved den Fest, hvor jeg overrakte ham Diplomet som Medlem af Videnskabernes Selskab.

I Overensstemmelse med sædvanlig Praksis her i Moskva var Festen efter Gesandtskabets Anmodning arrangeret af V.O.K.S.,[132] den Institution, som formidler de kulturelle Forbindelser mellem Sovjetunionen og Udlandet. Blandt de indbudte var flere af Akademiker Kapitzas nærmeste og Fagfæller*. Min Tale var ganske kort, holdt paa Engelsk med en latinsk Slutning, der af nærliggende Grunde indskrænkede sig til almindelige Gloser.

Det er for en Dansker en stor Oplevelse at se den Glæde, hvormed russiske Videnskabsmænd modtager enhver Hilsen fra Danmark og specielt fra Dem. Det aabner ellers lukkede Døre. Akademiker Kapitza har 2 store Ønsker, det ene at se Dem som Sovjetunionens Gæst, det andet at han selv kunde komme til Danmark. Om han kan rejse for Tiden, ved jeg ikke. Den første Forudsætning maatte sikkert være en officiel dansk Indbydelse.

Han beder mig hilse og sige Dem, at han og hans Familie har det godt. Institutet vokser, og Grunden er blevet udvidet med en stor Park med Udsigt over

---

[137] Danish translation of Kapitza's talk, reproduced on p. [310].

Floden og Moskva. Ogsaa den evigt unge Fru Kapitza sender de hjerteligste Hilsner.

Deres hengivne
Thomas Døssing.

* bl.a. Aleksander V. Fok,[138] som sender hilsner.

Hr. Professor, Dr.phil. Niels Bohr,
   Carlsberg,
      København, Valby.

Translation

J.No. 41.USSR.17.

1 Enclosure.

Moscow, 18 July 1946.

Dear Professor Bohr,

Academician Kapitza has asked me to send you the enclosed translation[137] of his speech at the celebration, where I presented him with the diploma as a member of the Academy of Sciences and Letters.

In accordance with the usual practice here in Moscow the celebration was arranged, at the request of the Embassy, by V.O.K.S.,[132] the institution that arranges the cultural ties between the Soviet Union and abroad. Among those invited were several of Academician Kapitza's family and colleagues*. My speech was quite short, given in English with a conclusion in Latin, which for obvious reasons was confined to simple words.

For a Dane it is a great experience to see the pleasure with which Russian scientists receive any greeting from Denmark, and especially from you. It opens doors otherwise closed. Academician Kapitza has 2 great wishes, the one to see you as a guest of the Soviet Union, the other that he could come to Denmark himself. I do not know whether he can travel at present. The first prerequisite must certainly be an official Danish invitation.

---

[138] Vladimir Aleksandrovich Fock (1898–1974), Russian theoretical physicist.

[316]

He asks me to send his regards and tell you that he and his family are well. The institute is growing, and the plot of land has been expanded with a large park with a view over the river and Moscow. The eternally young Mrs. Kapitza too sends her most cordial regards.

Yours truly
Thomas Døssing.

* including Aleksander V. Fock,[138] who sends his regards.

Professor, Dr.phil. Niels Bohr,
    Carlsberg,
        Copenhagen, Valby.

BOHR TO DØSSING, 10 August 1946
[Carbon copy]
[Source: NBA]

Tisvildelunde
Tisvildeleje

10 August 1946

Kære Minister Døssing.

Jeg takker hjertelig for Deres venlige Brev og behøver vist ikke at sige at jeg er meget taknemmelig for den af Gesandtskabet ordnede festlige Overrækkelse af Diplomet som Medlem af Videnskabernes Selskab til Akademikeren Kapitza ved hvilken Bestræbelserne for at opretholde og udbygge det internationale videnskabelige Samarbejde fandt saa smukt et Udtryk. Jeg var naturligvis ogsaa dybt interesseret i at læse den danske Oversættelse af Kapitzas Tale,[137] hvori han bl.a. kommer ind paa de store Problemer, som den fysiske Videnskabs sidste Fremskridt har stillet Verden overfor. Jeg har om dette Spørgsmaal skrevet nogle smaa Artikler,[110,111] som jeg allerede for længere Tid siden har sendt Kapitza uden at jeg dog fra ham har modtaget nogen Kommentar til de deri fremsatte Tanker. I Artiklerne, hvoraf jeg til Deres Underretning vedlægger Særtryk, har jeg søgt at give Udtryk for det Haab, at Videnskabens Udvikling, der paa saa gennemgribende Maade har ændret Vilkaarene for Folkeslagenes Samliv, i en nær Fremtid vil medføre afgørende Forbedringer i Forholdet mellem Nationerne og bringe den frie Adgang til alle Oplysninger vedrørende alle kulturelle Forhold i de forskellige Lande, uden hvilken gensidig Forstaaelse

[317]

og Tillid næppe kan opretholdes. Jeg haaber meget at Kapitza inden længe vil kunne følge den Indbydelse, som jeg paa Universitetets og Videnskabernes Selskabs Vegne allerede gentagne Gange har rettet til ham om at komme til København og tale om sine smukke og betydningsfulde fysiske Undersøgelser. Desværre har jeg endnu ikke faaet noget Svar derpaa og frygter, at et Brev, som han i et Telegram[139] for nogle Maaneder siden meddelte mig vilde blive afsendt, er gaaet tabt undervejs. Saasnart jeg om nogle Uger efter Ferien kommer til Byen skal jeg imidlertid efter Deres Raad sørge for, at denne Indbydelse paany sendes til Kapitza under de mest officielle Former.[140]

Baade min Kone og jeg haaber, at De selv og Fru Døssing er ved godt Helbred og glæder os til at faa Lejlighed til at træffe Dem, naar De igen kommer til København.

Med mange venlige Hilsner

Deres hengivne

[Niels Bohr]

Translation

Tisvildelunde
Tisvildeleje

10 August 1946

Dear Minister Døssing.

I cordially thank you for your kind letter and I probably do not need to say that I am very grateful for the festive presentation, arranged by the Embassy, of the membership diploma of the Academy of Sciences and Letters to Academician Kapitza, by which the endeavours to maintain and expand international scientific cooperation found such a beautiful expression. I was naturally also deeply interested in reading the Danish translation of Kapitza's speech,[137] in which among other things he touches upon the great problems that the latest advances in physical science have confronted the world with. I have written some small articles about this question,[110,111] which already quite some time

---

[139] See the introduction to Part I, p. [59], ref. 72.
[140] No such invitation has been found in the Niels Bohr Archive.

ago I sent to Kapitza without, however, having received any comment from him about the ideas presented in them. In the articles, of which I enclose offprints for your information, I have sought to express the hope that the development of science, which in such a radical way has changed the conditions for the coexistence of peoples, will in the near future lead to decisive improvements in the relationship between nations and result in free access to all information pertaining to all cultural conditions in the various countries, without which mutual understanding and trust can hardly be maintained. I hope very much that before long Kapitza will be able to accept the invitation that I have already sent him many times on behalf of the University and the Academy of Sciences and Letters to come to Copenhagen and talk about his beautiful and important investigations in physics. Unfortunately, I have not yet received any answer to this, and fear that a letter, which he informed me in a telegram[139] some months ago would be sent, has got lost on the way. As soon as I come to town in a few weeks, after the holidays, I will, however, according to your advice, make sure that this invitation be sent to Kapitza once more under the most official forms.[140]

Both my wife and I hope that you and Mrs. Døssing are in good health and look forward to having the opportunity to meet you when you come to Copenhagen again.

With many kind regards

Yours truly

[Niels Bohr]

BOHR TO ANDERSON, 23 July 1948.
[Typewritten]
[Source: BPP 116]
[Editor's comment: Handwritten by Aage Bohr on top of document: "Not sent off!"]
Gl. Carlsberg,
Copenhagen.

p.t. Lynghuset,
Tisvildeleje.   23th July, 1948.

Dear Sir John,
    Returned to Denmark after our visit to U.S.A., which became a good deal longer than anticipated when my wife and I saw you and Lady Anderson in

London on our way over, I wish for your personal information to tell you a little about my experiences.

I spent most of the time in Princeton as guest of the Institute for Advanced Study of which, as you may know, Oppenheimer is now the director. A few weeks after my arrival I had, however, the opportunity of meeting McCloy in New York and to talk with him about the situation. He advised me to write to Marshall and a few weeks later he submitted to Marshall in the way of introduction a short statement from me[141] together with my old memoranda[62,85] to President Roosevelt in the abbreviated form[142] of which I think I gave you copies during your stay in Copenhagen. I enclose therefore only a copy of the introductory statement of March 22nd, 1948.

It was then hoped that I should see Marshall immediately, but as you know it was suddenly decided that he should go for about six weeks to Bogota where also McCloy went. After McCloy's return he wished me to write a somewhat more explicit account of my views and I therefore sent him for forwarding to Marshall the enclosed general "comments" of May 17th,[143] where I tried to stress the necessity of making the stand for an open world a paramount issue but where, as you will see, I studiously avoided to enter on specific problems of security and procedure.

After Marshall had read the comments I met him in all privacy in the home of McCloy in Washington where we had a long and frank talk on many aspects of the matter and during which Marshall entered as well on difficulties of preparing the American public and Congress for a policy of openness, as on the anxieties as regards military security in case a proposal of universal openness should be accepted. He even touched upon the problem of consent on such policy from the English and Canadian partners in the atomic energy project.

To the best of my knowledge I tried to answer on all such questions and at the end of the talk Marshall invited me to write him a personal letter, for his closer consideration of the matter. After consultation with McCloy, who throughout has been most sympathetic and helpful, as well as with Frankfurter and Oppenheimer, I wrote the letter to Marshall of June 1st of which I enclose a copy.[144] As you will see, I entered in some detail on objections as regards security and I even ventured to comment on questions of procedure, but, as

---

[141] Bohr to Marshall, 22 March 1948, reproduced on p. [159].

[142] These abbreviated versions are not reproduced in the present volume.

[143] Comments of 17 May 1948, reproduced on p. [160].

[144] Bohr presumably means to refer to his letter to Marshall of 10 June 1948, reproduced on p. [163].

you will understand, I did not think it appropriate in the letter to discuss the attitude of the allied governments.

Although I did not have occasion to see Marshall again I understand that he occupies himself deeply with the matter, and through McCloy I obtained his consent to show the letter to his advisers in the State Department who have been especially connected with the consultations of the implications of atomic energy developments and to discuss all aspects of the matter with them. In the last weeks of my stay I had thus several long talks with Gullion,[145] who for years has served as liaison between the State Department and the American delegation to the Atomic Energy Commision of the United Nations, and with Gordon–Arneson[146] who hitherto has served within this delegation but now is going to be successor of Gullion, who leaves the State Department for a university career.

To the first of these talks I brought a rough draft of a few annotations which you will also find enclosed and in which I tried to make some of the remarks in the letter concerning security problems a little more clear.[147] I have the impression that both Gullion and Gordon–Arneson are now quite open to the view that the anxieties about security problems in case the suggested course should be followed ought not to be the main obstacle. To me this aspect of the matter appears a very minor part of the difficulties, but as you also emphasized yourself in Copenhagen,[148] the central issue is surely how the proposal will be received and what opportunities there should be to forestall and counteract reactions which might be anticipated.

I am of course deeply aware of the borders of my knowledge and of my lack of experience, and shall not comment upon the implications of the election campaign in America or on the intricacies of the momentary difficulties in Germany. Yet personally I feel convinced that the only way of overcoming the stalemate is afforded by an initiative of the kind suggested, and I feel sure that if the matter is brought up in a whole-hearted way it would be possible to achieve a great clarification of the situation and to follow up and retain the initiative in a manner which should bring about a hopeful and favourable development. This question was a subject of a long talk with Gordon–Arneson, who

---

[145] Edmund Asbury Gullion (1913–1998) was at this time executive secretary of the Combined Policy Committee, which had been established in 1943 to coordinate American, British and Canadian programmes with regard to atomic energy.

[146] R. Gordon Arneson (1916–1992), the State Department's liaison to the Atomic Energy Commission.

[147] Bohr's "Annotations" are reproduced on p. [169].

[148] Anderson visited Bohr in Copenhagen in September 1947.

is a most attractive and enlighted person and who came up from Washington for a long general talk with me in New York the day before we took the plane back to Denmark.

As regards my other experiences in America I have not much to tell. I did not take part in or attend any public discussions on political matters, and I kept away from any places where scientific or technical developments of military importance are taking place. On the whole I received a deep impression of the vigorous activities in pure scientific research and ordinary industrial developments, but at the same time I felt everywhere the real gloom which has taken hold of the minds of all internationally interested persons and the eagerness with which they look out for any idea which could open new prospects.

I hope you will find that I have acted with proper discretion in the delicate items of which I have written and shall be grateful to learn your views about the whole matter. Of course, you will appreciate the difficulties of my situation and know best yourself how to treat the information in this letter, but if you find it advisable I shall be glad if you would speak with Lord Halifax about it. I expect to hear about developments from McCloy, who will keep contact with me through the American Embassy in Copenhagen.

My wife joins me in sending Lady Anderson and yourself our kindest greetings and best wishes

Yours very sincerely

[Niels Bohr]

BOHR NOTES, 1949
[Typewritten with heading and note (last paragraph) in secretary's handwriting.]
[Source: BPP 112]

Notater til samtale med ambassadør Jos. Marvel

En afgørende Styrkelse af Frihedens Sag kunde opnaas, dersom der til Meddelelsen til Verden om Pagtens Undertegnelse fra Indbydernes Side knyttedes en Udtalelse af denne Art:

Ved denne Pagt, der har været følt nødvendig paa Grund af Usikkerheden i Verden og de voksende Bekymringer indenfor de demokratiske Lande, tilsigtes paa ingen Maade at forøge Spændingen mellem de Grupper af Nationer, hvori Menneskeheden i Øjeblikket beklageligvis er delt, men alene ved en Tilkendegivelse af Fællesskab i Syn paa den personlige Frihed som Grundlag for Menneskelivet og fælles Vilje til at forsvare denne Frihed, at sætte alt ind paa Fredens Opretholdelse.

Varig Fred kan imidlertid vanskeligt tænkes uden et Samarbejde mellem alle Nationer i gensidig Tillid, og den nødvendige Betingelse herfor maa være fri Adgang til Oplysning om alle Spørgsmaal, der angaar Nationernes indbyrdes Forhold. Kun dersom alle Medlemmer af enhver Nation kan danne sig begrundede Forestillinger om sociale Forhold og tekniske Fremskridt i ethvert andet Samfund, kan en Tilstand opnaas, hvor hæderlig Overbevisning og godt Eksempel alene bliver bestemmende for Civilisationens Udvikling og Folkenes Skæbne.

Samtidig med at vi bekendtgør det Samarbejde, der er etableret mellem et Antal demokratiske Nationer for Fredens Bevarelse, vil vi derfor fra vor Side tilbyde hele Verden fuldkommen fri Adgang til Oplysninger af enhver Art, der maatte ønskes, paa den Betingelse at tilsvarende Adgang til Oplysning gives af ethvert andet Land indenfor de forenede Nationer.

Tanker vedr. en mulig erklæring fra U.S.A. i sammenhæng med bekendtgørelse om Atlantpagtens oprettelse

Translation

Notes for conversation with ambassador Jos. Marvel

A decisive strengthening of the cause of freedom could be achieved if the conveners linked a statement of the following kind to the announcement to the world about the signing of the Pact:

With this pact, which has been deemed necessary because of the uncertainty in the world and the growing concerns within the democratic countries, there is no intention at all to increase the tension between the groups of nations into which humanity at the moment unfortunately is divided, but only, by a manifestation of a shared view of personal freedom as the foundation for human life and the shared will to defend this freedom, to make every effort towards the preservation of peace.

However, it is difficult to imagine lasting peace without cooperation between all nations in reciprocal trust, and the necessary requirement for this must be free access to information about all questions concerning the mutual relations between nations. Only if all members of any nation are able to form well-founded conceptions as regards social conditions and technical advances in any other society, can conditions be achieved where honest conviction and good example alone can be determining for the development of civilization and the fate of the peoples.

At the same time as we announce the cooperation that has been established between a number of democratic nations for the preservation of peace, we will therefore on our part offer the whole world completely free access to information of any kind that may be desired, on the condition that corresponding access to information be given by any other country among the united nations.

Thoughts concerning a possible declaration from the U.S.A. in connection with the announcement of the establishment of the Atlantic Pact

BOHR NOTES, undated
[Handwritten by Margrethe Bohr]
[Source: BPP 6.12]
[Editor's comment: Written at the end of 1950 or somewhat later.]

8 August 1945 blev Bomben kastet. I de følgende Dage var N[iels] og Aa[ge Bohr] optaget af at skrive en Artikel til Times.

N blev anmodet af den engelske Atomenergikommission[149] om at give sine views paa Følgerne af Atomenergiens Udvindelse political implications. Disse blev afleveret den Dag, N. rejste tilbage til Danmark. Der blev foreslaaet om det muligt vilde være det klogeste at foreslaa Aabenhed med det samme.

25 August rejste N hjem.   Smuts?

Blev anmodet af Frankfurter om at skrive en Artikel i Science.

4 Okt. Foredrag i Ingeniørforeningen. Forskellige Misforstaaelser opstod derved. Groves sendte en Mand over for at sige at der kun maatte omtales, hvad der stod i Smyth Rapporten.[150]

---

[149] The Atomic Energy Authority.
[150] See the introduction to Part I, p. [56].

Terletzsky kommer til København illegalt med Hilsen fra Kapitza.
7 Okt. N's 60 Aars Fødselsdag – Hilsen fra Anderson[117] etc.

Samtaler[151] med den amerikanske Ambassadør Davies[152] under Mødet[153] mellem Truman, Atlee[154] og Mackenzie King.[155]

Kapitza bliver indbudt til at give et Foredrag i Kbh.[156]

Liliental rapporten[128] – Breve sendt til O. Busch[157] – Brev til Baruch.[158]
Besøg i Amerika Sept–Nov. 1946. Møder i National Academy.[159]
L. Strauss[160] besøgte os paa Hotel Waldorff.
N. talte med Acheson – Benj. Cohen[161] – Mr. Herbert Marx.[162]

Besøg hos Baruch sammen med Kramers[163] og Tolmann[164] og Eberhard.[165]
Besøg hos Bronch[166] i Philadelphia.
Besøg hos Eberhard sammen med General MacNaughton.[167]

---

[151] In Copenhagen.

[152] Monnett B. Davis. See the introduction to Part I, p. [65].

[153] The meeting led to the Washington Declaration on atomic energy of 1945. See the introduction to Part I, p. [65].

[154] Clement Richard Attlee (1883–1967), British Prime Minister from August 1945 to 1951.

[155] William Lyon Mackenzie King (1874–1950), Prime Minister of Canada 1921–1926, 1926–1930 and 1935–1948.

[156] See Bohr to Kapitza, 17 November 1945 and 11 April 1946. Reproduced on pp. [302] and [307], respectively.

[157] From the chronology it seems that Bohr intended to refer to Vannevar Bush, to whom he sent the letter of 17 April 1946, reproduced on p. [308], thanking for the Acheson–Lilienthal Report.

[158] Bohr to Baruch, 1 June 1946, reproduced on p. [156].

[159] See the introduction to Part I, p. [52].

[160] Lewis Strauss (1896–1974). Appointed to the U.S. Atomic Energy Commission in 1946 of which he was chairman from 1953 to 1958.

[161] Benjamin V. Cohen (1894–1983), who worked in the State Department, is regarded as one the significant American voices in postwar policy planning.

[162] Herbert Marks (1907–1960), assistant to Dean Acheson.

[163] Hendrik Anthony Kramers (1894–1952), Dutch physicist, Bohr's first assistant and a particularly good friend. He was now chair of the scientific committee of the International Atomic Energy Commission under the United Nations (UNAEC). See the introduction to Part I, p. [65].

[164] Richard C. Tolman (1881–1948), American physical chemist and physicist.

[165] Probably Ferdinand Eberstadt (1890–1969), Baruch's assistant

[166] Detleff Wulff Bronk (1897–1975), prominent American biophysicist.

[167] Andrew G.L. McNaughton (1887–1966), Canadian representative to the UNAEC.

Ved Besøg i Canada blev Æresdoktor v. McGill i Montreal sammen med Sir John Anderson.

Rejste hjem paa Queen Mary med Sir John og Lady Anderson, traf Redaktør ... fra Times.[168]

Sir J Anderson og Lady Andersons Besøg i Kbh. 1947 Sept.

Marvel meget forstaaende og interesseret.

McCloy kommer til Danmark. N er sammen med ham hos Marvels og han kommer senere hjem til Carlsberg og ser N's Papirer og Manuskripter.

Febr. 1948 rejser N. til U.S.A. besøger paa Vejen Sir John og Lady Anderson i London, drak The hos dem og spiste Frokost i Ambassaden. Aage var rejst 1 Maaned før os.

Forbindelse med McCloy, Niels skrev Comments til Orientering for McCloy at bringe Marshall. Marshall rejste til Bogota og hans Ophold trak ud paa Grund af Oprøret.

N. spiste senere Middag hos McCloy sammen med Marshall – Far udarbejdede derefter Breve til Marshall ved Hjælp af Bang Jensen.[169] Drøftelser m. Marx og Benj. Cohen.

Besøgte Strauss paa Shoreham i Washington.

Rejser hjem c 25 Maj 1948. Sidste Dag kom Arneson op fra Washington til Northcourt for at tale med Niels.

Robert og K. Oppenheimer kommer til København i Sept 1948, vi alle rejser til Solvay mødet i Bruxelles.

Niels og jeg rejste til Paris for at N. kunde træffe Marshall. N. henvendte sig først til Benj. Cohen som satte N. i Forb. med Marshall, denne var optaget og bad General Osborne[170] tale med Niels. N. havde flere Samtaler med ham og med Arneson. Talte ogsaa med Tryggve Lie.[171]

---

[168] Ellipses in original.

[169] Povl Bang-Jensen (1909–1959), Danish official who helped Bohr in his political efforts.

[170] Frederick Henry Osborn (1889–1981) was at this time Deputy to the U.S. Representative to the UNAEC. He was also a pioneer in anthropology and population studies and a major proponent of eugenics.

[171] Trygve Halvdan Lie (1896–1968), the first Secretary General of the United Nations.

Foster Dulles[172] paa Besøg i Kbh. Drak The hos os paa Carlsberg.

N's Forb. med Dep[artementschef] H.H. Koch,[173] som fra første Øjeblik var dybt interesseret.

Marshall træder tilbage, Acheson blir hans Eftf.

Jan. 1949 rejser Aage til Amerika.

Efteraaret 1949 holder N. Gifford lectures[174] i Edinburgh, paa Tilbagerejsen besøgte vi Sir John Anderson i London. Boede hos Reventlows.[175] Var til Middag med Dep. Koch hos Mr. og Mrs. Morse,[176] hvem N. fortalte om ...

Samtaler med Stephen White.[177]

Rejste til Princeton Februar 1950.

Lykkedes ikke at træffe Acheson. N. traf Senator MacMahon[178]

Frankfurter – Cohen – Marx. Lewis Strauss – Marvel – Kennan.[179]

Det aabne Brev skrives ved Hjælp af Aage, Kauffmann,[63] Robert O, Bang Jensen og Hill.[180] Marietta[181] skriver Maskine. Courant.[182]

Venter med Udsendelsen til er kommen til Danmark.

Rejser over England for at tale med Reventlow og Sir John A. Sir John raader til at vente med Udgivelsen til Beg. af Juni, hvor Smuts kommer til Cambridge og muligvis vil støtte Tanken.

Smuts blir syg og kommer ikke.

Besøg i London.

---

[172] John Foster Dulles (1888–1959) would serve as Dwight D. Eisenhower's Secretary of State from 1953 to 1959.

[173] Hans Henrik Koch (1905–1987), prominent Danish government official who worked closely with Bohr in promoting the idea of an open world.

[174] See Vol. 10, pp. [172]–[181].

[175] Eduard Reventlow (1883–1963), Danish ambassador to the United Kingdom from 1939 to 1953.

[176] David A. Morse (1907–1990), Director–General of the International Labour Organization from 1948 to 1970 who showed genuine interest in Bohr's idea of an open world.

[177] Prominent American journalist.

[178] Brien McMahon (1903–1952), Democratic Senator from 1945 to 1952 and co-chairman of the Joint Committee of Atomic Energy from 1949.

[179] George F. Kennan (1904–2005), distinguished U.S. diplomat and historian.

[180] David Lewis Hill (1919–), nuclear physicist who was a student of Bohr's close collaborator John Archibald Wheeler (1911–). Hill had a close connection to Bohr and his institute and partook in discussions on political matters.

[181] (1922–1978), married to Aage Bohr.

[182] Richard Courant (1888–1972), eminent German–American mathematician and a close friend of Bohr. See the introduction to Part II, pp. [370] ff.

Brevet sendes ud. Dr. og Fru Rozental.[183] Paula[184] og Dir. Slebsagers[185] Hjælp

Meget lidt Reaktion i Offentligheden udenfor Danmark.

Peierls[186] skriver i M/C Guardian.[187]

Stephen White i New York Herald.[188]

– Land og Folk[189] –

Samarbejde med H.H. Koch i Tisv[ilde] i Sommeren 1950, og med Hill.

Tryggve Lie i Kbh, lykkes ikke at se ham.

Samtale med Halvard Lange.[190] Besøg af Mr. Berle.[191]

Samtaler med James Franck.[192] Besøg af Mary Shadow.[193]

Besøg i Stockh.

Artikel i Atomic Sc.[194] og i Nature.

Translation

8 August 1945 the bomb was dropped. During the following days N[iels] and Aa[ge Bohr] were occupied with writing an article for the Times.

N was asked by the British atomic energy commission[149] to give his views on the consequences of the exploitation of atomic energy – political impli-

---

[183] Stefan Rozental (1903–1994), Bohr's assistant, and Hanna Kobylinski (1907–1999), historian of China. Both were close friends of the Bohr family.

[184] Paula Strelitz (1892–1963), a relative of Bohr involved in publishing, who had to flee Germany as a refugee.

[185] Erik Slebsager (1909–1974), Managing Director of the publishing house Schultz in Copenhagen, which published the Open Letter and where Paula Strelitz worked.

[186] Rudolf Peierls (1907–1995), German–British nuclear physicist and close friend of Bohr who was instrumental in getting the atomic bomb project off ground.

[187] The Manchester Guardian, England.

[188] The New York Herald Tribune.

[189] "Land and People", Danish communist newspaper.

[190] (1902–1970), Norwegian Foreign Minister from 1946 to 1965.

[191] Adolf Augustus Berle (1885–1971) was influential in the U.S. State Department and had been part of Roosevelt's "brain trust".

[192] See ref. 24. Franck was famous for the "Franck report" of June 1945, in which a committee of physicists at the wartime "Metallurgical Laboratory" in Chicago reported on "Political and Social Problems" in the face of the atomic bomb.

[193] Probably Mary Merrill Shadow (1925–), member of the Tennessee State Legislature 1949–1953, who married David L. Hill (ref. 180) in December 1950.

[194] Bulletin of the Atomic Scientists, Chicago.

cations. These were delivered on the day N. went back to Denmark. It was suggested whether it would possibly be wisest to suggest openness right away.

25 August N went home. Smuts?

Was asked by Frankfurter to write an article in Science.

4 October. Lecture in the [Danish] Association of Engineers. Various misunderstandings resulted thereby. Groves sent a man over to say that only what was stated in the Smyth Report could be discussed.[150]

Terletzkii comes to Copenhagen clandestinely with greetings from Kapitza.

7 October. N's 60th birthday – Greetings from Anderson,[117] etc.

Conversations[151] with the American ambassador Davies[152] during the meeting[153] between Truman, Attlee[154] and Mackenzie King.[155]

Kapitza is invited to give a lecture in Cph.[156]

The Lilienthal Report[128] – letters sent to O. Busch[157] – letter to Baruch.[158]

Visit to America in Sep.–Nov. 1946. Meetings in the National Academy.[159]

L. Strauss[160] visited us at Hotel Waldorff.

N. spoke to Acheson – Benj. Cohen[161] – Mr. Herbert Marks.[162]

Visit to Baruch with Kramers[163] and Tolman[164] and Eberhard.[165]

Visit to Bronk[166] in Philadelphia.

Visit to Eberhard with General McNaughton.[167]

On a visit in Canada made honorary doctor at McGill in Montreal together with Sir John Anderson.

Went home on the Queen Mary with Sir John and Lady Anderson, met ..., editor from the Times.[168]

Sir John Anderson and Lady Anderson's visit to Cph. in 1947 Sep.

Marvel very understanding and interested.

McCloy comes to Denmark. N is with him at Marvels, and he later comes home to Carlsberg and sees N's papers and manuscripts.

Feb. 1948 N. travels to the U.S.A., visits Sir John and Lady Anderson in London on the way, had tea with them and had lunch at the Embassy. Aage had left 1 month before us.

Connection with McCloy, Niels wrote Comments for orientation for McCloy to take to Marshall. Marshall left for Bogota and his stay was prolonged because of the rebellion.

N. later had dinner at McCloy's together with Marshall – father thereafter prepared letters to Marshall with the help of Bang Jensen.[169] Discussions with Marks and Benj. Cohen.

Visited Strauss at Shoreham in Washington.

Travel home around 25 May 1948. On the last day Arneson came up from Washington to Northcourt to speak to Niels.

Robert and K. Oppenheimer come to Copenhagen in Sep. 1948, we all go to the Solvay meeting in Brussels.

Niels and I went to Paris so that N. could meet Marshall. N first approached Benj. Cohen, who put N. in touch with Marshall, he was busy and asked General Osborn[170] to speak to Niels. N. had several conversations with him and with Arneson. Also spoke to Trygve Lie.[171]

Foster Dulles[172] on a visit to Cph. Had tea with us at Carlsberg.

N's connection with Dept. Head H.H. Koch,[173] who was deeply interested from the first moment.

Marshall resigns, Acheson becomes his successor.

In Jan. 1949 Aage goes to America.

Autumn 1949 N. gives the Gifford lectures[174] in Edinburgh, on the way back we visited Sir John Anderson in London. Stayed at Reventlow's.[175] Dined with Dept. Head Koch at Mr. and Mrs. Morse's,[176] whom N. told of ...

Conversations with Stephen White.[177]

Journey to Princeton February 1950.

Did not succeed in meeting Acheson. N. met Senator McMahon[178]

Frankfurter – Cohen – Marks. Lewis Strauss – Marvel – Kennan.[179]

The Open Letter is written with the help of Aage, Kauffmann,[63] Robert O, Bang Jensen and Hill.[180] Marietta[181] types. Courant.[182]

Postpones distribution until coming to Denmark.

Goes via England to talk to Reventlow and Sir John A. Sir John advises to postpone publication until the beginning of June, when Smuts comes to Cambridge and possibly will support the idea.

Smuts falls ill and does not come.

Visit to London.

The letter is distributed. Dr. and Mrs. Rozental.[183] Paula[184] and Dir. Slebs-ager's[185] help

Very little reaction in the public outside Denmark.

Peierls[186] writes in the M/C Guardian.[187]

Stephen White in the New York Herald.[188]

– Land og Folk[189] –

Collaboration with H.H. Koch in Tisv[ilde] in summer 1950 and with Hill.
Trygve Lie in Cph., not able to meet him.

Conversation with Halvard Lange.[190] Visit by Mr. Berle.[191]

Conversations with James Franck.[192] Visit by Mary Shadow.[193]

Visit to Stockholm.

Article in Atomic Sc.[194] and in Nature.

PRESS RELEASE, 13 June 1950
[Published in Danish newspapers]

Statsminister Hans Hedtoft udtaler:

– Professor Bohrs Henvendelse til De forenede Nationer er fremkommet uden forudgaaende Drøftelse med den danske Regering. Uden her at kunne tage Stilling til Enkelthederne i Niels Bohrs Henvendelse, vil jeg sige, at naar Professoren fremhæver Betydningen af Samarbejde og kræver en aaben Verden med fri Adgang til Oplysninger og uhindret Lejlighed til Tankeudveksling over-alt som Middel til at fremme indbyrdes Tillid og garantere fælles Sikkerhed, kan vi fra dansk Side kun ønske, at disse Synspunkter maa give Stødet til indgaaende Overvejelser hos enhver, der har Medansvar for Verdens Fremtid.

Translation

Prime Minister Hans Hedtoft states:

– Professor Bohr's appeal to the United Nations has been formulated without prior discussion with the Danish government. Without here being able to con-sider the details of Niels Bohr's address, I will say that when the professor emphasizes the significance of cooperation and demands an open world with free access to information and unhindered opportunity for the exchange of ideas everywhere as means to advance mutual trust and guarantee common security, we can on the part of Denmark only wish that these viewpoints may give rise to careful considerations by everyone who shares the responsibility for the future of the world.

AAGE BOHR PRESS RELEASE, 28 April 1994
[Typewritten]
[Source: NBA]

Information concerning the visit of the Russian physicist Terletsky to Niels Bohr Institute in the autumn of 1945.

In connection with the attention of the media to the recent publication of a book by Sudoplatov,[195] in which he claims that my father, professor Niels Bohr, should have given information to a Soviet physicist Terletsky of importance for the Soviet atomic energy program, I can state that this assertion is completely incorrect and can provide the following information:
On the 1 or 2 November 1945, professor Mogens Fog, member of Parliament affiliated with the Danish communist party, came to my father asking him to meet a Soviet physicist who had come to Copenhagen with a letter from the Russian physicist Peter Kapitsa and wished to deliver it to my father in a confidential conversation that was to take place in secrecy. My father replied that, of course, he could not indulge in secret arrangements of any kind and told Mogens Fog that if any Russian scientist wished to speak to him, it would have to take place in full openness.[196]
On November 13, 1945 there came a letter[197] to my father from the Russian physicist, professor Terletsky who expressed a wish to visit the Institute. He was received here at this Institute, like other guests. My father met with Terletsky in his office, and, in accordance with my father's wish, I was present all the time.
Terletsky brought with him a letter of introduction (dated October 22, 1945)

---

[195] P. and A. Sudoplatov, with J.L. and L.P. Schechter, *Special Tasks: The memoirs of an unwanted witness – a Soviet spymaster*, Little, Brown and Company, Boston 1994.
[196] Bohr notes, 2 November 1945, reproduced on p. [301].
[197] The letter is deposited in the NBA.

from Kapitsa,[198] with whom my father had been acquainted for a long period of years. Kapitsa sent along recent scientific publications from his Institute. The conversation with Terletsky first dealt with Kapitsa and other personal acquaintances among Russian physicists. Terletsky then raised some technical questions concerning atomic energy, to which my father answered that he was not acquainted with details and referred Terletsky to the report recently published by the U.S. (Smyth Report), which gave an account of the principles on which chain reactions with the release of nuclear energy are based. My father did not make comments to any drawing.[199]

The Danish Intelligence Service was currently kept informed and undertook to take precautions to protect my father in connection with the visit to the Institute. Moreover, U.S. and British authorities were informed, and reports on the episode may therefore exist in archives at several places.

Aage Bohr

---

[198] Reproduced on p. [298].
[199] The drawing is referred to in Sudoplatov, *Special Tasks*, ref. 195, p. 207.

PART II

OTHER POLITICAL AND SOCIAL INVOLVEMENTS

INTRODUCTION

by

FINN AASERUD

Throughout his life Bohr involved himself in a variety of activities in politics and society besides those described in Part I. This introduction, and the reproduction of Bohr's contributions following it, document these. Bohr's popular writings not prepared for special political and social occasions, as well as his publications devoted to particular individuals, are the subject of Volume 12, the last volume of the Collected Works.

Some of the publications reproduced below were part of the long-term responsibilities Bohr had taken on in the domain of Danish science. As President of the Royal Danish Academy of Sciences and Letters and Chairman of the Society for the Dissemination of Natural Science he was expected to speak at the meetings and as a matter of form his words were subsequently published. They are presented in Sections 1 to 3, together with two published radio talks directly related to them.

The remainder of the publications reproduced below may be termed occasional writings. As his reputation grew, Bohr was frequently invited to give talks at conferences outside his own field of physics. He often consented, seeing them as an opportunity to present his views to a broader audience.

Some of Bohr's most important and best known philosophical contributions were written in response to such invitations. His lecture "Light and Life", famously extending the complementarity concept to biology, was given in August 1932 at the International Congress on Light Therapy in Copenhagen devoted to the role of light in "Biology, Biophysics, [and] Therapy".[1] Another classic lecture, in which Bohr extended complementarity to cultural studies,

---

[1] N. Bohr, *Light and Life*, Nature **131** (1933) 421–423, 457–459. Reproduced in Vol. 10, pp. [29]–[35].

constituted the opening of the International Congress of Anthropological and Ethnological Sciences, which took place in Elsinore and Copenhagen in August 1938.[2] Sections 4 to 7 contain those of Bohr's occasional writings that have not been published in earlier volumes of the Collected Works. Each of them relate to some more or less specific issue or activity specified in the section headings.

It should be noted that because the Niels Bohr Collected Works are based on Bohr's publications, some major efforts will remain unrepresented here. These include Bohr's involvement in the 1920s in the expansion and relocation of the Danish National Museum,[3] his efforts for Jewish refugee physicists under Hitler,[4] and his role in the 1950s in the establishment of, first, the European nuclear research establishment CERN[5] and, second, the Nordic Institute for Theoretical Physics (NORDITA) set up in conjunction with his own institute.[6]

The publications introduced below document the variety of Bohr's activities and interests as well as the care with which he presented them. Even when he spoke in an official capacity or wrote a brief foreword to a book or a pamphlet, the resulting publication was never trivial or indifferent. He would usually present the matter in a broad perspective drawing upon the general views he had developed on the various aspects of the human condition.

1. THE ROYAL DANISH ACADEMY

As his career developed, Bohr achieved a unique status in Danish science, and he was urged to accept a number of leading posts. He served as President of the Royal Danish Academy for Sciences and Letters from 1939 until his death in 1962. The Academy and its traditions meant a great deal to Bohr and he took very seriously the responsibilities and opportunities that his role

---

[2] N. Bohr, *Natural Philosophy and Human Cultures* in *Congrès international des sciences anthropologiques et ethnologiques, compte rendu de la deuxième session, Copenhague 1938*, Ejnar Munksgaard, Copenhagen 1939, pp. 86–95. Reproduced in Vol. 10, pp. [240]–[249].

[3] See Vol. 10, p. [250].

[4] The only publication from Bohr's hand providing some documentation of his work for the refugees is his obituary for Aage Friis, professor of history at the University of Copenhagen: N. Bohr, *He Stepped in Where Wrong Had Been Done: Obituary by Professor Niels Bohr*, Politiken, 7 October 1949. Being first and foremost a tribute to his colleague, it is reproduced and translated in Vol. 12. Bohr's efforts for the refugees are described in general terms in F. Aaserud, *Redirecting Science: Niels Bohr, philanthropy and the rise of nuclear physics*, Cambridge University Press, Cambridge 1990, pp. 105–164.

[5] A. Hermann *et al.*, *History of CERN*, North-Holland, Amsterdam, two volumes 1987 and 1990, provides an account of Bohr's role.

[6] The establishment of NORDITA is described briefly in *Nordita's oprettelse* in *NORDITA 1957–82*, the pamphlet printed in celebration of NORDITA's 25th anniversary, pp. 4–10.

as President afforded. He strongly valued the broad approach of the Academy, with members of two Classes, the Scientific–Mathematical and the Historical–Philosophical, attending the same meetings, thus making the Academy a vehicle for broad interaction between the various branches of the natural sciences as well as between these sciences and the humanities. To Bohr, the Danish word for "science" comprised the human as well as the natural sciences, the broad meaning of which term is retained in translations of Bohr's writings below. Bohr also appreciated the Academy's role in promoting international scientific cooperation and its potential for improving the material conditions for scientific research in Denmark.

Founded in 1742, the Royal Danish Academy was part of an international movement represented by the establishment of national institutions such as the Royal Society of London in 1660. In the first half of the nineteenth century, Hans Christian Ørsted played a prominent role in the Academy's activities. In the latter half of the century, the Academy became closely intertwined with the success of the Carlsberg Breweries. On 25 September 1876 the President of the Academy, Johan Nicolai Madvig,[7] received a letter from the founder of the breweries, Jacob Christian Jacobsen, announcing the establishment of the Carlsberg Foundation "in grateful appreciation of [Ørsted's] efforts to spread the light of knowledge to wider circles".[8] He asked the Academy to assume responsibility for the foundation. This was the beginning of a new era for the Academy, which continues to this day.

Bohr was elected member in 1917, one year after he had returned from Manchester to take up the professorship established for him at the University of Copenhagen. Like his younger brother, the mathematician Harald Bohr, who was elected the following year, Bohr attended the meetings frequently. Over the years he published some of his most comprehensive papers in the Academy's Communications (*Meddelelser*)[9] and gave many lectures on the status of his scientific work at the Academy's meetings.[10] He furthermore encouraged physi-

---

[7] Bohr is coauthor of a brief tribute to Madvig: N. Bohr and J. Pedersen, *Foreword* in *Johan Nicolai Madvig: Et mindeskrift*, Bianco Luno, Copenhagen 1955, p. vii. Reproduced and translated in Vol. 12.

[8] Jacobsen to Madvig, 25 September 1876, reproduced (in English translation) in O. Pedersen, *Lovers of Learning: A History of the Royal Danish Academy of Sciences and Letters, 1742–1992*, Munksgaard, Copenhagen 1992, p. 221.

[9] The first example is his seminal trilogy *On the quantum theory of line spectra* (1918–1922), which is reproduced in Vol. 3, pp. [65]–[184].

[10] Abstracts of many of these lectures were printed in the Academy's published Proceedings. The abstracts have been reproduced in the relevant volumes of the Niels Bohr Collected Works.

cists at his institute to publish in the Communications, recommending on the average more than one such publication per year to the Academy.

In 1927 Bohr was asked to become a candidate for President after the recently deceased Vilhelm Thomsen, an outstanding linguist who had been President since 1909. Bohr declined, referring to his heavy scientific work load. In 1934, three years after he had accepted the Academy's offer to move into the Carlsberg Mansion, he declined again, stating the same reason. Only when the presidency became vacant once more in 1939 did Bohr accept. He was to be reelected four times. In all, he presided for twenty-three years – the longest presidency since the early nineteenth century and the second longest ever.[11]

When Bohr accepted the presidency in the spring of 1939, he was visiting the United States where he was working on an important contribution to the understanding of the process of nuclear fission.[12] He was still reluctant to accept the post, less because of the time it would steal from his scientific work than, as he wrote to his wife in the first half of March, because he was "anxious about what the political development might bring".[13] Six days later, when he was closer to accepting the offer, he was more specific about the reasons for his doubt: "I did not know whether it was justifiable, in view of the dark and threatening background, to accept commitments with regard to representation at home and abroad."[14] He was grateful that the Academy had now decided to propose his election without asking him first. In this way, Bohr reasoned, "I could withdraw in a more reasonable and honourable way, if the development is such that I cannot continue my activity in Denmark." He thus anticipated the possibility that he would be forced to escape as a result of the Nazi threat.

Because of his absence, Bohr was not able to attend a meeting as President until 20 October 1939. Fifty-two members were present at the meeting, a typical number throughout Bohr's period as President. As documented in the Academy's published Proceedings (*Oversigt*),[15] he spoke briefly in memory

---

[11] The longest-serving President (1797–1831) was Ernst Heinrich Schimmelmann, who is described by Pedersen (ref. 8, p. 108) as "a Pomeranian upstart who proved to be a financial genius".

[12] See the introduction to Part I, p. [12].

[13] Niels to Margrethe Bohr, 10 March 1939. The correspondence between Niels Bohr and his wife is still in the possession of the Bohr family. The editor is grateful to Aage Bohr for permission to quote from this letter.

[14] Niels to Margrethe Bohr, 16 March 1939, *ibid.*

[15] The actions of the Academy President were of course frequently noted in the Proceedings. This section and the next introduce the passages which quote Bohr's own words and which have not been reproduced in earlier volumes of the Niels Bohr Collected Works. After this introduction the passages are reproduced in full followed by the other relevant publications by Bohr.

of his immediate predecessor S.P.L. Sørensen, prominent chemist and head of the Chemistry Division of the Carlsberg Laboratory, the research arm of the Carlsberg Breweries. After thanking his colleagues for the confidence shown in him, Bohr turned to the continued darkening of the world political situation, which had developed into World War II after Germany had invaded Poland on 1 September. He expressed "the anxiety that must be felt at this time for society in our country and for the scientific activity which it is the purpose of the Academy to protect." He added the hope "that the Academy may have success in offering its contribution to the restoration of the hitherto so fruitful cooperation in the domain of science between all peoples."[16]

Bohr's brief speech at the meeting of 15 March 1940, less than a month before the German occupation of Denmark, was given on the occasion of the attendance of Christian X, King of Denmark since 1912 and as such the Academy's Patron. As can be seen from this and later printed speeches to his King, Bohr took great care to compose an appropriate greeting on behalf of the Academy. He noted[17]

"... the many heavy burdens which rest upon Your Majesty just at this time, when such great storms sweep over humanity, and dangers and misfortunes also threaten our own country, when indeed one of our Nordic sister peoples is already so painfully afflicted."

Bohr was here referring to Finland, which had been invaded by the Soviet Union three months after the German invasion of Poland. Only three days prior to Bohr's speech, Finland had been compelled to sign the Treaty of Moscow, ceding territory to the Russians. Bohr expressed the hope that the King, "up to whom the whole people look with such great trust, may have the fortune to lead Denmark safely through these dangerous times in which we live." The King responded to Bohr's speech by promising "to do his best not to betray the trust that is shown him." Less than half a year later, sometime after the Germans had invaded Denmark, the seventieth birthday of Christian X gave Bohr another opportunity to address the King "in this for our country so fateful a time when the whole people are gathered around our King in love and trust".[18]

---

[16] *Meeting on 20 October 1939*, Overs. Dan. Vidensk. Selsk. Virks. Juni 1939 – Maj 1940, pp. 25–26. Reproduced and translated on pp. [379] ff.

[17] *Meeting on 15 March [1940]*, Overs. Dan. Vidensk. Selsk. Virks. Juni 1939 – Maj 1940, pp. 40–41. Reproduced and translated on pp. [385] ff.

[18] *Meeting on 20 September 1940*, Overs. Dan. Vidensk. Selsk. Virks. Juni 1940 – Maj 1941, pp. 25–26. Reproduced and translated on pp. [389] ff.

[341]

The German invasion of 9 April 1940 had little effect on the pursuit of science in Denmark as compared to other occupied lands.[19] In particular, the Academy continued its meetings despite the blackout, as Bohr stated three days later:[20]

"... both the Secretary and I personally have considered that it was right and important that we did not allow the meetings to cease, but that we continue our work on the tasks of the Academy in the service of science."

The meeting on 30 January 1942 was devoted to the centennial of the birth of Vilhelm Thomsen. Again the King was present, which invoked the following statement by Bohr:[21]

"... we see in the presence of Your Majesty here tonight a special expression of the warm sympathy with which our King follows all endeavours to protect the memories of the deeds that have been done by our countrymen in various fields of life throughout the ages, memories which in the future will be the inexhaustible source from which coming generations will draw strength to work in the same spirit, for the benefit of humanity and for the honour of our country."

In November 1942 the Academy celebrated the 200th anniversary of its establishment. As Bohr stated in his opening address:[22]

"Under other circumstances this event would have been celebrated with great publicity, and not least with extensive participation from the institutions all over the world with which the Academy has cooperated for so many years for the progress of science. Under the present circumstances, such participation is not possible, and we are compelled to commemorate the 200-year existence of our Academy in more modest and humble forms."

As Bohr noted, the King was absent owing to illness. He had sent a telegram, however, to which Bohr read his response before promptly submitting it.[23]

---

[19] See the introduction to Part I, p. [12].

[20] *Meeting on 12 April [1940], 4.15pm*, Overs. Dan. Vidensk. Selsk. Virks. Juni 1939 – Maj 1940, p. 55.

[21] *Meeting on 30 January 1942*, Overs. Dan. Vidensk. Selsk. Virks. Juni 1941 – Maj 1942, p. 32–34. Reproduced and translated on pp. [393] ff.

[22] *Meeting of 13 November 1942 on the 200th Anniversary of the Establishment of the Academy*, Overs. Dan. Vidensk. Selsk. Virks. Juni 1942 – Maj 1943, pp. 26–28, 31–32, 36, 40–41, 44–48. Reproduced and translated on pp. [397] ff.

[23] Translation on p. [407].

Many publications had been prepared for the celebration.[24] In particular, Johannes Pedersen, director of the board of the Carlsberg Foundation, presented his new book about the origins and activity of the foundation. This publication provided the occasion for an extensive speech by Bohr.[25] Taking J.C. Jacobsen's establishment of the Carlsberg Foundation and its role for the Academy as his point of departure, Bohr reflected on

"... the position the Academy has assumed in our society from its establishment until the present day, and on what kind of hopes we dare entertain for its future activity in the service of science and society."

He attached special significance to the Academy's role as a forum where "widely different branches of science are gathered at our meetings and take part in all the proceedings" and where "humanists and scientists ... tell one another what they each regard as important". He emphasized the Academy's role as "an ever more important link to international scientific cooperation". In this connection he noted in particular the establishment by the Danish State after World War I of the Rask–Ørsted Foundation – a pioneer institution in providing economic support for international cooperation in an especially difficult period. He viewed this development as

"... a manifestation of a successful cooperation between prominent figures within the political sphere and within the circle of scientists, who to a significant degree received their inspiration precisely from the debates in our Academy."

In other words, Bohr saw the Academy not merely as a forum where members discussed their scholarly interests, but as an institution encouraging scholars to work for the improvement of the material conditions for research.

Bohr then turned to science in its own right:

"What science brings us is not only skills which help us to master nature and to develop the organization of society; but inextricably connected with the increase of our knowledge is a deeper insight into our own being and our place in existence."

---

[24] See p. [408].
[25] Translation on pp. [411] ff.

In conclusion, he took up the present difficult situation for Denmark and its scholarship, raising the question, which he had dealt with in greater detail the year before,[26] of how Danish culture might best be characterized:

"As regards the special character of our culture here at home, it is first and foremost marked by the way in which the cultural movements, continually coming from the larger societies, are received and transformed here. When we ask whether our culture will continue to be able to flower and bear fruit, we face the same conditions that hold for the continued existence of any living organism. Even the most intimate knowledge of components and structure does not here lead to any answer, but we are left with the hope that a harmony, the full extent of which can only be glimpsed but never fully grasped, will still be retained. The expression for such hope we find, whether in individual living creatures or societies, in the will to live, and when we consider our own culture, we all surely, just in a time like this, have a vivid sense of how strong such a will can be."

He thus compared the understanding of a human society with that of a living organism, basing the hope for survival in the entire Danish people's will to live.

Bohr drew the statement about the "harmony ... which could only be glimpsed" almost verbatim from his lecture at the celebration in 1928 of his twenty-fifth anniversary of completing his pre-university education. In 1928, however, the background for the statement was not a threat to Danish culture but an interpretation of one of his favourite pieces of Danish literature.[27] Nor did the "harmony" yet apply to a living organism, but rather to "the conditions for human thinking".[28] This is a striking case of the continuity of Bohr's ideas under changing circumstances.

Bohr ended by observing that the Academy's future was

"... inextricably tied up with a happy destiny for our whole society; indeed,

---

[26] N. Bohr, *Danish Culture. Some Introductory Reflections* in *Danmarks Kultur ved Aar 1940*, Det Danske Forlag, Copenhagen 1941–1943, Vol. 1, pp. 9–17. Reproduced in Vol. 10, pp. [253]–[261] (Danish original) and [262]–[272] (English translation).

[27] P.M. Møller, *En dansk Students Eventyr* (The Adventures of a Danish Student), first published in 1843, five years after the author's death, on the basis of an unfinished manuscript. On the novel's influence on Bohr, see D. Favrholdt, *General Introduction: Complementarity Beyond Physics* in Vol. 10, pp. XXIII–XLIX, on pp. XXXI ff.

[28] N. Bohr, *Speech Given at the 25th Anniversary Reunion of the Student Graduation Class* [1928], private print, Copenhagen 1953. Reproduced in Vol. 10, pp. [226]–[232] (Danish original) and [233]–[236] (English translation). The passage referred to is on p. [235].

nor can it be detached from maintaining the cooperation of all nations for the advance of science. It is into these hopes we today put our trust."

Bohr's first period as President ended in April 1944, when he was in exile in Great Britain and the United States.[29] However, the Academy decided to postpone the election, allowing Bohr to continue to serve in the expectation that the matter would resolve itself. When Bohr returned, the Secretary informed him of the situation and asked whether he would accept reelection. According to the Proceedings, Bohr responded that "under normal circumstances he would have proposed that the post as President went to a member of the historical–philosophical class at the termination of his period. ... In view of the circumstances on his return he would be positively inclined to accept a request for reelection."[30] At an extraordinary meeting on 21 September Bohr was "reelected as President of the Academy for the rest of the present period, that is, until the election date in April 1949."[31]

Bohr's reelection took place just before the seventy-fifth birthday of King Christian X on 26 September. A greeting to the King was read at the Academy's first regular meeting of the season, on 19 October. It expressed "gratitude for the good conditions that enlightenment and science have had in Your Majesty's long reign". At the same meeting Bohr read a tribute on the occasion of Martin Knudsen's retirement as Secretary.[32]

Two weeks before its first regular meeting the Academy arranged a "Festive Meeting ... on the occasion of the President's Sixtieth Birthday".[33] It was one of many tributes on the occasion.[34] Bohr was handed an address, signed by all Danish members, honouring his scientific contributions in general as well as his role as "adviser and counsellor in connection with the efforts that in the

---

[29] Bohr's exile is described in the introduction to Part I, pp. [14]–[49].

[30] *Extraordinary Meeting on 21 September 1945*, Overs. Dan. Vidensk. Selsk. Virks. Juni 1945 – Maj 1946, pp. 30–31. The words are those of Johannes Hjelmslev, the mathematician and Chairman of the Mathematical–Scientific Class who during Bohr's absence had alternated in presiding over the meetings with Johannes Pedersen, Chairman of the Historical–Philosophical Class.

[31] *Ibid.*

[32] *Meeting on 19 October 1945*, Overs. Dan. Vidensk. Selsk. Virks. Juni 1945 – Maj 1946, pp. 30–31. The address to the King is reproduced and translated on pp. [415] ff. The speech for Knudsen is reproduced in Vol. 12.

[33] *Festive Meeting on 5 October 1945 on the Occasion of the President's Sixtieth Birthday*, Overs. Dan. Vidensk. Selsk. Virks. Juni 1945 – Maj 1946, pp. 27–29.

[34] See the introduction to Part I, p. [63].

most recent times have released energy sources of such gigantic might". The address expressed agreement with Bohr's hope that[35]

> "... science, precisely by its latest lesson of the necessity of concord, will contribute decisively to a harmonic coexistence between the peoples. ...
>
> If we direct our gaze upwards in this room, our eye meets the large picture of Prometheus, who steals the fire from the gods. No better symbol can be given for your work."

Johannes Pedersen, the Chairman of the Historical–Philosophical Class, then spoke in his capacity as Chairman of the Carlsberg Foundation's board of directors, announcing that the board had "decided to make available a sum of 100,000 kr. for the establishment of a trust which will bear Professor Bohr's name and which will serve to support research within a field decided by him." Bohr would dispose of the interest of the trust for as long as he lived. At the end of his talk Pedersen furthermore announced that the board had decided to pay for a painting of Bohr to be hung in the Academy building, subsequently commissioned from the Norwegian artist Henrik Sørensen.

Bohr gave his next published speech as President on 25 April 1947. The occasion was the death of King Christian X after a reign of thirty-five years. Bohr stressed in particular the King's role during World War II, when he[36]

> "... became the natural rallying point for the unshakeable adherence to the traditions and ideals which characterize our old culture and on which we build our lives, and awakened, with his fearless conduct and his encouraging example, the gratitude and admiration of the whole people, just as, out in the world, the King won a reputation of the greatest importance for our cause."

Toward the end of his speech Bohr announced plans for a special meeting devoted to the memory of Christian X.

The special meeting was held six months later, in the presence of the new King, Frederik IX. The President paid tribute to the deceased King Christian X for his never failing interest in the Academy, even in the dark times of occupation. He thanked the new King for wishing to take part in the present meeting as Patron of the Academy, as was the tradition. The King thanked for the President's kind words and announced that earlier in the day he had awarded

---

[35] *Festive Meeting*, ref. 33, p. 28.
[36] *Meeting on 25 April 1947*, Overs. Dan. Vidensk. Selsk. Virks. Juni 1946 – Maj 1947, pp. 53–54. Reproduced and translated on pp. [419] ff.

Bohr the highest Danish order of distinction, the Order of the Elephant, which was usually reserved for royalty and heads of state. "The intention", he said, " was not merely thus to give a distinction to Niels Bohr personally, but also to honour Danish science, of which Niels Bohr was such an outstanding representative."[37]

At the conclusion of the meeting Bohr made a point which he had also made to Christian X before the war: "the vocation of the scientist is remote from the administration of the state, but the prospering of science is of great importance for the life of society." Bohr now went further, implying that science had taken on a greater social role as a result of its achievements during the war:

"We trust that science, which through the ages has shown what kind of advances can be made through common human striving, by its renewed reminder of the necessity of concord must contribute to a brighter future for the whole of humanity."

In March 1949 the Academy's Secretary read the greeting that Bohr had presented to King Frederik IX on his fiftieth birthday, which expressed "the hope for keeping the peace and for a new flowering to which the results of the scientific work will perhaps be able to make significant contributions."[38]

Bohr's last three writings in the Academy Proceedings are all greetings to foreign institutions, namely the University of Glasgow, which celebrated its 500th anniversary in June 1951,[39] the *Akademie der Wissenschaften* in Göttingen, which celebrated its 200th anniversary the same year[40] and the Royal Society of London in connection with that institution's 300th anniversary in July 1960.[41] The latter greeting was the most extensive. The Academy was represented in London by Bohr, Niels Erik Nørlund (Academy President from 1928 to 1933) and Øjvind Winge (geneticist and head of the Academy's

---

[37] *Meeting on 17 October 1947 in Commemoration of King Christian X*, Overs. Dan. Vidensk. Selsk. Virks. Juni 1947 – Maj 1948, pp. 26, 28–29. Reproduced and translated on pp. [425] ff.

[38] *Meeting on 11 March 1949*, Overs. Dan. Vidensk. Selsk. Virks. Juni 1948 – Maj 1949, p. 46. Reproduced and translated on pp. [433] ff.

[39] *Meeting on 19 October 1951*, Overs. Dan. Vidensk. Selsk. Virks. Juni 1951 – Maj 1952, pp. 33–34. Reproduced on pp. [437] ff.

[40] *Meeting on 16 November 1951*, Overs. Dan. Vidensk. Selsk. Virks. Juni 1951 – Maj 1952, p. 39. Reproduced and translated on pp. [439] ff.

[41] *Meeting on 14 October 1960*, Overs. Dan. Vidensk. Selsk. Virks. Juni 1960 – Maj 1961, pp. 39–41. Reproduced on pp. [443] ff.

Mathematical–Scientific Class).[42] Among the Royal Society's early leaders, the address noted Newton in particular, who had "inaugurated a new era in the history of mankind". The address paid tribute to the Royal Society for having exerted a "never relaxing influence upon scientific life not only in Great Britain, but all over the globe" and emphasized the importance for the Danish Academy of its close contact with the British institution. As many as twenty-seven illustrious British scientists were listed, "whom we are proud to recall as having been our members". They included Ernest Rutherford, whose name "marks again the beginning of a new era in the history of mankind." The conclusion bears Bohr's unmistakeable stamp:

"The present times probably hold more dangers to the human race than any previous age, but also more hope of lasting progress, if good will can prevail. Our Academy is confident that the Royal Society of London, also in the future, will keep its leading position within scientific activity, and we know how great a benefit this will be to mankind."

2. ESTABLISHMENT OF THE DANISH NATIONAL SCIENCE FOUNDATION

After World War II, government support of basic science entered the agenda of several countries, taking different forms. In Denmark, the development was slow at first. A "Science and Society" study group, arising from the resistance movement during the war and consisting of about twenty-five prominent Danish scientists, agitated for the issue in the years after the war.[43] Toward the late 1940s August Krogh – prominent physiology professor, Nobel Prize winner and Academy member – expressed support for their cause by proposing that the Academy increase its membership to include younger scientists and otherwise make radical organizational changes in order to involve itself more actively in improving the conditions for scientific research in Denmark. When the Academy did not follow his recommendations, Krogh decided to resign from the Academy in protest. In January 1949, he submitted his resignation – the

[42] Bohr had represented the Academy at Royal Society celebrations before, notably at the celebration in 1946 of Isaac Newton's 300th birthday: N. Bohr, *Newton's Principles and Modern Atomic Mechanics* in *The Royal Society Newton Tercentenary Celebrations 15–19 July 1946*, Cambridge University Press, Cambridge 1947, pp. 56–61. Reproduced in Vol. 12.

[43] H. Knudsen, *Politik, penge og forskningsvilkår* in *Dansk Videnskabs Historie*, Vol. 4, scheduled for publication in 2006. The editor is grateful to Henry Nielsen, one of the editors of the series, for prepublication access to this article, as well as to Henrik Knudsen for illuminating discussions.

only one in the Academy's history – in the form of an open letter which he released to the Danish press.[44]

The Academy responded by sending Krogh a brief letter expressing regret about his decision. However, there was a growing awareness in the Academy of the acute need to improve the material conditions for science. In an interview in the Danish newspaper "Politiken" a few months later with the prominent Danish journalist Merete Bonnesen, who had also interviewed him upon his return from the Soviet Union in 1934,[45] Bohr expressed his concern in this regard. He talked about the need "to create conditions so that our traditions can continue in a way which can be of encouragement and benefit for the whole population."[46] After noting the important contributions of private foundations to Danish science, Bohr continued: "I should like to say loud and clear that these contributions can never be more than a supplement to the effort that the State must make for science." To Bohr it was particularly important "that young people who have shown talent and enthusiasm for scientific research will get such working conditions that there can be no doubt that here at home there is a basis for the realization of their hopes." As for the Academy, he pointed to the "many [members] who work most energetically for the improvement of the situation of science and the possibilities for its application in this country."

In early January 1951 the movement toward improving the conditions for Danish science accelerated further, with Politiken as a strong mouthpiece. On 18 January 1951 Bohr went further with the points from his newspaper interview, delivering a "deeply serious appeal" in the current affairs programme on Danish national radio which Politiken printed the following day.[47] He began by referring to the role of science in the development of society and the recognition in all civilized societies of the importance of participating "according to ability in this common great human work". "More than anything else", Bohr continued, "the international cooperation in the field of science shows us what can be achieved by joint striving towards common goals." He also referred to the traditions in Denmark for "successful participation in the

---

[44] B. Schmidt–Nielsen, *August and Marie Krogh: Lives in Science*, Oxford University Press, New York 1995, pp. 233–236.

[45] See the introduction to Part I, p. [9].

[46] M. Bonnesen, *Niels Bohr on jest and earnestness in science*, Politiken, 17 April 1949, p. 9. Reproduced and translated in Vol. 12 together with other publications by Bohr shedding light on aspects of his own life.

[47] *Niels Bohr's deeply serious appeal*, Politiken, 19 January 1951. Reproduced and translated on pp. [447] ff.

development in numerous fields of science" which "have been regarded by the Danish people with warm and selfless interest." He argued in general terms that because of its increasingly important role in society, science in Denmark required substantially more funding. He concluded:

> "This is a matter which not only concerns the academic world, but in the broadest sense the whole population. It is thus necessary that everyone is aware of the dangers involved for our society if provision is not made – in connection with the wide-ranging precautionary measures which are being implemented to consolidate Denmark's finances and to meet the serious problems facing the country – to keep, by relatively small sacrifices, the path open for ensuring the future in respects that can prove decisive for the prosperity of society and for the preservation of our traditions in the field of cultural life."

Although Bohr's appeal was part of a general effort in the world of Danish science to increase the funding for research, it was Bohr's words that weighed most for the majority of Danish newspapers.

On the radio, Bohr did not speak as President of the Danish Academy. However, at their meeting the following day the Academy members discussed Bohr's views and, after some debate, they agreed unanimously to Bohr's "proposal to address an appeal to the government from the Academy for larger means to be used for the advancement of science."[48] This was the same meeting at which Bohr gave a report on "The Epistemological Problem of Natural Science".[49]

Bohr and Jakob Niels Nielsen, Secretary of the Academy and a prominent mathematician active in matters of science policy, proceeded to write the appeal, which was submitted to the Minister of Education, Flemming Hvidberg, theologian and member of the Danish Conservative Party. The appeal was read at the next meeting of the Academy with explanatory comments by Bohr.[50]

Bohr and Nielsen's appeal was significantly more detailed than Bohr's speech on Danish radio. It stressed the Academy's concern that the conditions for science in Denmark were declining and the "the serious dangers that would result if our country was left behind". The need for added support was twofold:

---

[48] *Meeting on 19 January 1951*, Overs. Dan. Vidensk. Selsk. Virks. Juni 1950 – Maj 1951, p. 40.
[49] *Ibid*, p. 39. The brief published abstract of the talk is reproduced in Vol. 7, p. [384] (Danish original and English translation).
[50] The appeal was printed in *Meeting on 2 February 1951*, Overs. Dan. Vidensk. Selsk. Virks. Juni 1950 – Maj 1951, pp. 42–45. Reproduced and translated on pp. [453] ff.

"On the one hand there is a pressing need for improving conditions of study at the universities and the institutes of higher education, including the enlargement of laboratories, museums and similar educational and research institutions, as well as appreciable improvement of the conditions for young people doing their studies. The Academy wishes to give its warmest support to the statements received from wide circles earnestly requesting that remedy is made for the most obvious deficiencies in such respects at an early date.

On the other hand it is strongly required that significantly increased opportunities for scientific research are created through annual allocations on the part of the State."

Bohr and Nielsen distinguished between applied and fundamental science, suggesting that the former be taken care of by "increasing the funds made available to the research council for science and technology, [and] by allocating the necessary funds to an agricultural research council and a medical research council with corresponding spheres of activity." While the first-mentioned institution had been established in 1946 as *Det Teknisk–Videnskabelige Forskningsråd*, research councils concerning the other two activities were yet to be established.

However, it was "fundamental scientific research" which "by its nature corresponds most closely with the endeavours and activities of the [Academy] members, and here it is thus appropriate for the Academy to explain its viewpoint on the present situation and to put forward more concrete proposals." The address argued compellingly that the contributions of private foundations, in Denmark represented in particular by the Carlsberg Foundation, could no longer suffice to satisfy the pecuniary needs of fundamental science, suggesting that the Danish State might start with a budget comparable to that of the Carlsberg Foundation, gradually increasing it as necessary.

On 2 February 1951, the very day that the appeal was read to the Academy, the general movement for improving the conditions for Danish research and scholarship found expression in a large demonstration in Copenhagen. The demonstration, which was organized by the Danish student community, demanded more support for academic institutions in Denmark. According to the police, the procession was three kilometres long with as many as 10,000 professors, teachers and students taking part. The press reported that the procession was greeted by Education Minister Hvidberg who referred to Bohr's appeal on behalf of the Academy, explaining that he had negotiated with Bohr about it and stressing that he was working hard to see what might be done.

Within a year the Danish government proposed legislation for the establishment of a National Science Foundation to the Danish Parliament. Again Bohr went on the Danish national radio, strongly recommending the proposal. He concluded:[51]

"If circumstances allowed, it would be desirable that even much greater means could be made available for scientific research, as has happened in other countries of Denmark's size and cultural stage. All things considered, however, the suggested establishment of the National Science Foundation will mean that for a number of years conditions will have been created so that scientific research here at home can be maintained and continued to an extent corresponding to the country's traditions and which can secure us a position, in the common human striving in the field of culture, corresponding to our circumstances."

The foundation was established by legislation in June 1952 and the Academy's three representatives were elected in an extraordinary meeting three months later.[52] Its establishment constituted the beginning of a development leading to a true revolution in financial support for science in Denmark during the following decades.

3. THE SOCIETY FOR THE DISSEMINATION OF NATURAL SCIENCE

Hans Christian Ørsted had been inspired to establish the Society for the Dissemination of Natural Science (*Selskabet for Naturlærens Udbredelse*, SNU) in 1824 after a visit to England where he had been impressed by the activities of the Royal Institution. Like the Royal Institution, SNU saw it as its purpose to make science known outside the academic sphere. Its meetings were held at the Technical University of Denmark, which had been founded in part on Ørsted's initiative in 1829 under the name of the Polytechnical Institute (*Polyteknisk Læreanstalt*).

Bohr had been a regular member from his student days when in 1936 he joined SNU's board of directors. He attached great significance to SNU and its objective, and attended the meetings whenever possible, contributing himself

---

[51] N. Bohr, *Statement by Professor Niels Bohr in the current affairs programme on Danish National Radio, Monday 4 February 1952*, Videnskabsmanden: Meddelelser fra Foreningen til Beskyttelse af Videnskabeligt Arbejde **6** (No. 1, 1952), p. 3. Reproduced and translated on pp. [463] ff.

[52] *Extraordinary Meeting on 19 September 1952*, Overs. Dan. Vidensk. Selsk. Virks. Juni 1952 – Maj 1953, p. 31.

with popular lectures.[53] In 1939 he succeeded Martin Knudsen as Chairman,[54] devoting substantial energy to his added responsibility for the rest of his life.

At the beginning of his period as Chairman early in the twentieth century Knudsen took initiative to establish the H.C. Ørsted Medal, to be awarded to a Danish scientist for research of an international standard within physics or chemistry. The Danish King, who served as Patron of SNU, presented the medals. The first Ørsted Medal was awarded to S.P.L. Sørensen in 1909; the second to Bohr's physics teacher, Christian Christiansen in 1912; the third to Knudsen himself in 1916; and the fourth to Bohr in 1924.[55] As Chairman, Bohr was expected to say a few words at the medal award ceremony; the remarks were subsequently printed in the Danish journal "Fysisk Tidsskrift", published by SNU. In addition to Bohr's opening and closing remarks at each presentation of the medal, as reproduced below, the journal also printed the motivation for the award and the recipient's lecture on his or her scientific contribution.

The first medal given during Bohr's presidency was awarded in 1941 to Kai Linderstrøm–Lang, S.P.L. Sørensen's successor as head of the Carlsberg Laboratory's Chemistry Division. Bohr's opening remarks were largely directed to Christian X, expressing gratefulness for the King's continued interest in the Society, which, Bohr noted, was of particular importance during the present period of German occupation, "when the whole people is united in gratitude to and confidence in our King in the endeavour to protect the special and for us most precious culture that has flowered in this country for centuries."[56]

Bohr was going to present three more Ørsted medals, one in 1952[57] and two in 1959.[58] The former was awarded to the chemist Alex Langseth. The new King, Frederik IX, followed the royal tradition of presenting the medals. Prior to the meeting Bohr, together with among others H.C. Ørsted's grandson of the

---

[53] One of his popular lectures on scientific topics has been published: N. Bohr, *Recent Investigations of the Transmutations of Atomic Nuclei*, Fys. Tidsskr. **39** (1941) 3–32. Reproduced in Vol. 9, pp. [413]–[442] (Danish original) and [443]–[466] (English translation). The article is referred to in the introduction to Part I, p. [12].

[54] Bohr published several tributes to Knudsen, including a memorial talk given in SNU; see Vol. 12.

[55] N. Bohr, [Lecture], Fys. Tidsskr. **23** (1925) 10–17. Reproduced in Vol. 5, pp. [128]–[135] (Danish original) and pp. [136]–[142] (English translation).

[56] N. Bohr, *Eighth Presentation of the H.C. Ørsted Medal*, Fys. Tidsskr. **39** (1941) 175–177, 192–193. Reproduced and translated on pp. [469] ff.

[57] N. Bohr, *Ninth Presentation of the H.C. Ørsted Medal*, Fys. Tidsskr. **51** (1953) 65–67, 80. Reproduced and translated on pp. [477] ff.

[58] N. Bohr, *Tenth and Eleventh Presentation of the H.C. Ørsted Medal*, Fys. Tidsskr. **57** (1959) 145, 158. Reproduced and translated on pp. [487] ff.

same name, had shown the King the newly opened H.C. Ørsted Museum. The museum's premises had been provided by the Technical University of Denmark in connection with the 100th anniversary of Ørsted's death the year before, at the celebration of which Bohr had given a lecture on Ørsted's life and work.[59] The tenth and eleventh Ørsted Medals, the last to be awarded during Bohr's period as Chairman, were presented at a meeting in May 1959 to the chemist Jens Anton Christiansen and Paul Bergsøe, an industrial chemist and successful popularizer of science.

In 1942 SNU established an annual grant for particularly promising young Danish scientists.[60] The grant was set up in the memory of Kirstine Meyer, a pioneer woman physicist and historian of physics, who had died the year before.[61] The grant was given out every year for the rest of Bohr's life and is still being awarded.

4. SCANDINAVIAN SCIENTISTS, THE INSTITUTE AND THE UNIVERSITY OF COPENHAGEN

Bohr always attached great importance to Scandinavian cooperation in science. The first collaborators at his institute included Svein Rosseland from Norway and Oskar Klein from Sweden, with whom Bohr would stay in contact for the rest of his life. Over the years, Scandinavian scientists continued to add importantly to science at the institute and Bohr often went to the other Scandinavian countries to visit colleagues and attend conferences.

Institutionally, Scandinavian cooperation in science was represented especially by the Scandinavian Meetings of Natural Scientists dating back to 1839 when the first meeting was held in the Swedish town of Gothenburg.[62] At that time expectations were high for improving cooperation among Scandinavian scientists, broadly defined to include amateurs and medical practitioners. During the early years the meetings were led by three particularly prominent scientists – Hans Christian Ørsted from Denmark, Jöns J. Berzelius from Sweden and Christopher Hansteen from Norway – which gave them prestige and

---

[59] N. Bohr, *Hans Christian Ørsted*, Fys. Tidsskr. **49** (1951) 6–20. Reproduced in Vol. 10, pp. [341]–[356] (Danish original), [357]–[369] (English translation).

[60] N. Bohr, *Memorial Evening for Kirstine Meyer in the Society for Dissemination of Natural Science*, Fys. Tidsskr. **40** (1942) 173–175. Reproduced and translated on pp. [493] ff.

[61] N. Bohr, *Kirstine Meyer, n. Bjerrum: 12 October 1861 – 28 September 1941*, Fys. Tidsskr. **39** (1941) 113–115. Reproduced and translated in Vol. 12.

[62] N. Eriksson, *"I andans kraft, på sannings stråt...": De skandinaviska naturforskarmötena 1839–1936*, Gothenburg Studies in the History of Science 12, Acta Universitatis Gothoburgensis, Gothenburg 1991.

unity. As the years passed, the meetings, oriented as they were toward a broad audience, found it more and more difficult to compete with the increasing specialization within the natural sciences.

The seventeenth meeting in Gothenburg in 1923 was the first in which Finland joined as an equal partner, having gained full independence from Russia in 1918. By then, Bohr had become a member of the seven-man Danish board of the meetings. However, illness prevented him from taking part in the meeting itself.

Not until the eighteenth meeting, held in Copenhagen in 1929 in connection with the 100th anniversary of the Technical University of Denmark, did Bohr give a lecture. The topic was "The Atomic Theory and the Fundamental Principles Underlying the Description of Nature".[63] It was a seminal contribution, the published version of which was soon translated into several languages.

Bohr accepted the invitation to speak at the next meeting, held in Helsinki in 1936, which would prove to be the last. Besides giving a scientific address, "Properties of Atomic Nuclei", presenting his important new ideas on nuclear reactions,[64] he spoke as the Danish delegate at the banquet, which was attended by about 500 people, including dignitaries from the participating countries.

Bohr formed his banquet speech as an ode to small and vulnerable nations, referring implicitly to the threat of Nazi Germany:[65]

"Not least in times like these, when great dangers threaten our freedom and the harmonious development of the whole human culture, we feel in our small, so closely related, countries that we have to turn to each other for mutual support, not only for maintaining our own independence but also for giving even a small contribution to mutual understanding and fruitful cooperation of all peoples."

He ended by referring explicitly to the newly independent Finnish nation:

"I would therefore like to close by expressing the warmest wishes that the fresh shoot on the common human spirit flourishing here in Finland may continue to flower and set fruit for the benefit of humanity, shielded by expectant love for the special character of the new shoot as well as by

---

[63] Reproduced in Vol. 6, pp. [223]–[235] (Danish original) and [236]–[253] (English translation).

[64] Reproduced in Vol. 9, pp. [163]–[171] (Danish original) and [172]–[178] (English translation).

[65] N. Bohr, [*Speech at the Meeting of Natural Scientists in Helsinki, 14 August 1936*] in *Nordiska (19. skandinaviska) Naturforskarmötet i Helsingfors den 11.–15. augusti 1936*, Helsinki 1936, pp. 191–192. Reproduced and translated on pp. [501] ff.

unprejudiced understanding of the essential unity of human life under its eternally changing forms."

As we have seen, Bohr was also going to express his concern for the Finnish cause four years later in his second address as President of the Royal Danish Academy of Sciences and Letters.[66]

Bohr's involvement in the Scandinavian Meetings of Natural Scientists reflects his genuine belief in the Nordic cause which he sought to promote in many different ways throughout his life. In particular, he was instrumental in setting up the Nordic Institute for Atomic Physics (NORDITA) adjacent to his own institute in 1957.

* * *

The twenty-fifth anniversary of Bohr's publication in 1913 of his atomic model coincided with a substantial expansion of his institute's facilities. The events were duly celebrated at the institute with the participation of the staff and others who had contributed to the work, as well as representatives of the Danish cultural and political elite. The occasion received considerable attention in the Danish press. The address (in Danish) that Bohr gave to the invited audience represents an instructive continuation of the talk he had given at the opening of the Institute in March 1921.[67] The typescript, which gives the impression of an oral presentation and appears to be typed from shorthand notes of the talk itself, gives Bohr's perspective on the early history of his institute and documents the transition, experimentally and theoretically, from atomic to nuclear physics that had taken place there in the meantime.[68] It is a comprehensive account, in Bohr's personal style, which includes acknowledgement of the institute's many sources of support over the years and pays tribute to each and every collaborator at the institute for his contributions and activities. Bohr ended in a characteristic fashion:

"... it is the fruitful scientific work within science that all must consider as one of the most encouraging aspects at present within human civilization. It gives hope that a similar peaceful cooperation for the advancement of humanity's common interests can also be achieved in other fields."

---

[66] See above, p. [341].

[67] N. Bohr, *Dedication of the Institute for Theoretical Physics*, typescript, BMSS (9.4). Reproduced in Vol. 3, pp. [284]–[293] (Danish original) and [293]–[301] (English translation).

[68] N. Bohr, *Speech at the Inauguration of the Institute's High-Voltage Plant 5 April 1938*, typescript, BMSS (15.2). Reproduced and translated on pp. [509] ff.

Bohr's talks in 1921 and 1938 are of special interest in that the establishment and leadership of the Institute for Theoretical Physics was one of Bohr's major accomplishments, inseparable from his scientific and intellectual contributions.

A different address, prepared for the same event and given in English, was recorded for international radio broadcasting. As is characteristic for his publications, Bohr did not pay particular attention to his own or his institute's past accomplishments or future plans, but presented instead the recent development of physics in general terms, emphasizing the importance of international cooperation.[69] The address belongs among the several historical overviews that Bohr gave over the years of his own field of physics, the first being his Nobel Prize Lecture of 1922.[70]

*   *   *

On 1 June 1941 Bohr gave a lecture on the occasion of the anniversary of the University of Copenhagen. Given when Denmark was under German occupation, it may be considered in part as a political statement, corresponding to his introduction to the series "Denmark's Culture in the Year 1940"[71] and his wartime writings as President of the Royal Danish Academy of Science and Letters.[72] After discussing in a simple yet insightful manner the gradual erasure of prejudices resulting from the development of science, the unity of science and Denmark's historical role in scientific development, Bohr concluded:[73]

> "The immediate combination of openness towards all real advances in the field of culture and the adherence to the human values created here at home, sheltered by our own traditions, should perhaps be the most characteristic feature of Danish science, as of Danish intellectual life altogether. Let us hope that we will be able also in the future to keep the freedom to develop our own outlook, which is necessary to ensure that our cooperation with all the other peoples can continue to bear fruit."

---

[69] N. Bohr, *Science and its International Significance*, Danish Foreign Office Journal, No. 208 (May 1938) 61–63. Reproduced on pp. [527] ff. The tape recording of the radio talk is deposited at the NBA.

[70] N. Bohr, *Om Atomernes Bygning* in *Les prix Nobel en 1921–1922*, Norstedt, Stockholm 1923. Translated into English as *The Structure of the Atom*, Nature (Suppl.) **112** (1923) 29–44. Both versions are reproduced in Vol. 4, pp. [429]–[465] and [467]–[482], respectively.

[71] Bohr, *Danish Culture*, ref. 26.

[72] See above, pp. [340] ff.

[73] N. Bohr, *The University and Research*, Politiken, 3 June 1941. Reproduced and translated on pp. [533] ff.

The annual staff photograph at Bohr's institute, 1938. Sitting, from the left: E. Rasmussen, T. Bjerge, S. Hellmann, G.C. Wick, H. Levi, G. de Hevesy, N. Bohr, B. Schultz, J.C. Jacobsen, C. Møller, O.R. Frisch, T.S. Chang. Standing, from the left: M. Andersen, M. Sandersen, B.W. Jensen, Lorenzen, E. Dræby, K. Neisig, L.J. Laslett, B. Eriksen, N.O. Lassen, A. Jensen, H. Ostermann, S. Hoffer–Jensen, K.J. Brostrom, B. Bendt–Nielsen, N. Arley, L. Hahn, J. Koch. O. Rebbe, O. Müller, P. Beckmann Hansen, E. Bohr, H.W. Olsen.

[358]

5. PEACEFUL USES OF ATOMIC ENERGY

Preparations for peaceful applications of atomic energy were initiated in many countries soon after the end of the war, often under the auspices of a national Atomic Energy Commission. Denmark had no access to the main raw material, uranium, needed for taking part in this development. In preparation for the coming exploitation of the new energy sources Bohr emphasized instead the importance of training personnel and keeping abreast of scientific developments. He thus offered greater participation for young physicists and engineers in the work at his institute, in order to qualify them for later participation in the atomic energy field. Bohr was supported in this approach by the Thomas B. Thrige Foundation, which provided his institute with a grant for this particular purpose in 1952. Bohr's first publication belonging to this section stems from his close relationship with this foundation.[74]

This relationship originated before the war, when the foundation provided and installed the magnet for the institute's cyclotron, which started operating in late 1938. In the publication just mentioned, Bohr recalls that he learned of the foundation's decision to supply the cyclotron magnet in a conversation with Thomas B. Thrige a few years before Thrige's death in 1938. Thrige was a remarkable innovator, born in the Danish town of Odense and receiving his technical experience around 1890 in part in the laboratory of Thomas A. Edison, who gave him excellent recommendations. After returning to his hometown in 1893, his activities grew rapidly, and the Thomas B. Thrige Factories became the largest electrical enterprise in Denmark.

The Factories' encouraging attitude to Bohr's early efforts to prepare Denmark for the atomic age came to expression in the decision in 1953 to devote a full issue of its journal to "questions concerning atomic research"[75] in which Bohr's contribution constitutes the introduction. In a foreword, the Factories motivated its decision to publish such an issue as follows:[76]

"... we have thought that our readers might be interested in learning a little about the great work being done here at home in a field which today is science but which might become the basis for tomorrow's technology. We have approached Professor Niels Bohr and asked whether he and his collaborators would tell our readers a little about the work that is done at

---

[74] N. Bohr, [*Preface*], ...fra Thrige **6** (No. 1, February 1953), pp. 2–4. Reproduced and translated on pp. [555] ff.
[75] *Ibid*, p. 2.
[76] *Ibid.*

the University Institute for Theoretical Physics, an approach that has been met with great willingness."

This issue of the Thrige journal gives remarkable insight into the experimental work at Bohr's institute at this time. That one of the major Danish industries equated Bohr and his institute with the future use of atomic energy in Denmark testifies to Bohr's incomparable status in his home country.

In December 1953 the prospects for practical applications of atomic energy for the countries not involved in developing the atomic bomb changed with President Dwight D. Eisenhower's announcement of the "Atoms for Peace" programme in a public speech to the General Assembly of the United Nations. The programme was designed to open international exchange of nuclear materials, information and research for peaceful purposes, as opposed to military use of atomic energy.

Only four days after Eisenhower's announcement the Thrige Foundation granted 200,000 Danish kroner to the Academy of Technical Sciences (*Akademiet for de tekniske Videnskaber*, ATV) earmarked for an investigation of the opportunities for developing atomic energy in Denmark. The ATV had been established in 1937 as an independent institution with the purpose of promoting industrially relevant technical and scientific research. Its President was Robert Henriksen, the central figure in the Danish electrical supply industry during the immediate postwar years. Toward the end of February 1954, the ATV established an Atomic Energy Committee (*Atomenergiudvalget*) consisting of Bohr and three other members from science and industry. Bohr agreed to be Chairman, and the Committee went to work. Its close relationship with the Danish government consisted of Bohr giving regular briefings to the Danish Prime Minister, Hans Hedtoft.[77]

It was natural to seek cooperation with England and the United States. Bohr had close contacts in both countries. One of them was John Cockcroft, the director of the British Atomic Energy Research Establishment at Harwell since its creation in 1946. In May 1954 Bohr had discussions with Cockcroft in Copenhagen. In July two members of the Atomic Energy Committee visited Harwell in order to discuss the details of the establishment of a Danish programme, including the provision of enriched uranium and the building of reactors.

In the meantime Bohr had decided to make another extended visit to the United States where, as a member, he had a standing invitation from the Insti-

---

[77] F. Petersen, *Atomalder uden kernekraft: Forsøget på at indføre atomkraft i Danmark 1954–1985 set i et internationalt perspektiv*, Klim, Århus 1996, p. 63.

tute for Advanced Study in Princeton. The Institute's Director since 1947 was J. Robert Oppenheimer, Bohr's colleague with whom he had worked closely at Los Alamos when Oppenheimer was the scientific leader of the project to develop the atomic bomb.[78]

In early June, well before his departure for the United States, Bohr received a letter of invitation from John S. Sinclair, President since 1949 of the National Industrial Conference Board in New York which was organizing its third annual Conference on the Peaceful Uses of Atomic Energy. Sinclair explained that "the conference is designed to bring leaders of industry, science and government together for discussions of how to achieve maximum peacetime benefits." Bohr, who had been recommended the conference by Henrik Kauffmann, the Danish ambassador in Washington with whom he had established a close friendship during the war, responded: "It shall be a welcome experience to me to accept this invitation".

Sinclair to Bohr,
3 Jun 54
English
Full text on p. [717]

Bohr to Sinclair,
19 Jun 54
English
Full text on p. [719]

The conference took place from 13 to 15 October 1954, three weeks after Bohr, his wife and his assistant Aage Petersen had sailed to New York from Copenhagen where Bohr had just overseen another expansion of his institute's facilities, hailed by the Danish newspapers as the beginning of the atomic age in Denmark. Several nations, including Belgium, Brazil, Canada and France, participated in addition to Denmark. Germany was represented by its most prominent physicist, Werner Heisenberg. The newspapers reported that more than 2,000 leaders in government, science and industry were in attendance.

Bohr's role was to speak at the luncheon on the opening day. He was introduced by the former Chairman of the U.S. Atomic Energy Commission (AEC), Gordon Dean. In his address, entitled "Greater International Cooperation is Needed for Peace and Survival", Bohr began by thanking for the invitation "on behalf of the scientists from abroad".[79] He went on to give a spirited account, with an anecdotic bent, of the history of atomic and nuclear physics leading up to the application of nuclear energy.

Only towards the end of the lecture did Bohr make explicit the connection between its content and its title:

> "Scientific knowledge is surely one of the greatest assets of mankind, and
> the common search for truth presents at the same time a unique opportunity
> of furthering that understanding between nations which now is more nec-

---

[78] See the introduction to Part I, p. [17].

[79] N. Bohr, *Greater International Cooperation is Needed for Peace and Survival* in *Atomic Energy in Industry: Minutes of 3rd Conference October 13–15, 1954*, National Industrial Conference Board, Inc., New York 1955, pp. 18–26. Reproduced on pp. [561] ff.

Bohr speaking at the third annual Conference on the Peaceful Uses of Atomic Energy, New York,
1954. On the right, Henrik Kauffmann, the Danish ambassador in Washington.

essary than ever if unprecedented dangers to civilization are to be averted
and all people can reap together in peace the fruits which the progress of
science offers for the promotion of human welfare."

The American newspapers took note of Bohr's main message, many of them
quoting from the above passage. They also noted Bohr's reference to the recent
ratification by twelve European countries of the agreement to create a centre
for nuclear research (CERN) in Geneva, which gave reason for further hope
for international cooperation.

Bohr also accepted the invitation to take part in a private luncheon the following day for particularly distinguished guests. This gave him an opportunity to renew old contacts and establish new ones in the field of the application of atomic energy. On another occasion during his visit to the United States he met with Lewis Strauss, the current Chairman of the AEC. At their meeting, Strauss expressed interest in helping the Danes build a small test reactor. With regard to supplying enriched uranium, however, he was more doubtful.[80]

By November 1954, negotiations on the European side of the Atlantic had passed the preliminary, informal stage. It was time for the Danish government to set up a preliminary Danish Atomic Energy Commission which should take over the responsibilities of the non-governmental Committee and should prepare the establishment of a permanent Commission. Prime Minister Hedtoft died suddenly on 29 January 1955 and was succeeded by Hans Christian ("H.C.") Hansen, the former Foreign Minister. Continuity as regards the atomic energy question was assured by Finance Minister Viggo Kampmann, who was already active in the field. It was thus that a preliminary Danish Atomic Energy Commission, established on 8 March 1955 under Bohr's chairmanship, became the responsibility of the Ministry of Finance. Eleven members were added to the four in the original Committee. The Commission included representatives from science, industry and government. All but three of the members were chosen by Bohr. After the original grant from the Thrige Foundation to the ATV had practically been spent, the Danish government provided ten times that amount to the Atomic Energy Commission as a first instalment towards continuing the work.

On 16 and 17 March 1955 the Danish Industrial Council held its national meeting, an event taking place every five years. The application of nuclear energy was high on the agenda, and the main attraction was a concluding lecture by Niels Bohr in Denmark's largest cinema, the Palladium in Copenhagen. Bohr's lecture was a veritable show, with 1,200 people, mainly from industry, in attendance. The presentation included slides and even physical demonstrations of the detection of radiation from radioactive substances. It described the development of nuclear physics step by step from the discovery of radioactivity and the atomic nucleus through the exploration of induced nuclear transmutations to the discovery of nuclear fission and the possibility of a chain reaction. In the process, Bohr explained the main principles underlying the construction of reactors as energy resources. The article remains a very

---

[80] Petersen, *Atomalder uden kernekraft*, ref. 77, p. 68.

readable exposition of this exciting development written by someone closely associated with it.[81]

Toward the end of his talk Bohr stressed the need for drawing upon the experience and potential of a variety of Danish industrial enterprises. According to Bohr, the broadest approach possible was required:

"In order that we in this country may be able to keep up with the development of the industrial resources, which scientific and technical advances in this as well as in many other fields promise us, it is necessary not only that society supports endeavours to maintain and continue our traditions as regards education and research, but also that, through information and guidance at home and in school, we ensure for such studies the necessary ever larger recruitment of young people, upon whose abilities and fresh potential the future of our country must be built."

The meeting was closed by Robert Henriksen, who remarked on the necessity of introducing atomic energy in Denmark and discussed some practical questions in this respect.

The technical difficulties with the slide presentation during Bohr's lecture only added to the Danish newspapers' euphoric description of the event:[82]

"On the day, which may appear soon, and certainly within a generation, when atomic power plants provide Denmark with electricity, one will remember the day of 17 March 1955, and one will speak of it as a great and significant historical day."

The newly established Atomic Energy Commission gathered for its first meeting the day after Bohr's lecture.

In the course of planning the future of atomic energy in Denmark, contact was also made with Norway and Sweden, which had come further in developing reactor technology based on natural uranium. It was in this context that Bohr accepted the invitation to speak at the biennial meeting of Nordic parliamentarians, held in Copenhagen on 28 and 29 June. The letter of invitation was signed by the Danish representative of the Nordic Interparliamentary Council, Alsing Andersen, a prominent member of the Social Democratic Party and former minister in the Danish government:

---

[81] N. Bohr, *The Physical Basis for Industrial Use of the Energy of the Atomic Nucleus*, Tidsskrift for Industri, **7–8** (1955) 168–179. Reproduced and translated on pp. [571] ff.

[82] Politiken, 18 March 1955.

"... the Nordic Interparliamentary Council has decided to put the question of the application of atomic power on the agenda for the meeting on 29/6. ... I would like to add that it is of course first and foremost the possibilities for the practical application of atomic energy in civilian life and the possibilities for Nordic cooperation in this field, that the participants in the meeting are interested in."

Andersen to Bohr,
17 May 55
Danish
Full text on p. [707]
Translation on p. [708]

Because of the language differences, Bohr was asked to have a printed version of his talk distributed beforehand.

Bohr accepted the invitation and prepared the requested manuscript. The other main speakers were the Norwegian Gunnar Randers, reactor builder and adviser to the United Nations General Secretary, and Rickard Sandler, who had been Swedish prime minister in the mid-1920s and foreign minister in the 1930s. In his talk, "Atoms and Society", Bohr explained the physics behind atomic reactors, as he had done in a different form in his lecture to the Danish Industrial Council.[83] While stressing the need for Nordic cooperation, Bohr was now able to announce that the bilateral negotiations with Great Britain and the United States had reached a conclusion. Both countries would supply a test reactor, and provision of enriched uranium and heavy water was included in the agreements.[84] The facility would be built under the auspices of the Atomic Energy Commission on the peninsula of Risø on Roskilde Fjord some twenty miles west of Copenhagen.

Bohr concluded by expressing the hope that the forthcoming United Nations International Conference on Peaceful Uses of Atomic Energy, to be held in Geneva from 8 to 20 August the same year, would improve the international situation not only with regard to the application of atomic energy but, more broadly, with regard to his vision of an open world:[85]

"We hope first and foremost, however, that this [the Geneva] congress will contribute in a decisive way to the international cooperation for the peaceful use of atomic energy to the benefit of every single society and, through the removal of barriers to mutual information, will advance that understanding of the interests common to all peoples, which is the precondition for a

---

[83] Bohr, *Physical Basis*, ref. 81.

[84] *Til samfundets tarv: Forskningscenter Risøs historie* (ed. H. Nielsen), Forskningscenter Risø, Roskilde 1998, p. 52.

[85] N. Bohr, *Atoms and Society* in *Den liberale venstrealmanak*, ASAs Forlag, Copenhagen 1956, pp. 25–32. Reproduced and translated on pp. [609] ff. The photograph of Bohr is from his talk in New York the year before, ref. 79.

harmonious co-existence for the heightening of human culture in all its shapes and forms."

The conference in Geneva took place only weeks after Eisenhower had publicly proposed his "Open Skies" programme at a summit in the same city between the American, Soviet, British and French leaders. Although the Soviet Union accepted neither "Atoms for Peace" nor "Open Skies" as such, the Russian participation in the subsequent United Nations conference was substantial and impressive. In addition to Bohr, the Danish representatives included other members of the Atomic Energy Commission and several prominent figures from the electrical industry. As Bohr had expressed hope for, the conference represented a thaw in the relationship between East and West.

Bohr's contribution was the first of ten evening lectures on general topics given by particularly prominent scientists.[86] He concluded with the observation that[87]

"... contacts between different cultural communities may influence the attitude of each to an extent which may even lead to a common culture with a more embracing outlook."

On 21 December 1955, when all the necessary preparatory steps were completed, the Danish Parliament unanimously adopted the Act that established a permanent Atomic Energy Commission with the stated aim "to further the peaceful use of atomic energy for the benefit of society". Bohr agreed to continue as Chairman and did so with never failing attention until he died on 18 November 1962.[88]

On 18 October 1957 an impressive exhibition on "Electricity and the Atom" opened in Copenhagen's large "Forum" hall. Since the last such national exhibition, held in Tivoli twenty-five years earlier, the Danish consumption of electricity had increased tenfold, while the physicists, who had hardly been represented in 1932, had entered centre stage with the promise of nuclear energy. Bohr wrote a greeting to the exhibition.[89]

---

[86] *Proceedings of the International Conference on the Peaceful Uses of Atomic Energy, Volume 16, Record of the Conference*, United Nations, New York 1956.

[87] N. Bohr, *Physical Science and Man's Position* in *ibid.*, pp. 57–61. Reproduced (from Ingeniøren **64** (1955) 810–814) in Vol. 10, pp. [102]–[106], quotation on p. [105].

[88] The Danish Atomic Energy Commission was disbanded in May 1975 by the Act on Energy Policy (Lov om Energipolitiske Foranstaltninger).

[89] N. Bohr, *Greeting to the Exhibition from Professor Niels Bohr*, Elektroteknikeren **53** (1957) 363. Reproduced and translated on pp. [633] ff.

Just as the "Electricity and the Atom" exhibition opened, Bohr was on the way to the United States to receive the first Atoms for Peace Award.[90] In his talk at the award ceremony William Clay Ford explained that "the initial idea was sparked by President Eisenhower at Geneva in 1955".[91] The award became a reality after the Ford Motor Company, as its Vice President Ford stated, had "authorized a contribution of one million dollars to be administered by a non-profit organization." Ford noted that "an annual award would be made to the individual or individuals judged to have contributed most to the peaceful uses of atomic energy." The award was established in the memory of Henry Ford and his son, Edsel.

In the summer of 1956 Jerome B. Wiesner of the Massachusetts Institute of Technology had asked Bohr his opinion of whether to provide the award to the American physicist Isidore Isaac Rabi, "as a consequence of his role in organizing the Geneva Conference and his leadership in the initiation of the CERN Laboratory". Bohr responded that he could "most heartily support" such a motion. Two days after Wiesner's request, the President of Atoms for Peace Awards, Inc., James R. Killian, asked Bohr "to assist the trustees in their decision by submitting nominations". It may therefore have come as quite a surprise to Bohr when it was publicly announced on 13 March 1957 that it was he who was chosen as the first recipient of the Atoms for Peace Award of $75,000.

Bohr was awarded the prize at a ceremony on 24 October 1957, which included speeches by among others Dwight D. Eisenhower, President of the United States, who presented the award. The award citation noted several kinds of contributions made by Bohr, including the purely scientific ones and his accomplishment of building and maintaining the "intellectual and spiritual center" of the Copenhagen Institute for Theoretical Physics. It stressed in particular that Bohr had "exerted great moral force in behalf of the utilization of atomic energy for peaceful purposes." In his introductory speech Bohr's close colleague John A. Wheeler gave a spirited account of Bohr's personality and "the causes for which he stands". He pointed to Bohr's principle of complementarity as having provided deep insight into science and human relations. In the last half of his speech Wheeler paid tribute to Bohr's vision and courage in his fight for an open world, in which "Bohr asks our help".

Wiesner to Bohr,
16 Jul 56
English
Full text on p. [720]

Bohr to Wiesner,
10 Aug 56
English
Full text on p. [722]

Killian to Bohr,
18 Jul 56
English
Full text on p. [716]

---

[90] *The Presentation of the first Atoms for Peace Award to Niels Henrik David Bohr, October 24, 1957*, National Academy of Sciences, Washington, D.C. 1957. Reproduced on pp. [637] ff.

[91] W.C. Ford, *The Founding of the Award*, *ibid.*; Ford's article is not included in the reproduction below.

Niels Bohr is awarded the Atoms for Peace Award, 24 October 1957. Left to right: Lewis Strauss, Arthur Holly Compton, Bohr, Dwight D. Eisenhower and James R. Killian, Jr.

Without mentioning the open world explicitly, Bohr concluded in his response that[92]

"... the rapid advance of science and technology in our age, which involves such bright promises and grave dangers, presents civilization with a most serious challenge. To meet this challenge, which calls upon the highest human aspirations, the road is indicated by that world-wide cooperation which has manifested itself through the ages in the development of science."

The atomic test station announced by Bohr in his speech to the Nordic parliamentarians[93] was strongly represented at the "Electricity and the Atom" exhibition, drawing considerable attention. The Risø establishment was formally inaugurated on 6 June 1958. As Chairman of the Atomic Energy Commis-

[92] N. Bohr, *Response* in *Presentation*, ref. 90.
[93] Bohr, *Atoms and Society*, ref. 85.

sion, Bohr gave the main speech. Devoting the first half to the developments in physics constituting the background for the realization of nuclear energy, Bohr noted that "the subsequent eventful development ... has given humanity possession of resources whose application holds such rich promises but at the same time such great dangers."[94] He continued:

"It is of course obvious that any increase of our knowledge and ability brings with it increased responsibility, and in this respect the new development means nothing less than the most serious challenge for our entire civilization, which can only survive through cooperation between all peoples in mutual trust. Against this background, President *Eisenhower* created, with his speech in 1953 to the United Nations, new hopeful prospects through the initiative for international support for the peaceful uses of atomic energy, which found such great accord."

At the end of his speech Bohr presented his vision for the Risø facility:

"It has been our conviction that the implementation of research work, like that started here, was necessary in order to gain the experience and to create the opportunities for education here at home without which it would not be possible to realize the hopes shared by the whole population. Only in this way will we be able to maintain a position among other nations corresponding to our traditions, and together with them take part in the investigation of further perspectives which hold promises for creating inexhaustible resources of energy for the needs of humanity ..."

Bohr's broad approach, which included basic research and education in addition to development and application of nuclear energy, contributed no doubt to the long-term flexibility of Risø as a research institution and its thriving development to this day in spite of the fact that nuclear energy has yet to become part of the Danish energy supply.

6. CONTACT WITH MATHEMATICIANS

On 26 October 1945 Bohr's brother-in-law Niels Erik Nørlund celebrated his sixtieth birthday. Nørlund was active in geodesical work and had a background in astronomy as well as in his main field of interest, mathematics.

---

[94] *Professor Niels Bohr on Risø*, Elektroteknikeren **54** (1958) 238–239. Reproduced and translated on pp. [645] ff.

A Festschrift, consisting of contributions from as many as forty-eight "Danish mathematicians, astronomers and geodesists" was prepared for him. While hardly belonging in any of the categories named in the title of the Festschrift, Bohr accepted the invitation to contribute. Without mentioning Nørlund's name or providing any biographical information about him, Bohr was able to present the current status of atomic science from a perspective relevant to Nørlund's. In the area of atomic physics, Bohr thus wrote,[95]

"... there can be no question of reverting to a description meeting the demands for visualizability to which we are accustomed from everyday life, but rather of a constant further development of a mathematical formalism which is adapted to the measurement results obtainable in the new fields of observation."

During his visit to the United States in the late autumn of 1954, Bohr engaged himself not only in work in Princeton and the introduction of atomic energy in Denmark. At the end of October he received an honorary Doctor of Science degree from Columbia University in connection with the celebration of that university's 200th anniversary. The citation described him as "Distinguished son of Denmark, benefactor of all mankind",[96] and it was on this occasion that he gave the lecture that would become the basis for his seminal philosophical article, "Unity of Knowledge".[97]

On 29 November Bohr spoke at the dedication of the Institute of Mathematical Sciences at New York University. A revised version of Bohr's talk was published a little more than a year later.[98] The invitation had come from the mathematician Richard Courant, the Director of the new institute: "We would not want a big affair, but still I want to ask you whether it would be at all possible on this occasion for you to give a talk." Two weeks later Bohr responded that "it shall certainly be a great pleasure to me to give a talk". He continued:

Courant to Bohr,
13 May 54
English
Full text on p. [710]

Bohr to Courant,
28 May 54
English
Full text on p. [712]

[95] N. Bohr, *On the Problem of Measurement in Atomic Physics* in *Festskrift til N.E. Nørlund, i Anledning af hans 60 Aars Fødselsdag den 26. Oktober 1945, fra danske Matematikere, Astronomer og Geodæter, Anden Del*, Ejnar Munksgaard, Copenhagen 1946, pp. 163–167. Reproduced and translated on pp. [655] ff.

[96] New York Times, 1 November 1954.

[97] N. Bohr, *Science and the Unity of Knowledge* in *The Unity of Knowledge* (ed. L. Leary), Doubleday & Co., New York 1955, pp. 47–62. Reproduced in Vol. 10, pp. [83]–[98].

[98] N. Bohr, *Mathematics and Natural Philosophy*, The Scientific Monthly **82** (1956) 85–88. Reproduced on pp. [667] ff.

"... I might perhaps speak about the inherent connection between the appropriate mathematical tools and the epistemological problems in physical science. But perhaps we can talk it over when we meet."

Bohr and Courant had known each other since both were in their twenties.[99] A little over two years younger than Bohr, Courant had become close to Niels's mathematician brother Harald at the University of Göttingen in the years after 1910. Richard and Niels met for the first time in England in 1913. Their first correspondence stems from 1922, when Courant was involved in arranging Bohr's famous lecture series in Göttingen that year introducing his atomic theory. The year 1924 saw the publication of the book "Methoden der mathematischen Physik", which would influence the new physics and which Bohr characterized as "inspiring for the elucidation of the novel problems with which modern developments in physical science have confronted us."[100]

Through his connections with Rockefeller philanthropy, Bohr was instrumental in securing support for the establishment of a new mathematics institute for Courant in Göttingen, which was inaugurated in 1929. Courant's enjoyment of the new institute was going to be short-lived, however, for in 1934, like others among his Jewish Göttingen colleagues, he was compelled to leave the professorship. He accepted a position at New York University. Bohr and Courant continued to meet frequently, especially after the war. Courant took a keen interest in Bohr's efforts for an "open world", helping him make relevant contacts through his considerable network of political acquaintances. Courant also helped Bohr establish relations with the Ford Foundation, which would provide financial means to the Copenhagen institute in support of international cooperation.

In consideration of their long-term friendship, it is hardly surprising that Bohr accepted Courant's invitation to speak at the dedication of his new institute. Courant not only discussed the talk with Bohr before it was given; he was also involved in its subsequent preparation for publication, "sending, with apologies for the delay, a slightly revised manuscript of your talk." A month and a half later, Bohr sent Courant his corrected version, with the comment:

Courant to Bohr,
6 Apr 55
English
Full text on p. [713]

"... I have been approached from various sides as regards its publication, but I have not made any arrangements and shall leave it all to your better judgement and kind help."

Bohr to Courant,
24 May 55
English
Full text on p. [714]

---

[99] Some of the biographical information on Courant is taken from C. Reid, *Courant in Göttingen and New York: The Story of an Improbable Mathematician*, Springer, New York 1976.
[100] Bohr, *Mathematics and Natural Philosophy*, ref. 98.

Although Courant had originally suggested "Science" as the appropriate journal, the article finally appeared in "The Scientific Monthly".[101]

Entitled "Mathematics and Natural Philosophy" Bohr's article referred to a person's first experience of numbers in early childhood as well as to the historical development of the understanding of mathematics from that of the ancient Greeks to Newton's and Einstein's gradually more sophisticated mathematical concepts. Bohr went on to outline the novel relationships of physical phenomena encountered in the exploration of atomic structure and the decisive role of mathematical abstractions in establishing a consistent physical theory.

In 1957 Bohr's article appeared in a different context, as a contribution to the Technion Yearbook of the Israel Institute of Technology. Bohr had been present when this institution awarded honorary degrees to his fellow physicists and good friends Albert Einstein and James Franck in Princeton on 3 October 1954. In 1956 Bohr visited Technion as part of a second visit to Israel.[102]

7. ON FOOTBALL, CANCER, ISRAEL AND THE DANISH RESISTANCE

While in the United States in the spring of 1939 Bohr found time to make a voice-recording in celebration of the fiftieth anniversary of "Akademisk Boldklub" (AB – Academic Ball Club), of which his father had been a strong supporter and which was one of the leading Danish football teams. Although having played goalkeeper for the A team, Niels never reached the heights of his brother Harald, who played not only for AB but also some years in the Danish national football team, taking part in the 1908 Olympic Games in London where Denmark won the silver medal.

Bohr's recorded greeting was printed ten years later as an introduction to the published ten-year report of AB's accomplishments from 1939 to 1949. Although brief, it provides a glimpse into Bohr's experience in AB and presents his more general reflections on the "importance of sport in the life of study".[103]

In 1946 "Kjøbenhavns Boldklub" (KB – Copenhagen Ball Club), recognized as continental Europe's oldest football club, celebrated its seventieth anniversary. As part of the celebration eleven prominent Copenhageners were invited to answer the question: "What significance does sport have for society and

---

[101] See ref. 98. The manuscript for the talk and various drafts for the article, including Courant's corrections, as well as the corrected proofs, are in BMSS (20.4).
[102] The first visit is described below, p. [375].
[103] *Greeting from Niels Bohr*, in *Akademisk Boldklub 1939–1949*, Copenhagen 1949, pp. 7–8. Reproduced and translated on pp. [675] ff. The recording is deposited at the NBA.

why is football the most widespread sport in the world?"[104] Bohr responded by writing a brief note of congratulations to the club, emphasizing its "pioneer activity promoting the propagation and improvement of the game of football, which here at home has gradually won such great support on the part of youth from all levels of society."[105] The eleven men approached – including Hans Hedtoft who would become Danish prime minister the year after – were presented in the letter of invitation as a football team, with Bohr as goal keeper.

*   *   *

On 16 October 1953 Bohr gave an address on Danish national radio promoting the effort to combat cancer and encouraging the Danish population to contribute money for the cause. The address was printed the following day in the programme for the Danish Royal Theatre's production of "Pygmalion", held in support of the National Association for Combating Cancer in celebration of its twenty-fifth anniversary. Bohr gave his address in his capacity as President of the Danish Cancer Committee. He had accepted this responsibility in 1935 upon the death of his predecessor, the fifteen years older medical doctor J.P. Skot–Hansen. Bohr took on the Presidency the year after the refugee scientist George de Hevesy had found a sanctuary at Bohr's institute, where Hevesy immediately set out to develop his isotopic indicator method for medical purposes, including cancer research.

Indeed, Bohr personally, as well as his institute, had played a major role also in the earlier historical development that he outlined in his address. Thus, Bohr himself became a member of the Radium Foundation's national committee in 1921, the year in which the second national subscription for the cancer cause took place.[106] Moreover, the first doctoral thesis coming out of the Radium Station, "Studies concerning the radiation and transformation of radioactive substances", was completed in 1928 by Jacob Christian Jacobsen who had received employment as a radiophysicist there in 1921. Jacobsen was also an experimental physicist at Bohr's institute from its establishment and would work there until he died in 1965.

In his address, Bohr stated that during the interwar years[107]

---

[104] Einar Middelboe to Bohr, 1 April 1946, NBA.

[105] *Kjøbenhavns Boldklub, 1876 . 26. April . 1946, Program*, Arne Frost–Hansen Bogtryk, Copenhagen 1946, p. 7.

[106] The first national subscription had been organized in 1912.

[107] N. Bohr, [*Address*] in *Det kongelige Teater, Forestillingen lørdag den 17. oktober 1953*, pp. 4–6. Reproduced and translated on pp. [681] ff.

Alle er med i den nådeløse kamp. I en gigantisk anstrengelse på tværs af nationale grænser, hudfarver og ideologier har jordens folkeslag forenet sig i kampen mod denne frygtelige svøbe - denne sygdom, der brutalt rammer hver 5. medborger i samfundet. Verdens bedste hjerner, det fineste tekniske apparatur - alt er sat ind på at skaffe bedre resultater indenfor forskning og behandling. Men dertil kræves enorme summer. Hver eneste krone, De kan afse, tæller med. — Idag er det kræftdag. Køb mange kræftmærker. Sørg for at hele familien har mønter parat til vore sælgere.

The full-page newspaper advertisement for the cancer cause from 1960, featuring Niels Bohr. After the heading, "The fight against cancer", the text continues in English translation: *"Everybody is taking part in the merciless fight.* In a gigantic effort across national boundaries, skin colours and ideologies, the peoples of the Earth have united in the fight against this terrible scourge – this disease which brutally strikes every 5th citizen in society. The best brains in the world, the finest technical equipment – everything is concentrated on obtaining better results in research and treatment. But enormous amounts of money are required for this. Every single krone you can spare counts. Today is Cancer Day. Buy many cancer stamps. Make sure that the whole family has coins ready for our sellers." The text is signed "Niels Bohr, President of the Cancer Committee".

[374]

"... it became more and more obvious that it was necessary to a greater extent to initiate scientific investigations in order to be able to keep up with the development and utilize all paths that research might indicate."

In the summer of 1938 such a relationship formed between Bohr's institute (representing fundamental science) and the National Association for the purpose of developing and using sophisticated equipment for X-ray therapy. The records even indicate that for a short period during the war there was a hospital bed in the institute where people were treated with the new equipment. However, the equipment would have limited success at the Radium Station, where only few patients were treated owing to technical problems before the entire setup burnt down.

Bohr remained President of the Cancer Committee for the rest of his life. His commitment is epitomized by his consent to use his portrait in a full-page newspaper advertisement for the cancer cause in 1960. That year's campaign brought substantially more contributions than previous ones, a circumstance ascribed by insiders to this particular advertisement. In subsequent campaigns the same advertisement ran in newspapers even outside Copenhagen.

* * *

In the autumn of 1953 Bohr visited Israel together with his wife, his son Aage, and his assistant Stefan Rozental. The occasion was the laying of the foundation stone for the physics building at the Weizmann Institute in Rehovot, Israel, which had been established in 1949. The driving force behind the institute's establishment and development was Meyer Weisgal, for whom Bohr wrote a tribute which was published after Bohr's death.[108]

In connection with the visit Bohr accepted his election to the board of governors of the Weizmann Institute. On his return to Denmark he took the initiative, together with Flemming Hvidberg, who had just resigned as Minister of Education, to raise a subscription in Denmark to obtain funds in support of collaboration between Danish scientists and scientists from the Weizmann Institute. As part of this campaign, Hvidberg, Bohr and Rozental gave three consecutive talks on Danish national radio.

Bohr spoke about his strong impressions of the dedicated efforts in Israel to build a new society by "an adaptation which despite all difficulties is taking place at a pace which hardly has any parallel." He referred to his personal

---

[108] N. Bohr, [*Tribute*] in *Meyer Weisgal at Seventy: An Anthology* (ed. E. Victor), Weidenfeld and Nicolson, London 1966, pp. 173–174. Reproduced in Vol. 12.

meetings with Chaim Weizmann (prominent chemist and the first President of Israel) who "understood that the well-being of the new society would be contingent on the most vigorous endeavours to advance scientific research".

Bohr's talk was subsequently published in the Danish journal "Israel".[109] The campaign gave a fine result, and scientific contact between Denmark and Israel continued in a fruitful manner throughout Bohr's lifetime and after.[110]

*   *   *

The Comrades' Assistance Fund was established in Denmark after the war by a group who had been imprisoned during the occupation. On its ten-year anniversary in 1955 the Fund arranged a large gathering of previous concentration camp prisoners, in which the healthy former prisoners served as hosts to 500 comrades who had been injured during imprisonment. The programme included an address by Danish Prime Minister H.C. Hansen as well as Beethoven's Fifth Symphony and one of Tchaikovsky's piano concertos. The music was carefully chosen on the basis of its role during the war. The first four notes of Beethoven's symphony had been used as an introduction to broadcasts from the British Broadcasting Corporation to Nazi-occupied lands, while Tchaikovsky's concerto had been played on Swedish radio in the evening on 4 May 1945, when Danes listened intensely to hear the news leading up to the German surrender.

Bohr was asked, through Hvidberg, to contribute an article to the memorial booklet prepared in connection with the celebration. Bohr complied and also took part in the celebration itself together with his wife. In his article, Bohr emphasized that he was privileged during the last and hardest phase of the occupation to be able to follow the fate of the Danish population from a distance, where he[111]

> "... got the strongest impression ... of the admiration with which the events in occupied Denmark were followed all over the free world and of the

[109] N. Bohr, *The Rebuilding of Israel: A Remarkable Kind of Adventure*, Israel **7** (No. 2, 1954) 14–17. Hvidberg's and Rozental's talks are on pp. 12–13 and 18–19, respectively. Bohr's talk is reproduced and translated on pp. [689] ff. A tape-recording of Bohr's talk is held at the NBA.

[110] S. Rozental, *Niels Bohr: Memoirs of a Working Relationship*, Christian Ejlers, Copenhagen 1998, pp. 112–116.

[111] N. Bohr, *The Goal of the Fight: That we in Freedom May Look Forward to a Brighter Future* in *Ti år efter*, published by the Comrades' Assistance Fund as a memorial booklet including the programme for a reunion 19–20 March 1955. Reproduced and translated on pp. [701] ff.

importance the resistance against tyranny had for the respect for the people and the standing of the country when the liberation came."

Stressing the seriousness of the world situation in view of the existence of the atomic bomb, Bohr wrote that

"... despite all previous disappointments, we must have faith that the seriousness of the situation now facing humanity will find such wide understanding that new paths will be opened for the resolution of disputes and allow that the peoples together can fulfil the rich promises of progress in welfare that science holds out for us in so many fields."

The idea of an open world was still at the forefront of Bohr's thinking.

[377]

1. THE ROYAL DANISH ACADEMY

I. MEETING ON 20 OCTOBER 1939

MØDET DEN 20. OKTOBER 1939
Overs. Dan. Vidensk. Selsk. Virks. Juni 1939 – Maj 1940, pp. 25–26

TEXT AND TRANSLATION

See Introduction to Part II, p. [341].

Mødet den 20. Oktober 1939.

Præsidenten, NIELS BOHR, der for første Gang er tilstede i Selskabet efter Valget til Præsident, byder Medlemmerne Velkommen. Han mindes det store Tab, som Selskabet har lidt ved S. P. L. SØRENSEN's Død og erindrer om den smukke Maade, hvorpaa J. Hjelmslev i det første Møde efter S. P. L. Sørensen's Død gav Udtryk for Selskabets dybe Taknemmelighed.

Præsidenten udtaler derefter: »Professor Sørensen har ved sin rige, frugtbare Livsgerning som faa bidraget til at øge vort Lands Anseelse i videnskabelige Kredse Verden over, og for alt, hvad der vedrørte vort Selskab, nærede han en Kærlighed, der maatte gribe enhver af os. Alle havde vi glædet os til, at Professor Sørensen i en Aar-række skulde have ledet vort snart 200 Aar gamle Selskab og ikke mindst været vor Talsmand paa Mindedagen for dets Stiftelse, til hvilket netop han, der gennem sin lange Deltagelse i Selskabets Ledelse har bidraget saa meget til at vogte og hævde dets ærværdige Traditioner, havde Forudsætninger som næppe nogen anden. Det er med dyb Vemod, at vi maa tænke paa, at det kun i faa Maaneder blev Professor Sørensen forundt at virke som Selskabets Præsident; men samtidig har han med den usvækkede Energi og Begejstring, hvormed han, trods sin dødelige Sygdom, til det sidste saa virksomt tog Del i de mest betydningsfulde og krævende Arbejder i Videnskabens og vort Samfunds Tjeneste, givet os et Forbillede, der vil anspore enhver af os til uanset Skæbnens Tilskikkelser at yde vort bedste«.

Præsidenten giver dernæst Udtryk for sin Taknemmelighed for den Tillid, Selskabet har vist ham og udtaler Haabet om, at det maa lykkes at videreføre Selskabets Virksomhed i Overensstemmelse med dets store Traditioner. Dette Haab støttede han paa Forvisningen om at kunne søge Raad saavel hos Selskabets tidligere Præsidenter som hos de nuværende Klasseformænd og især hos Sekretæren, der i over 20 Aar har varetaget sit ansvarsfulde Hverv.

Præsidenten omtaler kort den Bekymring, som man for Tiden maa have for Samfundet i vort Land og for den videnskabelige Virksomhed, som det er Selskabets Formaal at værne om. Han haaber at Selskabet maa faa Lykke til at kunne yde dets Skærv til Genopbygningen af det hidtil saa frugtbare Samarbejde paa Videnskabens Omraade mellem alle Folkeslag.

TRANSLATION

Meeting on 20 October 1939.

...

NIELS BOHR, the *President*, who is present in the Academy for the first time since his election as president, bids the members welcome. He recalls the great loss the Academy has suffered with S.P.L. SØRENSEN's death and brings to mind the beautiful way in which J. Hjelmslev at the first meeting after S.P.L. Sørensen's death expressed the Academy's deep gratitude.[1]

The President then states: "Professor Sørensen, with his rich and fruitful lifework, has as few contributed to increasing the respect for our country in scientific circles all over the world, and for everything that had to do with our Academy he felt a love that every one of us must find moving. We had all looked forward to Professor Sørensen leading our nearly 200-year old Academy for many years and not least being our spokesman on the commemoration day of its establishment,[2] to which precisely he, who throughout his long-lasting participation in the leadership of the Academy has contributed so much to guard and uphold its honourable traditions, had qualifications as hardly any other. It is with deep regret that we have to remember that Professor Sørensen was permitted to act as President of the Academy for only a few months; but at the same time – with the unflagging energy and enthusiasm whereby he, despite his fatal illness, to the last so actively took part in the most important and demanding tasks in the service of science and Danish society – he has given us an example that will spur everyone of us to give his best irrespective of the acts of fate."

The President then expresses his gratitude for the trust the Academy has shown him and states the hope that there will be success in continuing the

---

[1] Søren Peter Lauritz Sørensen (1868–1939) died on 12 February, whereupon the mathematician Johannes Hjelmslev (1873–1950) gave his brief memorial speech for him at the meeting five days later: Overs. Dan. Vidensk. Selsk. Virks. Juni 1938 – Maj 1939, pp. 38–39.

[2] The commemoration took place on 13 November 1942. Bohr's words at the meeting are reproduced on pp. [397] ff.

Academy activity in accordance with its great traditions. He based this hope on the certainty that he would be able to seek the advice of the previous Presidents of the Academy as well as of the present Chairmen of the Classes and especially of the Secretary,[3] who has held his responsible post for more than twenty years.

The President briefly mentions the anxiety that must be felt at this time for society in our country and for the scientific activity which it is the purpose of the Academy to protect. He hopes that the Academy may have success in offering its contribution to the restoration of the hitherto so fruitful cooperation in the domain of science between all peoples.

---

[3] The physicist Martin Knudsen (1871–1949) was Secretary of the Academy from 1917 to 1945.

Completed in 1899, the building where the Royal Danish Academy of Sciences and Letters resides today has its origins in the Academy's close relationship with the Carlsberg Foundation. The Carlsberg Foundation occupies the ground floor, while the rest of the building is at the permanent disposal of the Academy.

[384]

II. MEETING ON 15 MARCH [1940]

MØDET DEN 15. MARTS [1940]
Overs. Dan. Vidensk. Selsk. Virks. Juni 1939 – Maj 1940, pp. 40–41

TEXT AND TRANSLATION

See Introduction to Part II, p. [341].

[385]

Mødet den 15. Marts.

Præsidenten byder HANS MAJESTÆT KONGEN Velkommen med følgende Ord:

»Deres Majestæt. Paa Det Kongelige Danske Videnskabernes Selskabs Vegne vil jeg gerne give Udtryk for vor Taknemmelighed over, at vor høje Protektor har villet beære vort Møde med sin Nærværelse. Vi ser heri et fornyet Vidnesbyrd om den velvillige Interesse, som Deres Majestæt stedse har vist Selskabets Virksomhed. Denne Interesse skatter vi ganske særligt, idet vi tænker paa de mange og tunge Byrder, der hviler paa Deres Majestæt netop i disse Tider, hvor saa store Storme gaar hen over Menneskeheden, og Farer og Ulykker ogsaa truer vort eget Land, ja, hvor allerede et af vore nordiske Brødrefolk er blevet ramt saa smerteligt. Vor lille Kreds her i Selskabet staar jo efter sit Virke og sin Sammensætning det aktive politiske Liv fjernt, men i saa fuld en Grad som nogen omfatter vi vor gamle Kultur med den dybeste Kærlighed og deler paa det inderligste Haabet om dens Bevarelse og fortsatte Blomstring. Vi vil derfor gerne have Lov at være med til at udtale de varmeste Ønsker om, at Deres Majestæt, til hvem hele Folket ser op med saa stor en Tillid, maa faa Lykke til at føre Danmark frelst igennem den farefulde Tid, hvori vi lever.«

HANS MAJESTÆT KONGEN takker for Præsidentens Velkomsthilsen, nævner, at det er første Gang, at han overværer et Møde i Selskabet i den nuværende Præsidents Tid og udtaler sin Overbevisning om, at Selskabets Ledelse er i gode Hænder. Kongen lover at gøre sit til ikke at forskertse den Tillid, som vises ham.

TRANSLATION

Meeting on 15 March [1940].

...

The *President* bids HIS MAJESTY THE KING[1] welcome with the following words:

"Your Majesty. On behalf of the Royal Danish Academy of Sciences and Letters I would like to express our gratitude that our noble Patron has wished to honour our meeting with his presence. We see in this a renewed testimony of the gracious interest that Your Majesty has always shown in the activity of the Academy. This interest we treasure especially, in that we think of the many heavy burdens which rest upon Your Majesty just at this time, when such great storms sweep over humanity, and dangers and misfortunes also threaten our own country, when indeed one of our Nordic sister peoples is already so painfully afflicted.[2] In view of its work and its composition, our small circle here in the Academy is, of course, remote from the active political life, but to an equal degree as any we embrace our old culture with the deepest love and most fervently share the hope of its preservation and continued flowering. We would therefore like to be allowed to utter the warmest wishes that Your Majesty, up to whom the whole people look with such great trust, may have the fortune to lead Denmark safely through these dangerous times in which we live."

HIS MAJESTY THE KING thanks for the welcome of the President, mentions that it is the first time that he attends a meeting in the Academy during the term of the present President and states his conviction that the leadership of the Academy is in good hands. The King promises to do his best not to betray the trust that is shown him.

---

[1] Christian X (1870–1947). He succeeded his father, Frederik VIII (1842–1912).
[2] Bohr here refers to Finland. See the introduction to Part II, p. [341].

III. MEETING ON 20 SEPTEMBER 1940

MØDET DEN 20. SEPTBR. 1940
Overs. Dan. Vidensk. Selsk. Virks. Juni 1940 – Maj 1941, pp. 25–26

TEXT AND TRANSLATION

See Introduction to Part II, p. [341].

Mødet den 20. Septbr. 1940 Kl. 16¹⁵.

Ved *Hs. Majestæt Kongens* 70-Aars Fødselsdag den 26. September vil *Præsidenten* overbringe en Adresse med følgende Ordlyd:

»Deres Majestæt.

Det Kongelige Danske Videnskabernes Selskab frembærer paa Deres Majestæts 70 Aars Fødselsdag sin Hyldest og ærbødige Lykønskning.

I denne for vort Land saa skæbnesvangre Tid samles hele Folket om vor Konge i Kærlighed og Tillid, og maaske stærkere end nogen Sinde forenes alle Danske i Bestræbelsen paa at værne om den særegne og for os dyrebare Kultur, der i Aarhundreder har blomstret her hjemme, og hvis Trivsel under Deres Majestæts lange Regering har haft saa gode Kaar.

Med dybfølt Taknemmelighed paaskønner Det Kongelige Danske Videnskabernes Selskab den Forstaaelse og Bevaagenhed, som Deres Majestæt, Selskabets høje Protektor, bestandig har tilkendegivet overfor dets Arbejde, og slutter sig paa det inderligste til Ønsket om at Deres Majestæt endnu i mange Aar maa bevare Sundhed og Kraft til fortsat at virke til Bedste for vort Land.«

TRANSLATION

Meeting on 20 September 1940 4.15pm.

...

On the occasion of *His Majesty the King*'s 70th birthday on 26 September, the *President* will hand over an address with the following words:

"*Your Majesty*.

On your Majesty's 70th birthday the Royal Danish Academy of Sciences and Letters brings its homage and respectful congratulations.

In this for our country so fateful a time the whole people are gathered around our King in love and trust, and, perhaps more strongly than ever before, all Danes are united in the endeavour to guard the special, and for us precious, culture which has flowered for centuries here at home and whose prosperity in the course of the long reign of Your Majesty has had such good conditions.

The Royal Danish Academy of Sciences and Letters appreciates with deeply felt gratitude the understanding and attention that Your Majesty, the noble Patron of the Academy, has always displayed towards its work, and joins most sincerely in the wish that Your Majesty for many years more may enjoy health and strength to continue to work for the good for our country."

Vilhelm Thomsen, 1842–1927.

[392]

IV. MEETING ON 30 JANUARY 1942

MØDET DEN 30. JANUAR 1942
Overs. Dan. Vidensk. Selsk. Virks. Juni 1941 – Maj 1942, pp. 32–34

TEXT AND TRANSLATION

See Introduction to Part II, p. [342].

[393]

Mødet den 30. Januar 1942.

Mødet holdes i Erindringen om Selskabets tidligere mange-aarige Præsident, Professor Vilhelm Thomsen, i Anledning af 100-Aarsdagen for hans Fødsel, den 25. Januar.

Præsidenten byder *Hans Majestæt Kongen* Velkommen med følgende Ord:

»*Deres Majestæt.*

Paa Det Kongelige Danske Videnskabernes Selskabs Vegne vil jeg gerne give Udtryk for vor dybe Taknemmelighed over, at Selskabets høje Protektor ved sin Nærværelse har villet kaste Glans over dette Møde, hvor vi er samlet for at mindes den store Forsker, hvis Gerning har bidraget saa meget til vort Lands Anseelse overalt i Verden, og ikke mindst har været til saa stor Ære for dette gamle Selskab, som han i saa mange Aar ledede. Idet vi tænker paa, hvor mange Byrder der hviler paa Deres Majestæt, og med hele det danske Folk deler Beundringen og Taknemmeligheden for alt, hvad Kongen i disse trange Aar har udrettet for Landet, ser vi i Deres Majestæts Nærværelse her i Aften et særligt Udtryk for den varme Deltagelse, hvormed vor Konge følger alle Bestræbelser for at værne Minderne om de Bedrifter, der paa Livets forskellige Omraader gennem Tiderne er øvet af vore Landsmænd, Minder, som i Fremtiden vil være den uudtømmelige Kilde, hvoraf kommende Slægter kan øse Kraft til at virke i samme Aand til Gavn for Menneskeheden og til Ære for vort Land.«

Hans Majestæt Kongen takker for Præsidentens Velkomsthilsen. Kongen giver Udtryk for sin Taknemmelighed over, at han ikke staar alene og lover at bestræbe sig paa ikke at svigte den Tillid, som vises ham, hverken i Dag eller i Fremtiden.

Præsidenten erindrer om, at Vilhelm Thomsen's Minde fra forskellig Side var blevet hædret i Anledning af 100-Aarsdagen. Saavel hans Fødeby Randers, som København har i disse Dage vedtaget at knytte hans Navn til et offentligt Sted, en Hæder hvortil Randrusianersamfundet har taget Initiativet og er blevet støttet af Anbefalinger fra Københavns Universitet og Det Konge-lige Danske Videnskabernes Selskab. — Fra Selskabet blev der paa 100-Aarsdagen ved en Deputation nedlagt en Krans paa hans Grav, og paa dette første Møde efter Mindedagen har man ønsket i Selskabet at mindes Vilhelm Thomsen i Gerning.

TRANSLATION

Meeting on 30 January 1942.

The meeting is held in memory of the previous President for many years, professor VILHELM THOMSEN, on the occasion of the centenary of his birth on 25 January.

...

The *President* bids *His Majesty the King* welcome with the following words: "*Your Majesty.*

On behalf of the Royal Danish Academy of Sciences and Letters I would like to express our deep gratitude that, by his presence, the noble Patron of the Academy has wished to lend lustre to this meeting, where we are gathered to commemorate the great researcher, whose deeds have contributed so much to the esteem of our country all over the world, and have not least been to such great honour to this old Academy, which he led for many years. As we think of the many burdens that rest upon Your Majesty, and together with the whole Danish people share the admiration and the gratitude for all that the King has accomplished for the country in these difficult years, we see in the presence of Your Majesty here tonight a special expression of the warm sympathy with which our King follows all endeavours to protect the memories of the deeds that have been done by our countrymen in various fields of life throughout the ages, memories which in the future will be the inexhaustible source from which coming generations will draw strength to work in the same spirit, for the benefit of humanity and for the honour of our country."

His Majesty the King thanks for the welcome by the President. The King expresses his gratitude that he does not stand alone, and promises to endeavour not to betray the trust that is shown him, neither today nor in the future.

The *President* recalls that the memory of Vilhelm Thomsen had been honoured from various quarters on the occasion of the centenary. Randers, his place of birth, as well as Copenhagen, have at this time resolved to link his name to a public place, an honour to which the Society of Randrusians has taken the initiative and has been supported by recommendations from the University of Copenhagen and the Royal Danish Academy of Sciences and Letters.[1] – On the day of the centenary a deputation from the Academy laid a wreath on his grave, and on this first meeting after the day of commemoration it has been the wish of the Academy to make an act of commemoration for Vilhelm Thomsen.

[1] Vestertorv (Western Market Place), opposite the railway station of the town of Randers in north-eastern Jutland was now renamed Vilhelm Thomsens Plads (Vilhelm Thomsen Square). In Valby, Copenhagen, a street was named Vilhelm Thomsens Allé.

V. MEETING ON 13 NOVEMBER 1942
ON THE 200TH ANNIVERSARY OF THE ESTABLISHMENT
OF THE ACADEMY

MØDET DEN 13. NOVEMBER 1942
PAA 200-AARSDAGEN FOR SELSKABETS STIFTELSE
Overs. Dan. Vidensk. Selsk. Virks. Juni 1942 – Maj 1943,
pp. 26–28, 31–32, 36, 40–41, 44–48

The original page numbers are indicated
in the margin of the translation

TEXT AND TRANSLATION

See Introduction to Part II, p. [342].

[397]

Mødet den 13. November 1942
paa 200-Aarsdagen for Selskabets Stiftelse.

Præsidenten aabnede Mødet og udtalte følgende:

»I Aften er det netop 200 Aar siden, at det Møde fandt Sted, ved hvilket vort Selskab stiftedes. Under andre Forhold vilde denne Begivenhed være blevet fejret under stor offentlig Opmærksomhed og ikke mindst med omfattende Deltagelse fra de Institutioner Verden over, med hvilke Selskabet i saa mange Aar har samarbejdet paa Videnskabens Fremskridt. Som Forholdene ligger, er jo en saadan Deltagelse udelukket, og vi er henvist til at mindes vort Selskabs 200-aarige Bestaaen under mere beskedne og tilbagetrukne Former. Tilmed er vor høje Protektor, *Hans Majestæt Kongen,* der allerede for nogle Maaneder siden havde givet Tilsagn om at bære vort Møde med sin Nærværelse, paa Grund af Kongens beklagelige Uheld og derpaa følgende alvorlige Sygdom forhindret i at komme til Stede i Aften. Det er derfor med den største Taknemmelighed, at Selskabet for et Øjeblik siden har modtaget dette Telegram:

»Min hjertelige Lykønskning i Anledning af Jubilæet.

Christian R.«

Jeg foreslaar, at Selskabet sender følgende Telegram til *Hans Majestæt Kongen:*

»Det Kongelige Danske Videnskabernes Selskab, samlet til Møde paa 200-Aarsdagen for dets Stiftelse, har med største Glæde modtaget Deres Majestæts Lykønskning og vil gerne over for Deres Majestæt give Udtryk for Selskabets dybe Taknemmelighed for den Bevaagenhed og Interesse, hvormed vor høje Protektor har fulgt Selskabets Virksomhed igennem hele Deres Majestæts lange Regeringstid, under hvilken alle Bestræbelser paa Kulturlivets Omraade her i Landet har haft saa gode Kaar. I Tilknytning til de Følelser, der bevæger hele det danske Folk, føjer Det Kongelige Danske Videnskabernes Selskab til sin ærbødige Tak de inderligste Ønsker for Deres Majestæts snarlige fuldstændige Helbredelse.«

Idet jeg gaar ud fra, at Selskabet slutter sig dertil, vil dette Telegram umiddelbart blive afsendt.

I Anledning af Jubilæet har vi endvidere modtaget en smuk Adresse med følgende Hilsen:

»Det Kongelige Norske Videnskabers Selskab bringer på
200-Årsdagen Det Kongelige Danske Videnskabernes Selskab sin
søsterlige Hilsen med Ønsket om fortsatt fruktbringende Arbeid
for felles Mål til Fremme av nordisk Videnskap og Kultur, idet
vi med Takk minnes at to av dets eldste Medlemmer, Schöning
og Suhm, i 1760 var vårt Selskaps Medstiftere.

Trondheim 13de November 1942.

RAGNVALD IVERSEN   S. SCHMIDT-NIELSEN
Preses   Generalsekretær.«

Det er vort Selskab overordentlig kært at modtage denne
Hilsen fra et Søsterselskab, med hvilket vi er forbundne med
saa mange fælles dyrebare Minder. Vi deler af Hjertet de Ønsker,
der er udtrykt i den smukke Adresse fra Det Kongelige Norske
Videnskabers Selskab, og vi vil i vort Svar udtale vor dybtfølte Tak«.

Præsidenten meddelte, at der, som angivet paa Mødesedlen,
vilde blive forelagt forskellige Skrifter i Anledning af Jubilæet,
og han gav først Ordet til Redaktøren, L. L. HAMMERICH, der forelagde det første Bind af et af Selskabet selv udgivet Skrift,
indeholdende Samlinger til dets Historie.

Præsidenten takkede Redaktøren for hans Redegørelse for
Formaalet og Indholdet af det forelagte Værk, i hvis Tilblivelse
han selv havde taget saa virksom Del. »Vi skylder hele Redak-
tionsudvalget Tak for det store Arbejde, men først og fremmest
skylder vi, saaledes som det fremgaar af Redaktørens Beretning,
Fuldmægtig A. LOMHOLT stor Taknemmelighed for den Omhu
og Indsigt, han har lagt for Dagen ved Udarbejdelsen og ved
sine aarelange forberedende Undersøgelser og Ordning af Selskabets Arkiv, uden hvilke Værket ikke kunde være blevet til.
Det er Selskabets Haab, at disse Samlinger til dets Historie vil
tiltrække sig Opmærksomhed hos alle, der er interesserede i det
danske Samfunds kulturelle Udvikling, og vil blive en værdifuld Kilde til Oplysning for fremtidige Forskere af Videnskabens
Historie i vort Fædreland«.

Præsidenten gav dernæst Ordet til AAGE FRIIS, der forelagde
et af Det Kongelige Danske Selskab for Fædrelandets Historie
udgivet Skrift om Selskabets Medstifter, Hans Gram: *Vita
Johannis Grammii.*

Præsidenten udtalte:

»Det er med største Taknemmelighed, at Det Kongelige Danske Videnskabernes Selskab modtager denne overmaade smukke Hilsen og Gave fra et Søsterselskab, med hvilket det gennem de mange Aar har staaet i saa nær Forbindelse. Den Ære, der er vist vort Selskab ved denne Udgivelse netop i Dag af et Værk om den fremragende Videnskabsmand og Samfundsborger, som for 200 Aar siden tog Initiativet til vort Selskabs Stiftelse, paaskønner vi dybt, og paa Selskabets Vegne vil jeg gerne bede Professor AAGE FRIIS om at overbringe Det Kongelige Danske Selskab for Fædrelandets Historie vor hjerteligste Tak og Genhilsen tillige med de varmeste Ønsker for vort Søsterselskabs Virksomhed og fortsatte Trivsel.«

Præsidenten gav dernæst Ordet til N. E. NØRLUND, der forelagde det første Bind af sit store Værk: *Danmarks Kortlægning. En historisk Fremstilling.*

Præsidenten takkede med følgende Ord:

»Paa Selskabets Vegne vil jeg gerne bringe Professor N. E. NØRLUND en hjertelig Tak for den Hyldest, der er vist vort gamle Selskab gennem Forelæggelsen netop i Dag af dette store Værk om Danmarks Kortlægning, hvori der gøres Rede for den meget betydelige Indsats, som Selskabet under de første 100 Aar af dets Bestaaen har ydet til Fremme af dette Formaal. Som Professor Nørlund har fremdraget, har vort Land paa dette Omraade allerede tidlig ydet store Bidrag til Videnskaben, og vi glæder os over, at disse Traditioner ogsaa i senere Tider har været opretholdt paa ypperlig Maade, og at Danmark atter gennem Professor Nørlunds egne frugtbare Undersøgelser indtager saa fremtrædende Plads i Arbejdet paa den geodætiske Videnskabs Udvikling.«

Præsidenten gav endelig Ordet til Formanden i Carlsbergfondets Direktion, JOHANNES PEDERSEN, der forelagde et Skrift: *Carlsbergfondet* om Fondets Tilblivelse og Virksomhed.

Præsidenten udtalte:

»Paa Selskabets Vegne vil jeg gerne bringe Professor JOHAN-
NES PEDERSEN en hjertelig Tak for hans Forelæggelse af sit op-
lysende Skrift om det store Fond, der siden dets Oprettelse
paa lykkeligste Maade har virket som en Livskilde for dansk
Videnskab. Udgivelsen af dette Skrift netop paa Selskabets
200-Aarsdag vil paa den mest velkomne Maade samle Opmærksomheden om det vidunderlige Æventyr i Selskabets Historie,
som Carlsbergfondets Tilblivelse har været. J. C. Jacobsens
Henvendelse til Selskabet med Meddelelsen om hans store Gave
og hans Forslag om Selskabets Medvirken til dennes Frugtbargørelse er utvivlsomt den mest betydningsfulde Hændelse i
Selskabets Historie i det sidst forløbne Aarhundrede.

Det Kongelige Danske Videnskabernes Selskab tænker i Dag
med den største Taknemmelighed paa den Tillid, Carlsbergfondets Stifter viste det, og vi vil dertil føje vor Tak til de
Mænd, valgt af vor Kreds, som paa saa lykkelig Vis har for-
staaet at fordele og forvalte de store Midler, som ved J. C.
Jacobsens enestaaende Gave er skænket dansk Videnskab. Vi
føler ogsaa Taknemmelighed for den Maade, hvorpaa Direktionen, der efter Fondets Statuter i alle sine Beslutninger er
uafhængig af Selskabet og kun bundet af de Retningslinier,
som Stifteren har angivet, stedse i J. C. Jacobsens Aand har
støttet Videnskabernes Selskab og fremfor alt givet Selskabet
Raadighed over disse smukke Lokaler, hvor vi kan mødes under
saa værdige Former. Et nyt Vidnesbyrd om denne Indstilling
ser vi i den store Gave, som Fondet i Dag har skænket
Selskabet, og som vi modtager med en dybtfølt Tak og vil
bestræbe os for paa bedste Maade at anvende til Selskabets
Formaal.

Paa en Dag som denne mindes vi ogsaa med Taknemmelighed de store Gaver, Legater og Stiftelser, Selskabet i Aarenes
Løb har modtaget fra forskellig Side, og ikke mindst tænker vi
paa, at Statsmyndighederne i de seneste Aar ved betydelige Bevillinger har muliggjort Opretholdelsen af vor Publikationsvirk-
somhed og derved handlet i Overensstemmelse med de gamle
Traditioner, der gaar tilbage til Selskabets allerførste Tid, hvor

det fra Kongens egen Kasse modtog den for dets Virksomhed uundværlige Understøttelse.

De Værker og Skrifter, der her i Dag er blevet forelagt, og de Ord, der har været knyttet til de forskellige Forelæggelser, giver hver for sig saa indgaaende Oplysninger om væsentlige Træk af Selskabets Historie og danner tilsammen saa alsidigt et Billede af vor Virksomhed i de forløbne 200 Aar, at vi næppe paa nogen smukkere og mere levende Maade kunde være blevet mindet om vore ærværdige Traditioner og om de mange fremragende Mænd, som gennem Tiderne har kastet saa megen Glans over vort gamle Selskab. Ved denne Lejlighed vil det imidlertid være naturligt, at vi endnu i nogle Øjeblikke samler vore Tanker om den Stilling, Selskabet har indtaget i vort Samfund fra sin Stiftelse til i Dag, og om hvilke Haab vi tør nære til dets fremtidige Virksomhed i Videnskabens og Samfundets Tjeneste.

Den umiddelbare Anledning til Selskabets Stiftelse var jo, som vi ved, Kravet om Løsningen af visse bestemte Opgaver, der fordrede almen videnskabelig Indsigt og i det første Aarhundrede af dets Bestaaen, hvor der her i Landet kun fandtes faa Organer til Støtte for Videnskaben og dens Anvendelser, havde vort Selskab i begge disse Henseender en vigtig Mission at udføre. Navnlig paatog det sig, ofte efter direkte Opfordring fra Kongemagten, Varetagelsen af forskellige Opgaver, blandt hvilke Landets Kortlægning, saaledes som vi netop har hørt, var den største og heldigst gennemførte. Omkring Midten af det forrige Aarhundrede havde dette Forhold dog væsentligt ændret sig, idet mange selvstændige danske Institutioner, under hvilke de fleste af disse Opgaver mere naturligt henhørte, efterhaanden var vokset op. Ved denne Udvikling spillede vel ikke Selskabet selv, men mange af dets ledende Mænd en afgørende Rolle. Jeg skal blot minde om, at det var paa Initiativ af H. C. Ørsted, der i hele 37 Aar var vort Selskabs Sekretær, at den polytekniske Læreanstalt stiftedes, ligesom allerede tidligere Grundlæggeren af Landbohøjskolen, P. C. Abildgaard, i sine sidste Leveaar havde varetaget dette vigtige Hverv inden for Selskabet.

Betydningen af, at der i vort Samfund stadig fandtes et Centrum, om ikke saa meget for videnskabelige Forskningers Udførelse, saa dog for deres almindelige Værdsættelse, kan imidlertid næppe overvurderes. Det mest slaaende Vidnesbyrd herom er J. C. Jacobsens saa smukt formulerede Henvendelse til Selskabet om Carlsbergfondets Oprettelse, som Fondets Formand lige har fremdraget. Netop i denne Forbindelse tænker vi ogsaa paa, hvor meget det den Gang betød, at videnskabelig Virksomhed i dens mangeartede Afskygninger paa saa harmonisk Maade var knyttet sammen inden for vort Selskab. At Repræsentanter for vidt forskellige Grene af Videnskaben samles til vore Møder og deltager i alle Forhandlinger, giver vort Selskab dets særlige Præg. Den enestaaende Lejlighed, som Humanister og Naturvidenskabsmænd derved faar til at meddele hverandre, hvad de hver for sig betragter som væsentligt inden for deres Forskning, udelukker en Ensidighed, der ofte følger med større Forhold, hvor en skarpere Adskillelse af praktiske Grunde har været nødvendig. Omend ikke alle Kundskabs- og Forskningsomraader, der fortjener Navn af Videnskab, er repræsenteret inden for Selskabet, har de Forbindelser, der her knyttes, sikkert haft og har stadig overordentlig Betydning for den harmoniske Udvikling af de videnskabelige Bestræbelser i vort Land, ja for hele Kulturlivet herhjemme.

Som Forholdene udviklede sig, er vort Selskab tillige blevet et stedse vigtigere Bindeled ved det internationale videnskabelige Samarbejde. Paa dette Omraade kan Danmark med Rette være stolt af den af hele Samfundet delte Forstaaelse for den vigtige Opgave, det her drejede sig om. Især staar Oprettelsen af Rask-Ørsted Fondet — ved hvilken der fra dansk Side gjordes saa betydningsfuld en Indsats paa et Tidspunkt, hvor mellemfolkelige videnskabelige Bestræbelser var særlig vanskelige og særlig paakrævede — som Udtryk for et lykkeligt Samarbejde mellem ledende Personligheder fra det politiske Liv og fra Videnskabsmændenes Kreds, der i væsentlig Grad fik sine Impulser netop fra Forhandlingerne inden for vort Selskab. Ikke alene har dette Fond muliggjort en Deltagelse her fra Landet i det internationale videnskabelige Samarbejde i et Omfang, der ellers næppe havde været tænkeligt, men tillige har Fondets Virksomhed dannet Mønster for Institutioner, skabt i større Lande med tilsvarende rigere Midler. Anerkendelsen af dansk Initiativ og Virke paa dette Omraade er maaske ikke den mindst vigtige af de Omstændigheder, hvorpaa vi bygger vort Haab om, at Danmark, naar fredelige Tider vender tilbage, atter vil finde sin Plads som aktiv Deltager i de mellemfolkelige kulturelle Bestræbelser.

Intetsteds finder vi stærkere Vidnesbyrd om Frugtbarheden af menneskeligt Samarbejde end paa den videnskabelige Forsknings Omraade. Naar vi tænker paa, hvordan Slægter har hjulpet hverandre med at bære Sten til Videnskabens stolte Bygning og i Fællesskab har højnet dens Tinder og styrket dens Fundamenter, føler vi os stillet over for et Værk, hvortil den enkelte kun formaar at bidrage i Kraft af den Rigdom, som rummes i det hele fælles Eje. Hvad Videnskaben bringer os, er ikke alene Kundskaber, der hjælper os til at beherske Naturen og til at udbygge Samfundenes Organisation; men uadskillelig forbundet med Forøgelsen af vor Viden er en dybere Indsigt i vort eget Væsen og vor Stilling i Tilværelsen. Dette fælles Træk for alle Tiders Erkendelsesstræben er i vor egen Tidsalder kommet i Forgrunden som næppe nogensinde før og møder os ligefuldt ved saadanne Yderpunkter som Udforskningen af Naturen omkring os og Studiet af de menneskelige Kulturer. Paa det første Omraade har vi jo lært, at selv vore allersimpleste Begreber som Rum og Tid og Aarsag og Virkning ikke er altomfattende Former, men Idealisationer, hvis Anvendelse hviler paa hidtil upaaagtede Forudsætninger. Paa det andet Omraade har Erkendelsen af gensidig Befrugtning mellem tilsyneladende uafhængige Kulturer og af den Arv, de hver for sig har overtaget fra Kulturer, om hvilke vi indtil for kort Tid siden intet Kendskab havde, givet os ny Belæring om Kulturernes eget Liv og deres dybe indbyrdes Afhængighedsforhold.

Hvad vor hjemlige Kulturs Egenart angaar, er denne fremfor noget præget af den Maade, hvorpaa de kulturelle Rørelser, der stadig kommer fra de større Samfund, hos os modtages og omdannes. Naar vi spørger, om vor Kultur stadig vil kunne sætte Blomster og bære Frugt, møder vi de samme Forhold, som gælder ethvert levende Væsens fortsatte Bestaaen. Selv det mest indgaaende Kendskab til Bestanddele og Opbygning fører her ikke til noget Svar, men vi er henvist til Haabet om, at en Harmoni, der i sit fulde Omfang kun kan anes, men aldrig helt kan gribes, stadig vil bevares. Udtrykket for et saadant Haab finder vi, hvad enten det gælder de enkelte levende Væsener eller Samfundene, i deres Livsvilje, og naar vi tænker paa vor egen Kultur, har vi vist alle netop i en Tid som denne en levende Følelse af, hvor stærk en saadan Vilje kan være.

Vender vi til sidst tilbage til Spørgsmaalet om vort Selskabs Fremtid, ved vi kun det ene, at vore Ønskers Opfyldelse er uadskillelig forbundet med en lykkelig Skæbne for vort hele Samfund, ja, kan ej heller løses fra Opretholdelsen af alle Folkeslags Samarbejde paa Videnskabens Fremgang. Det er til disse Haab, vi i Dag sætter vor Lid.

Dermed er Dagsordenen udtømt for dette Møde paa Selskabets 200-Aarsdag, som — trods de jævne ydre Former — hos os alle har kaldt paa mange af vore rigeste Minder. Selv om Mødet er hævet, er vor Sammenkomst imidlertid ikke til Ende, og vi vil i Aftenens Løb faa Lejlighed til atter at fremkalde disse Minder.«

The 200th anniversary meeting of the Royal Danish Academy of Sciences and Letters, 13 November 1942. Left row, front to back: N. Bjerrum, E. Biilman, N. Bohr (President) and M. Knudsen (Secretary). Then, sitting from left to right: E. Arup, J.N. Brønsted, K. Fabricius, H. Pedersen, K. Sandfeld, A. Christensen, A.W. Langseth, L.L. Hammerich (Editor), A. Friis, J.A. Christiansen, H.O. Lange, B. Strömgren, J. Nielsen, C. Blinkenberg, B. Jessen, K. Jessen, N.E. Nørlund, E. Dyggve, H. Bohr, F. Poulsen, P. Tuxen and A. Jensen. Standing from left to right: J. Ørskov, S. Orla–Jensen, J.C. Bock, M. Thomsen and P. Rubow. In the foreground to the right: E. Lundsgaard, J. Pedersen (looking to the left) and M. Kristensen.

TRANSLATION

Meeting on 13 November 1942   pp. 26–28
on the 200th Anniversary of the Establishment of the Academy.

...

The *President* opened the meeting and stated the following:
"This evening it is exactly 200 years ago that the meeting took place at
which our Academy was established. Under other circumstances this event
would have been celebrated with great publicity, and not least with extensive
participation from the institutions all over the world with which the Academy
has cooperated for so many years for the progress of science. Under the present
circumstances, such participation is not possible, and we are compelled to
commemorate the 200-year existence of our Academy in more modest and
humble forms. Additionally, our noble Patron, *His Majesty the King*, who
already some months ago had agreed to honour our meeting with his presence,
has been prevented from coming here tonight because of the King's unfortunate
accident and subsequent severe illness.[1] It is therefore with the most sincere
gratitude that the Academy a moment ago received the following telegram:

"My cordial congratulations on the occasion of the jubilee.

Christian R."

I propose that the Academy send the following telegram to *His Majesty the
King*:

"The Royal Danish Academy of Sciences and Letters, gathered for a meeting
on the bicentenary of its establishment, has with the greatest pleasure received
Your Majesty's congratulations and would like to express the deep gratitude

[1] The King had fallen off his horse on 19 October that year.

[407]

of the Academy towards Your Majesty for the attention and interest, whereby our noble Patron has followed the activity of the Academy throughout the long reign of Your Majesty, during which all efforts in the domain of cultural life in this country have had such favourable conditions. In connection with the feelings which move the whole Danish people, the Royal Danish Academy of Sciences and Letters adds to its respectful thanks the most sincere wishes for the swift complete recovery of Your Majesty."

On the assumption that the Academy agrees, this telegram will be sent immediately.

On the occasion of the jubilee we have furthermore received a beautiful address with the following greeting:

"The Royal Norwegian Academy of Science brings the Royal Danish Academy of Sciences and Letters on the bicentenary its sisterly greeting with wishes of continued fruitful work towards common goals for the advancement of Nordic science and culture, as we gratefully remember that two of its oldest members, Schöning and Suhm, were the coestablishers of our Academy in 1760.[2]

Trondheim 13 November 1942.

RAGNVALD IVERSEN
Chair

S. SCHMIDT–NIELSEN
Secretary General."[3]

Our Academy greatly cherishes the receipt of this greeting from a sister academy, with which we are connected through so many common precious memories. We cordially share the wishes expressed in the beautiful address from the Royal Norwegian Academy of Science, and we will in our reply express our deeply felt gratitude."

––––––––––

The *President* reported that, as indicated in the programme, various publications would be presented in connection with the jubilee, and he gave the

[2] Gerhard Schöning (1722–1780) and Peter Frederik Suhm (1728–1798) contributed to the writing of Norwegian and Danish history, respectively. While living in Trondheim, they established, together with the Norwegian bishop Johan Ernst Gunnerus (1718–1773), the Royal Norwegian Academy.
[3] Ragnvald Iversen (1882–1960), philologist of Norwegian; Sigval Schmidt–Nielsen (1877–1956), chemist.

floor first to the Editor, L.L. Hammerich,[4] who presented the first volume of a publication issued by the Academy itself, containing collections as regards its history.[5]

...

The *President* thanked the Editor for his account of the purpose and content   pp. 31–32 of the submitted publication, in the creation of which he personally had taken so active a part. "We owe the entire editorial board thanks for the great effort, but first and foremost, as it appears from the account of the Editor, we owe section head A. LOMHOLT[6] much gratitude for the care and insight that he has shown in the preparation and in his preliminary investigations and arrangement over many years of the Academy archive, without which the publication could not have been realized. It is the hope of the Academy that these collections as regards its history will attract the attention of everyone who is interested in the cultural development of Danish society and will become a valuable source of information for future researchers in the history of science in our country."

The *President* thereafter gave the floor to AAGE FRIIS[7] who presented a text, published by the Royal Danish Society for National History,[8] about the founder of the Academy, Hans Gram:[9] *Vita Johannis Grammii*.[10]

...

The *President* stated: p. 36
"It is with the most sincere gratitude that the Royal Danish Academy of Sciences and Letters receives this exceedingly beautiful greeting and gift

---

[4] Louis Leonor Hammerich (1892–1975), Germanic philologist.
[5] *Det Kongelige Danske Videnskabernes Selskab 1742–1942: Samlinger til Selskabets Historie*, Volume 1 (ed. A. Lomholt), Munksgaard, Copenhagen 1942. The series was going to consist of five volumes, the last of which was published in 1973.
[6] Asger Lomholt (1901–1990).
[7] (1870–1949), Danish historian and chairman of the Danish Committee for the Support of Refugee Intellectual Workers, established in 1933, of which Bohr was also a founding member. Bohr's obituary for Friis, *He Stepped in where Wrong Had Been Done*, Politiken, 7 October 1949, is reproduced in Vol. 12.
[8] Founded in 1745 by Jacob Langebek (1710–1775).
[9] (1685–1748), philologist and historian.
[10] (ed. B. Kornerup), Det Kongelige Danske Selskab for Fædrelandets Historie, Copenhagen 1942.

from a sister academy, with which it through many years has been so closely connected. We deeply appreciate the honour bestowed upon our Academy by this publication, precisely today, of a work about the excellent scientist and citizen, who 200 years ago took the initiative towards the establishment of our Academy, and on behalf of the Academy I would like to ask professor AAGE FRIIS to give the Royal Danish Society for National History our most heartfelt thanks and greetings in return, as well as to bring the warmest wishes for the activity and continued prosperity of our sister academy."

———————

The *President* thereafter gave the floor to N.E. NØRLUND[11] who presented the first volume of his great treatise: *The Mapping of Denmark. A historical account.*[12]

...

pp. 40–41

The *President* expressed his thanks with the following words:

"On behalf of the Academy I would like to thank professor N.E. NØRLUND cordially for the tribute shown to our old Academy through the presentation, just today, of this great work about the mapping of Denmark, in which is described the very important effort that the Academy has made for the advancement of this purpose during the first 100 years of its existence. As professor Nørlund has pointed out, our country made great contributions to science in this field already at an early stage, and we are pleased that these traditions have been maintained in an excellent way also in more recent times, and that Denmark once again, through professor Nørlund's own fruitful investigations, occupies such a prominent position in the work to develop the science of geodesy."

———————

The *President* finally gave the floor to JOHANNES PEDERSEN,[13] chairman of the Carlsberg Foundation's board of directors, who presented a text: *The*

[11] Niels Erik Nørlund (1885–1981), mathematician and Bohr's brother-in-law. See the introduction to Part II, p. [370].

[12] N.E. Nørlund, *Danmarks Kortlægning: En historisk Fremstilling, bind 1: Tiden til Afslutningen af Videnskabernes Selskabs Opmaaling*, Munksgaard, Copenhagen 1942.

[13] (1883–1977), orientalist and a close acquaintance of Bohr who wrote an article in Pedersen's honour: *Physical Science and the Study of Religions* in *Studia Orientalia Ioanni Pedersen Septuagenario A.D. VII. id. Nov. Anno MCMLIII*, Munksgaard, Copenhagen 1953, reproduced in Vol. 10, pp. [275]–[280]. Pedersen was going to succeed Bohr as President of the Royal Academy upon the latter's death in 1962.

Carlsberg Foundation[14] about the establishment and activity of the Foundation.

...

The *President* stated: pp. 44–48

"On behalf of the Academy I would like to convey heartfelt gratitude to Professor JOHANNES PEDERSEN for his presentation of his informative treatise about the great Foundation which since its establishment has served in the most fortunate way as a life-giving source for Danish science. The publication of this treatise on precisely the 200-year anniversary of the Academy will in the most welcome fashion focus attention on the marvellous adventure in the history of the Academy constituted by the creation of the Carlsberg Foundation. J.C. Jacobsen's[15] approach to the Academy with the announcement of his large donation and his proposal of the Academy's involvement in its utilization is without doubt the most important event in the Academy's history over the last hundred years.

Today the Royal Danish Academy of Sciences and Letters thinks with the most profound gratitude of the trust shown to it by the founder of the Carlsberg Foundation and we will add to this our thanks to those men, chosen from our circle, who in such a fortunate way have understood to distribute and manage the large funds which through J.C. Jacobsen's unique gift have been bestowed on Danish science. We also feel gratitude for the way that the board of directors, which according to the Foundation statutes is independent of the Academy in all its decisions and only bound by guidelines of conduct stated by the founder, has in the spirit of J.C. Jacobsen always supported the Royal Academy and, first and foremost, has given the Academy the use of these beautiful premises where we can meet under such dignified circumstances. New testimony of this outlook can be seen in the large gift which the Foundation has donated to the Academy today and which we receive with heartfelt gratitude and will strive to use for the purposes of the Academy in the best way.[16]

On a day like today we also recollect with gratitude the large gifts, legacies and scholarships which the Academy has received over the years from various quarters, and not least do we remember that the State authorities in recent years

---

[14] J. Pedersen, *Carlsbergfondet*, Munksgaard, Copenhagen 1942. Published in English in a revised and abbreviated form as J. Pedersen, *The Carlsberg Foundation*, Bianco Luno, Copenhagen 1956.

[15] Jacob Christian Jacobsen (1811–1887).

[16] The Academy had received 200,000 Danish kroner "to be spent for the advancement of the Academy's purpose in accordance with the arrangements that the Academy might make itself."

[411]

have made possible, by considerable grants, the continuation of our publication activities and have thereby acted in accordance with the old traditions that go back to the very beginnings of the Academy, when it received the indispensable support for its activities from the King's own treasury.

The works and treatises that have been presented today, and the words that have been linked to the various presentations, give each on their own such detailed information about important traits of the Academy's history, and form together such a comprehensive picture of our activities in the past 200 years that we could hardly be reminded in any more beautiful or vital way of our honourable traditions and of the many excellent men who throughout the ages have reflected so much glory on our old Academy. However, it would be fitting on this occasion to gather our thoughts for a few more moments on the position the Academy has assumed in our society from its establishment until the present day, and on what kind of hopes we dare entertain for its future activity in the service of science and society.

The immediate cause for the foundation of the Academy was, as is well known, the demand for the solution of certain definite tasks requiring general scientific insight, and in the first century of its existence, when in this country there existed only few bodies for the support of science and its applications, our Academy had an important mission to fulfil in both these respects. In particular, it assumed the responsibility, often at the direct behest of the King, for various tasks among which, as we have just heard, the mapping of the country was the largest and the best executed. Around the middle of the last century, this state of affairs had altered significantly in that, little by little, many independent Danish institutions had sprung up, to which the majority of these tasks more naturally belonged. In this development it was presumably not the Academy itself, but rather many of its leading personalities, that played a decisive role. I will mention that it was on the initiative of H.C. Ørsted, who was the Secretary of our Academy for as many as 37 years, that the Technical University was established,[17] just as, even earlier, P.C. Abildgaard, founder of the Agricultural College, in his later years had filled this important post in the Academy.[18]

However, it is hardly possible to overestimate the significance of the fact that in our society there still existed a centre, if not so much for carrying

[17] Hans Christian Ørsted (1777–1851). The Technical University was established in 1829.

[18] The veterinarian Peter Christian Abildgaard (1740–1801) founded a private veterinary school in 1773, which was taken over by the Danish state three years later. This was the forerunner of the Royal Danish Veterinary and Agricultural College, established by legislation in 1856. Abildgaard served as Secretary of the Royal Danish Academy from 1795 to 1801.

out scientific investigations then at least for their general appreciation. The most striking testimony of this is J.C. Jacobsen's so beautifully formulated approach to the Academy about the establishment of the Carlsberg Foundation, to which the Foundation's chairman has called attention. Just in this connection we also remember how much it meant at that time that scientific activity in all its multifarious shades was linked together in such a harmonious way within our Academy. The fact that representatives of widely different branches of science are gathered at our meetings and take part in all the proceedings gives our Academy its special character. The unique opportunity thus offered to humanists and scientists to tell one another what they each regard as important within their research excludes a narrowness which often follows with larger scale, in which a sharp division has been necessary for practical reasons. Even though not all areas of knowledge and research deserving to be called science are represented in the Academy, the ties forged here have certainly had, and still have, utmost significance for the harmonious development of the scientific endeavours in our country, indeed for the whole cultural life here at home.

As circumstances developed, our Academy has, in addition, become an ever more important link to international scientific cooperation. In this field Denmark can rightly be proud of the understanding, shared by society as a whole, for the important task that was involved here. Especially the establishment of the Rask–Ørsted Foundation[19] – by which such an important effort was made on the part of Denmark at a time when international scientific endeavours were especially difficult and especially necessary – stands as a manifestation of a successful cooperation between prominent figures within the political sphere and within the circle of scientists, who to a significant degree received their inspiration precisely from the debates in our Academy. Not only has this Foundation made possible participation from this country in international scientific cooperation to an extent otherwise hardly imaginable, but, in addition, the activity of the Foundation has served as a model for institutions established in larger countries with correspondingly richer resources. The recognition of Danish initiative and activity in this field is perhaps not the least important of the circumstances upon which we build our hope that Denmark, when peaceful times return, will once again find its place as an active participant in the international cultural endeavours.

Nowhere do we find a stronger testimony of the fruitfulness of human cooperation than in the field of scientific research. When we consider how generations have helped each other with carrying stones to the proud edifice of

---

[19] The Rask–Ørsted Foundation was established in 1919.

science and have together heightened its peaks and strengthened its foundations, we feel we are faced with an achievement to which the individual can only contribute by virtue of the wealth that is held in the whole common property. What science brings us is not only skills which help us to master nature and to develop the organization of society; but inextricably connected with the increase of our knowledge is a deeper insight into our own being and our place in existence. This trait common for the quest for knowledge through all the ages has in our own time come to the fore as hardly ever previously and meets us equally at such extreme points as the investigation of the nature around us and the study of human cultures. In the first field we have learned that even our very most simple concepts, such as space and time and cause and effect, are not universal forms but idealizations whose use rests on hitherto unheeded assumptions. In the second field the recognition of mutual germination between seemingly independent cultures, and of the legacy they each on their own have adopted from cultures about which we until quite recently had no knowledge, has given us a new lesson about the cultures' own life and their deep mutual state of dependence.

As regards the special character of our culture here at home, it is first and foremost marked by the way in which the cultural movements, continually coming from the larger societies, are received and transformed here. When we ask whether our culture will continue to be able to flower and bear fruit, we face the same conditions that hold for the continued existence of any living organism. Even the most intimate knowledge of components and structure does not here lead to any answer, but we are left with the hope that a harmony, the full extent of which can only be glimpsed but never fully grasped, will still be retained. The expression for such hope we find, whether in individual living creatures or societies, in the will to live, and when we consider our own culture, we all surely, just in a time like this, have a vivid sense of how strong such a will can be.

If in conclusion we return to the question of the future of our Academy, we are certain of only one thing, that the fulfilment of our wishes is inextricably tied up with a happy destiny for our whole society; indeed, nor can it be detached from maintaining the cooperation of all nations for the advance of science. It is into these hopes we today put our trust.

Thus the agenda is exhausted for this meeting on the day of the Academy's bicentenary, which – despite the modest outer forms – has brought forth in us all many of our richest memories. Even though the meeting is adjourned, our gathering is not at an end, however, and in the course of the evening we will have the opportunity to call forth these memories yet again."

VI. MEETING ON 19 OCTOBER 1945

MØDET DEN 19. OKTOBER 1945
Overs. Dan. Vidensk. Selsk. Virks. Juni 1945 – Maj 1946, pp. 29–31

TEXT AND TRANSLATION

See Introduction to Part II, p. [345].

[415]

Mødet den 19. Oktober 1945.

Til *Hs. Majestæt Kongens* 75-Aars Fødselsdag den 26. September 1945 havde Selskabet sendt følgende Adresse, for hvilken Kongens Tak var modtaget:

»Deres Majestæt,

Det Kongelige Danske Videnskabernes Selskab frembærer sin ærbødige Lykønskning til Deres Majestæts fem og halvfjerdsindstyve Aars Fødselsdag, der efter tunge Aar oprinder i en lykkelig Tid, da vort Fædreland igen er frit, og hele det danske Folk atter kan samle sig om Fredens Opgaver.

Som Danmark har været skaanet for Krigens værste Ødelæggelser, saaledes har ogsaa Videnskaben herhjemme som Helhed haft bedre Kaar end mange andre Steder. Trods Vanskeligheder, hvorunder nogle af vore Medlemmer har været fanget af Fjenden og andre blev jaget af Lande, har Selskabet haft den Lykke at kunne fortsætte sin i Stilhed virkende Gerning i Bevidstheden om den Betydning, som videnskabelig Forskning indebærer for menneskeligt Fremskridt.

Det Kongelige Danske Videnskabernes Selskab tænker med Taknemmelighed paa de gode Vilkaar, som Oplysning og Videnskab har haft i Deres Majestæts lange Regeringstid, og paa, hvad Danmarks Konge har ydet for sit Land i Ulykkens mørke Aar, da Kulturværdier, der danner Baggrund ogsaa for fri Forsknings Udfoldelse, truedes som ingensinde før.

I Hengivenhed mod vor høje Protektor ønsker Selskabet Deres Majestæt lyse Fremtidsdage med Fremgang for vort Fædreland.«

TRANSLATION

Meeting on 19 October 1945.

...

On the occasion of *His Majesty the King*'s 75th birthday on 26 September the Academy has sent the following address, for which the King's thanks have been received:

"Your Majesty,

The Royal Danish Academy of Sciences and Letters brings its respectful congratulations on the occasion of Your Majesty's seventy-fifth birthday, which after dark years takes place in a joyous time, when our native country is free again and the whole Danish people can once more concentrate on the tasks of peace.

Just as Denmark has been spared from the worst destruction of war, also science here at home as a whole has enjoyed better conditions than many other places. Despite difficulties, during which some of our members were taken prisoner by the enemy and others were forced to flee the country, the Academy has had the good fortune to be able to quietly continue its active work, conscious of the importance that scientific research implies for human progress.

The Royal Danish Academy of Sciences and Letters remembers with gratitude the good conditions that education and science have had in Your Majesty's long reign, and what the King of Denmark has done for his country in the dark years of misfortune, when cultural values which form the background also for the free conduct of research, were threatened as never before.

In devotion to our noble Patron, the Academy wishes Your Majesty bright days of the future with progress for our native country."

VII. MEETING ON 25 APRIL 1947

MØDET DEN 25. APRIL 1947
Overs. Dan. Vidensk. Selsk. Virks. Juni 1946 – Maj 1947, pp. 53–54

TEXT AND TRANSLATION

See Introduction to Part II, p. [346].

Mødet den 25. April 1947.

Præsidenten meddelte, at Selskabets høje Protektor *Hans Majestæt Kong* CHRISTIAN DEN TIENDE efter 35-Aars Kongegerning var afgaaet ved Døden paa Amalienborg den 20. April 1947.

Præsidenten udtalte i den Anledning følgende:

»Siden vort sidste Møde har en dyb Sorg ramt hele det danske Folk ved Kong Christian den Xdes Død. Store og for vort Land dybt indgribende Begivenheder fandt Sted under Kongens lange Regeringstid. Kort efter Tronbestigelsen udbrød den første Verdenskrig, i hvilken Danmarks Inddragelse blev lykkelig undgaaet, og ved hvis Afslutning saa dybt et Saar i det danske Folkelegeme heledes ved de sønderjydske Landsdeles Genforening med Moderlandet. Ved den Værdighed og Umiddelbarhed, hvormed Kongen i de Dage tolkede alle Danskes Følelser, vandt Kong Christian den Xde den Plads i Folkets Hjerte, som gennem Aarene skulde blive dybere og dybere fæstet. Samhørighed mellem Konge og Folk skulde dog naa den største Inderlighed og Styrke i de tunge Tider under den sidste store Krig, hvor for første Gang i vor tusindaarige Historie Danmark for en Tid mistede sin Frihed og blev besat af fjendtlige Tropper. For den ubrydelige Vedhængen ved de Traditioner og Idealer, der kendetegner vor gamle Kultur, og paa hvilke vi bygger vort Liv, blev Kongen det naturlige Samlingsmærke og vakte ved sin ranke Færd og sit opmuntrende Forbillede hele Folkets Taknemmelighed og Beundring, lige som Kongen ude i Verden vandt en Anseelse af største Betydning for vor Sag. At det blev forundt vor gamle Konge at opleve Lykken over Danmarks Befrielse, føltes af alle som en væsentlig Del i den store Glæde. I sine sidste Aar omgaves Kongen fra hele Folkets Side af en Ærbødighed og Kærlighed, der vanskeligt kan finde sin Lige, og under den fremadskridende alvorlige Sygdom, der skulde gøre Ende paa Kongens Liv, samledes alle i Bekymring og Medfølelse.

Her i Videnskabernes Selskab, hvor vi ved saa mange Lejligheder har følt den varme Interesse, vor høje Protektor nærede for vort Arbejde, som for alle Bestræbelser for at højne Kulturen i vort Land, og hvor Kongen saa ofte har kastet Glans over vore Møder, deltager vi inderligt i Sorgen over Kong Christian den Xdes Død og i Taknemmelighed for alt, hvad Kongen har været for vort Land.

I Overensstemmelse med Selskabets Traditioner paatænkes det at hellige et særligt Møde til Mindet om Kong Christian den Xde, et Møde, der antagelig vil blive det første i den kommende Sæson, og vi haaber ved den Lejlighed, at Hans Majestæt Kong Frederik den IXde vil beære Selskabet ved sin Nærværelse. Jeg er sikker paa at have hele Selskabets Tilslutning til, at vi over for Hans Majestæt giver Udtryk for vor Deltagelse og vor Hyldest, og at vi anmoder Kongen om at vise Selskabet den Ære og i Lighed med sine Forgængere yde os den Støtte at være Det Kongelige Danske Videnskabernes Selskabs Protektor.«

Selskabets Medlemmer sluttede sig dertil ved at rejse sig.

TRANSLATION

Meeting on 25 April 1947.

...

The President announced that the Academy's noble Patron *His Majesty King* CHRISTIAN X, after reigning for 35 years had died at Amalienborg[1] on 20 April 1947.

The *President* stated the following on this occasion:

"Since our last meeting the whole Danish people has been struck by a deep sorrow on the death of King Christian X. Great, and for our country extremely vital, events took place during the King's long reign. Shortly after his accession to the throne the First World War broke out, in which Denmark's involvement was fortunately avoided, and at the end of which such a deep wound in the Danish national body was healed on the reunion of the provinces of Southern Jutland with the mother country.[2] Through the dignity and spontaneity with which the King during those days gave expression to the feelings of all Danes, King Christian X won that place in the heart of the people which through the years was to become ever more deeply entrenched. The affinity between king and people was, however, to reach its greatest intensity and strength during the dark times during the last great war, when Denmark for the first time in our thousand-year history lost for a time its freedom and was occupied by enemy troops. The King became the natural rallying point for the unshakeable adherence to the traditions and ideals which characterize our old culture and on which we build our lives, and awakened, with his fearless conduct and his encouraging example, the gratitude and admiration of the whole people, just

---

[1] Built around 1750, Amalienborg came into the possession of the Danish royal family in 1794.

[2] The southern part of the Danish mainland had been lost to Germany in the Danish–German War of 1864. The northern part of the area was reunited with Denmark in 1920 after a plebiscite.

as, out in the world, the King won a reputation of the greatest importance for our cause. That our old King lived to experience the happiness of Denmark's liberation, was felt by everyone as an important part of the great joy. In his last years, the King was surrounded by a respect and love from the whole people, the like of which would be difficult to find, and during the progressive severe illness, which was to put an end to the King's life, all united in concern and sympathy.

Here in the Academy of Sciences and Letters, where on so many occasions we have felt the warm interest that our noble Patron held for our work, as for all endeavours to raise the culture in our country, and where the King so often has lent lustre to our meetings, we sincerely share in the grief at King Christian X's death and in gratitude for all that the King has been for our country.

In accordance with the traditions of the Academy the dedication of a special meeting in commemoration of King Christian X is contemplated, a meeting that probably will be the first in the coming season, and we hope on that occasion that His Majesty King Frederik IX[3] will honour the Academy with his presence.[4] I am certain that I have the approval of the whole Academy that we express our sympathy and pay our homage to His Majesty, and that we ask the King to show the Academy the honour and, like his predecessors, give us the support of being the Patron of the Royal Danish Academy of Sciences and Letters."

The members of the Academy showed their approval by rising.

---

[3] (1899–1972). He was succeeded by Queen Margrethe II (1940–).

[4] The new King did take part in the commemoration meeting, which took place on 17 October 1947. The proceedings are reproduced on pp. [425] ff.

VIII. MEETING ON 17 OCTOBER 1947
IN COMMEMORATION OF KING CHRISTIAN X

MØDET DEN 17. OKTOBER 1947
TIL MINDE OM KONG CHRISTIAN X
Overs. Dan. Vidensk. Selsk. Virks. Juni 1947 – Maj 1948, pp. 26–29

TEXT AND TRANSLATION

See Introduction to Part II, p. [347].

Mødet den 17. Oktober 1947 til Minde om Kong Christian X.

Selskabets Protektor, *Hans Majestæt Kong* FREDERIK IX, overværede Mødet.

Præsidenten takkede paa Selskabets Vegne Kongen, fordi denne i Lighed med en lang Række af tidligere danske Konger, havde villet vise Selskabet den Ære og yde det den Støtte at være dets Protektor. Præsidenten udtalte derefter bl. a.:

»Lige fra Det Kongelige Danske Videnskabernes Selskabs Stiftelse, hvori Kong Christian VI tog saa virksomt Del, har Selskabet nydt godt af Kongemagtens Bevaagenhed, hvad der gennem Tiderne paa mange Maader har været af største Betydning for Selskabets Virke. I Dag tænker vi især paa den aldrig svigtende Interesse, som Deres Majestæts Fader igennem sin lange Regeringstid viste Selskabet. Ikke alene har Kong CHRISTIAN DEN TIENDES Nærværelse ved festlige Lejligheder kastet Glans over vore Møder, men ogsaa ved vore almindelige Forhandlinger har Kongen ofte beæret Selskabet med sin Tilstedeværelse. Med særlig Taknemmelighed mindes vi, at Kong Christian den X fandt Tid og Lejlighed til at komme til Stede iblandt os selv i

Besættelsens tunge Tid, hvor Kongen ved sit ranke Forbillede blev Samlingspunktet for hele Folkets Bestræbelser for at opretholde de Idealer og Traditioner, der kendetegner vor gamle Kultur. De ligefremme og alvorlige Ord, som Kongen ved denne Lejlighed rettede til Selskabet, virkede dybt paa alle Medlemmerne og føltes som en stor Bestyrkelse af vort Forsæt. Kongen havde ogsaa udtalt sit Ønske om at overvære Mødet paa 200-Aarsdagen for Selskabets Bestaaen, men paa Grund af den Sygdom, der bragte saa megen Ængstelse, men som ikke kuede Kongens eget Mod, maatte vi ved vort Jubilæum dybt savne vor høje Protektor. Med hele det danske Folk delte Selskabet Glæden over, at det blev Kong Christian X forundt at opleve Danmarks Befrielse, og vi deltog inderligt i Landets Sorg over Kongens Bortgang.«

Præsidenten omtalte videre, at Selskabet, undtagen gennem sine Publikationer, i Almindelighed ikke henvendte sig til Offentligheden. Lige siden Selskabets Stiftelse havde det imidlertid været en ubrudt Tradition, at det efter en Konges Død paa højtidelig Maade mindedes sin bortgangne Protektor. Præsidenten udtalte Selskabets Taknemmelighed for, at Kong Frederik IX, som Selskabets nuværende Protektor, havde villet bivaane Mødet.

Kongen takkede for de Ord, Præsidenten havde henvendt til ham og udtalte, at han med Glæde havde overtaget Protektoratet for Selskabet. Kongen bemærkede, at han jo ikke var Videnskabsmand, men haabede, at han maatte faa Lejlighed til at komme til Stede ogsaa paa almindelige Mødeaftener, naar Emnerne særlig maatte tiltrække sig hans Opmærksomhed.

Kongen meddelte, at han samme Dag havde tildelt Selskabets Præsident Elefantordenen. Det var Hensigten dermed ikke blot at udmærke NIELS BOHR personlig, men tillige at hædre dansk Videnskab, for hvilken Niels Bohr var saa ypperlig en Repræsentant. Kongen anmodede Medlemmerne om at rejse sig og hylde Niels Bohr, der havde kastet Glans over Danmarks Navn.

Præsidenten takkede Kongen for de hjertelige Ord og udtrykte sin Taknemmelighed for den store Ære, som Kongen havde ladet ham blive til Del. Paa hele Selskabets Vegne ønskede han ogsaa at takke Kongen for hans Udtryk for den varme Interesse, hvormed Kongen omfattede Videnskabens Trivsel, og udtalte Haabet om, at det skulde lykkes Selskabet at videreføre dets minderige Traditioner til Fremme for Videnskaben og til Gavn for vort Land.

Præsidenten gav derefter Ordet til Professor Knud Fabricius, der gennem sin historiske Forskning havde haft særlig Anledning til at beskæftige sig med Begivenhederne i Kong Christian den Tiendes Regeringstid.

Professor KNUD FABRICIUS holdt derefter følgende Foredrag:

»*Træk af Christian den Tiendes Kongegerning*«.

Præsidenten takkede Professor Fabricius for hans Foredrag, der ved de Oplysninger, som han havde samlet og fremdraget, havde føjet nye Træk til det Billede af Kong Christian den Tiendes Personlighed, der vilde omfattes med Taknemmelighed og Beundring, saa længe vort Land bestod, og som samtidig paany havde belyst det inderlige Forhold mellem Kongemagten og hele det danske Folk.

Præsidenten rettede til Slut følgende Ord til Kongen:

»Jeg ved, at jeg taler paa hele Selskabets Vegne, naar jeg overfor Deres Majestæt giver Udtryk for vore dybtfølte Ønsker om Lykke i Kongens ansvarsfulde Gerning til vort Lands Bedste. Videnskabsmændenes Kald ligger jo Statens Styrelse fjernt, men Videnskabens Trivsel er af væsentlig Betydning for Samfundets Liv. Studiet af Kulturen og Historien, som Deres Majestæt tidligt har vist saa aktiv Interesse, minder os ikke alene om vor egen Fortid og Arv, men holder os tillige bestandig hele Menneskeslægtens Samhørighed for Øje, og Naturens Udforskning viser os stedse nye Veje til Fremme af menneskeligt Velfærd og til Forstaaelse af vor Stilling i Tilværelsen. Den seneste Udvikling, der atter stiller en stor Fremgang i Udsigt, har samtidig maattet gøre det klart for alle, at Civilisationens fortsatte Bestaaen vil kræve et Samarbejde mellem alle Folkeslagene i gensidig Tillid. Vi stoler paa, at Videnskaben, som gennem Tiderne har vist, hvad Fremskridt der kan opnaas ved fælles menneskelig Stræben, ved dens fornyede Paamindelse om Sammenholdets Nødvendighed maa bidrage til en lysere Fremtid for hele Menneskeheden.

Paa denne Baggrund og i denne Aand skal vi efter beskeden Evne bestræbe os for, at Deres Majestæt maa finde Glæde og Tilfredsstillelse af Fortsættelsen af den traditionsrige Forbindelse mellem Monarken og Det Kongelige Danske Videnskabernes Selskab.«

Før Kongen forlod Mødet, hilste han paa hver enkelt af de tilstedeværende Medlemmer.

TRANSLATION

Meeting on 17 October 1947
in Commemoration of King Christian X.

The Patron of the Academy, *His Majesty King* FREDERIK IX,
attended the meeting.

...

The *President* thanked the King on behalf of the Academy, because, like a long line of earlier Danish kings, he had wished to show the Academy the honour, and grant it the support, of being its Patron. The President thereafter said, among other things:

"Ever since the establishment of the Royal Danish Academy of Sciences and Letters, in which King Christian VI took such an active part, the Academy has enjoyed the favour of the Crown, which through the ages has in many ways been of the greatest importance for the activity of the Academy. Today we think in particular of the never failing interest that Your Majesty's father showed the Academy throughout his long reign. Not only has the presence of King CHRISTIAN X on festive occasions lent lustre to our meetings, but also at our ordinary meetings the King has often honoured the Academy with his presence. We remember with special gratitude that King Christian X found the time and the opportunity to be present among us even in the dark time of the occupation, when the King, with his fearless example, became the rallying point for the endeavours of the whole people to maintain the ideals and traditions that characterize our old culture. The direct and serious words the King addressed to the Academy on this occasion, affected all the members deeply and were felt as a great strengthening of our purpose. The King had also pronounced his wish to attend the meeting on the 200th anniversary of the Academy's existence, but due to the illness which caused so much anxiety, but which did not daunt the King's own courage, we were sorely obliged to miss our noble Patron at our jubilee. With the whole Danish people the Academy shared the joy that King Christian X lived to experience the liberation of Denmark, and we shared sincerely in the sorrow of the country at the death of the King."

The President further said that the Academy did not normally address the public except through its publications. Ever since the establishment of the Academy it had, however, been an uninterrupted tradition that upon the death of a king it commemorated its departed Patron in a solemn manner. The President expressed the gratitude of the Academy that King Frederik IX, as the present Patron of the Academy, had wished to attend the meeting.

The *King* thanked for the words that the President had addressed to him and stated that he had accepted the patronage for the Academy with pleasure. The King remarked that he was of course not a scientist, but hoped that he would have opportunity to be present also on ordinary meeting evenings, when the topics might particularly attract his attention.

The King announced that on the same day he had bestowed the Order of the Elephant[1] on the President of the Academy. The intention was not merely thus to give a distinction to NIELS BOHR personally, but also to honour Danish science, of which Niels Bohr was such an outstanding representative. The King asked the members to rise and acclaim Niels Bohr, who had lent lustre to the name of Denmark.

The President thanked the King for the cordial words and expressed his gratitude for the great honour which the King had bestowed upon him. On behalf of the whole Academy he also wished to thank the King for his expression of the warm interest with which the King embraced the prosperity of science, and pronounced the hope that the Academy would succeed in continuing its memorable traditions to the advancement of science and to the benefit of our country.

The President thereupon gave the floor to Professor Knud Fabricius,[2] who through his historical research had had special occasion to concentrate on the events of King Christian X's reign.

Professor KNUD FABRICIUS thereupon gave the following lecture:

"*Features of Christian X's reign*".

The President thanked professor Fabricius for his lecture, which, through the information that he had collected and presented, had added new features to the image of King Christian X's personality which would be regarded with gratitude and admiration for as long as our country existed, and which at the same time had once again illuminated the sincere relationship between the Crown and the whole Danish people.

In conclusion, the President addressed the following words to the King:

---

[1] See the introduction to Part II, p. [347].
[2] (1875–1967), historian.

"I know that I speak for the whole Academy when I express to Your Majesty our deeply felt wishes for success in the responsible task of the King for the good of our country. Of course, the vocation of the scientist is remote from the administration of the state, but the prospering of science is of great importance for the life of society. The study of culture and history, in which Your Majesty early has shown such an active interest, does not only remind us of our own past and heritage, but also constantly displays the interdependence of the human race, and the exploration of nature always shows us new paths towards the advancement of human welfare and towards the understanding of our place in existence. The latest development, which once again holds out the prospect of a great advance, must at the same time make it clear to everyone that the continued existence of civilization will require cooperation between all peoples in mutual confidence. We trust that science, which through the ages has shown what kind of advances can be made through common human striving, by its renewed reminder of the necessity of concord must contribute to a brighter future for the whole of humanity.

On this background and in this spirit we must, according to our humble ability, endeavour that Your Majesty may find pleasure and satisfaction in the continuation of the connections, rich in tradition, between the Monarch and the Royal Danish Academy of Sciences and Letters."

Before the King left the meeting, he greeted each of the members present.

IX. MEETING ON 11 MARCH 1949

MØDET DEN 11. MARTS 1949
Overs. Dan. Vidensk. Selsk. Virks. Juni 1948 – Maj 1949, pp. 45–46

TEXT AND TRANSLATION

See Introduction to Part II, p. [347].

[433]

Mødet den 11. marts 1949.

Sekretæren oplæste den lykønskningsadresse, som præsidenten den 11. marts havde overrakt Hans Majestæt kongen i anledning af 50-årsdagen.

Adressen havde følgende ordlyd:

»Deres Majestæt.

Det Kongelige Danske Videnskabernes Selskab frembærer i ærbødighed på Deres Majestæts halvtreds års fødselsdag sin lykønskning i taknemmelighed for det tilsagn om velvilje, der har fundet udtryk i vor midte fra vor høje protektors side. Derved fortsættes en tradition, i hvilken Danmarks konger gennem to århundreder med interesse og sympati har fulgt selskabets arbejde, omend dette til tider kunne synes at ligge fjernt fra dagens nærmeste krav.

Selv om videnskaben ofte har modtaget tilskyndelser fra udviklingens aktuelle problemer, har dens stærkeste drivfjeder altid været en dybere sandhedssøgen såvel i menneskeåndens selverkendelse og udtryksform som i eftersporing af naturens love.

Efter ufreds år, i hvilke hele vor kultur og vor frihedsarv stededes i den største fare, nærer vor tid håbet om fredens bevarelse og om en ny blomstring, til hvilken det videnskabelige arbejdes resultater måske vil kunne yde væsentlige bidrag. Måtte Deres Majestæt få lykke til i disse betydningsfulde år i manddomsgerningens fulde kraft at lede vort land mod en rig udfoldelse og en tryg fremtid.«

TRANSLATION

Meeting on 11 March 1949.

...

The Secretary read out the congratulatory address the President had handed over on 11 March to His Majesty the King on the occasion of his 50th birthday. The address read as follows:

"Your Majesty.

On the occasion of Your Majesty's fiftieth birthday, the Royal Danish Academy of Sciences and Letters respectfully brings its congratulations in gratitude for the assurance of good-will that has been expressed in our midst by our noble Patron. Thereby a tradition is upheld in which Denmark's kings throughout two centuries have followed with interest and sympathy the work of the Academy, although at times this might seem remote from the immediate needs of the day.

Even though science has often received impulses from the problems of current interest caused by the turn of events, its strongest mainspring has always been a deeper search for truth, in self-knowledge and mode of expression of the human mind as well as in exploration of the laws of nature.

After years of strife, during which our whole culture and heritage of freedom were placed in the greatest danger, our time nourishes the hope for keeping the peace and for a new flowering to which the results of the scientific work will perhaps be able to make significant contributions. May Your Majesty succeed in these important years, in the full strength of the prime of life, in leading our country towards a rich blossoming and a secure future."

[435]

The fifth centenary of the University of Glasgow. From left to right: Bohr, Philip Ivor Dee, John Douglas Cockcroft and Ernest Orlando Lawrence.

X. MEETING ON 19 OCTOBER 1951

MØDET DEN 19. OKTOBER 1951
Overs. Dan. Vidensk. Selsk. Virks. Juni 1951 – Maj 1952, pp. 33–34

See Introduction to Part II, p. [347].

Mødet den 19. oktober 1951.

»The Royal Danish Academy of Sciences and Letters present their most sincere congratulations to the University of Glasgow on her fifth centenary.

The Academy is happy to take part in the celebration of this most remarkable event through its delegation.

In the history of the universities of Europe the University of Glasgow is one of the oldest, and it has to its record an unbroken line of merits which bears witness both to the ingenious contributions of many scientists and scholars to the world's inheritance of science and culture, and to the never-failing support of the Scottish people. To quote only one example, the activity of Lord Kelvin at this seat of learning and research during half a century has been an inspiration to investigators without number in his time and in ours. The role which scientific inquiry is to play in the life of man through its application was foreseen at an early date and is reflected in the flourishing development of numerous institutions of applied science.

Deeply rooted in the national traditions of a gifted people, and with its doors wide open for the searching mind of our day, the University of Glasgow draws her line from a glorious past to a rich future part in the best endeavours of mankind.«

XI. MEETING ON 16 NOVEMBER 1951

MØDET DEN 16. NOVEMBER 1951
Overs. Dan. Vidensk. Selsk. Virks. Juni 1951 – Maj 1952, p. 39

TEXT AND TRANSLATION

See Introduction to Part II, p. [347].

Mødet den 16. november 1951.

Til 200-års jubilæet i Akademie der Wissenschaften, Göttingen, den 9.—10. november 1951 havde selskabet sendt en adresse med de to delegerede, Niels Bohr og Johs. Pedersen.

Adressen havde følgende ordlyd:

»Die Königliche Dänische Akademie der Wissenschaften entbietet der Akademie der Wissenschaften in Göttingen zur Feier Ihres zweihundertjährigen Bestehens Gruss und Glückwunsch.

Wir gedenken der langen Reihe glänzender Namen, die als Vertreter der Natur- und Geisteswissenschaften Ihrer Akademie angehört haben, und der tiefgreifenden Wirkung so vieler von Göttingen ausgegangener schöpferischer Ideen. Wir gedenken auch des regen Gedankenaustausches, den Ihre Akademie mit Gelehrten aller Länder unterhielt, und der in verschiedenen Epochen auch zu engen Beziehungen mit unserem Lande führte, sowie der bereitwilligen Aufnahme, die ausländische Forscher, häufig auch dänische, von jeher in Göttingen fanden.

Der Sitz Ihrer Akademie wurde dadurch zu einem Brennpunkt wissenschaftlicher Zusammenarbeit über alle nationalen Schranken hinweg.

Möge es ihr auch in dieser Hinsicht vergönnt sein, grosse Traditionen weiterzuführen und zu jenem Verständnis zwischen den Völkern beizutragen, das im Wesen der Wissenschaft liegt, und dessen Notwendigkeit wir alle heute stärker als je empfinden.«

TRANSLATION

Meeting on 16 November 1951.

...

On the occasion of the 200-year anniversary of the Akademie der Wissenschaften, Göttingen, 9–10 November 1951, the Academy had sent an address with the two delegates, Niels Bohr and Johannes Pedersen.

The address read as follows:

"The Royal Danish Academy of Sciences and Letters sends the Akademie der Wissenschaften in Göttingen greetings and congratulations on the celebration of its two hundred years of existence.

We recall the long line of brilliant names who have belonged to your Academy as representatives of the sciences and the humanities, and the profound effect of so many creative ideas originating from Göttingen. We also recall the lively exchange of ideas which your Academy maintains with scholars from all countries and which, at various times, also led to close relations with our country, as well as the welcoming reception which researchers from abroad, often also from Denmark, always found in Göttingen.

The seat of your Academy thereby became a focus of scientific cooperation across all national barriers.

May you be allowed also in this respect to continue the great traditions and to contribute to the understanding between the peoples which is part of the nature of science and whose necessity we all feel more strongly than ever today."

[441]

XII. MEETING ON 14 OCTOBER 1960

MØDET DEN 14. OKTOBER 1960
Overs. Dan. Vidensk. Selsk. Virks. Juni 1960 – Maj 1961, pp. 39–41

Communication on behalf of the Royal Danish Academy to the
Royal Society's 300th Anniversary where the
Academy was represented by N. Bohr, N.E. Nørlund and Ø. Winge

See Introduction to Part II, p. [347].

[443]

Mødet den 14. oktober 1960.

»The Royal Danish Academy of Sciences and Letters sends its most cordial congratulations to the Royal Society of London on the occasion of the 300th anniversary of its foundation, and begs to express its very best wishes for the future.

As one of the oldest existing societies for the promotion of the sciences, the Royal Society of London has throughout three centuries exerted a never relaxing influence upon scientific life not only in Great Britain, but all over the globe.

Under the leadership of prominent presidents, one of whom, Isaac Newton, inaugurated a new era in the history of mankind, the Royal Society of London has been the initiator, promotor, and sponsor of a long series of important and successful enterprises within nearly all fields of the sciences, and the publications of the Royal Society of London have in a great measure contributed to scientific progress.

Our Academy is happy that we have always been in close connection with the Royal Society of London. In planning our activities, in working out our statutes, we have on several occasions throughout two centuries been able to follow the wise example of the Royal Society of London. Before our Academy came into existence, the results of a Danish expedition to Egypt were published in a preliminary form by the Royal Society of London, to be completed a few years later by our Academy. From the very beginning, members of our Academy were elected members of the Royal Society of London. Two of the presidents of the Royal Society of London were among our first honorary members. Indeed, during the whole existence of our Academy, there has probably never been a year, certainly no prolonged period, in which no member of the Royal Society of London was a member of our Academy.

Among the many illustrious men whom we are proud to recall as having been our members, we may name: Joseph Banks, Thomas Young, Humphry Davy, John Herschel, Michael Faraday, Charles Lyell, Roderick Impey Murchison, William J. Hooker, Joseph D. Hooker, Richard Owen, William Huggins, James Prescott Joule, Arthur Cayley, Thomas Henry Huxley, William Thomson Kelvin, Charles Robert Darwin, John William Strutt Rayleigh, William Ramsay, Charles Scott Sherrington, Joseph John Thomson, William Bateson, Joseph Barcroft, William Maddock Bayliss, William Henry Bragg, Ernest Henry Starling, Frederick Gowland Hopkins, and Ernest Rutherford – to end with a name which marks again the beginning of a new era in the history of mankind.

When towards the end of the last century, international cooperation became more urgent than before, the Royal Society of London took a decisive initiative in founding the International Catalogue of Scientific Literature and in supporting it for years, even at excessive expenditure.

The Royal Society of London played a prominent part in creating the International Association of Academies and, after its collapse during the First World War, in recreating this important instrument of international scientific cooperation as the International Council of Scientific Unions.

Again, after the collapse of international cooperation during the Second World War, the Royal Society of London resumed the thread as soon as possible: our Academy will always be profoundly grateful in remembering the visit, in 1945, of the Senior Secretary of the Royal Society of London, Archibald Hill, who came to us with greetings which warmed our hearts, and with practical plans for future international cooperation.

The present times probably hold more dangers to the human race than any previous age, but also more hope of lasting progress, if good will can prevail. Our Academy is confident that the Royal Society of London, also in the future, will keep its leading position within scientific activity, and we know how great a benefit this will be to mankind.«

2. ESTABLISHMENT OF THE DANISH NATIONAL SCIENCE FOUNDATION

XIII. NIELS BOHR'S DEEPLY
SERIOUS APPEAL

NIELS BOHRS DYBT ALVORLIGE APPEL
Politiken 19 January 1951

Broadcast on the current affairs programme,
Danish National Radio, 18 January 1951

TEXT AND TRANSLATION

See Introduction to Part II, p. [349].

Niels Bohrs dybt alvorlige appel

„Det er paatrængende nødvendigt, at der ydes større støtte end hidtil for videnskabens fremme"

„En løsning kan ikke udskydes under henvisning til landets øjeblikkelige vanskeligheder"

I radioens aktuelle kvarter i aftes henvendte professor, dr. *Niels Bohr* sig til den danske befolkning. Indlæget formede sig som en dybt alvorlig appel til folk og øvrighed om at yde større støtte til videnskabens fremme. Vi bringer her talen i dens fulde ordlyd:

FRA kulturens ældste tider har bestræbelserne for at udvide grænserne for menneskets viden været af største betydning for samfundslivets udvikling, og i vor tid har videnskaben paa flere og flere maader sat sit præg paa vor tilværelse. Samtidig med, at det stadig dybere indblik i naturens lovmæssigheder har skabt uanede muligheder for levevilkaarenes forbedring, har vi vundet et stedse rigere kendskab til mange sider af den menneskelige kultur og dens udvikling gennem tiderne. I alle civiliserede samfund har man da ogsaa erkendt det som en vigtig opgave efter evne at deltage i dette store menneskelige fællesværk, og mere end noget andet holder det mellemfolkelige samarbejde paa videnskabens omraade os for øje, hvad der kan naas ved forenet stræben mod fælles maal.

Det danske samfund har gamle traditioner for fremgangsrig deltagelse i udviklingen paa mangfoldige af videnskabens omraader og ikke mindst for samarbejde med andre folkeslag til forøgelse af egne erfaringer og den fælles viden. Disse bestræbelser har gennem tiderne sat sig dybe spor i dansk aandsliv og har if det danske folk været omfattet med varm og offervillig interesse. Udover den støtte, som den

Professor Niels Bohr.

videnskabelige undervisning og forskning har fundet fra statens side, har de betydelige fondsmidler, der af enkelte borgere efter *J C. Jacobsens* store forbillede er blevet stillet i samfundets tjeneste, bidraget væsentlig til at muliggøre den indsats, der herhjemme kunne ydes paa de forskellige forskningsomraader. Danmarks deltagelse i det internationale samarbejde paa videnskabens omraade har endvidere paa forbilledlig maade været fremmet gennem oprettelsen af *Rask-Ørstedfondet*, der ved afslutningen af den første verdenskrig skabtes af staten i erkendelse af betydningen af det mellemfolkelige kulturelle samarbejde.

PAA denne baggrund forefindes der her i landet saavel en betydelig sagkundskab paa videnskabens forskellige omraader som en værdifuld erfaring med hensyn til forskningsarbejdets formaalstjenlige planlæggelse, og ikke mindst staar en veluddannet og evnerig yngre generation af videnskabsdyrkere rede til at videreføre udviklingen. De krav, der maa stilles baade til de nødvendige hjælpemidler til forskningen og til det videnskabelige arbejdes organisation er imidlertid i de seneste aar undergaaet en gennemgribende ændring som følge af den rivende udvikling inden for videnskabens mest forskellige grene, der har aabnet helt uventede muligheder for forskningen og anvendelsen af dens resultater.

I erkendelse af disse forandrede forhold er der verden over, saavel fra store som fra smaa samfunds side, ikke alene sat energiske bestræbelser i gang for udnyttelsen af videnskabens resultater til samfundets tarv, men tillige stillet meget forøgede midler til raadighed for forskningen til uddybelse af vor viden om naturen og mennskelivet i alle dets afskygninger. *Dersom vort land ikke hurtigt skal blive ladt tilbage og miste en stilling blandt nationerne lige vigtigt for samfundets fremgang som for vor kulturs betryggelse, er det derfor paatrængende nødvendigt, at der ogsaa herhjemme ydes større støtte end hidtil for videnskabens fremme.* Som forholdene ligger, betyder dette ikke blot væsentlig forøgede hjælpemidler til forskningen paa alle omraader, men tillige, at der for den ungdom, der føler kald til en gerning i videnskabens tjeneste, skabes saadanne kaar, at den kan faa lejlighed til at udvikle sine evner og betingelser for at udnytte disse til gavn for samfundet.

DE synspunkter, der har gjort sig gældende ved beslutningerne om den store indsats, der selv i smaa og økonomisk vanskeligt stillede lande i disse aar gøres til fremme af videnskabelig forskning paa alle omraader, gælder ogsaa for vort land. *Det maa paa det stærkeste fremhæves, at det ikke drejer sig om en sag, hvis løsning kan udskydes med henvisning til landets øjeblikkelige økonomiske vanskeligheder, men man maa gøre sig klart, at hvis vi paa videnskabens omraade sakker agterud i forhold til andre nationer, vil vi have afskaaret os fra muligheder, der i væsentlig grad vil være bestemmende for levevilkaarene i vort land og for folkets tillid til vor kulturs fremtid.* Det er en sag, der angaar ikke alene den akademiske verden, men i videste forstand hele befolkningen. Det er derfor nødvendigt, at enhver er op mærksom paa de farer, det indebærer for vort samfund, dersom man ikke — i forbindelse med de vidtgaaende forholdsregler, der iværksættes for at underbygge Danmarks økonomi og møde de alvorlige problemer, landet staar overfor — sørger for med forholdsvis mindre ofre at holde ejen aaben til sikring af fremtiden i henseender, der kan blive afgørende for samfundets trivsel og for bevarelsen af vore traditioner paa det kulturelle livs omraade.

TRANSLATION

Niels Bohr's deeply serious appeal

"It is urgently necessary that greater support than previously should be provided for the advancement of science"

"A solution cannot be delayed with reference to the country's immediate difficulties"

On the radio's current affairs programme yesterday evening Professor Dr. *Niels Bohr* addressed the Danish people. The contribution was formulated as a deeply serious appeal to the public and the authorities to provide greater support for the advancement of science. We bring here the speech in its entirety:

From the oldest times of culture the endeavours to extend the limits of human knowledge have been of greatest significance for the development of society, and in our time science has in more and more ways made its mark on our existence. At the same time as the steadily deeper insights into the laws of nature have created unsuspected possibilities for improving the conditions of life, we have gained an ever deeper insight into the many facets of human culture and its development throughout the ages. In all civilized societies it has also been recognized as an important task to participate according to ability in this common great human work, and more than anything else the international cooperation in the field of science shows us what can be achieved by joint striving towards common goals.

Danish society has old traditions for successful participation in the development in numerous fields of science and, not least, for cooperation with other peoples for the increase of own experience and shared knowledge. Throughout the ages these endeavours have left profound traces on intellectual life in Denmark and have been regarded by the Danish people with warm and selfless interest. In addition to the support that scientific education and research have found on the part of the State, the considerable means from foundations that have been made available for the service of society by individual citizens following *J.C. Jacobsen*'s great example, have contributed significantly to making possible the effort that could be extended here at home in the various fields of research. Denmark's participation in international cooperation in the field of science has furthermore been advanced in an exemplary way through the establishment of the *Rask–Ørsted Foundation*, which was created by the State at the close of the first world war in recognition of the importance of international cultural cooperation.

On this background, there is in this country not only considerable expertise in the various fields of science, but also valuable experience as regards expedient planning of research work, and, not least, there is a well-educated and talented young generation of scientific workers ready to continue the development. The requirements that must be set, both for the necessary resources for research and for the organization of the scientific work, have, however, in recent years undergone a radical change as a result of the rapid development within the multifarious branches of science, which has opened quite unexpected possibilities for research and the application of its results.

In recognition of these altered circumstances, all over the world, in societies large and small, not only have vigorous endeavours been initiated for the application of scientific results for the benefit of society, but, in addition, greatly increased resources have been made available for research towards deepening our knowledge of nature and human life in all its forms. *If our country is not to be rapidly left behind and lose a position among the nations just as important for the advance of society as for the conservation of our culture, it is therefore urgently necessary that also here at home greater support than previously should be provided for the advancement of science.* Under present circumstances this means not only considerably increased resources for research in all fields, but also that such conditions are created for the young people who feel called to work in the service of science that they can have the opportunity of developing their talents and conditions for using these for the benefit of society.

The viewpoints behind the decisions about the great effort which is being made in these years, even in small and financially weak countries, for the advancement of scientific research in all fields, also hold for our country. *It must be emphasized most strongly that this is not a matter whose solution can be delayed with reference to the country's immediate financial difficul-*

ties, but we must be aware that if we lag behind in the field of science in relation to other nations, we will have cut ourselves off from opportunities which to a considerable degree will be decisive for the living conditions in our country and for the confidence of the people in our culture's future. This is a matter which not only concerns the academic world, but in the broadest sense the whole population. It is thus necessary that everyone is aware of the dangers involved for our society if provision is not made – in connection with the wide-ranging precautionary measures which are being implemented to consolidate Denmark's finances and to meet the serious problems facing the country[1] – to keep, by relatively small sacrifices, the path open for ensuring the future in respects that can prove decisive for the prosperity of society and for the preservation of our traditions in the field of cultural life.

[1] The Social Democratic minority government under Hedtoft had stood down on 25 October 1950 as a result of the opposition's criticism of its handling of Denmark's economic problems. It was succeeded by a Liberal–Conservative government led by Erik Eriksen under which severe new taxes, compulsory savings and a defence loan were introduced by mid-November.

XIV. MEETING ON 2 FEBRUARY 1951

MØDET DEN 2. FEBRUAR 1951
Overs. Dan. Vidensk. Selsk. Virks. Juni 1950 – Maj 1951, pp. 42–45

With J.N. Nielsen

TEXT AND TRANSLATION

See Introduction to Part II, p. [350].

Mødet den 2. februar 1951.

I fortsættelse af drøftelsen i det foregående møde af spørgsmålet om at rette en henvendelse fra selskabet til regeringen om tilvejebringelsen af forbedrede muligheder for videnskabelig forskning havde præsidenten og sekretæren overbragt undervisningsministeren en skrivelse, hvori der nærmere blev gjort rede for de synspunkter, som fremkom under drøftelsen i selskabet, og hvortil selskabets medlemmer havde givet deres enstemmige tilslutning.

Skrivelsen, der oplæstes, og hvortil præsidenten gjorde nogle bemærkninger, havde følgende ordlyd:

»Det Kongelige Danske Videnskabernes Selskab, hvis formål er at fremme naturvidenskaben og de humanistiske videnskaber, har gennem sin 200-årige historie været nøje forbundet med videnskabens udvikling i vort land, og det er derfor naturligt, at det med vågen opmærksomhed følger de betingelser, der fra dansk side bydes deltagelsen i den videnskabelige forskning. Inden for selskabet har man derfor i længere tid med bekymring iagttaget, hvorledes disse betingelser i materiel henseende i stedse ringere grad svarer til betydningen af de opgaver, der trænger sig på, og man har derfor ofte overvejet, hvordan man bedst kunne støtte bestræbelserne for at forbedre forholdene.

Der skal ikke her i enkeltheder gøres rede for de foreliggende omstændigheder, men der kan henvises til en udtalelse af selskabets præsident i radioen den 18. januar, hvoraf teksten vedlægges. I sit møde den efterfølgende dag gav Videnskabernes Selskab enstemmigt sin tilslutning til de i udtalelsen fremsatte synspunkter, og efter en påfølgende forhandling besluttede selskabet at rette en henvendelse til statsmyndighederne om at søge tilvejebragt forbedrede muligheder for videnskabelig forskning som et nødvendigt led i bestræbelserne for at bringe vort land igennem tidens vanskeligheder. Hertil kræves der skridt i to retninger:

På den ene side er der påtrængende behov for forbedring af studieforholdene ved universiteterne og de højere læreanstalter, herunder udbygning af laboratorier, museer og lignende undervisnings- og forskningsinstitutioner, samt en følelig lettelse af vilkårene for den studerende ungdom. Til de fra vide kredse fremkomne tilkendegivelser med indtrængende opfordring til, at der ved snarlige gennemgribende foranstaltninger rådes bod på de mest fremtrædende mangler i sådanne henseender, ønsker selskabet at give sin varmeste tilslutning.

På den anden side er det stærkt påkrævet, at der gennem årlige bevillinger fra statens side skabes betydeligt udvidede muligheder for videnskabelig forskning. Størrelsen af sådanne bevillinger må fra tid til anden tages op til ny overvejelse, da den afhænger af, hvilke opgaver der ligger for, og hvilke kræfter der kan sættes ind på deres løsning. Angående nødvendigheden af en sådan forøgelse af statens støtte til forskningen, som er gennemført i en lang række lande, hvor den fra første færd har været betragtet som et væsentligt led i genopbygningsarbejdet efter krigen, skal man påny henvise til ovennævnte radioudtalelse, i hvilken der peges på de alvorlige farer, det vil medføre, dersom vort land bliver ladt tilbage i udviklingen. Man kan her skelne mellem sådanne videnskabelige undersøgelser, der mere eller mindre direkte tager sigte på løsningen af bestemte opgaver, der til enhver tid stilles af samfundets funktion og behov, og den grundvidenskabelige forskning, der i første række tilstræber at uddybe vor viden om naturen og menneskelivet og derigennem at skabe grundlaget for og give impulser til løsning af samfundsopgaver.

Undersøgelser af den førstnævnte art vil antagelig mest formålstjenligt kunne støttes ved, at man foruden at forøge de midler, der er stillet til rådighed for det teknisk-videnskabelige forskningsråd, bevilger de fornødne midler til et landbrugsvidenskabeligt og et lægevidenskabeligt forskningsråd med tilsvarende virkefelter. Planerne for sådanne foranstaltninger må naturligvis udformes i samråd med de på disse områder virkende institutioner, inden for hvilke en større sagkundskab er til stede end den, der repræsenteres af Videnskabernes Selskabs medlemskreds. Hvad den grundvidenskabelige forskning angår, drejer det sig derimod om et område, der efter sin art falder på det nøjeste i tråd med medlemmernes bestræbelser og virksomhed, og her er det da nærliggende for selskabet at gøre rede for sit syn på den nuværende situation og at fremkomme med mere konkrete forslag.

Når det på mange områder i tidligere årtier har været muligt her i landet at følge med i udviklingen og at gøre en betydelig indsats til videnskabens fremme, skyldes dette, udover den støtte, som videnskabelig undervisning og forskning har fået fra statens side, i første linie Carlsbergfondet, ved hvis oprettelse der fra stifteren, J. C. Jacobsens side vistes selskabet så stor tillid. Det må her fremhæves, at Carlsbergfondet i henhold til statutmæssig bestemmelse virker for formål, der ikke direkte indbefattes i de forpligtelser, der må skønnes naturligt at påhvile staten. Dog har fondet, der også ved tidligere lejligheder har givet ekstraordinære tilskud til egentlige universitetsinstitutioner og f. eks. har bekostet bygningen af matematisk institut, under hensyntagen til, at arbejdet inden for grundvidenskaberne i de senere år har mødt stedse stigende vanskeligheder ved at opfylde tidens krav, søgt at bøde på forholdene ved at stille betydelige midler til rådighed for udvidelsen af universitetets institut for teoretisk fysik samt oprettelsen af et tidssvarende astronomisk observatorium og indretningen af laboratorier til biologiske undersøgelser på forskellige felter, hvis sammenspil stiller særlig fremgang i udsigt.

Selv om dansk videnskab står i dyb taknemmelighedsgæld til Carlsbergfondet og ligeledes til de forskellige andre fonds, der senere er kommet til som et udslag af privat offervilje, må det erkendes, at fremtiden ikke kan bygges på disse fonds alene, der bl. a. på grund af pengenes synkende værdi ikke vil kunne magte opgaven, selv om deres medvirken fortsat vil være af stor betydning. Man anerkender med stor taknemmelighed, at staten ved oprettelsen af Rask-Ørsted fondet og den senere forøgelse af dettes midler har ydet en meget betydningsfuld hjælp til dansk videnskabelig virksomhed i dens forbindelse med udlandet. Som forholdene har udviklet sig, vil det imidlertid ikke længere være muligt her hjemme at følge med i den rivende udvikling af videnskaben, der stadig åbner nye muligheder for resultater af største betydning for samfundets liv, med mindre staten som led i bestræbelserne for at genoprette landets trivsel påtager sig den opgave mere effektivt at fremme grundvidenskaberne og deres anvendelser.

Efter de erfaringer, der er vundet ikke mindst gennem Carlsbergfondets virksomhed, vil det efter selskabets skøn være nødvendigt, at de midler, der fra statens side årlig stilles til rådighed til grundvidenskabelig forskning, til at begynde med er af en lignende størrelse som det beløb, Carlsbergfondet årlig anvender på dette område, og at beløbets størrelse senere reguleres i takt med udviklingen og under hensyntagen til de ved midlernes anvendelse indhøstede erfaringer. Man regner med, at disse nye midler ikke skal dække egentlige anlægsudgifter som bygning af laboratorier og anskaffelse af permanent udstyr, men at de benyttes til støtte for anskaffelse af specielle hjælpemidler og apparater samt til at sikre den nødvendige arbejdskraft til gennemførelsen af de videnskabelige undersøgelser.

Ved fremsættelsen af denne henvendelse føler det Kongelige Danske Videnskabernes Selskab det i den foreliggende situation som sin pligt overfor samfundet at henstille til statsmyndighederne, at der snarest muligt søges opnået en forsvarlig løsning af denne uopsættelige og for vort land livsvigtige sag.«

TRANSLATION

Meeting on 2 February 1951.

...

In continuation of the discussion at the previous meeting[1] about the question of directing an application from the Academy to the Government about the provision of improved opportunities for scientific research, the President and the Secretary had handed a letter to the Minister of Education, which described in greater detail the viewpoints that arose during the discussion in the Academy and to which the members of the Academy had given their unanimous support.

The letter, which was read aloud and about which the President made some remarks, ran as follows:

"The Royal Danish Academy of Sciences and Letters, whose mission is to advance the natural and human sciences, has throughout its 200-year history been closely connected to the development of science in this country, and it is therefore natural that it follows with lively attention the conditions that are offered in Denmark for the participation in scientific research. For a long time the Academy has observed with anxiety how these conditions as regards resources correspond to an ever lesser extent to the importance of the urgent tasks, and it has thus often been considered how it would be best to support the endeavours for improving the situation.

Here the circumstances in question will not be described in detail, but reference may be made to a statement by the President of the Academy on the radio on 18 January, the text of which is enclosed.[2] At its meeting the following day the Royal Academy gave unanimous support to the viewpoints made in the statement, and after a subsequent discussion the Academy decided

---

[1] The meeting on 19 January 1951, the proceedings of which are not reproduced in this volume. See the introduction to Part II, p. [350].

[2] The talk is reproduced on pp. [447] ff.

to present an appeal to the State authorities about trying to provide improved possibilities for scientific research as a necessary part of the endeavours to bring our country through the current difficulties. To achieve this, moves are required in two directions:

On the one hand there is a pressing need for improving conditions of study at the universities and the institutes of higher education, including the enlargement of laboratories, museums and similar educational and research institutions, as well as appreciable improvement of the conditions for young people doing their studies. The Academy wishes to give its warmest support to the statements received from wide circles earnestly requesting that remedy is made for the most obvious deficiencies in such respects at an early date.

On the other hand it is strongly required that significantly increased opportunities for scientific research are created through annual allocations on the part of the State. The size of such allocations must now and again be taken up for reconsideration, as it depends on which problems are now before us and which efforts can be applied for their solution. As regards the necessity of such an increase in State support for research, which has been carried out in many countries where, from the very start, it has been regarded as an important component of reconstruction after the war, reference must once again be made to the aforementioned radio statement, in which attention is drawn to the serious dangers that would result if our country was left behind in the development. Here it is possible to distinguish between scientific investigations aiming more or less directly at the solution of concrete tasks which at any one time are posed by the functions and needs of society, and fundamental scientific research which in the first instance endeavours to deepen our knowledge of nature and human life and thereby to create the basis for, and give impulses to, the solution of tasks incumbent on society.

Investigations of the first kind could probably be most suitably supported, in addition to increasing the funds made available to the research council for science and technology, by allocating the necessary funds to an agricultural research council and a medical research council with corresponding spheres of activity.[3] The plans for such arrangements must naturally be drawn up in consultation with the institutions active in these fields, within which greater expertise is to be found than that represented by the circle of Royal Academy members. As regards fundamental scientific research, on the other hand, we

---

[3] The Technical–Scientific Research Council had been established in 1946. Research Councils covering the fields of agricultural and medical research were yet to be established. See the introduction to Part II, p. [351].

have to do with a sphere which by its nature corresponds most closely with the endeavours and activities of the members, and here it is thus appropriate for the Academy to explain its viewpoint on the present situation and to put forward more concrete proposals.

When in earlier decades it has been possible in this country to keep up with developments and to make a significant contribution to the advancement of science in many fields, this was primarily due to, in addition to the support given to scientific education and research on the part of the State, the Carlsberg Foundation, at whose establishment J.C. Jacobsen, its founder, showed the Academy such great trust. It must here be emphasized that, according to its statutes, the Carlsberg Foundation works for objectives that are not directly included in the obligations that must be thought to rest naturally with the State. Nevertheless, the Foundation, which also on earlier occasions has given extraordinary support to university institutions proper and, for example, paid for the building of the mathematical institute,[4] has sought, under consideration of the fact that work within the fundamental sciences in recent years has faced ever increasing difficulties in meeting current requirements, to improve conditions by making considerable funds available for the enlargement of the university's institute for theoretical physics,[5] as well as the establishment of a modern astronomical observatory[6] and the equipping of laboratories for biological investigations in various fields whose interaction promises particular progress.[7]

Even though Danish science owes a great debt of gratitude to the Carlsberg Foundation, and similarly to the various other foundations which have arisen later as a result of private generosity, it must be acknowledged that the future cannot be built upon these foundations alone which, because of among other things the depreciation of the currency, will not be able to fulfil the task, even though their participation will continue to be of great importance. It is recognized with great appreciation that, by the establishment of the Rask–Ørsted

[4] A new Institute of Mathematics at the University of Copenhagen was built in 1933 adjacent to Bohr's institute and inaugurated in February 1934. It was directed by Niels Bohr's brother, Harald.
[5] The allocations to Bohr's institute after the war for expansions and apparatus constituted a significant part of the Carlsberg Foundation's budget.
[6] Bohr probably refers to the astronomical observatory in the village of Brorfelde outside Copenhagen, which was formally inaugurated in 1953.
[7] In the years after the war the Carlsberg Foundation provided several relatively small grants for such interdisciplinary work. Just in the year that Bohr spoke, for example, grants had been given to George de Hevesy's (1885–1966) continued contribution in Denmark (he had lived in Stockholm since 1943) to the application of his indicator method, and to Hevesy's former assistant Hilde Levi's (1909–2003) work on carbon dating. On Hevesy, see the introduction to Part II, p. [373] and pp. [522] ff.

Foundation and the subsequent increase of its funds, the State has provided very significant help to Danish scientific activity as regards its relations with other countries. As the situation has developed it will, however, no longer be possible in this country to keep up with the rapid development of science which continually opens new opportunities for results of the greatest importance for the life of society, unless the State as part of the endeavours to reestablish the prosperity of the country assumes the task to advance more efficiently the fundamental sciences and their applications.

On the basis of the experience gained not least by the activity of the Carlsberg Foundation, the Academy deems that it will be necessary that the funds made available annually by the State for fundamental scientific research are from the start of a size similar to the amount the Carlsberg Foundation expends annually in this field, and that the size of the amount is subsequently regulated in accordance with the turn of events and with reference to the experience gained by the use of the funds. It is taken for granted that these new funds should not cover building expenses proper such as for the construction of laboratories and the acquisition of equipment, but that they should be used to help in the acquisition of special equipment and apparatus as well as to ensure the availability of necessary personnel for carrying out the scientific investigations.

In presenting this appeal, the Royal Academy of Sciences and Letters feels in the present situation that it is its obligation towards society to suggest to the State authorities that a proper solution of this urgent matter, so vital for our country, should be sought achieved as soon as possible."

XV. STATEMENT BY PROFESSOR NIELS BOHR
IN THE CURRENT AFFAIRS PROGRAMME ON
DANISH NATIONAL RADIO, MONDAY 4 FEBRUARY 1952

UDTALELSE AF PROFESSOR NIELS BOHR
I RADIOENS AKTUELLE KVARTER,
MANDAG DEN 4. FEBRUAR 1952
Videnskabsmanden: Meddelelser fra Foreningen til
Beskyttelse af Videnskabeligt Arbejde **6** (No. 1, 1952), p. 3

TEXT AND TRANSLATION

See Introduction to Part II, p. [352].

UDTALELSE AF PROFESSOR NIELS BOHR
i radioens aktuelle kvarter, mandag den 4. februar 1952

Det lovforslag om oprettelsen af »Statens almindelige videnskabsfond«, som undervisningsministeren har forelagt og om hvilken rigsdagens forhandlinger begynder i morgen, vil, om det — som man inderligt må håbe — bliver gennemført, betyde et overordentligt fremskridt for kulturelle bestræbelser i vort land.

Da der i de bemærkninger, der er knyttet til lovforslagets forelæggelse, er omtalt en skriftlig henvendelse rettet i januar 1951 af Det kgl. danske Videnskabernes Selskab til undervisningsministeren, vil jeg gerne i nogle få ord gøre rede for de betragtninger, der ligger til grund for denne henvendelse.

Forståelsen af videnskabelig forsknings betydning for åndslivet og den tekniske udvikling deles af alle kredse i den danske befolkning, og gang på gang har der været givet udtryk for en sådan forståelse på rigsdagen og af de skiftende regeringer. Imidlertid har videnskabens rivende udvikling og de for forskningen nødvendige stedse mere bekostelige hjælpemidler bragt med sig, at en ny anstrengelse må gøres fra landets side, dersom vi ikke skal sakke agterud i forhold til andre nationer og afskære os fra muligheder, der i væsentlig grad vil være bestemmende for levekårene i vort land og for folkets tillid til vor kulturs fremtid.

Særlig i årene efter krigen er ønsker om forbedringen af vilkårene for dansk videnskabeligt studium og forskning kommet stærkt til orde fra forskellig side, og til overvejelse af hele spørgsmålet har undervisningsministeriet for nogle år tilbage nedsat en kommission til behandling af statens forhold til videnskaben.

Med baggrund af den sagkundskab og erfaring, som Videnskabernes Selskabs medlemmer repræsenterer, har man i henvendelsen til undervisningsministeren særlig ønsket at henlede opmærksomheden på betydningen af, at de krav, som opretholdelsen og videreførelsen af forskningen stiller, og om hvilke man tør håbe på fuld enighed fra alle sider, løses umiddelbart, inden uoprettelig skade er sket.

Det synspunkt, det her drejer sig om, og som har fundet udtryk i lovforslaget, er nødvendigheden af, at der stilles midler til rådighed, som kan sættes ind på løsningen af betydningsfulde videnskabelige opgaver, der fortsat melder sig under videnskabens stadige udvikling på de forskellige områder. Den fremgangsmåde, hvortil

man uden sådanne midler er henvist at søge regeringens og rigsdagens tilslutning til hver enkelt bevilling, ville nemlig være forbundet med overvejelser fra disse myndigheders side, der kan foranledige forsinkelser og udskydelser, som stiller sig ivejen for en aktiv indsats i løsningen af de foreliggende forskningsopgaver.

I erkendelsen af disse omstændigheder har man da også — navnlig efter den sidste verdenskrig — i de fleste lande skabt videnskabelige forskningsfonds, og stillet meget betydelige midler til deres rådighed.

Her i landet blev der netop i denne henseende skabt et forbillede ved det af brygger J. C. Jacobsen for 75 år siden stiftede Carlsbergfond, der har kunnet yde en støtte til forskningen såvel inden for åndsvidenskaberne som naturvidenskaberne, der har været af uvurderlig betydning for det relativt høje stade, som dansk forskning har indtaget gennem de sidste generationer.

Videnskabernes Selskab, hvem Carlsbergfondets stifter viste den tillid, blandt dets medlemmer at vælge fondets direktion, har haft lejlighed til at følge fondets frugtbare virksomhed på nært hold. Glæden over denne har imidlertid i de senere år været blandet med voksende bekymring over midlernes utilstrækkelighed til at imødekomme de nødvendige krav for at opretholde dansk videnskabs stade og sikre landets stilling på dette område blandt de andre kulturlande.

Ansættelsen af det årlige beløb, som i Videnskabernes Selskabs henvendelse skønnedes nødvendigt til den grundvidenskabelige forskning, hviler såvel på den ved Carlsbergfondets virksomhed vundne erfaring om det umiddelbare behov, som hensyntagen til det grundlag, der er skabt ved denne virksomhed, og den støtte, som fondet også i fremtiden vil kunne yde.

Om forholdene tillod det, ville det være ønskeligt, at endnu langt større midler kunne stilles til rådighed for den videnskabelige forskning, således som det er sket i andre lande af Danmarks størrelse og kulturstade. Alt taget i betragtning vil imidlertid den foreslåede oprettelse af statens videnskabsfond betyde, at der for en årrække vil være skabt betingelser for, at den videnskabelige forskning her hjemme kan opretholdes og videreføres i et omfang, der svarer til landets traditioner, og som kan sikre os en til forholdene svarende plads i de fællesmenneskelige bestræbelser på kulturens område.

TRANSLATION

STATEMENT BY PROFESSOR NIELS BOHR
in the current affairs programme on
Danish National Radio, Monday 4 February 1952

The proposal for legislation regarding the establishment of the National Science Foundation, which the Minister of Education has submitted and about which the debates in parliament are to begin tomorrow, will, if – as one must sincerely hope – it is passed, signify an extraordinary advance for the cultural endeavours in our country.

As the remarks which are linked to the presentation of the proposal contain reference to a written application made in January 1951 by the Royal Danish Academy of Sciences and Letters to the Ministry of Education,[1] I would like to explain in a few words the considerations forming the basis for this application.

The understanding of the importance of scientific research for intellectual life and technological development is shared by all circles in the Danish population, and time and time again expression for such an understanding has been made in Parliament and by the changing Governments. However, the rapid development of science and the ever more costly facilities required for research have caused the need for a new effort on the part of the country, if we are not to lag behind in relation to other nations and cut ourselves off from opportunities that to a considerable degree will be decisive for living conditions in our country and for the people's trust in the future of our culture.

Especially in the postwar years wishes for the improvement of conditions for Danish scientific study and research have been strongly voiced from various quarters and, in order to consider the whole question, the Ministry of Education set up a commission some years ago to deal with the relationship of the state to science.[2]

On the background of the expert knowledge and experience which the members of the Academy of Sciences and Letters represent, one has, in the ap-

---

[1] Reproduced on pp. [453] ff.

[2] This was "The Commission of 22 August 1946 for the investigation of the question of the advancement of science in Denmark through measures by the state", known as the Science Commission. It consisted of politicians, civil servants, and representatives from Danish institutions of higher education such as H.M. Hansen (see p. [513], ref. 5) and J. Pedersen (see p. [410], ref. 13). It was established after pressure from among others the "Science and Society" study group, on which see the introduction to Part II, p. [348].

plication to the Minister of Education,[3] wished in particular to direct attention to the importance of immediately meeting, before irreparable damage is done, the demands which are posed by the sustenance and continuation of research, and about which one dare hope for full agreement from all parties.

The viewpoint involved here, and which has found expression in the proposal for legislation, is the necessity of making means available which can be applied for the resolution of important scientific tasks that continue to materialize during the constant development of science in the various fields. The procedure whereby one is relegated, without such means, to seek the approval of Government and Parliament for every single grant, would thus involve consideration on the part of these authorities which can cause delays and postponements standing in the way of an active effort in the resolution of the research tasks at hand.

In recognition of these circumstances, scientific research foundations have accordingly been established in most countries – especially after the last world war – and very significant means have been placed at their disposal.[4]

In this country an example was created precisely in this regard by the Carlsberg Foundation – established 75 years ago by J.C. Jacobsen, the brewer – which has been able to extend support to research within both the humanities and the natural sciences, a support that has been of invaluable importance for the relatively high position that Danish research has assumed throughout the last generations.

The Academy of Sciences and Letters, to which the founder of the Carlsberg Foundation showed the trust of choosing the foundation's board of directors from among its members, has had the opportunity to follow the foundation's fruitful activity at close hand. The joy over this activity, however, has in recent years been mixed with growing anxiety about the insufficiency of the means to meet the necessary demands for maintaining the position of Danish science and ensuring the standing of the country in this field among the other civilized nations.

The determination of the annual amount that in the application from the Academy of Sciences and Letters was deemed necessary for fundamental scientific research, rests on the experience about the immediate needs gained through the activity of the Carlsberg Foundation, as well as the consideration of the

---

[3] Flemming Friis Hvidberg (1897–1959). See the introduction to Part II, pp. [350] and [351].

[4] For example, National Science Foundations had recently been established in Sweden and the United States.

basis that this activity has created and the support that the foundation will be able to extend also in the future.

If circumstances allowed, it would be desirable that even much greater means could be made available for scientific research, as has happened in other countries of Denmark's size and cultural stage. All things considered, however, the suggested establishment of the National Science Foundation will mean that for a number of years conditions will have been created so that scientific research here at home can be maintained and continued to an extent corresponding to the country's traditions and which can ensure us a position, in the common human striving in the field of culture, corresponding to our circumstances.

3. THE SOCIETY FOR THE DISSEMINATION OF NATURAL SCIENCE

XVI. EIGHTH PRESENTATION OF THE H.C. ØRSTED MEDAL

OTTENDE UDDELING AF H.C. ØRSTED MEDAILLEN
Fys. Tidsskr. **39** (1941) 175–177, 192–193

To K. Linderstrøm–Lang

TEXT AND TRANSLATION

See Introduction to Part II, p. [353].

Ottende Uddeling af H. C. Ørsted Medaillen.

Onsdag d. 19. Marts 1941 afholdt Selskabet for Naturlærens Udbredelse et festligt Møde i Anledning af, at H. C. Ørsted Medaillen tildeltes Professor, Dr. phil. K. Linderstrøm-Lang.

Mødet afholdtes i Den polytekniske Læreanstalts store Foredragssal, der Kl. 20, da Selskabets Protektor, Hans Majestæt Kongen, ankom, var fyldt af en festklædt Forsamling.

Direktionens Formand, Professor, Dr. Niels Bohr indledede med følgende Tale:

Deres Majestæt, mine Damer og Herrer!

Paa Selskabet for Naturlærens Udbredelses Vegne vil jeg gerne byde velkommen til denne festlige Sammenkomst i Anledning af den 8. Uddeling af den Æresbevisning til en dansk Videnskabsmand, der er knyttet til Mindet om Selskabets Stifter. Fremfor alt ønsker Selskabet at give Udtryk for sin dybe Taknemmelighed over, at Deres Majestæt, Selskabets høje Protektor, har villet komme til Stede ligesom ved de tidligere Uddelinger af H. C. Ørsted Medaillen. Lige fra Selskabets Stiftelse for mere end 100 Aar siden har den Interesse og Støtte for dets Bestræbelser, der stadig er blevet det vist fra Kongehusets Side, været af største Betydning for Selskabets Virksomhed. I Dag, hvor hele Folket i Taknemmelighed og Tillid staar sammen om vor Konge i Bestræbelsen for at værne om den særegne og for os saa dyrebare Kultur, der i Aarhundreder har blomstret her i Landet, er Deres Majestæts Bevaagenhed os om muligt mere værdifuld end nogensinde. Vi er jo netop samlede her for at mindes de store Bidrag til Naturvidenskabens Udvikling og til dens Anvendelse i Samfundets Tjeneste, som er knyttet til Navnet paa Selskabets Stifter, og for at glæde os over, at de stolte Traditioner siden har kunnet opretholdes, saaledes at der ogsaa i Dag i vort Samfund arbejder Videnskabsmænd, hvis Fortjenester gør dem værdige til Tildelingen af den Æresbevisning, der bærer H. C. Ørsteds Navn. Ved det Valg blandt disse, som Direktionen ved denne Lejlighed efter bedste Skøn har truffet, understreges danske videnskabelige Traditioner ogsaa paa ganske særlig Maade.

Professor Kai Linderstrøm-Lang har jo paa saa fremragende Maade arbejdet videre paa det Grundlag, der herhjemme blev skabt af den første Modtager af H. C. Ørsted-Medaillen, Professor S. P. L. Sørensen, gennem hans mangeaarige frugtbare Forskervirksomhed paa Carlsberglaboratoriets kemiske Afdeling, hvis Ledelse Professor Linderstrøm-Lang efter ham har overtaget. Selve Tilblivelsen af dette i sin Art enestaaende Laboratorium skyldes det Borgersind og den Begejstring for Videnskaben, som Brygger J. C. Jacobsen maaske fandt det allersmukkeste Udtryk for, da han siden skænkede hele sin store Virksomhed til dansk Videnskab og i sin første Meddelelse til det kgl. danske Videnskabernes Selskab om sin Hensigt at op- rette Carlsbergfondet netop fremhævede sin Beundring og Tak- nemmelighed for alt, hvad H. C. Ørsted har udrettet for vort Land. Hvad Carlsbergfondet har betydet for dansk Forskning og dens Anseelse i Verden, behøver jeg her ikke nærmere at uddybe; men i Dag ønsker vi især at give Udtryk for vor Glæde over, at der stadig i Spidsen for Carlsberglaboratoriet staar Mænd, der viderefører dets traditionsrige Virksomhed paa en Maade, der saa fuldkomment op- fylder Stifterens Forventninger.

Før jeg paa Selskabets Vegne overrækker Professor Linderstrøm- Lang H. C. Ørsted-Medaillen, skal jeg efter Selskabets Skik oplæse den medfølgende Skrivelse, hvormed Direktionen har motiveret sit Valg ved Uddelingen af denne Æresbevisning.

Sluttelig udtalte Professor Bohr:

Paa hele Selskabets Vegne vil jeg gerne give Udtryk for den Beundring, hvormed vi alle har fulgt Professor Linderstrøm-Langs Redegørelse for sine saa betydningsfulde Undersøgelser og smukke Arbejdsmetoder. Især har han jo ved den glimrende Forevisning af disse Metoder givet os det stærkeste Indtryk af den Eksperimenterkunst, hvorpaa hans Undersøgelser er bygget. Een enkelt Forevisning endnu tror jeg vil glæde alle tilstedeværende, nemlig to Lysbilleder af den Medaille, som Professor Linderstrøm-Lang med saa megen Ære har modtaget.

Her ser vi H. C. Ørsteds smukke Ansigtstræk og hans karakteristiske Underskrift, og her ser vi de i Medaillen prægede og indgraverede Ord: Fra Selskabet for Naturlærens Udbredelse til K. Linderstrøm-Lang.

Til Slut vil jeg gerne paa Selskabets Vegne atter udtrykke vor Taknemmelighed over, at Deres Majestæt har villet komme til Stede og dermed kaste Glans over vor Fest, samt forlene den med den højeste Værdighed. Idet Selskabet slutter sig til hele Folkets Ønske om, at Deres Majestæt endnu i mange Aar maa bevare Sundhed og Kraft til Gavn for vort Land, vil vi gerne dertil knytte Selskabets Forhaabning om at turde se Deres Majestæt iblandt os ved mange senere lignende Lejligheder under den fremtidige Virksomhed, gennem hvilken Selskabet haaber stadig at kunne opretholde sine gamle Traditioner.

TRANSLATION

Eighth Presentation of the H.C. Ørsted Medal.

On Wednesday 19 March 1941 the Society for the Dissemination of Natural Science held a festive meeting on the occasion of the awarding of the H.C. Ørsted Medal to Professor K. Linderstrøm–Lang,[1] D.Phil.

The meeting was held in the large lecture hall at the Technical University which was filled with an audience in festive dress when His Majesty the King, Patron of the Society, arrived at 8 o'clock.

Professor Niels Bohr, chairman of the board of directors, opened the meeting with the following words:

Your Majesty, Ladies and Gentlemen,

On behalf of the Society for the Dissemination of Natural Science I wish to welcome you to this festive gathering on the occasion of the 8th awarding to a Danish scientist of the honour linked to the memory of the Society's founder. Above all, the Society wishes to express its deep gratitude that Your Majesty, the Society's Royal Patron, has agreed to be present just as at the previous awardings of the H.C. Ørsted Medal. Ever since the founding of this Society more than 100 years ago, the interest and support for its endeavours steadfastly shown by the royal family have been of the greatest importance for the activities of the Society. Today, when the whole people is united in gratitude to and confidence in our King in the endeavour to protect the special and for us most precious culture that has flowered in this country for centuries, Your Majesty's attention is, if possible, more valuable to us than ever before. We are gathered here precisely to remember the great contributions to the development of science and its application in the service of society that is linked to the name of the founder of the Society, and to rejoice that it has been possible to maintain these proud traditions, so that in our country also today there are scientists working whose merits make them worthy of the awarding of the honour that bears H.C. Ørsted's name. With the choice among these, which the

---

[1] Kai Ulrik Linderstrøm–Lang (1896–1959).

board has made to the best of its ability, Danish scientific traditions are also underlined in a very special way.

Professor Linderstrøm–Lang has of course continued work in an outstanding manner on the basis created here in Denmark by the first recipient of the H.C. Ørsted Medal, Professor S.P.L. Sørensen, through the latter's fruitful research activity over many years at the chemical section of the Carlsberg Laboratory, whose leadership Professor Linderstrøm-Lang had taken over after him.[2] The very creation of this laboratory, which is unique of its kind, is due to the public spirit and the enthusiasm for science, for which the brewer J.C. Jacobsen perhaps found the most beautiful expression when he subsequently donated his entire great enterprise to Danish science and, in his first report to the Royal Danish Academy of Sciences and Letters about his intention to establish the Carlsberg Foundation, precisely emphasized his admiration and gratitude for everything H.C. Ørsted has accomplished for our country. I need not here elaborate further upon what the Carlsberg Foundation has meant for Danish research and its esteem in the world; but today we wish in particular to express our joy that at the head of the Carlsberg Laboratory there are still men who carry on its traditional activity in a manner which so perfectly fulfills the expectations of the founder.

Before I hand over the H.C. Ørsted Medal to Prof. Linderstrøm–Lang on behalf of the Society I shall read, in accordance with the custom of the Society, the accompanying text whereby the board has motivated its choice for awarding this honour.

...

In conclusion Professor Bohr stated:

On behalf of the whole Society, I would like to express the admiration with which we all have followed Professor Linderstrøm–Lang's exposition of his very important investigations and beautiful working methods. In particular

---

[2] The Carlsberg Laboratory, the research branch of the Carlsberg Breweries, was established by J.C. Jacobsen (p. [411], ref. 15), the founder of the Breweries, in 1875. The following year it was placed under a board of trustees elected by the Royal Academy. It consisted from the outset of a chemical and a physiological division. Sørensen was the second leader of the chemical division from 1901 to 1938, when Linderstrøm–Lang took over.

he has given us, with his excellent display of these methods, the strongest impression of the experimental technique upon which his investigations are built. There is yet one more display which I think will please everybody present, that is, two slides of the Medal that Professor Linderstrøm–Lang has received with so much honour.

Here we see the handsome features of H.C. Ørsted and his characteristic signature, and here we see the words moulded and engraved in the medal: From the Society for the Dissemination of Natural Science to K. Linderstrøm–Lang.

Finally, on behalf of the Society, I would again like to express our gratitude to Your Majesty for having agreed to be present and thus lending lustre to our celebration, as well as endowing it with the greatest dignity. As the Society joins with the wish of the whole people that Your Majesty may for many years yet keep your health and strength for the benefit of our country, we would like to add the Society's hope to see Your Majesty among us on many later similar occasions during the future activity, through which the Society hopes to continue be able to maintain its old traditions.

XVII. NINTH PRESENTATION OF THE H.C. ØRSTED MEDAL

NIENDE UDDELING AF H.C. ØRSTED MEDAILLEN
Fys. Tidsskr. **51** (1953) 65–67, 80

To A. Langseth

TEXT AND TRANSLATION

See Introduction to Part II, p. [353].

[477]

Niende uddeling af H. C. Ørsted medaillen.

Onsdag den 3. december 1952 afholdt Selskabet for Naturlærens Udbredelse et festligt møde i anledning af, at H. C.
Ørsted medaillen tildeltes professor, dr. phil. *A. Langseth.*

Mødet indledtes paa særlig smuk maade ved, at Selskabets
protektor, Hans Majestæt Kongen, umiddelbart inden mødet
i festsalen paa Danmarks tekniske Højskole, aflagde sit første besøg i H. C. Ørsted museet. Præsidenten for Selskabet
for Naturlærens Udbredelse, professor, dr. *Niels Bohr* fore-
viste den smukke samling af Ørsteds apparater. Til stede
ved forevisningen var direktionens medlemmer samt H. C.
Ørsteds sønnesøn, direktør *H. C. Ørsted* med frue.

Fra besøget på Ørsted museet. Kongen og professor Niels Bohr. Kongen står
med Ørsteds kompas i hånden. (Pol. Phot.)

Efter besøget paa museet gik Kongen, ledsaget af Selskabets
præsident og de øvrige besøgende, over i festsalen, hvor de
festklædte mødedeltagere var samlede.

Professor Bohr bød velkommen med følgende tale:

1

[478]

Deres Majestæt, mine Damer og Herrer.

Paa Selskabet for Naturlærens Udbredelses vegne vil jeg gerne give udtryk for vor dybe taknemmelighed for, at Deres Majestæt som Selskabets høje protektor har villet komme til stede og foretage overrækkelsen af H. C. Ørsted-medaillen. Selskabet mindes ogsaa med taknemmelighed, at Deres Majestæts fader, Kong Christian X, har overværet de tidligere tildelinger af medaillen, og især at den gamle konge paa denne maade viste Selskabet saa varm en interesse i krigens mørke tid, da medaillen i 1941 forrige gang tildeltes en dansk videnskabsmand.

Paa Selskabets vegne byder jeg dernæst de tilstedeværende repræsentanter for danske institutioner, der hver paa sin maade virker for fremme af naturvidenskaben i vort samfund, velkommen. Ligeledes er det Selskabet en glæde, at saa mange medlemmer af Danmarks naturvidenskabelige Samfund, Fysisk Forening og Kemisk Forening har fulgt indbydelsen til sammen med dets egne medlemmer at hædre den fremragende videnskabsmand, som Selskabet har besluttet at tildele den æresbevisning, der er knyttet til mindet om Selskabets stifter.

Efter dets navn er det et hovedformaal for vort gamle Selskab at medvirke til udbredelsen af kendskabet til naturlærens stadige fremskridt. Adskillige af de opgaver, som H. C. Ørsted med saa stor styrke tog op i Selskabets første tid, er jo efterhaanden overtaget af institutioner, der paa forskellig vis virker for dette formaal. Ikke alene har den radiofoni, hvis vidunderlige udvikling har sin første rod i Ørsteds opdagelse af elektromagnetismen, i vor tid skabt uanede muligheder for udbredelse af kundskaber til hele befolkningen, men mange af de naturvidenskabelige landvindinger, hvorom Ørsted i sin tid med begejstring talte til kredsen af Selskabets første medlemmer, indgaar jo nu i det lærestof, der meddeles i den almene skoleundervisning eller ved Universitetet og de højere læreanstalter.

I denne forbindelse maa vi særlig tænke paa, at det er Ørsteds eget initiativ, som vi skylder oprettelsen af den stadigt blomstrende og nu saa vidt forgrenede institution, inden for hvis mure vi er samlede her i aften. De drømme, der fyldte Ørsteds sind om, hvad de fysiske og kemiske videnskaber ville komme til at betyde for hele samfundslivet, har jo først og fremmest den polytekniske Læreanstalt eller Danmarks tekniske Højskole bragt til virkeliggørelse her i landet.

Denne højskole har lige siden sin oprettelse gæstfrit givet husly for Ørsteds gamle Selskab og har i fjor i anledning af hundredeaarsdagen for Ørsteds død givet fornyet udtryk derfor ved at overlade Selskabet et smukt lokale til anbringelsen af det H. C. Ørsted-museum, hvori saa mange af de instrumenter, som Ørsted benyttede ved sine berømte undersøgelser, og mange af de minder fra hans hjem, som familien saa trofast har værnet om, er opstillet og gjort tilgængelige for offentligheden. Umiddelbart før dette møde har Selskabets direktion haft den ære og glæde at kunne forevise det smukke og minderige museum for Hans Majestæt Kongen.

Foruden den foredragsvirksomhed, som Selskabet stadig søger at videreføre i Ørsteds aand og tilpasse efter tidens forhold, har Selskabet altid følt det som en særlig kærkommen pligt at holde minderne om Ørsteds rige forskerliv og frugtbare gerning i det danske samfund i ære, og det var netop som led i saadanne bestræbelser, at den medaille, som i dag for 9. gang uddeles, for mere end en menneskealder siden stiftedes paa initiativ af Direktionens daværende formand Martin Knudsen. Medaillens uddeling giver Selskabet en mulighed for at give en velfortjent anerkendelse til danske forskere inden for de fysiske og kemiske videnskaber, der har formaaet at opretholde og videreføre de traditioner, som har rødder i aarhundreder herhjemme, og til hvilke Ørsteds navn er uløseligt knyttet.

Det har for Selskabets direktion været en glæde til den række af fremragende danske kemikere, der har været hædret paa denne maade, og blandt hvilke S. P. L. Sørensen modtog den første H. C. Ørsted-medaille og K. Linderstrøm-Lang den sidste, at føje Alex Langseths navn. Den skrivelse, hvormed Direktionen har motiveret denne uddeling af H. C. Ørsted-medaillen, lyder saaledes:

Efter foredraget udtalte professor Bohr:

På Selskabets vegne vil jeg gerne bringe professor Langseth en hjertelig tak for den rige oplevelse, det har været at lytte til hans foredrag, som har givet os et så levende indtryk af den tankerigdom og experimenteringskunst, hvormed han i så høj grad har øget vort kendskab til molekylernes verden og til de lovmæssigheder, der åbenbarer sig i det små, og er bestemmende for egenskaberne hos de stoffer, hvoraf vore redskaber og vore legemer er opbygget. Det er næppe nødvendigt nærmere at uddybe vor tak og vor beundring for, hvad vi har hørt og set, men det turde måske være af interesse endnu at vise nogle billeder af den medaille, som professor Langseth har modtaget.

Her ses medaillens ene side med H. C. Ørsteds smukke ansigtstræk og karakteristiske underskrift, og her ses medaillens anden side med inskription: Fra Selskabet for Naturlærens Udbredelse til A. Langseth.

Til slut vil jeg gerne endnu en gang på Selskabets vegne takke Deres Majestæt, hvis tilstedeværelse og overrækkelse af medaillen har forlenet dette møde med den højeste værdighed og understreget, hvor dybt det ligger hele det danske folk og dets styre på sinde, at videnskabelig forskning herhjemme stadig må blomstre og sætte frugt som bidrag til menneskehedens fælles bestræbelser for forøgelsen af kendskabet til og forståelse af den natur, der omgiver os, og som vi selv tilhører.

TRANSLATION

Ninth Presentation of the H.C. Ørsted Medal.

On Wednesday 3 December 1952, the Society for the Dissemination of Natural Science held a festive meeting on the occasion of the presentation of the H.C. Ørsted medal to Professor *A. Langseth*,[1] D.Phil.

The meeting was begun in an especially beautiful way with His Majesty the King, Patron of the Society, immediately before the meeting in the ceremonial hall of the Technical University of Denmark, visiting the H.C. Ørsted Museum for the first time. Professor Dr *Niels Bohr*, President of the Society for the Dissemination of Natural Science, showed the beautiful collection of Ørsted's apparatuses.[2] The members of the board of directors together with Director *H.C. Ørsted*,[3] H.C. Ørsted's grandson, and his wife were present at the showing.

After the visit to the museum the King, accompanied by the Society's President and the other guests, walked over to the ceremonial hall where the festively dressed meeting participants were assembled.

Professor Bohr bid welcome with the following address:

Your Majesty, Ladies and Gentlemen.

On behalf of the Society for the Dissemination of Natural Science I would like to express our profound gratitude that Your Majesty, as the noble Patron of the Society, has agreed to be here and to present the H.C. Ørsted Medal. The Society also remembers with gratitude that King Christian X, Your Majesty's father, has been present at the previous presentations of the Medal, and in particular that in this way the old king showed such a warm interest for the Society in the dark days of war, the last time the Medal was awarded to a Danish scientist in 1941.

---

[1] Alex Langseth (1895–1961), professor of chemistry at the University of Copenhagen.

[2] The caption for the photograph on p. [478] reads, in English translation: "From the visit to the Ørsted Museum. The King [Frederik IX] and Professor Niels Bohr. The King is holding Ørsted's compass."

[3] Hans Christian Ørsted (1876–1962), Director of the Office of the Nordic Employers' Confederation in Brussels (1922–1939) and in Geneva (1939–1946.)

[482]

On behalf of the Society I now welcome the representatives of Danish institutions present here, which each in its way works for the advancement of science in our society. Similarly it is a pleasure for the Society that so many members of the Science Society of Denmark,[4] the Physical Society[5] and the Chemical Society[6] have accepted the invitation to honour, together with its own members, the outstanding scientist to whom the Society has decided to award the honour linked to the memory of the Society's founder.

As expressed in its name, it is a major objective of our old Society to contribute towards the dissemination of knowledge about the constant advances of natural science. Several of the tasks that H.C. Ørsted took up with such great energy in the first years of the Society have in time, of course, been taken over by institutions serving in various ways for this purpose. Not only radio broadcasting, whose wonderful development has its first root in Ørsted's discovery of electromagnetism, has in our time created unimagined possibilities for dissemination of knowledge to the whole population, but many of the scientific conquests, which Ørsted in his lifetime spoke about with enthusiasm to the circle of the Society's first members, are now included in the syllabus taught in general school education or at the University and the institutions of higher learning.

In this connection we must remember in particular that it is to Ørsted's own initiative that we owe the establishment of the still flowering and now so widely branched institution within whose walls we are gathered here this evening. The dreams that filled Ørsted's mind about what the physical and chemical sciences would come to mean for the whole life of society have been realized in this country first and foremost by the Polyteknisk Læreanstalt or Technical University of Denmark.

Ever since its establishment this institution has generously provided shelter for Ørsted's old Society and last year, on the occasion of the 100th anniversary of Ørsted's death, gave renewed expression for this by handing over to the Society a beautiful room for the placement of the H.C. Ørsted Museum, in which so many of the instruments Ørsted used with his famous investigations, and many of the memorabilia from his home, which his descendants have

---

[4] *Danmarks naturvidenskabelige Samfund* was established in 1911 by the industrialist Gustav Adolph Hagemann (1842–1916) and Bohr's physicist colleague Martin Knudsen (1871–1949) as a lecture forum for prominent scientists, primarily from abroad. Bohr gave two lectures in the Society, in 1917 ("Recent results regarding the internal constitution of atoms") and 1938 ("Transmutations of elements").

[5] Founded 1908.

[6] Founded 1879.

looked after so faithfully, have been displayed and made accessible to the public. Immediately prior to this meeting the Society's board of directors has had the honour and the pleasure of being able to show the beautiful museum, rich in memories, to His Majesty the King.[7]

Apart from the lecturing activity, which the Society still seeks to carry on in Ørsted's spirit and adapt to contemporary circumstances, the Society has always felt it as an especially welcome duty to honour the memories of Ørsted's rich life of research and fruitful work in Danish society, and it was precisely as part of such efforts that the Medal, being presented today for the ninth time, was established more than a generation ago on the initiative of Martin Knudsen, then chairman of the board of directors. The presentation of the Medal gives the Society an opportunity to give a well-deserved mark of appreciation to Danish researchers in the physical and chemical sciences who have been able to maintain and continue the traditions that have centuries-old roots here at home, and to which Ørsted's name is inextricably linked.

It has been a pleasure for the Society's board of directors to add the name of Alex Langseth to the series of outstanding Danish chemists who have been honoured in this way, and among whom S.P.L. Sørensen received the first H.C. Ørsted medal and K. Linderstrøm–Lang the last one. The text whereby the board of directors has motivated this presentation, reads as follows:

...

After the lecture Professor Bohr stated:

On behalf of the Society I would like to thank Professor Langseth cordially for the rich experience it has been to listen to his lecture, which has given us such a lively impression of the richness of thought and art of experimentation whereby he has to such a high degree increased our knowledge of the world of molecules and the regularities which reveal themselves on a small scale and are determining for the properties of the substances of which our tools and our bodies are built. It is hardly necessary to elaborate further our thanks and our admiration for what we have heard and seen, but it might all the same be of interest to show some pictures of the medal that Professor Langseth has received.

Here can be seen the one side of the medal with the handsome features of H.C. Ørsted and his characteristic signature, and here can be seen the other

[7] The original H.C. Ørsted Museum has since become a part of Denmark's Technical Museum in Elsinore, north of Copenhagen.

side of the medal with the inscription: From the Society for the Dissemination of Natural Science to A. Langseth.

In conclusion, I would like once again on the Society's behalf to thank Your Majesty, whose presence and presentation of the medal have endowed this meeting with the greatest dignity and have underlined how deeply the whole Danish people and its government feel that scientific research here at home should continue to flower and set fruit, contributing to the common endeavours of humanity for the increase in knowledge and understanding of the nature that surrounds us and of which we ourselves are part.

XVIII. TENTH AND ELEVENTH PRESENTATION OF THE H.C. ØRSTED MEDAL

TIENDE OG ELLEVTE UDDELING AF
H.C. ØRSTED MEDAILLEN
Fys. Tidsskr. **57** (1959) 145, 158

To J.A. Christiansen and P. Bergsøe

TEXT AND TRANSLATION

See Introduction to Part II, p. [353].

[487]

Tiende og ellevte uddeling af H. C. Ørsted medaillen.

Tirsdag d. 19. maj 1959 afholdt Selskabet for Naturlærens Udbredelse et festligt møde, hvor H. C. Ørsted medaillen tildeltes professor, dr. phil. *J. A. Christiansen* og dr. techn. *Paul Bergsøe*. Selskabets protektor, Hans Majestæt Kongen, foretog overrækkelsen af medaillerne.

Formanden for selskabets direktion, professor, dr. *Niels Bohr*, bød velkommen med følgende tale:

Deres Majestæt, mine Damer og Herrer.

På Selskabet for Naturlærens Udbredelses vegne vil jeg gerne give udtryk for vor dybe taknemmelighed for, at Deres Majestæt som Selskabets høje protektor har villet komme til stede og foretage overrækkelsen af H. C. Ørsted medaillen.

På Selskabets vegne retter jeg dernæst en velkomst til de tilstedeværende repræsentanter for danske institutioner, der på hver sin måde virker for fremme af naturvidenskaben i vort samfund. Blandt disse institutioner føler Selskabet sig i særlig gæld til Danmarks tekniske Højskole, der ligesom Selskabet skylder H. C. Ørsted sin tilblivelse, og som igennem så mange år med broderlig gæstfrihed har ydet husly for vort gamle selskabs virksomhed. Ligeledes er det Selskabet en glæde, at så mange medlemmer af Danmarks naturvidenskabelige Samfund, Fysisk Forening, Kemisk Forening og Astronomisk Selskab har fulgt indbydelsen til sammen med vore egne medlemmer at overvære denne højtidelighed, hvor man i tilknytning til mindet om H. C. Ørsteds store virke i videnskabens og det danske samfunds tjeneste har lejlighed til at hædre to landsmænd, der hver på sin vis har virket i Ørsteds ånd for fremme af den videnskabelige forskning og udbredelsen af kendskabet til dens fremskridt.

Ved den tildeling af H. C. Ørstedmedaillen, der vil foregå i dag, vil den blive modtaget af professor J. A. Christiansen og dr. Paul Bergsøe med den motivering, som vil fremgå af de skrivelser, som jeg nu skal oplæse.

Efter foredragene udtalte professor Bohr:

På alle tilhørernes vegne vil jeg gerne give udtryk for vor tak for den rige oplevelse, som modtagerne af H. C. Ørsted medaillen ved deres foredrag har givet os, og udtale de varmeste ønsker for en fortsættelse af deres virke til yderligere fremskridt for videnskaben og fremme af Selskabets formål.

Til slut vil jeg gerne endnu en gang på Selskabets vegne takke Deres Majestæt, hvis tilstedeværelse og overrækkelse af medaillen har forlenet dette møde med den højeste værdighed og understreget, hvor dybt det ligger hele det danske folk og dets styre på sinde, at videnskabelig forskning herhjemme stadig må blomstre og sætte frugt som bidrag til menneskehedens fælles bestræbelser for forøgelsen af kendskabet til den natur, der omgiver os, og vor beherskelse af de kræfter, den rummer.

TRANSLATION

Tenth and Eleventh Presentation of
the H.C. Ørsted Medal.

On Tuesday 19 May 1959, the Society for the Dissemination of Natural Science held a festive meeting, where the H.C. Ørsted Medal was awarded to Professor *J.A. Christiansen*,[8] D.Phil., and *Paul Bergsøe*,[9] D.Tech. His Royal Majesty, the Society's Patron, presented the medals.

The chairman of the Society's board of directors, Professor Dr *Niels Bohr*, bid welcome with the following address:

Your Majesty, Ladies and Gentlemen.

On behalf of the Society for the Dissemination of Natural Science I wish to express our profound gratitude that Your Majesty, as the noble Patron of the Society, has agreed to be here and to present the H.C. Ørsted Medal.

On behalf of the Society I now welcome the representatives of Danish institutions present here, which each in its way works for the advancement of science in our society. Among these institutions the Society feels in particular debt to the Technical University of Denmark, which like the Society owes its establishment to H.C. Ørsted, and which throughout so many years, with brotherly hospitality, has offered shelter for the activities of our old Society. Similarly it is a pleasure for the Society that so many members of the Science Society of Denmark, the Physical Society, the Chemical Society and the Astronomical Society have accepted the invitation to be present, together with our own members, at this ceremony where, in connection with the memory of H.C. Ørsted's great work in the service of science and society, we have the opportunity to honour two countrymen who each in his way have worked in the spirit of Ørsted for the advancement of scientific research and the dissemination of knowledge about its progress.

At the awarding of the H.C. Ørsted Medal, which will take place today, it will be received by Professor J.A. Christiansen and Dr Paul Bergsøe with the

---

[8] Jens Anton Christiansen (1888–1969), professor of physical chemistry at the University of Copenhagen.

[9] (1872–1963), chemist, metallurgist and popularizer of science.

motivation given by the texts I shall now read aloud.

...

After the lectures Professor Bohr stated:

On behalf of the whole audience, I would like to express our gratitude for the rich experience which the recipients of the H.C. Ørsted Medal have given us by their lectures, and pronounce the warmest wishes for a continuation of their work towards further progress for science and furtherance of the objectives of the Society.

In conclusion, I would like once again on the Society's behalf to thank Your Majesty, whose presence and presentation of the medal have endowed this meeting with the greatest dignity and have underlined how deeply the whole Danish people and its government feel that scientific research here at home should continue to flower and set fruit, contributing to the common endeavours of humanity for the increase in knowledge of the nature that surrounds us and our command of the forces it holds.

[491]

XIX. MEMORIAL EVENING FOR KIRSTINE MEYER IN THE SOCIETY FOR DISSEMINATION OF NATURAL SCIENCE

MINDEAFTEN FOR KIRSTINE MEYER
I SELSKABET FOR NATURLÆRENS UDBREDELSE
Fys. Tidsskr. **40** (1942) 173–175

TEXT AND TRANSLATION

See Introduction to Part II, p. [354].

Mindeaften for Kirstine Meyer
i Selskabet for Naturlærens Udbredelse.

Onsdag den 14. Oktober 1942 afholdt Selskabet for Naturlærens Udbredelse et Møde til Minde om Lektor Fru Dr. phil. Kirstine Meyer, f. Bjerrum.

Formanden, Professor Dr. phil. Niels Bohr indledede Mødet med at udtale:

Ved vort første Møde her i Selskabet i Fjor, der fandt Sted kort efter Kirstine Meyers Død og faa Dage før at hun vilde have fyldt 80 Aar, meddeltes det, at Selskabets Direktion, som Led i den Hyldest, der fra saa mange Sider var forberedt hende, havde truffet Beslutning om, paa denne Dag at udnævne Kirstine Meyer til Selskabets Æresmedlem i Anerkendelse af hendes store Virke for Højnelsen af Fysikundervisningen her i Landet og især for hendes dybtgaaende og frugtbare historisk-fysiske Studier, ved hvilke hun som ingen anden har kastet Lys over Baggrunden for nogle af de største danske Fysikeres Virke og først og fremmest over vort Selskabs Stifter, H. C. Ørsteds grundlæggende Opdagelser og enestaaende folkeopdragende Virksomhed. Denne Hyldest skulde Skæbnen til vor dybe Beklagelse hindre vort Selskab i at bringe Kirstine Meyer. Det var derfor med en saa meget større Glæde, at Direktionen i Løbet af Aaret modtog en Forespørgsel fra en Kreds af Kirstine Meyers Venner og Slægt, om Selskabet vilde paatage sig Bestyrelsen og Uddelingen af et Mindelegat for Kirstine Meyer, som man paatænkte at oprette. Naturligvis blev denne Henvendelse modtaget med største Taknemmelighed, og Svaret fra Direktionen, hvori man erklærede sig rede til at paatage sig Bestyrelsen og Uddelingen af Legatet, sluttede med følgende Sætning, som jeg skal oplæse:

»Paa Selskabets Vegne vil man samtidig gerne give Udtryk for Taknemmelighed over den Ære, der herved er vist Selskabet, og hvorved det ikke alene vil faa en meget forønsket Lejlighed til for kommende Slægter at fremhæve den Betydning, som Fru Dr. phil.

Kirstine Meyers Forskning har haft for dansk historisk-fysisk Videnskab og især for Mindet om Selskabets Stifter, H. C. Ørsted, men
tillige vil give Selskabet et yderligere Virkefelt til Opmuntring af
unge danske Videnskabsmænd paa Naturlærens Omraade.«

Af Legatets Fundats skal jeg oplæse følgende Hovedpunkter:

1.

Legatets Navn er
Lektor, Dr. phil. Fru Kirstine Meyer født Bjerrums Mindelegat.

2.

Legatets Grundkapital andrager for Tiden 13.000 Kr. Grundkapitalen kan forøges ved Bidrag udefra eller ved Henlæggelser af
Indtægterne efter Bestyrelsens Bestemmelse.

3.

Grundkapitalen maa ingensinde benyttes til Uddeling.

4.

Legatet bestyres af Direktionen for Selskabet for Naturlærens
Udbredelse. Det aarlige Regnskab revideres af dette Selskabs Revi-
sorer og forelægges paa Selskabets aarlige ordinære Generalforsam-
ling til Efterretning.

5.

Af Renteindtægten af Legatets Kapital uddeles Portioner paa
mindst 500 Kr. til unge lovende danske Studerende eller Kandidater
til Fremme af deres Uddannelse eller deres Forskning paa Naturlærens Omraade. Legatet bør uddeles mindst hvert andet Aar og
saa vidt muligt paa Kirstine Meyers Fødselsdag, den 12. Oktober.

Det har været en stor Glæde for Direktionen, at der allerede forefindes en tilstrækkelig Renteydelse af Legatets Kapital til, at den
første Uddeling af Legatet har kunnet finde Sted paa Kirstine Meyers
Fødselsdag for to Dage siden, og jeg skal her meddele, at det ved
denne første Uddeling er givet til Hr. cand. mag. N. O. Lassen,
der igennem sin Deltagelse i videnskabelige Undersøgelser paa Universitetets Institut for teoretisk Fysik har lagt saadanne Evner og
Dygtighed for Dagen, at der maa næres meget betydelige Forventninger til hans fremtidige videnskabelige Virksomhed. Jeg kan endvidere meddele, at cand. mag. Lassen paa et af vore Møder i dette

Efteraar her i Selskabet vil holde et Foredrag om de Undersøgelser, hvori han har taget saa virksom Del.

Efter denne Beretning om Mindelegatet og dets første Uddeling skal vi gaa over til Aftenens Foredrag om Kirstine Meyers store Opdrager- og Forskergerning, som vil blive holdt af Hr. Lektor, Dr. phil. Mogens Pihl, der selv paa saa udmærket Maade er traadt i Kirstine Meyers Fodspor igennem sine historisk-fysiske Forskninger. Det er mig en stor Glæde hermed at give Ordet til Dr. phil. Mogens Pihl.

TRANSLATION

Memorial Evening for Kirstine Meyer
in the Society for Dissemination of Natural Science.

On Wednesday 14 October 1942 the Society for the Dissemination of Natural Science held a meeting in memory of Lecturer Kirstine Meyer,[1] D.Phil., n. Bjerrum.

The President, Professor Niels Bohr, introduced the meeting by stating:

At our first meeting here in the Society last year, which took place shortly after Kirstine Meyer's death and a few days before her 80th birthday, it was reported that the board of the Society, as part of the celebrations that had been arranged for her from so many quarters, had made the decision to appoint Kirstine Meyer on that day as an honorary member of the Society in recognition of her great work for the improvement of physics education in this country, and in particular for her thorough and fruitful studies in the history of physics, by which as none other she has cast light on the background for the work of some of the greatest Danish physicists, and first and foremost on the fundamental discoveries and unique popular education activity of our Society's founder H.C. Ørsted. To our deep regret, fate was to prevent our Society from bringing this tribute to Kirstine Meyer. It was therefore with all the more pleasure that in the course of the year the board received a request from a circle of Kirstine Meyer's friends and relations as to whether the Society would take responsibility for the administration and presentation of a memorial grant for Kirstine Meyer, which one considered establishing. This application was naturally accepted with the greatest gratitude, and the reply from the board, which declared readiness to take responsibility for the administration and presentation of the grant, concluded with the following sentence which I shall read aloud:

"On behalf of the Society, one would at the same time like to express gratitude for the honour which is hereby shown to the Society and whereby it will not only have a much desired opportunity to emphasize for coming generations the importance that the research of Mrs Kirstine Meyer, D.Phil.,

[1] Kirstine Meyer (1861–1941). Bohr's obituary for Meyer, *Kirstine Meyer, n. Bjerrum: 12 October 1861 – 28 September 1941*, Fys. Tidsskr. **39** (1941) 113–115, is reproduced in Vol. 12.

has had for research in Danish history of physics and especially for the memory of the Society's founder, H.C. Ørsted, but will additionally give the Society a further sphere of action for the encouragement of young Danish scientists in the field of natural science."

From the charter of the grant I shall read out the following main points:

1.

The name of the grant is
The Memorial Grant of Lecturer Kirstine Meyer, n. Bjerrum, D.Phil.

2.

The original capital amounts to 13,000 kroner at present. The original capital may be augmented with external contributions or by setting aside income according to the decision of the board of directors.

3.

The original capital must never be used for awards.

4.

The grant is administered by the board of directors of the Society for the Dissemination of Natural Science. The annual accounts are audited by the auditors of this Society and are presented at the annual ordinary general assembly of the Society for general information.

5.

From the interest accrued from the capital of the grant, portions of at least 500 kroner are awarded to young promising Danish students or graduates for the advancement of their education or their research in the field of natural science. The grant should be awarded at least every second year and as far as possible on Kirstine Meyer's birthday on 12 October.

It has been a great pleasure for the board that there already exists a sufficient return of interest on the original capital as to have allowed the first award of the grant to be made on Kirstine Meyer's birthday two days ago, and I shall here announce that this first award has been given to N.O. Lassen,[2] M.Sc., who through his participation in scientific investigations at the University Institute

---

[2] Niels Ove Lassen (1914–), experimental nuclear physicist at Bohr's institute involved in the installation of the first cyclotron in the late 1930s. He remained active at the institute well beyond his retirement in 1984.

for Theoretical Physics has displayed such talents and ability that quite considerable expectations can be entertained as regards his future scientific activity. I can further announce that Lassen, M.Sc., will give a lecture at one of our meetings in the Society this autumn about the investigations in which he has so actively taken part.

After this report on the memorial grant and its first presentation, we shall proceed to the evening's lecture about Kirstine Meyer's great work in education and research, which will be given by Lecturer Mogens Pihl,[3] D.Phil., who himself in such an excellent fashion has followed in Kirstine Meyer's footsteps through his research in the history of physics. It is a great pleasure for me to give herewith the floor to Mogens Pihl, D.Phil.

[3] (1907–1986), physicist and historian of physics.

4. SCANDINAVIAN SCIENTISTS, THE INSTITUTE AND THE UNIVERSITY OF COPENHAGEN

XX. [SPEECH AT THE MEETING OF NATURAL SCIENTISTS IN HELSINKI, 14 AUGUST 1936]

[TALE VED NATURFORSKERMØDET I HELSINGFORS,
14. AUGUST 1936]
"Nordiska (19. skandinaviska) Naturforskarmötet i Helsingfors
den 11.–15. augusti 1936", Helsinki 1936, pp. 191–192

TEXT AND TRANSLATION

See Introduction to Part II, p. [355].

Nordic (19th Scandinavian) Meeting of Natural Scientists, Helsinki, August 1936. Bohr stands in the middle.

På danskarnas vägnar svarade prof. NIELS BOHR:

Mine Damer og Herrer,

Af de Hilsener, ligesaa forskellige i Udtryksform som ens i Sindelag, der ved Aabningshøjtideligheden forleden bragtes af Repræsentanterne for hvert af de 4 nordiske Lande, der har fulgt Finlands Indbydelse til dette skandinaviske Naturforskermøde, vil De alle have faaet det stærkeste Indtryk af den enstemmige Glæde, denne Indbydelse har vakt overalt i Norden. Med hvor store Forventninger vi end er kommet, er de dog langt overtrufne af de rige Oplevelser, vi har haft i disse Dage, hvor vi har været samlede til Arbejde og Fest her i Finlands skønne levende Hovedstad. Ikke alene har Mødet i videnskabelig Henseende haft et Forløb fuldt værdigt finsk Videnskabs store Traditioner, der paa saa begejstret og frugtbar Maade videreføres af den nulevende Slægt, men hele det finske Folk med dets Regering i Spidsen har vist os en saa enestaaende Gæstfrihed og lagt saa stor en Forstaaelse for vor fælles Sag for Dagen, at dette Møde altid vil staa for os som et af vore smukkeste Minder. En Tak for den nye Belæring og den Opmuntring til fortsatte Bestræbelser, som det videnskabelige Samarbejde har givet os, vil Gæsterne jo faa Lejlighed til at bringe ved den Afslutning paa vort Møde, der nu desværre saa stærkt nærmer sig. Her ved Festlighedernes Højdepunkt i denne straalende Ramme er det derimod naturligt, at det mere er den rent menneskelige Side af vort Samvær, der ligger os paa Sinde. Ikke blot føler vi den mest umiddelbare Glæde ved her at være sammen med vore finske Værter og Værtinder, i hvis Hjem mange af os har faaet saa gæstfri en Modtagelse, men først og fremmest er vi alle grebne af den Følelse af dybere Samhørighed, som har lydt gennem hvert af de hjertelige finske Velkomstord, og for hvilken Finlands Undervisningsminister netop har givet saa smukt og kraftigt et Udtryk.

Ikke mindst i Tider som disse, hvor store Farer truer vor Frihed og den hele menneskelige Kulturs harmoniske Udvikling, føler vi os i vore smaa saa nært beslægtede Lande henvist til hverandre for gensidig Støtte, ikke blot til at opretholde vor egen Selvstændighed, men ogsaa til at yde selv et ringe Bidrag til Folkeslagenes indbyrdes Forstaaelse og frugtbare Samvirke. Den stærkeste Opmuntring til en saadan Stræben rummes jo i Tanken om, hvad i svundne Tider smaa Samfund, truede af Fjenders Overmagt eller Naturens Vælde, som de gamle Israeliter eller de stolte græske Stammer eller de frie Bønder paa Island i Sagatiden, i Kampen for deres Selvstændighed har formaaet at bidrage til hele Menneskehedens og vor egen Kultur. Den samme Selvhævdelsens Aand giver sig for Øjeblikket maaske sit allerstærkeste Udtryk i den frie finske Nations Kamp for at gøre Brug af sine unge Kræfter i Arbejdet paa Fædrelandets rivende materielle og kulturelle Udvikling og for at finde Udtryk for sit eget Væsen i Ord og Toner og i Farver og Form paa en Maade, der har givet Genklang i Menneskehjerter Verden over og først og fremmest hos de nordiske Brødrefolk, hvor de fælles rige Minder giver saa følsom en Sangbund. Det finske Folks Indsats er os tillige en Lære om, hvordan Folkeelementer af forskelligste Herkomst og Sprog kan hjælpe hverandre paa gensidig befrugtende Maade i Arbejdet paa samme Maal. Enhver, der har haft den uforglemmelige Oplevelse at rejse i dette dejlige Land med de mange stolte Mindesmærker fra Nordens fælles Historie i Strid og Fred, forstaar jo, hvordan der her gror op en egenartet Form for menneskelig Kultur, i hvilken svensk Ridderaand paa uadskillelig Maade indgaar i en Folketradition med urgamle, vidt forgrenede Rødder, og hvis varme Menneskevenlighed og Lune lyser ud af Øjnene paa ethvert Barn og gammel Kvinde eller Mand, som vi møder paa Landevejen, hvor den snor sig gennem de grønne opdyrkede Pletter, der ligesom de blaa Søer omkranses af Finlands endeløse Skove.

Jeg ejer ikke en Digters Evne til at finde Udtryk for, hvad der fylder mit Sind, men som en, der har fulgt en Naturforskers Kald, ligger det mig paa Hjerte at betone, at vor Videnskab ikke alene er os en Hjælp til at gøre os til Herrer over Naturens Kræfter, men at den som Kunsten, omend paa forskellig Maade, ogsaa er os en Kilde til dybere Kendskab til os selv. Af Naturvidenskabens Udvikling er vi jo Gang paa Gang blevet belært om, hvordan selv de mest dagligdags Begreber kan rumme upaagtede Fordomme, hvis Bortrydning er Betingelsen for, at en større Sammenhæng kan fremtræde i hele sin Skønhed. Allermest skæbnesvanger kan dog vore Fordomme blive, naar det gælder at bedømme Menneskelivets egne Udviklingsmuligheder, hvis Rigdom ingen Fantasi kan udtømme. Jeg vil derfor gerne slutte med at udtale de varmeste Ønsker om, at det friske Skud paa den fælles Menneskeaand, som trives her i Finland, fortsat maa blomstre og sætte Frugt til Menneskehedens Bedste, værnet saavel af forventningsfuld Kærlighed til det nyes Egenart som af fordomsfri Forstaaelse af Menneskelivets væsentlige Enhed under dets evig skiftende Former.

TRANSLATION

Professor Niels Bohr responded on behalf of the Danes:

Ladies and gentlemen,

From the greetings, just as different in form of expression as identical in attitude, which were brought at the opening ceremony recently by the representatives of each of the four Nordic countries who have accepted Finland's invitation to this Meeting of Scandinavian Scientists, you will all have received the strongest impression of the unanimous pleasure this invitation has given rise to everywhere in the Nordic countries.[1] No matter how great the expectations with which we have come here, they have been exceeded by far by the rich experiences we have had in these days when we have been gathered for work and celebration here in Finland's beautiful lively capital city. Not only as regards science has the meeting had a course fully worthy of the great traditions of Finnish science, which are being continued in such an enthusiastic and fruitful way by the present generation, but the whole Finnish people, with its Government to the fore, has shown us such outstanding hospitality and displayed so great an understanding for our common cause, that this meeting will forever remain with us as one of our most beautiful memories. At the conclusion of our meeting, which is now unfortunately so rapidly approaching, the guests will of course have the opportunity to express their gratitude for the new lessons and the encouragement for continued endeavours that the scientific cooperation has given us. Here at the peak of the festivities, in this magnificent venue, it is natural, on the other hand, that it is rather the purely human aspect of our gathering that we have at heart. Not only do we feel the most immediate pleasure by being together here with our Finnish hosts and hostesses, in whose homes many of us have had so hospitable a reception, but first and foremost we are all gripped by the feeling of deeper belonging, which has sounded through each of the heartfelt Finnish words of welcome and of which Finland's Minister of Education[2] has just given so beautiful and forceful expression.

---

[1] In the Scandinavian (or Nordic) languages, "Scandinavia" refers to Denmark, Norway and Sweden, whereas the "Nordic countries" (or "Norden") also include Finland and Iceland. For this reason, the Finns found it pertinent to denote the meeting the "Nordic (19th Scandinavian) Meeting of Natural Scientists".

[2] Oskari Mantere (1874–1942), who had been Prime Minister of Finland for a brief period from 1928 to 1929.

Not least in times like these, when great dangers threaten our freedom and the harmonious development of the whole human culture, we feel in our small, so closely related, countries that we have to turn to each other for mutual support, not only for maintaining our own independence but also for giving even a small contribution to mutual understanding and fruitful cooperation of all peoples. The greatest encouragement to such a striving is held of course in the thought of what small societies in times past, threatened by the superior force of enemies or the power of nature, such as the old Israelites or the proud Greek tribes or the free peasants on Iceland in the times of the Sagas, in the fight for their independence have managed to contribute to the culture of all humanity and that of our own. The same spirit of self-assertion is at present perhaps being expressed most strongly of all in the fight of the free Finnish nation to make use of its youthful strength in the work towards the rapid material and cultural development of the fatherland and to find expression for its own identity in words and music and in colours and form in a way that has resonated in human hearts all over the world and first and foremost in the Nordic sister nations, where the rich shared memories provide so sensitive a sounding-board. The effort of the Finnish people is in addition a lesson for us about how groups of people of the most different origin and language can help each other in a mutually stimulating way in the work towards the same goal. Anyone who has had the unforgettable experience of travelling in this lovely country, with the many proud monuments from the shared history of the Nordic countries in war and peace, understands, of course, how a singular form of human culture is developing here, in which the Swedish spirit of chivalry in an inseparable way is part of a folk tradition with ancient widely branching roots, and whose warm benevolence and humour shine out of the eyes of every child or old woman or man we meet on the high road, where it winds its way through the green cultivated patches which, like the blue lakes, are encircled by Finland's endless forests.

I do not have a writer's ability to find expression for what fills my mind, but as one who has followed a scientist's calling, I have at heart to emphasize that our science is not only a help for us to make ourselves masters over the forces of nature, but that, like art, if in a different way, it is also a source for us of deeper knowledge of ourselves. We have indeed been taught over and over again by the development of natural science how even the most mundane concepts can hold undiscerned prejudices, whose removal is the precondition for allowing a greater coherence to appear in all its beauty. Our prejudices can be the most disastrous of all, however, when it comes to judging human life's own potentialities, whose richness no imagination can exhaust. I would therefore like to close by expressing the warmest wishes that the fresh shoot on the common human spirit flourishing here in Finland may continue to flower and set fruit for the benefit of humanity, shielded by expectant love for the special character of the new shoot, as well as by unprejudiced understanding of the essential unity of human life under its eternally changing forms.

XXI. SPEECH AT THE INAUGURATION OF THE INSTITUTE'S HIGH-VOLTAGE PLANT, 5 APRIL 1938

TALE VED INDVIELSEN AF INSTITUTTETS HØJSPÆNDINGSANLÆG,
5. APRIL 1938
Typescript

Probably transcript of stenographic notes
taken at Bohr's talk.

TEXT AND TRANSLATION

See Introduction to Part II, p. [356].

Bohr speaks at his institute on 5 April 1938, celebrating the twenty-fifth anniversary of his atomic model as well as the expansion of his institute. On his jacket can be seen the Danish Medal of Merit in Gold, with which King Christian X had honoured him earlier in the day for twenty-five years of meritorious employment by the Danish state. (Courtesy: The Royal Library, Copenhagen.)

[510]

Tale ved Indvielsen af Instituttets Højspændingsanlæg
5. April 1938.

Jeg er meget taknemmelig for de venlige Velkomstord og ikke mindst for den store Anerkendelse, Hs. Majestæt Kongen har ladet mig blive til Del.[1] Det var jo ellers min Pligt at byde Dem alle hjertelig velkommen herude i Dag, og vi er jo særlig taknemmelige for at se saa mange Repræsentanter for Regeringen og de Stats- og kommunale Myndigheder, der lige fra Begyndelsen af har vist vort Arbejde saa stor Interesse og har ydet os saa stor en Støtte ligesom ogsaa ved denne Lejlighed. Vi er ogsaa meget glade for at se saa mange Repræsentanter for de forskellige private Institutioner, der har ydet os saa megen Støtte og gennem saa store Gaver til Instituttet, at de nye Anlæg, som i Dag skal fremvises, er blevet mulige. Førend vi gaar over til selve Fremvisningen, tænkte jeg, at det maaske kunde interessere Dem at høre en kort Beretning om Instituttets Historie og Udvikling og nogle faa Ord om de Op-gaver, vi særlig har beskæftiget os med, og om de nye Opgaver som vi haaber at faa Held til at angribe med de nye Midler, vi har faaet. Det, der gav Anledning til Oprettelsen af et saadant Institut med Navnet "Teoretisk Fysik", det var den Situation, der var opstaaet indenfor Fysikken paa det Tidspunkt, hvorvidt at det var særlig paakrævet at muliggøre det intimeste Samarbejde mellem teoretiske og eksperimentelle Undersøgelser paa Atomfysikkens Omraade.[2] Vi havde jo paa det Tidspunkt gennem de store Opdagelser vedrørende Materiens Sammensætning, Opbygning, Slutstene, vundne af Rutherfords Opdagelse af Atomernes Kerne, faaet et fuldstændig bestemt Billede af Atomernes Op-bygning, i hvert Tilfælde et indgaaende Kendskab til Atomernes Byggestene. Samtidig var det klart, at den Model af Atomet, den saakaldte "Rutherford'ske Atommodel", hvorefter et Atom er bygget paa lignende Maade som et Sol-system, d.v.s. bestaar af et Antal lette Partikler, der er bundne gennem den elektriske Tiltrækning, Elektronerne, til Atomkernen, som er en Partikel, der er ganske overordentlig lille i Forhold til Atomet selv, altsaa et Billede, som fuldkomment minder om et Solsystem. Udfra en saadan Model var det klart, at det var ganske umuligt udfra de sædvanlige fysiske Betragtninger, der for … at forstaa Atomernes karakteristiske Egenskaber, men vi havde været saa lykkelige at faa en Nøgle til Forstaaelsen af denne Gaade gennem Plancks

[1] The King, Christian X, had honoured Bohr with the prestigeous Danish Medal of Merit in Gold.
[2] Cf. Bohr's speech at the inauguration of the institute in 1921: *Dedication of the Institute for Theoretical Physics*, typescript, BMSS (9.4). Reproduced in Vol. 3, pp. [284]–[293] (Danish original) and [293]–[301] (English translation).

Opdagelse ved Aarhundredeskiftet af det saakaldte Virkningskvantum, hvor man lærte at kende et helt nyt Træk af atomistisk Natur i Naturlovene, et Træk der var helt forskelligt fra den gamle Erkendelse af Stoffernes endelige Delelighed, og man opdagede saa at sige et Træk af Udelelighed i selve Atomprocessernes Forløb og finder en Mulighed til at forstaa, hvordan man i det mindste kunde komme frem ved Hjælp af den Rutherfordske Atommodel, idet man blev tvungen til at antage, at Atomernes Stabilitet hidrørte fra dette nye Træk, og at f. Eks. enhver Ændring i et Atoms Tilstand var en individuel Proces, hvorved Atomet foretog en Overgang fra en stationær Tilstand til en anden. Det var en meget skematisk teoretisk Betragtning og kunde komme til Nytte ved at give os en Mulighed til at faa direkte Oplysninger om den Maade, hvorpaa Elektronerne er bundne i Atomet ved Studie af deres saakaldte Spektre ved de enkelte Overgangsprocesser. Ved at undersøge Lysets Farve kan man finde ud af, hvor stor Energi der udsendes og lære at kende Bindingsforholdene i Atomerne. Selv om disse Resultater, som man kunde faa paa Grundlag af Plancks og Rutherfords Opdagelser, viste, at man var paa rette Vej, saa var der langt igen, førend man kunde faa en virkelig sammenhængende, i sig selv sammenhængende og afsluttet Beskrivelse af Atomernes Forhold, og det var klart, at man kunde prøve sig frem, og at man havde ikke anden Vej end at prøve at se, hvad man kunde hitte ud af ved Hjælp af sine Erfaringer (teoretiske Undersøgelser) og dels ved at spørge Naturen selv om nye Oplysninger (experimentelle Undersøgelser). Den Tanke at oprette et Institut for Arbejdet paa dette Gebet mødte en meget stor Forstaaelse og Tilslutning fra Universitetet og Regeringsmyndighederne. Gennemførelsen af Planen blev meget væsentlig lettet derved, at en Kreds af Interesserede Privatmænd paa Direktør Berlèmes[3] Initiativ indsamlede en større Sum, hvorfor Grunden blev købt, hvorpaa vi nu befinder os, og det var den første Udflytning af videnskabelige Institutter til denne Del af Byen, hvor nu den nye Universitetsby vokser op.[4] Ligeledes var det af den største Betydning, at Carlsbergfondet satte os i Stand til at erhverve kostbare spektroskopiske Apparater. Arbejdet i Begyndelsen hvilede først og fremmest paa de Mennesker, der kunde knyttes til Instituttet, og jeg tænker i Dag med Taknemmelighed paa den Støtte og

[3] Aage Berlème (1886–1967), Danish business man who had been Bohr's schoolmate.

[4] Before the inauguration of Bohr's institute in 1921, the entire University had been located in central Copenhagen. In the 1930s, institutes for physical chemistry and mathematics were built adjacent to Bohr's institute, while the physiological activities of the faculties of medicine and science were joined together in a large complex of buildings nearby with support from the Rockefeller Foundation.

Indsats, som Professor H.M. Hansen[5] i Begyndelsen var, før han fik sit eget Institut. Vi har ogsaa fra Begyndelsen af haft grumme meget dygtige og fremragende eksperimentelle Medarbejdere, bl.a. Dr. Jacobsen,[6] som har været ved Instituttet fra første Begyndelse, og nuværende Professor Werner,[7] der en Tid var Assistent i Begyndelsen, og som nu har sin egen Virksomhed i Aarhus, og med hvem vi haaber i de kommende Aar at komme til at staa i intimt Samarbejde. Det var meget væsentligt for os, at vi fra Udlandet fik Besøg af meget fremragende unge Mennesker. Den første af disse var nuværende Professor Kramers i Leiden, som vi har den store Glæde at se i Dag, som først kom som ung Student, men blev knyttet til Instituttet og virkede her som Lektor i 10 Aar og har haft saa stor Betydning for Instituttets Udvikling og Undervisningen.[8] En anden, jeg gerne vil nævne i en senere Forbindelse, det var det Besøg vi fik af Professor Hevesy,[9] en ungarsk Fysiker, som jeg havde lært at kende i Rutherfords Laboratorium i Manchester, og Muligheden for hans Tilknytning til Instituttet var Oprettelsen af Rask–Ørstedfondet.[10] Jeg vil gerne minde om den Indsats, som man fra dansk Side gjorde umiddelbart efter Krigen, den store Forstaaelse for Betydningen af Samarbejdet mellem Folkene ogsaa paa Videnskabens Omraade, som havde lidt saa stort Afbræk under den store Krig, og vi fik et stort Antal ogsaa i de Aar, baade Professor Klein,[11] der nu er ansat i Stockholm, Rosseland,[12] der nu er Professor i Astrofysik i Oslo,

---

[5] The Danish physicist Hans Marius Hansen (1886–1956) was a close colleague of Bohr with a prominent career at the University of Copenhagen. Bohr's greeting on Hansen's 60th birthday, *A Personality in Danish Physics*, Politiken, 7 September 1946, as well as his two obituaries, *A Shining Example for Us All*, Politiken, 14 June 1956, and *Obituary for Hans Marius Hansen*, Fys. Tidsskr. **54** (1956) 97, are reproduced in Vol. 12.

[6] Jacob Christian Jacobsen (1895–1965).

[7] Sven Werner (1898–1984) worked at Bohr's institute from 1924 to 1927. He took up his professorship at the University of Aarhus in 1938 after a period at the Technical University of Denmark, during which time he retained contact with Bohr's institute.

[8] Hendrik Anthony Kramers (1894–1952) became Bohr's first assistant in 1916, the same year as Bohr was appointed professor and five years before the inauguration of Bohr's institute. Bohr wrote two obituaries for Kramers, one in Danish, *On the Death of Henrik Anton Kramers*, Politiken, 27 April 1952, and a more comprehensive one in English, *Hendrik Anthony Kramers †*, Ned. T. Natuurk. **18** (1952), 161–166. Both obituaries are reproduced in Vol. 12.

[9] George de Hevesy (1885–1966).

[10] The Rask–Ørsted Foundation was established in 1919.

[11] Having spent much time in Copenhagen from 1919 on, Oskar Klein (1894–1977) succeeded Werner Heisenberg as Bohr's assistant in 1927. After his move to Stockholm in 1931, he remained one of Bohr's closest friends and collaborators.

[12] Svein Rosseland (1894–1985) was a visitor to Bohr's institute from its establishment to 1924 and from 1926 to 1927. He worked closely with Oskar Klein.

Dr. Nishina,[13] Japan, og Dr. Ray[14] fra Indien, og under det store Samarbejde der var indenfor Instituttet og indenfor forskellige Institutter Verden over, lykkedes det virkelig at naa til en saadan foreløbig Afrunding af det første Angreb paa Atomernes Bygning. Man naaede at faa et Overblik over den Maade, hvorpaa Elektronerne er bundne i Atomerne, et Overblik der tillod at forstaa det Slægtskabsforhold mellem Grundstoffernes fysiske og kemiske Egenskaber, det periodiske System. I den Forbindelse var jo Professor Hevesys Tilknytning til Instituttet af den største Betydning, idet det lykkedes ham gennem Opdagelsen af et Grundstof, som han og hans Medarbejder, Professor Coster[15] i Groningen, kaldte Hafnium, at faa en Støtte for de teoretiske Synspunkter, vi dengang arbejdede med paa Instituttet. Paa den Tid kom der nye Muligheder for Samarbejdet paa Videnskabens Omraade gennem Rockefellerfondet, gennem Oprettelsen af International Education Board som en Del af Rockefellerfondet,[16] og først og fremmest tog de den Opgave op at lette det internationale Samarbejde, og i Betragtning af den Indsats, vi kunde gøre her ved Instituttet, skænkede de os et stort Beløb, saa Instituttets Bygninger kunde udvides og blive dobbelt saa stort som før. Den nødvendige Udvidelse af Grunden hertil blev skænket af Københavns Kommune, men dette International Education Board tog ogsaa andre Opgaver op, og det vil maaske interessere Dem at minde om, at her var netop Rask–Ørstedfondets Oprettelse Forbillede. Ved Besøg af Direktøren[17] for International Education Board i København saa han med største Sympati paa de Bestræbelser, som man her i Landet havde taget Initiativet til. Og i de følgende Aar kom som Rockefeller-Fellows til Instituttet lovende unge Videnskabsmænd fra forskellige Lande og fik Lejlighed til en Tid at udvide deres egen Horisont og bringe til dette Institut nye Impulser, og det er interessant at minde om, at den allerførste af saadanne Rockefeller-Stipendiater var nuværende Professor Heisenberg[18] i Leipzig, som gennem adskillige Aar arbejdede her ved Instituttet, og som har gjort saa overordentlig en Indsats til Opbygningen og

---

[13] Yoshio Nishina (1890–1951) stayed at the institute from 1925 to 1928, during which time he worked closely with Oskar Klein. He subsequently became a main force in the introduction of nuclear physics in Japan and retained contact with Bohr and his institute throughout his life.

[14] Bidhu Bhusan Ray (1894–1944) visited Bohr's institute for a year from October 1924.

[15] Dirk Coster (1889–1950) worked at Bohr's institute from 1923 to 1924.

[16] The Rockefeller Foundation and the International Education Board were two independent granting agencies within Rockefeller philanthropy.

[17] Wickliffe Rose (1862–1931).

[18] Werner Heisenberg (1901–1976).

Afklaringen af en ny Atommekanik, som gav Anledning til Virkningskvantet,[19] som i Skønhed og Fuldstændighed i enhver Henseende kan sammenlignes med Newtons Mekanik.

Jeg skal slet ikke komme nærmere ind paa saadanne Enkeltheder mere, men blot sige, at denne Udvikling af den nye Atommekanik kom til at danne en foreløbig Afslutning paa Bestræbelserne paa at finde ud af, hvordan Elektronerne er bundne i Atomet. Man oversaa efterhaanden baade paa Grundlag af den Erfaring, der var samlet, og ved Hjælp af de nye Hjælpemidler de Fænomener i en meget vidtstrakt Grad, og Interessen blev naturligvis afledt til andre Sider af Atomforskningen, nemlig til Studiet af Atomkernen selv. Arbejdet herpaa var grundet paa en anden af Rutherfords store Opdagelser, nemlig Paavisningen af at det er muligt under særlige Omstændigheder at frembringe Sønderdelinger af Atomerne, Atomkernesønderdelinger. Under sædvanlige Forhold er dens Uforanderlighed Grundlaget for selve Grundstoffernes Uforanderlighed, men ved Bombardement af Atomkernen med meget energirige Partikler, der udsendes fra Radium, der hidrører fra Eksplosioner af Radiumatomkernen, er det muligt at frembringe Sønderdelinger af andre Stoffers Kerner og frembringe en Forandring af et Grundstof, hvad der kan betragtes som en helt ny Æra i Naturvidenskabens Historie. Disse Opdagelser er efterfulgt af en ganske overordentlig Udvikling til at begynde med i Cambridge. Det næste store Fremskridt var Paavisningen af, at det var muligt at frembringe Atomkernesønderdelinger ved Partikler, som man selv giver store Hastigheder ved Hjælp af store elektriske Spændinger. Udviklingen tog umaadelig Fart, og ikke mindst lærte man at kende saadanne Sønderdelinger, ved hvilke der udsendes Neutroner, saa man lærte en ny Bestanddel af Materien at kende og finder et overordentlig virksomt Hjælpemiddel til at foretage Eksperimenter med Atomerne, fordi Neutronerne ikke frastødes fra den elektriske almindelige Kerne. De kan trænge ind i den og ramme den, og man har faaet helt nye Muligheder til saadanne Undersøgelser. Opdagelsen af allerstørste Betydning var Opdagelsen af den saakaldte kunstige Radioaktivitet af Madame Curies[20] Datter[21] og Svigersøn[22] for faa Aar siden, at det er muligt ved Atomkerneforvandlinger at frembringe nye radioaktive Stoffer, og det har givet Muligheder for at gøre fysiske Eksperimenter, men det har især givet Mulighed for at

---

[19] "som gav Anledning til Virkningskvantet" ("which gave rise to the quantum of action") should probably read "som Virkningskvantet gav Anledning til" ("which the quantum of action gave rise to").

[20] Marie Curie (1867–1934).

[21] Irène Curie (1897–1956).

[22] Frédéric Joliot (1900–1958).

benytte denne kunstige Radioaktivitet til at studere kemiske Omsætninger, til at studere Stofskiftet i levende Organismer, idet man efter en Metode, som for ca. 25 Aar siden blev foreslaaet af Professor Hevesy, benytter radioaktive Atomer til Indikatorer for Forvandlinger af Atomer af et eller andet Grundstof. Man kan saa at sige sætte Mærke ved enkelte Atomer og definere gennem Atomer, at de gaar ind i Organismen til en Tid og gaar ud til en anden Tid. Oprindelig havde Professor Hevesy foretaget saadanne Undersøgelser ved sit første Besøg i København[23] i Forbindelse med Biologer knyttet til Landbohøjskolen[24] og Finseninstituttet,[25] men de nye Muligheder gav Anledning til Samarbejde mellem Instituttet her og mange forskellige videnskabelige Institutioner, de to Afdelinger af Universitetets fysiologiske Institutter, der bestyres af Professor Krogh[26] og Professor Lundsgaard,[27] og Carlsberglaboratoriet[28] og Tandlægeskolen.[29] Man har ogsaa undersøgt paa nye Maader særlige Forhold ved Stofskiftet indenfor Tænderne. Dette Arbejde stillede jo store Krav til Instituttet, fordi vi jo alle forstod dets store Betydning, samtidig med de fysiske Undersøgelser. Vi fik en meget værdifuld Støtte en Tid fra Insulinfondet,[30] og vi fik hurtigt Rockefellerfondet[31] interesseret i at støtte det hele Arbejde og give os en stor Bevilling til Anskaffelse af et Højspændingsanlæg. Det var først gjort muligt ved en Bevilling fra Carlsbergfondet[32] ... og haaber at faa større Midler i Hænde til virkelig at give danske Biologer de fornødne Midler til at foretage saadanne Undersøgelser, der aabner helt nye Udsigter indenfor den biologiske Forskning. Man kan jo kun se et lille Stykke ind i Fremtiden baade med Hensyn

---

[23] Hevesy worked at Bohr's institute from 1920 to 1926, and again from 1934 to 1943.

[24] See p. [412], ref. 18.

[25] The Finsen Institute was established in 1896 with the purpose of continuing the treatment with light instigated by Niels Finsen (1860–1904), a pioneer of modern phototherapy, and the first Dane to receive the Nobel Prize – in physiology or medicine for 1903. In the 1920s it added to its activities by incorporating a hospital for cancer treatment, the so-called Radium Station, which was further expanded in 1937.

[26] August Krogh (1874–1949), professor of physiology at the Faculty of Science.

[27] Einar Lundsgaard (1899–1968), professor of physiology at the Faculty of Medicine.

[28] See p. [474], ref. 2.

[29] The School of Dentistry in Copenhagen, established in 1888, was the first such school in Denmark. It obtained university status, with an obligation to conduct research, in 1941.

[30] The Nordic Insulin Foundation was established in 1926 with proceeds from Danish insulin production, which August Krogh had been instrumental in initiating from 1922. It provided its first grant to Bohr's institute in 1935.

[31] The substantial support from the Rockefeller Foundation, the first part of which was approved only weeks after the grant from the Insulin Foundation, was given as part of its newly established "experimental biology" programme.

[32] Support from the Rockefeller Foundation was conditional upon support from other sources.

til Opgaver og Hjælpemidler, og netop under dette Arbejdes Planlæggelse viste
det sig, at nye Hjælpemidler maa fremskaffes, nye Hjælpemidler, der tillader
at opnaa endnu større Hastigheder for Partiklerne ved Beskydning ved i Stedet
for at lade dem gaa igennem høje Felter at lade dem løbe rundt i et Magnetfelt
og gradvis accelerere dem op ved Hjælp af svage Felter (Professor Lawrence
i Berkeley, Cyclotron).[33] Det vilde ganske overstige vore Midler at anskaffe
et saadant Hjælpemiddel; men Ingeniør Thrige[34] tog den Tanke op med stor
Venlighed og Forstaaelse, at Thomas B. Thriges Fond vilde skænke os en stor
Magnet som Grundlag for saadanne Undersøgelser, og en saadan Magnet er
virkelig forfærdiget af Direktør Meyer[35] i Odense og er nu opstillet i Instituttet.
Vi har endda i de sidste Dage faaet den yderligere, meget kærkomne Underretning, at Thrigefondet ikke alene har skænket os Magneten, men ogsaa vil
skænke os Dækning af Udgifter ved den tekniske elektriske Installation, som er
udført i Forbindelse med Cyclotronens Opstilling. Jeg vil ogsaa nødigt undlade
at nævne, hvor stor en Hjælp det har været, at vi for to Aar siden fra en Kreds
af danske Institutioner modtog en Gave af et kraftigt Radiumpræparat paa 600
mg,[36] og det har været hele Grundlaget for de Undersøgelser, som hidtil har
været foranstaltet, ogsaa de biologiske Undersøgelser under Professor Hevesys
Ledelse, og det vil ogsaa, naar de nye Anlæg fungerer, blive et uvurderligt
Hjælpemiddel for vore Undersøgelser. Naar man indenfor beskedne Rammer
vil tage Del i Undersøgelser af denne Art, kræves der betydelige Midler, og vi
haaber i Fremtiden baade fra Myndighedernes og private Institutioners Side at
møde en lignende Venlighed og Gavmildhed.

Jeg maa maaske lige inden jeg slutter, nævne et Par Ord om selve An-
læggets Tilvejebringelse. Vi havde den Lykke endnu at kunne have Arkitekt
Varmings[37] Støtte. Han har udført de andre Udvidelser af Bygningen her, som
Martin Borch[38] oprindelig havde [forestået]. Før sin Død lykkedes det Arkitekt
Varming at fuldende Bygninger … . Det store Tab, som Arkitekt Varmings
Død betød …Ingeniør Varming og Ingeniør Steensen,[39] som har baade hjulpet

---

[33] Ernest Orlando Lawrence (1901–1958), the inventor of the cyclotron.

[34] Bohr describes his relationship with Thrige on pp. [555] ff.

[35] Viggo Meyer (1884–1952).

[36] The "Radium Gift" for Bohr's fiftieth birthday on 7 October 1935 was organized by Hevesy and involved sixteen Danish firms and foundations, including the Carlsberg Breweries, the Carlsberg Foundation and the Thrige Foundation. The radium was bought for 100,000 Danish kroner.

[37] Kristoffer Nyrop Varming (1865–1936) had also been responsible for the expansion of Bohr's institute beginning in 1924.

[38] Martin Borch (1852–1937), Kristoffer Varming's uncle, had designed the original institute.

[39] Jørgen Varming (1906–), Kristoffer Varming's son, and Niels Steensen (1904–1986) had established the consulting engineering firm Steensen & Varming in 1933.

os med selve Konstruktionen og de tekniske Anlæg paa saa udmærket Maade. Jeg vil ogsaa gerne udtale vor Taknemmelighed overfor Haandværkerne, der ikke mindst i denne Tid har gjort et meget stort Arbejde, og jeg tænker særlig paa det utrættelige Arbejde af Montør Petersen[40] fra Thrige, der har arbejdet her under alle de senere Installationer. Saa er der ogsaa gjort et meget stort Arbejde indenfor Instituttets egne Værksteder, som under den fortrinlige Ledelse af Laboratoriemester Olsen[41] og Vagtmester Jensen[42] har udført ganske overordentlig smukke Snedkerarbejder, som vi vil faa at se; men først og sidst er det Instituttets videnskabelige Medarbejdere, der maatte bære Byrderne af Arbejdet, og det væsentlige Arbejde med Cyclotronen skylder vi Dr. Jacobsen, der gennem sine Undersøgelser i de mange Aar paa Instituttet har saa enestaaende Erfaringer. Vi har haft den Lykke, at Professor Lawrence har overladt os en af sine udmærkede Medarbejdere,[43] der har været os til største Hjælp ved Indretningen af Højspændingslaboratoriet. Der har Arbejdet hvilet væsentligt paa Assistent Bjerge,[44] og vi har ogsaa haft stor Hjælp af nuværende Professor i Boston v. Hippel,[45] der har særligt Kendskab til Højspændingsfænomener. Jeg kan ikke nævne alle, men vil ikke forglemme Assistent Rasmussen,[46] der er Leder af vor spektroskopiske Afdeling og har overvaaget hele Arbejdet paa den mest utrættelige Maade. Den nøje Forbindelse, vi har haft med andre Institutter, har været af største Betydning for os alle og ogsaa for de unge danske Medarbejdere, der har haft alle Muligheder for paa ganske anderledes Maade at komme ind i det store Arbejde paa Fysikens Omraade, men det har endda været saadan, at vi har haft den Lykke ved Hjælp af de Midler, man fra Udlandet har stillet til vor Raadighed, at knytte unge danske Videnskabsmænd for en Tid til Instituttet, saa at de har kunnet faa den Uddannelse, som gør

[40] There were several fitters named Petersen at Thrige in 1938, and it has been impossible to establish which one of them took part in the work at Bohr's institute.

[41] Holger Olsen (1889–1958) worked at Bohr's institute from 1920 until his retirement in 1954.

[42] In 1924, August Jensen (1890–1964) was chosen among 227 applicants to the position of non-scientific assistant. He worked at Bohr's institute until his retirement in 1946.

[43] Lawrence Jackson Laslett (1913–1993).

[44] Torkild Bjerge (1902–1974) worked as assistant at Bohr's institute from 1937 to 1939, when he was appointed professor of physics at the Technical University of Denmark.

[45] Arthur Robert von Hippel (1898–2003) worked in Copenhagen from January 1935 to August 1936, en route from Istanbul to the Massachusetts Institute of Technology, where he spent the rest of his long career.

[46] Ebbe Kjeld Rasmussen (1901–1959), Danish physicist. Rasmussen worked at Bohr's institute from 1928 to 1942, when he accepted a professorship at the Royal Veterinary and Agricultural College. Bohr's obituary for him, *Ebbe Kjeld Rasmussen 12 April 1901 – 9. October 1959*, Fys. Tidsskr. **58** (1960) 1–2, is reproduced in Vol. 12.

det muligt for dem senere at bære Byrden ved Arbejderne. Endnu vil jeg blot sige, at naar vi overhovedet tænker paa denne Side af Sagen, da er det frugtbare videnskabelige Arbejde indenfor Videnskaben det, alle maa tænke paa som en af de mest opmuntrende Sider i Øjeblikket indenfor den menneskelige Civilisation. Det giver Haab om, at lignende fredeligt Samarbejde til Fremme for Menneskehedens fælles Interesser ogsaa kan opnaas paa andre Omraader.

TRANSLATION

Speech at the inauguration of the Institute's high-voltage plant
5 April 1938.

I am very grateful for the kind words of welcome and not least for the great distinction that His Majesty the King has bestowed upon me.[1] It was of course my duty otherwise to cordially welcome all of you here today, and we are especially grateful to see so many representatives of the Government and the State and municipal authorities, which from the very beginning have shown such great interest in our work and offered such great support, just as also on this occasion. We are also very pleased to see so many representatives of the various private institutions that have offered us so much support and through so large donations to the Institute that the new apparatus to be presented today has become possible. Before we proceed to actually looking around, I thought it might interest you to hear a brief account of the history and development of the Institute and a few words about the tasks we have concentrated on in particular, as well as the new tasks that we hope to have the good fortune to attack with the new tools we have acquired. The occasion for the establishment of an institute like this, with the name "theoretical physics", was the situation that had arisen within physics at that time, as to whether it was specially necessary to make possible the closest cooperation between theoretical and experimental investigations in the field of atomic physics.[2] We had of course at that time, through the great discoveries about the composition, constitution, of matter – keystones, won by Rutherford's discovery of the atomic nucleus – obtained a completely defined picture of the structure of the atoms, in any case detailed knowledge of the building blocks of the atoms. At the same time, it was clear that the model of the atom, the so-called "Rutherford model of the atom", according to which an atom is built in a similar way to a solar system, that is, consists of a number of light particles, the electrons, which are bound through the electric attraction to the atomic nucleus, which is a particle that is quite extremely small compared to the atom itself, thus a picture which perfectly resembles a solar system. On the basis of such a model it was obvious that it was quite impossible, on the basis of the usual physical considerations, which for ..., to understand the characteristic properties of the atoms, but we have had such good fortune to get a key to the understanding of this riddle through Planck's discovery, at the turn of the century, of the so-called quantum of action, when one learned about a completely new feature of an atomistic kind in the laws of nature, a feature quite different from the old understanding of the finite divisibility of matter, and one discovered, so to speak, a feature of indivisibility in the sequence of the atomic processes themselves and finds an opportunity to understand how one at least could proceed with the help of the Rutherford model of the atom, in that one was forced to assume that the stability of atoms stemmed from this new feature, and that, for example, any change in the state of the atom was an individual process whereby the atom made a transition from one stationary state to another. It was a very schematic theoretical consideration and could be useful by giving us an opportunity to get direct information about the way whereby the electrons are bound in the atom by studying their so-called spectra at the individual transition processes. By investigating the colour of the light, one can establish how much energy is emitted and get to know the conditions of binding in the atoms. Although these results, which could be obtained on the basis of Planck's and Rutherford's discoveries, showed that one was on the right path, there was a long way to go before one could obtain a really coherent – coherent in itself – and complete description of the conditions of the atoms, and it was obvious that one could feel one's way and that there was no other way than to try and see what could be found out with the help of one's experience (theoretical investigations) and partly by asking nature itself for new information (experimental investigations). The idea of establishing an institute for the work in this domain met with very great understanding and support from the University and the government authorities. The execution of the plan was made considerably easier by a circle of interested private persons, on the initiative of Director Berlème,[3] collecting a large sum of money for which the plot of land, where we are now, was bought, and it was the first moving-out of scientific institutes to this part of the city, where the new university town is now growing up.[4] Similarly, it was of the greatest significance that the Carlsberg Foundation enabled us to acquire expensive spectroscopic instruments. The work at the beginning rested first and foremost on those people who could be attached to the Institute, and today I think with gratitude of the support and effort of Professor H.M. Hansen[5] at the beginning, before he got his own institute. From the beginning we have also had exceedingly clever and outstanding experimental collaborators, among others Dr Jacobsen,[6] who has been at the Institute from the very beginning, and Werner,[7] now Professor, who was an assistant for a time at the beginning, and who now has his own activity in Aarhus, and with whom we hope, in the coming years, to have a very close cooperation. It was very important for us to receive visits of very outstanding young people from abroad. The first of these was Kramers, now Professor in Leyden, whom we have the great pleasure of seeing today, who first came as a young student, but was attached to the Institute and worked here as a lecturer for 10 years and has had such great importance as regards both the development of the Institute and the teaching.[8] Another visit I would like to mention in a later connection was the one by Professor Hevesy,[9] a Hungarian physicist whom I had got to know in Rutherford's laboratory in Manchester, and the possibility for his attachment to the Institute was the establishment of the Rask–Ørsted Foundation.[10] I would like to recall the effort that was made on the part of Denmark immediately after the war, the great understanding for the importance of cooperation between the peoples also in the field of science, which had suffered such great interruption during the Great War, and we had a large number of visitors also in those years, Professor Klein[11] who is now employed in Stockholm, Rosseland[12] who is now Professor in astrophysics in Oslo, Dr Nishina,[13] Japan, and Dr Ray[14] from India, and during the great cooperation taking place within the Institute and within various institutes worldwide, one really succeeded in achieving such a preliminary rounding-off of the first attack on the structure of atoms. One managed to achieve an overview of the way whereby the electrons are bound in the atoms, an overview that allowed an understanding of the relationship between the physical and chemical properties of the elements, the periodic system. In this connection, Professor Hevesy's attachment to the Institute was, of course, of the greatest significance, in that he succeeded, through the discovery of an element which he and his collaborator, Professor Coster[15] in Groningen, called Hafnium, to obtain support for the theoretical viewpoints we were working with then at the Institute. At that time new opportunities for cooperation in the field of science appeared through the Rockefeller Foundation, through the establishment of the International Education Board as a part of the Rockefeller Foundation,[16] and first and foremost they took up the task of making international cooperation easier, and in consideration of the effort we could make here at the Institute, they donated a large sum to us, so that the Institute buildings could be expanded and become twice as large as before. The necessary extension of the ground for this was donated by the Copenhagen municipality, but this International Education Board also took up other tasks, and it will perhaps be of interest to recall that here the model was precisely the establishment of the Rask–Ørsted Foundation. On a visit to Copenhagen, the Director[17] of the International Education Board observed with greatest sympathy the endeavours to which the initiative had been taken in this country. In the following years, promising young scientists from various countries came to the Institute as Rockefeller Fellows and had the opportunity to broaden their own horizon for a time and bring new impulses to this institute; and it is interesting to recall that the very first of such Rockefeller stipendiates

[522]

was Heisenberg,[18] now Professor in Leipzig, who worked for several years here at the Institute, and who has made such an extraordinary effort towards the building and the clarification of a new atomic mechanics, which gave rise to the quantum of action,[19] which in beauty and completeness in every respect can be compared with Newton's mechanics.

I shall not at all go further into such details any longer, but only say that this development of the new atomic mechanics came to form a preliminary completion of the endeavours to find out how the electrons are bound in the atom. Both on the basis of the experience gained and with the help of the new tools, one came to understand, as time passed, the phenomena to a very great degree, and interest was naturally diverted to other aspects of atomic research, that is, to the study of the atomic nucleus itself. The work on this was based on another of Rutherford's great discoveries, that is, the demonstration that it is possible under special conditions to produce disintegration of the atoms, nuclear disintegrations. Under ordinary conditions, its immutability is the basis for the immutability of the elements themselves, but by bombardment of the atomic nucleus with highly energetic particles, which are emitted from radium [and] which originate from the explosions of the radium atomic nucleus, it is possible to bring about disintegrations of the nuclei of other substances and to produce a change of an element, which may be regarded as a completely new era in the history of natural science. These discoveries are followed by a quite extraordinary development which started in Cambridge. The next great advance was the demonstration that it was possible to bring about nuclear disintegrations with particles which are given great velocities with the help of great electric voltages. The development took up enormous speed and not least was knowledge acquired of those disintegrations by which neutrons are emitted, so that one got to know a new constituent of matter and finds an extraordinarily effective tool for making experiments with atoms, because the neutrons are not repelled from the electric normal nucleus. They can penetrate into it and hit it and quite new opportunities for such investigations have been obtained. The discovery of the very greatest significance was the discovery of so-called artificial radioactivity by Madame Curie's[20] daughter[21] and son-in-law[22] a few years ago, that it is possible by transmutations of the atomic nucleus to produce new radioactive substances, and this has provided opportunities for making physical experiments, but it has in particular given the opportunity of using this artificial radioactivity to study chemical transformations, to study the metabolism in living organisms, since according to a method suggested about 25 years ago by Professor Hevesy, radioactive atoms are used as indicators for transmutations of atoms of one or another element. It is possible, so to

[523]

speak, to label individual atoms and define through atoms that they enter into the organism at one time and depart from it at another time. Professor Hevesy had originally made such investigations during his first visit to Copenhagen[23] together with biologists attached to the Agricultural College[24] and the Finsen Institute,[25] but the new opportunities gave occasion for cooperation between the Institute here and many different scientific institutions – the two departments of the University physiological institutes, directed by Professor Krogh[26] and Professor Lundsgaard,[27] and the Carlsberg Laboratory[28] and the School of Dentistry.[29] The special circumstances regarding the metabolism inside the teeth have also been investigated in new ways. This work made great demands, of course, on the Institute, because we all realized its great importance, at the same time as the physical investigations. We received very valuable support for a time from the Insulin Foundation,[30] and we quickly got the Rockefeller Foundation[31] interested in supporting the whole work and to give us a large grant for the acquisition of a high-voltage generator. It was only made possible with a grant from the Carlsberg Foundation[32] ... and hope to get large means at our disposal to really give Danish biologists the necessary means to carry out such investigations, which open quite new perspectives within biological research. It is of course only possible to see a little way into the future both as regards tasks and tools, and precisely during the planning of this work it turned out that new tools have to be acquired, new tools that make it possible to reach still greater velocities for the particles at bombardment by letting them run around in a magnetic field and gradually accelerate them with the help of weak fields (Professor Lawrence in Berkeley, Cyclotron),[33] instead of letting them pass through strong fields. It would of course quite exceed our means to acquire such a tool; but with great kindness and understanding, Engineer Thrige[34] took up the idea that the Thomas B. Thrige Foundation would give us a large magnet as the basis for such investigations, and such a magnet has indeed been manufactured by Director Meyer[35] in Odense and is now set up at the Institute. We have even in recent days received the further very welcome news that the Thrige Foundation has not only given us the magnet but will also cover the expenses for the technical electrical installations that have been made in connection with setting up the cyclotron.

I should not fail to mention how great a help it has been that two years ago we received a gift of a strong radium preparation of 600 mg from a circle of Danish institutions,[36] and this has been the whole basis for the investigations carried out until now, including the biological investigations led by Professor Hevesy, and it will also be an invaluable tool for our investigations when the new apparatus is in function. When one wants to take part in investigations of

At the presentation of the Radium Gift to Bohr in 1935. Left to right: –, Martin Knudsen, H.M. Hansen (partly hidden), Niels Neergaard, S.P.L. Sørensen, N. Bohr, Niels Bjerrum, –, –, Johannes Pedersen (in front), Benny Dessau and Frederik Sander.

this kind within a modest framework, considerable means are required, and we hope in the future to meet similar kindness and generosity on the part of both the authorities and private institutions.

Just before I finish, I may perhaps say a few words about how the plant itself was set up. We had the good fortune still to have the assistance of architect Varming.[37] He has carried out the other extensions of the building here, for which Martin Borch[38] was originally responsible. Architect Varming succeeded in completing the buildings ... before his death. The great loss that Architect Varming's death meant ... . Engineer Varming and Engineer Steensen,[39] who both have helped us in such an excellent way with the construction itself and the technical plant. I would also like to express our gratitude to the workmen, who not least during this time have done a very big job, and I think especially of the tireless work by Petersen,[40] the fitter, from Thrige, who has worked here during all the latest installations. Then a very big job has also been done in the Institute's own workshops which, under the excellent leadership of Laboratory Master Olsen[41] and Janitor Jensen,[42] have made quite extraordinarily beautiful carpentry that we are going to see, but first and last it is the Institute's scientific collaborators who had to carry the load of the work, and the considerable work with the cyclotron we owe to Dr Jacobsen, who through his investigations in the many years at the Institute has such unique experience. We have had the good fortune that Professor Lawrence has let us have one of his excellent collaborators,[43] who has been of the greatest help in the equipping of the high-voltage laboratory. There the work has mainly rested on Assistant Bjerge,[44] and we have also had great assistance from von Hippel,[45] now professor in Boston, who has special knowledge of high voltage phenomena. I cannot mention everyone, but I will not leave out Assistant Rasmussen,[46] who is the head of our spectroscopic department and who has monitored the whole work in the most tireless way. The close connection we have had with other institutes has been of the greatest importance for all of us and also for the young Danish collaborators, who have had every opportunity in quite a different way to get into the great work in the field of physics, but it has even been the case that we have had the good fortune, with the help of the means put at our disposal from abroad, to attach young Danish scientists to the Institute for a time, so that they have been able to get the education that makes it possible for them to carry the load of the work later. Finally, I will just say that when we at all consider this aspect of the matter, then it is the fruitful scientific work within science that all must consider as one of the most encouraging aspects at present within human civilization. It gives hope that a similar peaceful cooperation for the advancement of humanity's common interests can also be achieved in other fields.

XXII. SCIENCE AND ITS INTERNATIONAL SIGNIFICANCE

Danish Foreign Office Journal, No. 208 (May 1938) 61–63

See Introduction to Part II, p. [357].

Copenhagen University Institute for Theoretical Physics with the University Institute for Mathematics adjoining on the left. The new high-voltage hall is seen on the right. Copenhagen c. 1940.

DANISH FOREIGN OFFICE

**COMMERCIAL
& GENERAL REVIEW**

JOURNAL

NUMBER 208 · MAY 1938

SCIENCE *and its*

INTERNATIONAL SIGNIFICANCE

Broadcast lecture by

Professor NIELS BOHR, Copenhagen

This is Copenhagen, Denmark, calling.
The announcer:

This programme is being sent from the Theoretical Physics Institute of the University. It is dedicated to the head of the institute, Professor Niels Bohr, the famous Danish physicist and Nobel Prize winner. An American guest in our institute, Professor Louis Turner of Princeton University, has consented to give a few introductory words. Will you please take the microphone, *Professor Turner?* . . . Yes, with the greatest pleasure:

Twenty-five years ago, on the fifth of April, nineteen hundred and thirteen, Professor Niels Bohr finished his first paper on the Bohr theory of the structure of the atom. On the basis of the late Lord Rutherford's discovery of the nucleus of the atom, and Max Planck's discovery that radiation is emitted in separate quanta, Niels Bohr created his fundamental theory of the atom. For the past twenty five years this theory has been the guide in all progress in atomic physics.

The institute of Theoretical Physics was built for Professor Bohr in nineteen hundred and twenty in order to bring about an intimate collaboration between theoretical and experimental physics. To-day a new department of this institute has been officially opened in the presence of representatives of the Danish government and other notables.

In the last few years the greatest progress in physics has been achieved in the study of the structure of the atomic nuclei and of their disintegrations. Professor Bohr's recent contributions in this field are also of fundamental significance. This newly opened part of the laboratory contains modern equipment for experimental studies of this sort. The buildings and their equipment are gifts of the Rockefeller Foundation and of various Danish foundations. The more than one hundred colleagues from America and other foreign lands who have had the happy experience of working here with Professor Bohr, will envy me my good fortune in being here for this celebration. I am sure that I speak for them as well as for myself in offering congratulations and best wishes for another quarter century of splendid work. Professor Bohr will now speak to you.

[529]

Professor Bohr:

I am very thankful for the great honour shown me by the invitation to say a few words about the exploration of the wonderful world of atoms to listeners in the large English speaking communities on both sides of the Atlantic. It especially offers me a most welcome opportunity to give expression for the indebtedness all scientists in this field owe to the pioneer work of men from these great nations which have contributed so much to human civilisation. Nothing can indeed remind us more strongly of the unity of mankind than the evolution of science which, from primitive endeavours to collect and order experience, has grown to be a most powerful source of progress for humanity.

To this great work all civilized nations have given their share, and, approaching our subject, we need only think of men like *Franklin, Volta, Ørsted, Ampère, Faraday* and *Hertz*, to realize how international cooperation has created the basis for the wonders of electrical engineering, which have reshaped in an undreamt measure our daily life and at the same time given us new means to unveil the secrets of nature. The names of *Röntgen* and *Curie* will here suffice to recall, how even most beneficial aids to the relief of human suffering may be the immediate result of pure scientific inquiry into the physical nature of our world.

Of course, in a short speech it would be quite impossible to give any detailed account of the marvellous development of atomic physics in our days, and I shall therefore only recall a few points especially suited to illustrate the decisive role cooperation here has played. As everyone knows, a corner- stone in this development was the discovery of the electron, which was traced as a constituent element of all matter by the pioneer-work of *Sir J. J. Thomson* and his school in the old University of Cambridge, and the charge and mass of which was finally measured with such astounding accuracy through *Millikan's* ingenious experiments in Chicago. It is also of common knowledge, that *Lord Rutherford's* fundamental research on radioactivity led him to the discovery of the atomic nucleus which in so unsuspected manner completed our knowledge of the constitution of the atom. This discovery, however, revealed at the same time the impossibility of explaining many characteristic properties of atoms by means of the ordinary laws of physics, which apply to the behaviour of matter in bulk.

Still, due of the fundamental discovery, a few years earlier, of the elementary quantum of action, to which *Planck* was led through his penetrating analysis of black body radiation, physicists were in no way unprepared for such a situation. Planck's discovery meant indeed the disclosure of an essential new feature of atomicity in the laws of nature, quite foreign to the point of view of classical physics and supplementing most unsuspectedly the old doctrine of the restricted divisibility of matter. Thus was initiated a new epoch in natural philosophy, and it needs hardly be recalled that *Einstein,* in the very same year in which he published his first paper on relativity, also contributed most fruitfully to the foundations of atomic physics by pointing out, how the existence of the quantum of action offered an immediate explanation of a number of phenomena which hitherto had frustrated the efforts of scientists.

One had therefore not to look far for a clue to the solution of the new dilemma with which Rutherford's discovery confronted us, and it is a special pleasure for me here to stress that if, twentyfive years ago, I had the good fortune to give a modest contribution to this development, it was above all thanks to the hospitality I then as a young man, enthusiatic for the great promises of the new theoretical outlooks, enjoyed in the famous laboratories of England, where the fundamental experimental disco-

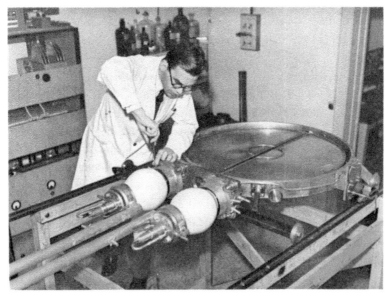

Electrodes for the Copenhagen Cyclotron granted to Professor Bohr
by the Thomas B. Thrige Fund, Denmark

veries were just being made. In particular I
think with grateful emotion of the unique friend-
liness and straightforwardness with which *Rutherford,* in the mindst of his unceasing creative
activity, was always prepared to listen to any
student, behind whose youthful inexperience he
perceived a serious interest.

To develope on the foundation laid by *Planck*
and *Rutherford* a rational theory of atomic constitution claimed, however, a clarification of the
principles of atomic mechanics, which was
gradually accomplished by a number of experimental as well as theoretical investigations, in
which physicists from countries all over the
world have taken effective part. The astounding
rapidity of this brilliant development has in fact
only been made possible by a most intensive
international cooperation that has been facili-
tated not least by the liberality with which *the
Rockefeller Foundation* has permitted thousands
of promising young men to complete their
education and enlarge their outlook by temporary visits to institutes abroad, where they at
the same time have brought new impulses. Also
in this Institute we have had the pleasure and
benefit of the visit of a large number of Rockefeller fellows, and it is perhaps not without
interest to mention that the first among them
and, I think, actually the very first regular
Rockefeller fellow, was *Heisenberg,* whose fundamental contributions to atomic mechanics are
so well-known. Likewise it is a special pleasure
for us to think of the stay here, as a fellow of
the American-Scandinavian Foundation, of Professor *Urey,* whose later discovery of the heavy
hydrogen has opened such important new fields
of research.

In recent years we have entered a new era
in physical and chemical science initiated by
another of the great discoveries with which
Rutherford's genius presented us. Thus, in nineteen hundred and nineteen *Rutherford* succeeded
in proving that the nucleus of the atom, the
intangibility of which under all ordinary conditions is the very basis for the constancy of the
elements, may under the extreme circumstances afforded by the bombardement with high-
speed particles, itself undergo disintegrations.
This modern revival of the old dreams of the
alchemists has, in the last few years, led to a
succession of most startling discoveries regarding the ultimate constituents of matter and has
already found very important practical applications, especially through the discovery of
artificial radioactivity by the daughter and son-
in-law of the immortal Madame Curie. Above
all this discovery has given us entirely new
means of studying the incessant exchange of
matter within living organisms by means of the
indicator method originated by Professor
Hevesy, the privilege of whose collaboration we
enjoy in this Institute.

Part of accelerating tube for nuclear disintegration experiments.

The intense world-wide interest in this young
nuclear physics has given rise to great efforts in
developing ever more powerful tools for this
kind of research, ranging from the construction
on a gigantic scale of electrostatic machines
akin to those of Franklin's days to the design
of electromagnetic accelerators of atomic particles on quite new principles. On my recent
visit to America I received an unforgetful impression of the success of these courageous
efforts, and it is a most pleasant duty on this
occasion to express my special indebtedness towards the ingenious inventor of the cyclotron,
Professor *Lawrence* of Berkeley, who has so
generously put his precious experience at the
disposal of this Institute as well as of many
other laboratories all over the world. In building
such complicated and costly apparatus, the mutual aid of scientific investigation and practical
industrial activity is felt particularly strongly
on account of the technical perfection required
as well as of the necessary financial support. In
our case this support has been most liberally
granted by the *Rockefeller Foundation* and by
Danish foundations which likewise owe their
sources to industries which again depend essentially on science. In concluding this address I
wish to give expression for the hope, that the example of the fruitful cooperation in science,
independent of national borders and prejudices,
may stimulate efforts to contribute by peaceful
collaboration also in other fields of human
culture to the common progress of mankind.

[531]

XXIII. THE UNIVERSITY AND RESEARCH

UNIVERSITETET OG FORSKNINGEN
Politiken, 3 June 1941

Talk on Danish national radio, 1 June 1941,
on the anniversary of the founding of the University of Copenhagen

TEXT AND TRANSLATION

See Introduction to Part II, p. [357].

[533]

NIELS BOHR

UNIVERSITETET OG FORSKNINGEN

*Tale ved Statsradiofoniens Udsendelse
den 1. Juni 1941 paa Aarsdagen for
Grundlæggelsen af Københavns Universitet*

Særtryk af „Politiken"s Kronik 3. Juni 1941

FORUDEN at være et Midtpunkt for Ungdommens Indførelse i den samlede Viden, man paa ethvert Tidspunkt var naaet frem til, maatte vort Universitet fra første Færd tillige blive et Hjemsted for videnskabelig Forskning. Mellem Forskning og Meddelelse af Viden kan der jo slet ingen streng Grænse trækkes, saa vist som al dybere Kundskab efter sin Art selv har en søgende Karakter.

Videnskabens Formaal, at forøge og ordne vore Erfaringer, stiller os stadig over for Opgaver, der bringer Spørgsmaalet om menneskelig Erkendelses Væsen i skarp Belysning. Det er jo langtfra saaledes, at de Rammer, hvori Erfaringerne kan indordnes, er givet os paa Forhaand. Tværtimod er det kun gennem Erfaringerne selv, at de Lovmæssigheder, der muliggør vort Overblik over Fænomenernes Mangfoldighed, gradvis kommer til vort Kendskab

3

Haand i Haand med Indhøstningen af Erfaringer paa alle Kundskabsom- raader har der da ogsaa op igennem Tiderne fundet en bestandig Udvikling af vor Begrebsbygning Sted, kendeteg- net lige fuldt ved Skærpelsen af vor Tænknings Vaaben som ved Uddybelsen af vort Syn paa Menneskets egen Stil- ling i Tilværelsen.

Hvad har det i denne Henseende ikke betydet, at vi ikke længere som en Selvfølge gaar ud fra, at vi befinder os i Altets Midtpunkt, men siden den Sejr over dybt rodfæstede Fordomme, som Renæssancetidens Videnskab vandt, er blevet klar over, at vi lever paa en lille Klode, der sammen med de andre Planeter er Drabanter for vor Sol i dens Bevægelse mellem Universets My- riader af glødende og udslukte Stjer- ner.

Ogsaa den Klarlæggelse af saa mange af de for Naturfænomenerne gældende almindelige Love, der danner Grundlaget for hele vor Tids vidunder- lige tekniske Udvikling, hviler jo, selv om det ofte glemmes, paa Videnskabs- mændenes Aarvaagenhed over for de saa let oversete Fingerpeg, som nye Erfaringer kan give os om Utilstrække- ligheden af de Synspunkter, der havde

4

været fyldestgørende for de tidligere Erfaringers Sammenfatning.

Nutildags er det maaske ogsaa svært tilstrækkeligt at paaskønne, hvor gennemgribende den ved sin Fremkomst saa haardt bekæmpede Udviklingstanke har ændret vort Syn paa den Plads, vi selv indtager blandt Jordens forunderlig rige Mangfoldighed af levende Væsener. Hvor mange Skranker har der heller ikke været at overvinde, før de inden for Naturvidenskaberne udviklede Metoder fandt rationel Anvendelse paa Studiet af Livets Foreteelser og derved muliggjorde den for Menneskeslægtens Velfærd saa betydningsfulde Opblomstring af Lægekunsten i vor Tid.

Hvor dyb og indtrængende en Belæring om Betingelserne for Menneskelivets Udfoldelse har vi endelig ikke faaet gennem de humanistiske Videnskaber, der har deres Rod i den gamle Interesse for Folkeslagenes Sprog og Historie. Studiet af Menneskesamfundenes Liv paa Jordens forskellige Egne i Fortid og Nutid aabner jo maaske mere end noget andet vort Øje for de Muligheder for Kulturens Udvikling, som netop Arbejdet paa at forøge vor Viden under stadig Kamp mod gamle

5

og nye Fordomme stiller hele Menneske-
slægten i Udsigt.

✳

Den rivende Udvikling paa alle Videnskabens og Teknikkens Omraader,
der i højere Grad end nogensinde hindrer den enkelte i at sammenholde den
hele menneskelige Viden, har ingenlunde fordunklet, men langt snarere
fremhævet al Videnskabs Enhed, som
netop Universiteterne staar som Sym-
bolet for. Ikke alene har et Samarbejde
mellem Dyrkere af tilsyneladende vidt
forskellige Kundskabsgrene vist sig stadig mere frugtbart, men ogsaa Grænserne mellem mange Omraader af Videnskaben, der hidtil var strengt adskilte, er ved Udviklingens Gang blevet
flydende, ja er ofte helt forsvundet.

Saadanne Forhold møder vi paa alle
Videnskabens Virkefelter. Herhjemme
ligger det os vel særlig nær at tænke
paa, i hvor høj Grad den af hele Folket med saa stor Interesse fulgte
Granskning af vore rige Oldtidsminder
er fremmet ved det stedse voksende
Samarbejde mellem Arkæologer, Geo-
loger, Zoologer, Botanikere og Kemikere, der har bidraget saa meget til den
alsidige Belysning af de Forhold, hvor-

6

under vore Forfædre i fjerne Tider levede

Udviklingen af nye Forskningsmetoder har paa alle Videnskabens Omraader beriget vor Viden og udvidet vor Horisont. Inden for Naturvidenskaben har især det Indblik i Atomernes hidtil for menneskelig Erfaring stængte Verden, som Eksperimentalteknikkens uanede Fremgang har givet os, medført en fuldstændig Sammensmeltning af store Kundskabsomraader og vidtgaaende tilladt os at virkeliggøre de gamle græske Tænkeres dristige Drømme om at føre al Naturbeskrivelse tilbage til Betragtninger af rene Tal.

Paa Grundlag af vore Forestillinger om Atomernes Bygning erkender vi, hvorledes simple fysiske Naturlove afspejler sig i Kemiens rige, for Studiet af saavel den uorganiske Natur som Organismernes Funktioner saa vigtige Erfaringer. Samtidig har Astronomien, som saa ofte tidligere, bragt os de mest slaaende Vidnesbyrd om disse Loves Gyldighed i hele Universet. Selv de fjerneste Himmellegemer har vist sig at være opbygget af de samme Atomer som Stofferne her paa Jorden, og i Stjernernes Udviklingsgang genfinder vi Virkningerne af de samme

7

Atomomdannelser, som Forsøgene i vore Laboratorier først har lært os at kende.

Som det Gang paa Gang er gaaet, naar vor Viden er blevet forøget paa uventet Maade, har Begreber og Synspunkter, der hidtil havde staaet som urørlige Fundamenter, alligevel vist sig for snævre for Erfaringernes Sammenfatning, og før tilsyneladende Modsigelser i de nye Erfaringer kunde opklares, har det endda været nødvendigt i vore Dage at tage det urgamle filosofiske Problem om Forholdet mellem Fænomen og Iagttagelse op til fornyet Diskussion.

Jeg behøver i denne Forbindelse næppe at minde om den Afrunding, som hele vort Verdensbillede har opnaaet igennem Relativitetsteorien, der har klarlagt, hvorledes Brugen af selv de mest elementære Begreber som Rum og Tid paa en hidtil upaaagtet Maade er betinget af Iagttagerens Standpunkt. Inden for de ved Atomforskningen aabnede nye Erfaringsomraader kunde Orden dog først tilvejebringes, efter at ogsaa Betingelserne for selve Aarsagsbe-

8

grebets formaalstjenlige Anvendelse var stillet i ny Belysning.

Spørgsmaalet om Aarsag og Virkning kan jo kun med Rette stilles, naar vi, som det er Tilfældet ved alle Dagliglivets Foreteelser, kan se ganske bort fra Iagttagelsens Indflydelse paa Objekternes Opførsel. Efter selve Arten af de Love, som gælder for de individuelle Atomprocesser, kan vi imidlertid slet ikke skelne mellem Atomernes egen Opførsel og deres Vekselvirkning med de for Fænomenernes Beskrivelse uundværlige Iagttagelsesmidler. Sammenfatningen af Erfaringerne ligger derfor uden for Aarsagssætningens Rækkevidde og har krævet Indførelse af mere almengyldige Synspunkter, ved hvis Udformning Matematikkens til saa høj en Kunst udviklede Abstraktioner atter har været af uvurderlig Hjælp for Naturbeskrivelsen.

Erkendelsen af Aarsagsbegrebets begrænsede Anvendelighed betyder langtfra noget Afkald paa Naturfænomenernes dybere Forstaaelse, men understreger tværtimod al menneskelig Tænknings fælles Forudsætninger. Ikke mindst er det jo fra Selviagttagelse bekendt, hvor umuligt det er, skarpt at skelne mellem Objektet for Undersø-

9

gelsen og selve det iagttagende Subjekt. Den Beskrivelsesmaade, som vi ved psykologiske Studier er henvist til, viser derfor ogsaa en iøjnefaldende Lighed med det inden for Atomfysikken udviklede saakaldte Komplementaritetssynspunkt.

Allerstærkest minder den nye Belæring os dog om den Visdom, som mange af de store Tænkere allerede i Civilisationens Barndom lagde for Dagen, naar de i Forbindelse med deres Grublen over Tilværelsens Gaader atter og atter fremhævede, at kun Erkendelsen af, at vi i Livets vekslende Spil er saavel Deltagere som Iagttagere, lader os ane en Tilværelsens Harmoni, der efter sin Art forsvinder, saa snart vi prøver at presse den ind i snævrere Synspunkters Rammer.

*

Hvor verdensfjernt end en Videnskabsmands Arbejde ofte maa synes, naar det kun har ringe Forbindelse med Dagliglivets Krav og bevæger sig paa Omraader, der ligger langt borte fra menneskelige Lidenskabers Tumleplads, er det dog Forskeren, der ikke lader sig binde af andre Hensyn end

Sandhedens Erkendelse, som opmurer og underbygger det faste Stade, hvorfra man ene kan haabe stedse klarere at overse Menneskets Plads i Tilværelsen og Kulturens Udviklingsmuligheder.

Hvad den enkelte kan bidrage til dette Formaal er uendelig beskedent, og vi maa ofte tænke paa, hvorledes Newton selv sammenlignede sit Værk med en Drengs Leg paa Strandbredden med at samle de smukke Sten, som Havet opskyller. Paa næppe noget andet Felt føler vi saa stærkt som inden for videnskabelig Forskning, at det drejer sig om i Forening at fremelske noget, der lever sit eget Liv, uafhængigt af den Enkelte, og faar sin hele Kraft fra Menneskenes fælles Erkendelsesstræben.

For Opretholdelsen af den Aand, der præger et Universitet som vort, er det jo ikke mindst Ungdommens Videtrang og Virkelyst, der udgør den Jordbund, hvorfra stadig Fornyelse hentes. Følelsen af, at Ung og Gammel hver paa sin Vis arbejder mod samme Maal og tjener samme Sag, er maaske den allermest værdifulde Side ved Livet. som det former sig i Universiteternes Høresale og Laboratorier, og den, der efterlader det dybeste Indtryk hos

11

enhver, der i kortere eller længere Tid
har deltaget deri.

Som et Resultat af Samarbejde
fortsat gennem Tiderne og udstrakt til
alle Folkeslag holder Videnskabens
stolte Bygning os bestandig det almen-
menneskelige Fællesskab for Øje. Især
i de mindre Samfund, hvor vi i
højere Grad end i de større er henvist
til at søge Belæring uden for vor egen
Kres, har denne Følelse være vaagen
og har dannet Baggrunden for de betydningsfulde Bidrag til Videnskabens
Fremskridt, som de smaa Kultursam-
fund ofte har kunnet give.

Om den ærefulde Indsats, hvormed
vort eget Land gennem Tiderne har
bidraget til den videnskabelige Forskning paa de forskelligste Omraader,
vidner Navne som Tycho Brahe, Steno,
Rømer, Rask, Ørsted og Finsen. Hvad
saadanne Mænd har kunnet yde, hviler
igen paa den Interesse for Videnskaben, der saa længe har været levende
i det danske Samfund, og den Offervilje, der stedse har været lagt for
Dagen, naar det gjaldt Deltagelsen i
Arbejdet paa Kulturens Fremskridt.

Ethvert dansk Barn har med Stolthed lært om den enestaaende Gavmildhed, som Kong Frederik II udviste,

12

da han for at formaa Tycho Brahe, vort Universitets første Søn af Verdensry, til herhjemme at fortsætte sit for Astronomiens nyere Udvikling grundlæggende Arbejde, stillede Hjælpemidler til hans Raadighed i et Omfang, der er næsten uden Sidestykke i Videnskabens Historie og vel kun kan sammenlignes med de Midler, for hvilke de rige amerikanske Institutioner i vore Dage lader bygge kæmpemæssige Observatorier til Udforskningen af Himmelrummets fjerneste Egne.

Med Taknemmelighed maa vi ogsaa tænke paa de efter vore Forhold betydelige Midler til Støtte af videnskabelig Forskning, vort lille Land er kommet til at raade over især ved det Forbillede paa Borgersind, som J. C. Jacobsen gav, da han skænkede hele sin store Virksomhed til dette Formaal og i Gavebrevet fandt saa smukt Udtryk for den Gæld, hvori det danske Samfund staar til H. C. Ørsted, der ikke alene ved sine Opdagelser kastede Glans over vort gamle Universitet, men hvis Begejstring og Virketrang vi tillige skylder Tilblivelsen af saa mange af vore andre Institutioner til Kundskabs Udbredelse og Udnyttelse.

Hvad dansk Indsats paa vidt for-

13

skellige Omraader har betydet for Videnskabens Udvikling, blev ogsaa paa lykkeligste Maade understreget, da den danske Stat efter Verdenskrigen oprettede Rask-Ørstedfondet med det Formaal yderligere at fremme det Samarbejde mellem danske og udenlandske Videnskabsmænd, der gennem lange Tider har været saa afgørende for Opretholdelsen af et højt Stade herhjemme paa alle Videnskabens Felter.

Den umiddelbare Forening af Aabenhed over for alle virkelige Fremskridt paa Kulturens Omraade og Fastholden ved de menneskelige Værdier, der i Ly af vore egne Traditioner er skabt herhjemme, turde maaske være det mest betegnende for dansk Videnskab som for dansk Aandsliv overhovedet. Lad os haabe, at vi ogsaa i Fremtiden vil kunne bevare den Frihed til at udvikle vor egen Indstilling, der er nødvendig for, at vort Samarbejde med alle de andre Folkeslag stadig kan bære Frugt.

Niels Bohr.

TRANSLATION

Besides being a centre for the introduction of young people to the sum of knowledge one had obtained at any time, our university should from the very start also be an abode for scientific research. It is impossible to draw a sharp boundary between research and dissemination of knowledge in so far as all deep knowledge by its nature has itself a seeking character.

The aim of science, to increase and order our experience, constantly confronts us with tasks that bring the question of the nature of human knowledge into sharp focus. It is very far from the case that the framework into which experience can be arranged is given us in advance. On the contrary, it is only through experience itself that the regularities making possible our comprehensive view of the multiplicity of the phenomena gradually come to our knowledge.

Thus, hand in hand with the harvesting of experience in all fields of knowledge, a constant development of the structure of our concepts has taken place through the ages, characterized just as much by the sharpening of the weapon of our thinking as by the deepening of our view of man's own position in existence.

What has it not meant in this regard that we no longer assume as a matter of course that we find ourselves in the centre of the universe but, after the victory over deeply rooted prejudices gained by the science of the Renaissance, have come to realize that we live on a small globe which, like the other planets, is a satellite of our sun in its movement among myriads of glowing and extinguished stars of the universe.

Also the clarification of so many of the general laws valid for natural phenomena, which form the basis of the whole wonderful technical development of our time, rests, although this is often forgotten, on the vigilance of scientists as regards the so easily overlooked hints that new experience can give us about the insufficiency of the viewpoints which had been adequate for the comprehension of earlier experience.

Nowadays it is perhaps also difficult to appreciate sufficiently how fundamentally the idea of evolution, so strongly opposed when it appeared, has changed our view of the place we ourselves occupy among the Earth's wonderfully rich multiplicity of living beings. How many barriers had not to be overcome until the methods developed in the natural sciences found rational application in the study of the phenomena of life and thus made possible the flowering of medicine in our time so essential for the welfare of humanity.

How deep and penetrating a lesson have we not gained in the end about the conditions for the flourishing of human life through the humanities, which have their roots in the old interest in the language and the history of the various peoples. The study of the life of human societies in diverse areas on Earth in the past and in the present opens our eyes perhaps more than anything else to the possibilities for cultural development which precisely the work to increase our knowledge, in constant battle against old and new prejudices, holds in prospect for all mankind.

*

The rapid development in all fields of science and technology, which more than ever prevents the individual from holding together all human knowledge, has in no way obscured, but much rather emphasized, the unity of all knowledge for which precisely universities stand as a symbol. Not only has cooperation between the cultivators of seemingly vastly different branches of knowledge proved itself ever more fruitful, but also the boundaries between many fields of science, which have hitherto been strictly separate, have in the course of development become indefinite, indeed have often disappeared entirely.

We meet this situation in all spheres of science. Here at home it would seem particularly obvious for us to consider to how high a degree the study of our rich ancient relics, followed by the whole population with such great interest, has been advanced by the steadily growing cooperation between archaeologists, geologists, zoologists, botanists and chemists, which has contributed so much to the comprehensive elucidation of the circumstances under which our ancestors lived in distant times.

The development of new methods of research has enriched our knowledge and widened our horizon in all fields of science. Within natural science especially the insight into the world of atoms, hitherto closed to human experience, which the unsuspected advance of experimental technique has given us, led to a complete melting together of large fields of knowledge and allowed us in a far-reaching manner to realize the old Greek thinkers' bold dreams of bringing all description of nature back to contemplations of pure numbers.

On the basis of our conceptions of the structure of atoms we recognize how simple physical laws of nature are mirrored in the rich experience of chemistry, so important for the study of inorganic nature as well as of the functions of organisms. At the same time, as so many times before, astronomy has brought us the most striking evidence of the validity of these laws in the whole universe. Even the most distant celestial bodies have proved to be built up of the same atoms as the substances here on Earth and in the course of the development of the stars we rediscover the effects of the same atomic transformations which we first learnt about from the experiments in our laboratories.

*

As has happened time after time, when our knowledge has been increased in an unexpected fashion, concepts and viewpoints hitherto standing as inviolable foundations have nevertheless proved too narrow for the comprehension of experience, and before apparent contradictions in the new experience could be explained, it has even been necessary in our time to take up the age-old philosophical problem of the relationship between phenomenon and observation for renewed discussion.

In this connection I hardly need remind you of the rounding off of our world picture gained through the theory of relativity, which has clarified how the use of even the most elementary concepts such as space and time, in a hitherto unrecognized way, are conditional upon by the standpoint of the observer. In the new areas of experience opened by atomic research, order could only be brought about after the conditions for the suitable application of the very concept of causality were also placed in a new light.

The question of cause and effect can only rightly be asked when we, as is the case in all everyday events, can disregard entirely the influence of the observation on the behaviour of the objects. By the very nature of the laws valid for the individual atomic processes we cannot, however, distinguish at all between the behaviour of the atoms themselves and their interaction with the means of observation indispensable for the description of the phenomena. The comprehension of the experience lies therefore outside the reach of the principle of causality and has necessitated the introduction of more general viewpoints, on whose formulation the abstractions of mathematics, developed to such a high art, have once more been of invaluable help for the description of nature.

The recognition of the limited applicability of the concept of causality does not at all imply a renunciation of a deeper understanding of natural phenomena, but, on the contrary, underlines the general presuppositions of all human thought. Not least is it known from self-observation how impossible it is to distinguish sharply between the object of study and the observing subject itself. The mode of description which we have to use in psychological studies

[549]

therefore shows a conspicuous similarity with the so-called complementarity viewpoint developed in atomic physics.

Most strongly, however, the new lesson reminds us of the wisdom that, already in the childhood of civilization, many of the great thinkers displayed when, in connection with their pondering over the riddles of existence, they emphasized again and again that only the recognition that, in the changing play of life, we are actors as well as observers, lets us glimpse a harmony of existence which by its nature disappears as soon as we try to press it into the framework of narrower viewpoints.

*

However unworldly the work of a scientist may often seem when it has only little connection with the demands of everyday life and moves in fields that lie far away from the arena of human passions, it is nevertheless the researcher who does not let himself be bound by other considerations than the recognition of truth, that builds up and supports the solid base wherefrom one alone can hope to see ever more clearly man's place in existence and the possibilities for the development of culture.

What the individual can contribute to this purpose is infinitely modest, and we must often think of how Newton himself compared his work with a boy's game on the beach collecting the beautiful stones that the ocean washes ashore. In hardly any other field do we feel so strongly as in scientific research that what is involved is jointly to foster something which lives its own life, independent of the individual, and gets its whole strength from humanity's common striving for knowledge.

For the sustenance of the spirit marking a university such as ours, it is, of course, not least the thirst for knowledge and the industry of the young people which form the basis wherefrom constant renewal is gained. The feeling that young and old, each in their way, work towards the same goal and serve the same cause is perhaps the most valuable aspect of the life taking place in the lecture halls and laboratories of the universities and the aspect leaving the deepest impression on anyone who has taken part in it for a shorter or longer time.

As a result of cooperation continued through the ages and extended to all peoples, the proud edifice of science constantly reminds us of the fellowship of all mankind. Especially in the smaller societies, where to a greater extent than in the larger ones we have to look for lessons outside our own circle, this feeling has been vigilant and has formed the background for the important contributions to the advance of science that small civilized societies have often been able to provide.

Names[1] such as Tycho Brahe,[2] Steno,[3] Rømer,[4] Rask,[5] Ørsted and Finsen[6] testify to the honourable effort whereby our own country has contributed throughout the ages to scientific research in the most diverse fields. What such men have been able to contribute depends in turn on the interest in science which so long has been alive in Danish society, and the willingness to sacrifice which has always been manifested when it came to participation in the work for the advance of culture.

Every Danish child has learnt with pride about the unique generosity shown by King Frederik II[7] when, in order to persuade Tycho Brahe, our university's first son of world renown, to continue his work here at home, fundamental for the recent development of astronomy, he placed facilities at his disposal to an extent which is almost unparalleled in the history of science, and which probably can only be compared with the funds for which wealthy American institutions nowadays build gigantic observatories for the investigation of the most distant regions of the universe.

We must also think with gratitude of the, by our standards, considerable means for the support of scientific research that our small country has come to have at its disposal especially through the example of public spirit shown by J.C. Jacobsen, when he donated his whole great enterprise for this purpose, and in the deed of gift found such a beautiful expression for the indebtedness of Danish society to H.C. Ørsted, who not only through his discoveries lent lustre to our old university, but to whose enthusiasm and industry we also owe the establishment of so many of our other institutions for the dissemination and utilization of knowledge.

What Danish efforts in widely different fields have meant for the development of science was also underlined in the most fortunate manner when the Danish State established the Rask–Ørsted Foundation after the World War

---

[1] Bohr dwells in some more detail on these names in *Danish Culture. Some Introductory Reflections* in *Danmarks Kultur ved Aar 1940*, Det Danske Forlag, Copenhagen 1941–1943, Vol. 1, pp. 9–17. Reproduced in Vol. 10, pp. [253]–[261] (Danish original) and [262]–[272] (English translation).

[2] (1546–1601).

[3] Nicolaus Steno, or Niels Stensen, (1638–1686), Danish geologist and philosopher. See Vol. 10, p. [268].

[4] Ole Rømer (1644–1710).

[5] Rasmus Rask (1787–1832), prominent Danish language scholar regarded as a founder of comparative linguistics.

[6] See p. [516], ref. 25.

[7] (1534–1588), King of Denmark and Norway 1559–1588.

with the purpose of furthering even more the cooperation between Danish and foreign scientists, which over long periods of time has been so decisive for the maintenance of a high standard in all fields of science here at home.

The immediate combination of openness towards all real advances in the field of culture and the adherence to the human values created here at home, sheltered by our own traditions, should perhaps be the most characteristic feature of Danish science, as of Danish intellectual life altogether. Let us hope that we will be able also in the future to keep the freedom to develop our own outlook, which is necessary to ensure that our cooperation with all the other peoples can continue to bear fruit.

Niels Bohr.

Niels Bohr in 1941.

5. PEACEFUL USES OF ATOMIC ENERGY

XXIV. [PREFACE]

[FORORD]
...fra Thrige **6** (No. 1, February 1953) 2–4

Untitled preface to a special issue of the
journal of the Thomas B. Thrige Factories
dedicated to activities at the Niels Bohr Institute

TEXT AND TRANSLATION

See Introduction to Part II, p. [359].

[555]

THOMAS B. THRIGE

ODENSE · KØBENHAVN

Eftertryk kun tilladt med Kildeangivelse

Nærværende Hefte af „… fra Thrige" behandler udelukkende Spørgsmaal i Forbindelse med Atomforskningen, idet vi har ment, at vore Læsere kunde være interesseret i at erfare lidt om det store Arbejde, der udføres herhjemme paa et Omraade, som i Dag er Videnskab, men som maaske bliver Grundlaget for Morgendagens Teknik. Vi har henvendt os til Professor Niels Bohr og spurgt, om hans og hans Medarbejdere vilde fortælle vore Læsere lidt om de Arbejder, der udføres paa Universitetets Institut for teoretisk Fysik, en Henvendelse, som blev imødekommet med stor Beredvillighed.

Vi vil gerne paa dette Sted rette en varm Tak til Professor Bohr og hans Medarbejdere for det store Arbejde, der er nedlagt i at gøre et vanskeligt Stof tilgængeligt for Læsere uden for de sagkyndiges snævre Kreds. Det er med Stolthed og Glæde, at vi udsender dette Hefte, hvis Indhold vi betragter som endnu et Vidnesbyrd om den fremskudte Stilling, dansk Atomforskning under *Niels Bohrs* Førerskab indtager paa et Omraade, der i Dag betegner Brændpunktet for den internationale Forskning Jorden over.

THOMAS B. THRIGE

Bestræbelserne på efter evne her hjemme at følge med i den løfterige udvikling inden for atomfysikken har ikke alene fra de offentlige myndigheder men også fra private institutioners side mødt velvillig forståelse og højst værdifuld støtte. Denne interesse og offervilje har skabt det nødvendige grundlag for forskningsarbejdet på dette område og har tillige været en stadig kilde til opmuntring for enhver, som har deltaget i dette arbejde.

Et uforglemmeligt minde om en sådan opmuntrende oplevelse var en samtale, som jeg havde med fabrikant Thomas B. Thrige i et af hans sidste leveår. Til min store glæde meddelte fabrikant Thrige mig, at det af ham stiftede fond til fremme af elektroteknikken her i landet ville yde støtte til erhvervelsen af en cyklotron til Universitetets Institut for Teoretisk Fysik, samt at man på Thrigefabrikkerne var beredt til at fremstille den meget store elektromagnet, der danner hovedbestanddelen i dette sindrige og mægtige hjælpemiddel for udforskningen af atomkernernes egenskaber.

Den store og vanskelige opgave med magnetens konstruktion løstes på fortræffelig måde af Thrigefabrikkernes erfarne medarbejderstab under direktør Meyers ledelse, og få år efter stod virkelig cyklotronen færdig på Instituttet og kunne tages i brug for undersøgelser, hvori man i Europa hidtil ikke havde kunnet deltage. Dette var den første begyndelse til den forbindelse med Thrigefabrikkerne, som skulle blive en stadig mere afgørende hjælp for hele den eksperimentelle virksomhed på Instituttet.

I alle disse år har man fra Thrigefabrikkernes side stedse på den mest beredvillige måde stillet erfaring og arbejdskraft til rådighed for apparatkonstruktioner, der efter deres art ofte lå uden for fabrikkernes sædvanlige arbejdsfelt og fabrikationsprogram. Overhovedet ville det næppe have været muligt her i landet på virksom måde at følge med i udviklingen på atomforskningens område uden den effektive støtte fra en stor elektroteknisk fabrikationsvirksomhed, og uden at der bag denne virksomhed stod et fond med det formål at virke i samfundets interesse.

Af artiklerne i dette hefte, hvori en række af Instituttets medarbejdere har gjort rede for dettes organisation og instrumentelle hjælpemidler, vil man få et stærkt indtryk af, hvor uadskilleligt fremskridtene på elektroteknikkens og atomforskningens område er forbundet, og i hvor dyb en taknemmelighedsgæld Instituttet står til den af Thomas B. Thrige stiftede fond og til den af ham skabte store danske industrielle virksomhed.

29. december 1952.

Niels Bohr

TRANSLATION

The current issue of "...fra Thrige" deals solely with questions concerning atomic research, as we have thought that our readers might be interested in learning a little about the great work being done here at home in a field which today is science but which might become the basis for tomorrow's technology. We have approached Professor Niels Bohr and asked whether he and his collaborators would tell our readers a little about the work that is done at the University Institute for Theoretical Physics, an approach that has been met with great willingness.

Here we would like to direct a warm thanks to Professor Bohr and his collaborators for the great work that has been put into making difficult material accessible to readers outside the narrow circle of experts. It is with pride and pleasure that we publish this issue, whose contents we regard as one more token of the advanced position that Danish atomic research, under the leadership of *Niels Bohr*, holds in a field that constitutes today the focal point for international research all over the world.

THOMAS B. THRIGE

The endeavours here at home to keep up, according to our ability, with the promising development within atomic physics have met accommodating understanding and most valuable support not only on the part of public authorities but also from private institutions. This interest and readiness to give has created the necessary basis for research work in this field and has in addition been a constant source of encouragement for anyone who has participated in this work.

An unforgettable memory of such an encouraging experience was a conversation I had with Thomas B. Thrige, the manufacturer, in one of the last years of his life. To my great pleasure, manufacturer Thrige informed me that the foundation established by him for the advancement of electrical engineering in this country would provide support for the acquisition of a cyclotron for the University Institute for Theoretical Physics, and that the Thrige factories were prepared to construct the very large electromagnet that constitutes the main element of this ingenious and powerful tool for the investigation of the properties of atomic nuclei.

The large and difficult task of constructing the magnet was solved in an excellent way by the experienced staff at the Thrige factories under the leadership of Director Meyer,[1] and a few years later the cyclotron was actually standing ready at the Institute and could be put to use for investigations in which it hitherto had not been possible to participate in Europe. This was the first beginning of the connection with the Thrige factories, which was to become an ever more decisive help for the whole experimental activity at the Institute.

In all these years, the Thrige factories have always, in the most accommodating way, made experience and labour available for construction of apparatus, which by its nature often lay outside the usual field of work and programme of manufacture for the factories. It would hardly have been possible in this country to keep up at all in an active way with the development in the field of atomic research without the efficient support from a large electrotechnical manufacturing enterprise, and without the existence behind this enterprise of a foundation with the purpose of acting in the interest of society.

From the articles in this journal, in which several of the Institute's collaborators have described its organization and the instrumental resources, one will receive a strong impression of how inseparably the advances in the fields of electrotechnology and atomic research are intertwined, and how deep a debt of gratitude the Institute owes to the foundation established by Thomas B. Thrige and to the great Danish industrial enterprise created by him.

29 December 1952.

Niels Bohr

---

[1] See p. [517], ref. 35.

XXV. GREATER INTERNATIONAL COOPERATION
IS NEEDED FOR PEACE AND SURVIVAL

"Atomic Energy in Industry:
Minutes of 3rd Conference October 13–15, 1954",
National Industrial Conference Board, Inc., New York 1955, pp. 18–26

See Introduction to Part II, p. [361].

International Developments in Atomic Energy: I

Chairman: GORDON DEAN, Lehman Brothers, New York City, and Former Chairman, United States Atomic Energy Commission

Welcome: JOHN S. SINCLAIR, President, National Industrial Conference Board

Address: <u>Greater International Cooperation is Needed for Peace and Survival</u> NEILS BOHR, Director, Institute for Theoretical Physics, Copenhagen, Denmark

Doctor Bohr

WELCOME BY JOHN S. SINCLAIR

ON BEHALF of The Conference Board, its Trustees, and Staff I want to extend a
brief but warm welcome to all of you. And in particular we want to welcome
our friends and guests from abroad and from South America and Canada. It is
a pleasure and a real privilege to have you participate in this Third Conference of ours devoted to the "Peaceful Uses of Atomic Energy." I sincerely
hope that all of you, speakers and those attending the sessions during the
next three days, will find it a rewarding and informative experience.

INTRODUCTION BY THE CHAIRMAN

IT IS a pleasure to preside here today at this opening luncheon of the National
Industrial Conference Board's 3rd Annual Conference on "Atomic Energy in Industry."

At the outset, I shall introduce to you some of our distinguished
guests at the head table. First of all I should like to introduce Mrs. Niels
Bohr; Mr. Clyde Rogers, Vice-President of The Conference Board; Dr. R. G.
Shuttleworth of the Union of South Africa; Mr. Robert LeBaron, formerly
Assistant to the Secretary of Defense for Atomic Energy; Mr. Maxil Ballinger,
who plans these conferences for The Conference Board; Dr. Werner Heisenberg,
from West Germany; Dr. Edward Hess, also of West Germany; Dean E. Blyth Stason,
Dean of the Law School of the University of Michigan; Gov. Pierre Ryckmans,
who is President of The Belgian Atomic Energy Center; Admiral Alvaro Alberto,
who heads the Brazilian Atomic Energy project; His Excellency, Ambassador
Hendrik de Kaufmann of Denmark; Dr. Francis Perrin, who is High Commissioner
of the French Atomic Energy Commission; Dr. Giorgio Valerio, of C.I.S.E. and
Societa Edison of Italy, and Dr. Mario Sylvestri, also of Italy; Dr. H. D.
Smyth, former Commissioner of the United States Atomic Energy Commission;
Dr. Anker Hansen, of the Danish Consul General's office in New York City;
Dr. Marvin Fox of the Brookhaven National Laboratory, and Mr. G. Clark Thompson,
Director of the Division of Business Practices of The Conference Board.

I have had few honors in my life which even begin to approach the
honor of introducing Niels Bohr to an American audience. His work is known to
all of you, and yet I'm afraid that I would be quite remiss if I simply con-
tented myself with saying, "Here is Niels Bohr." It is just possible that you
may not know that as far back as 1907 he received an award from the Royal
Danish Academy of Sciences. His studies with Thomson and Rutherford in England,
I feel sure, marked a high point in his life, and yet that was in 1911. In
1921, he became head of the Institute for Theoretical Physics at the University
of Copenhagen, and in 1922 he was awarded the Nobel Prize in Physics.

The Conference Board

Dean - 2

Dr. Bohr has been on many visits to America since he first gave lectures at Yale in 1923. After visits in 1933 and in 1937, he worked at Princeton in 1939 on the theory of nuclear fission with Professor John E. Wheeler. After his escape from German-occupied Denmark in 1943, he went to England and then came to the United States, where for a time he worked at the Los Alamos Scientific Laboratory.

Dr. Bohr has been awarded many scientific and civilian assigns, the most exclusive among them being the Danish Knighthood of the Elephant in 1947. He is a member of fifty-four scientific societies, five of these are in the United States. He has received honorary degrees from eighteen universities including Providence, California, and Princeton, and is to be awarded another by Columbia University within a few weeks. Dr. Bohr is President of the Royal Danish Academy of Sciences.

Those who were at Los Alamos during the war said to me, "The biggest event was the arrival of Bohr. It was as though the great white father had appeared." One of the Los Alamos scientists said to me recently, "I may be prejudiced or I may not have all the facts, but in my own opinion, Bohr is the greatest man of science of all time."

To think straight, to think boldly, to think unafraid - what a treat this is, and yet these are the things that Dr. Bohr has always done. So you bring to us, Dr. Bohr, here in America, some things that we need and things that we cherish, and we hope that you will speak out. Whatever you may say we shall listen to it, for we look upon you as an intelligent man, a wise man, a kind and inspiring man and a brave man. We welcome you to this conference and to these shores. Dr. Bohr.

The Conference Board

GREATER INTERNATIONAL COOPERATION IS NEEDED FOR PEACE AND SURVIVAL

Niels Bohr, Director, Institute for Theoretical Physics, Copenhagen, Denmark

ON BEHALF of the scientists from abroad who have been so kindly invited to take part in this Conference on the "Peaceful Uses of Atomic Energy," arranged by the National Industrial Conference Board, I want to express our deep appreciation and gratitude. It is indeed a privilege and a welcome opportunity even modestly to contribute to this great development which entails such unique promises for the welfare and prosperity of mankind.

I must confess that I felt somewhat embarrassed by being asked to speak to you at this great assembly. You cannot expect from me expert knowledge about technological developments, nor any considered opinion about the economic problems which are developing. However, I thought it might be appropriate on this occasion, where a beginning is being made for further international cooperation in the new field of atomic energy, for me to say something about the decisive part international collaboration has played in the development of modern physical science and perhaps to recall some treasured personal remembrances.

The exploration of the world of atoms which we have witnessed in the last fifty years is a true adventure. It has been a journey on untried ground where scientists from many countries have traveled together and constantly met with new marvels, but also with surprises which have given us a deeper appreciation of our position as observers of that nature of which we are a part ourselves. In such developments where new experimental evidence constantly inspired imagination which, in turn, offered guidance in the search for further progress, it has been most helpful to have the cooperation of scientists with different experiences, and even belonging to different schools of thought.

In such efforts where everyone, so to speak, stands on the shoulders of others, the question of individual merit and accomplishment is, of course, a dubious one. We can at most speak only of different degrees of preparedness to grasp and utilize the opportunities offered us. I say all this because, when I shall mention a few names I am fully aware that I am leaving out far more, and my intention is only to point to the different origins of common progress.

Modern development of atomic science grew indeed from many roots, and, as everybody knows, the idea that matter is composed of atoms goes back to antiquity. Still, in spite of all fruitfulness of atomic ideas in physics and chemistry, the existence of atoms was generally considered as hypothetical until fairly recent times. The situation changed, however, with the fundamental discoveries which followed each other so quickly at the close of the last and the beginning of the present century. I need not remind you how the exploration of the beautiful phenomena accompanying electric discharges in rarefied gases led Roentgen to his discovery of the penetrating radiation. That proved to be not only a highly efficient tool in physical investigations but, by allowing us so to say to see through human bodies, also soon became invaluable in medicine and surgery. We also recall how shortly afterwards Becquerel discovered the natural radioactivity of heavy elements, and especially the impression created when Mme. Curie in 1898 succeeded in isolating from uranium minerals the wonderful substance named radium.

The Conference Board

Bohr - 2

From these years also came the discovery of the electron as a constituent of the atoms of all elements. It was recognized not only that in metallic conduction we have to do with a flow of electrons between the far heavier metal ions, but also that evidence of the binding of electrons within the atoms could be ascertained in many other properties of matter. Especially instructive in this respect was the Zeeman effect, which showed that the mechanism of spectral emission was influenced by magnetic fields, in a way reflecting the properties of the beams of free electrons produced in discharge tubes.

At the beginning of the century, the problem of the electronic constitution of matter was explored by scientists from many countries, and rapid progress was achieved not least in the internationally composed group working in the famous Cavendish Laboratory in Cambridge under the inspired leadership of J. J. Thomson. In particular, I may recall how Thomson succeeded in the first approximate estimate of the number of electrons in the neutral atoms of the elements.

Soon, however, the whole development was to be furthered beyond all expectations by the researches of Ernest Rutherford who, first as a member of the Cambridge group and later leading groups himself in Montreal and in Manchester, with unfailing intuition step by step unraveled the remarkable laws of radioactive transmutations and the properties of the radiation accompanying such transmutations. The climax of these researches was reached in 1911, when Rutherford by the discovery of the atomic nucleus created a new epoch in physical science.

Rutherford's discovery presented us with a picture of the atom which at once offered the explanation of many fundamental properties of matter. Thus, the invariability of the elements in ordinary physical and chemical processes is immediately accounted for by the circumstance that the nucleus, which, within an extremely small volume contains almost the whole mass of the atom, is not affected by such processes where we have to do only with changes in the binding of the extra-nuclear electrons. Since this binding is essentially governed by the electric charge of the nucleus we understand, moreover, the existence of so-called isotopes containing nuclei with equal charge but different masses and exhibiting very closely the same physical and chemical properties. The general appearance of such isotopy, which was first discovered in Soddy's researches on the chemical properties of radioactive elements, was to be brought out in the succeeding years, especially by Aston's refinement of the methods of mass spectroscopy.

Everybody in the group in Manchester which it was my good fortune to join just at that time, was of course eagerly occupied with the great program opened by Rutherford's discovery. Already in Cambridge, where I worked in 1911, I had heard much of Rutherford's genius and vigorous personality, and I remember how an old assistant to Thomson told that he had seen many young men working in the Cavendish laboratory but nobody who could swear at his apparatus like Rutherford. Still, the background for Rutherford's unique gifts as a pioneer and leader in research was not only his vision and perseverance, but also the true human attitude which in spite of such impatience with apparatus, or rather with himself, always made him prepared to take a helpful interest in the work of his collaborators and to listen patiently to ideas however vaguely expressed by any young man in whom he perceived a serious striving.

The Conference Board

Bohr - 3

     With Rutherford in those years worked physicists from many different countries. Foremost among the British was Moseley, who in the short span of years that was left to him before he fell in the first World War, carried out the famous investigations on the high frequency spectra which allowed the unambiguous determination of the charge of the nucleus of any element. Another young Englishman who started his work in Manchester was Chadwick who, years after, when he with Rutherford had gone to Cambridge, discovered the neutron, which is not only an essential constituent of atomic nuclei but has also proved a powerful tool in nuclear research.

     Among the physicists from other countries who worked with Rutherford in those years was Geiger, whose name is well-known due to the counting device now so widely used and the tickering of which in any physical laboratory makes us vividly aware of the atomic world, until recently so remote from human experience. Another now-famous scientist was Hevesy who, as a quite young man in Rutherford's laboratory, conceived the idea of using radioactive isotopes as tracers of stable elements in physical and chemical processes. This ingenious method of so-called marked atoms was soon in the hands of Hevesy himself and many other scientists, not least in this country, and was to become a most fruitful source of information about organic metabolism.

     The greatest fascination of Rutherford's model of the atom for a theoretical physicist was the challenge it presented of a simple comprehension of the immense amount of experimental evidence concerning the properties of the elements collected through the years by physicists and chemists. Again contributions from the most various sources proved decisive, and above all a way to progress was indicated by the discovery of the universal quantum of action by which Planck in the first year of this century initiated a new approach to natural philosophy. In fact, the obvious impossibility of accounting for the inherent stability of atoms on the basis of Newtonian mechanics suggested that in any change in the state of the atom we have just to do with a complete transition between two among a discrete sequence of so-called stationary quantum states. In particular, it was found possible, in a way reminding of Einstein's original explanation of atomic photo effects, to account for the remarkable laws governing line spectra, unraveled especially by Rydberg. I need also hardly recall how these ideas soon after were most convincingly supported by the ingenious experiments of Franck and Hertz on the excitation of spectra by impact of electrons on atoms.

     Even if steady progress as regards the quantum theory of atomic constitution was obtained in the following years, by the elaboration of quantization rules by Sommerfield and his school and not least by Pauli's introduction of the exclusion principle, the difficulties of developing a consistent description of atomic phenomena was only gradually overcome by the establishment of a rational generalization of the classical physical theories involving a radical revision of the very observational problem. This achievement was a result of the concerted efforts of a whole generation of theoretical physicists, among whom names like Broglie, Heisenberg, Schroedinger, Born, and Dipac will be familiar to you all.

     Many more, indeed, contributed to the development, and using a humorous distortion of a metaphor of a great statesman speaking about how much a few brave men had done for a whole people, one may perhaps say that hardly in the history of science have so many worked assiduously together on a point related to something so small. In Copenhagen, where in the years between the wars we had the privilege of the collaboration of some of the most eminent of the younger theoretical physicists from the different countries, it was an

Bohr - 4

unforgettable experience to witness how, when the problems became ripe for clarification, a wide range of phenomena embracing intricacies of spectral structure, chemical bonds, permanent magnetism, and even radioactive nuclear transformations step by step found detailed explanation.

Returning to the specific problems of nuclear physics, we recall above all how Rutherford in spite of his heavy obligations during the first World War succeeded in 1918 in demonstrating the possibility of inducing nuclear transmutations by the energetic particles emitted by radium, and thus initiated the revolutionary development often picturesquely called modern alchemy. Continual progress in such researches was achieved, especially in Cambridge, where Rutherford followed Thomson as the leader of the Cavendish Laboratory, and a milestone was passed when Cockroft and Walton in 1932 succeeded in splitting lithium nuclei by bombardment with protons accelerated by means of a high-voltage generator constructed for the purpose.

This was indeed a most remarkable event, and attention was not least focused on the fact that the kinetic energy of the nuclear fragments was about a hundred times greater than that of the impinging protons. The source of this energy is the reorganization of the interplay between the nuclear constituents which are held together by great forces acting at small distances, and the effects observed were found to be in complete agreement with the universal relation between energy and mass predicted by Einstein as one of the many consequences of the relativity theory which has so greatly widened our horizon and given our world picture a unity surpassing all previous expectations. Incidentally, we may recall that the apparent constancy of mass in ordinary chemical processes is due to the comparatively small energy changes accompanying the reorganization of the binding of the extra-nuclear electrons in such processes.

The experiments in the Cavendish Laboratory initiated discussions of the prospects of releasing nuclear energy on a large scale, and I remember the stir in the public press and the anxiety voiced as regards catastrophic consequences of the continuation of the experiments. Still, such anxieties were premature since under the circumstances by far the greater amount of the energy of the bombarding proton beams as well as of the emerging high-speed nuclear fragments was lost in collisions with the atomic electrons and the matter through which they passed and in which they, after capture of electrons, would settle down as neutral atoms taking part in the ordinary heat motion. Nuclear reactions of such type can therefore only maintain themselves in bodies of exceedingly high temperature like the sun, where the protons remain free and keep large average velocities. Indeed, today, through the work initiated by Gamow and completed by Bethe, we know about the nuclear transmutations which are the sources of that radiation which through a billion years has upheld organic life on the earth but the origin of which until a few years ago was a complete mystery.

Such results were the outcome of a development again based on contributions from many different sides. I need hardly recall how much nuclear investigation has been furthered by the construction of powerful accelerators of atomic particles, like van de Graff generators based on the same principles as the electrification machines used in Franklin's time, or the cyclotron, first constructed by Lawrence and in which charged particles circulating in a magnetic field are speeded up by an alternating electric tension synchronous with the revolution of the particles. These devices permitted the extension

The Conference Board

Bohr - 5

of the study of nuclear reactions by proton bombardment. Further, most interesting results were achieved when new atomic projectiles became available after Urey's discovery of the heavy isotope of hydrogen whose nucleus, the so-called deuteron, consists of a proton and a neutron.

A most important step in nuclear research was also the discovery of artificial radioactivity by Mme. Curie's daughter and her husband Joliot. Especially we remember how Fermi, by means of neutrons liberated in nuclear reactions, was able to produce radioactive isotopes of almost all elements which have promoted the development of the tracer methods in biological research to such great degree. Above all, however, the study of neutron-induced radioactivity in the heaviest elements like uranium led Hahn and Strassmann to the discovery of nuclear fission, by which the uranium nucleus is split in two fragments ejected with energies amounting to about a hundred million electron volts.

This discovery initiated an intense development, again furthered by international cooperation. As is well known, the essential features of the mechanism of uranium fission were elucidated by Meitner and Frisch, and on my visit to Princeton in the spring of 1939, where I had the pleasure of collaborating with Wheeler, the theory was further developed and led to the conclusion that fission by slow neutrons was possible only in the rare uranium isotope with mass number 235. Of most decisive importance, however, was the discovery, made independently by Joliot and his co-workers in Paris and by Fermi and collaborators at Columbia, that the fission process was accompanied by the liberation of neutrons in sufficient number to induce further fission processes. The actual realization of such chain reactions, by which energy is released on a large scale was, as all know, accomplished under Fermi's direction in Chicago in 1942.

I need not comment on the far-reaching consequences of this great achievement, which often is referred to as the initiation of the atomic age in the history of mankind, but before I conclude I wish to say a few words about the continuation of international cooperation in the fields of atomic physics which, although they do not point to immediate practical prospects, involve great promises as regards the clarification of the fundamentals of physical science. I think of the discovery of new elementary particles with astounding properties that are revealed by the study of the so-called cosmic rays. The suggestion of the existence of such particles, known as mesons and first observed by Anderson and Blackett, came indeed from Yukawa's ingenious attempt to account for the strong forces which keep nuclei together and possess no analogue in ordinary physical experience.

In the years after the war new types of mesons with different properties have been observed, especially by Powell and his collaborators in Bristol, and the experimental evidence rapidly grew when it was found that mesons could be directly produced by means of the great cyclotrons in this country, and further advances have recently been obtained with the completion of the huge cosmotron in Brookhaven. In Europe a great desire has been felt for participation in the investigation of these fundamental problems, and as you may know twelve European countries have ratified a convention with the purpose of creating a center for nuclear research in Geneva equipped with the necessary costly facilities. In the preparation of the plans, expert advice from American scientists has been invaluable and we look forward to a most intimate future collaboration.

The Conference Board

Bohr - 6

Scientific knowledge is surely one of the greatest assets of mankind, and the common search for truth presents at the same time a unique opportunity of furthering that understanding between nations which now is more necessary than ever if unprecedented dangers to civilization are to be averted and all people can reap together in peace the fruits which the progress of science offers for the promotion of human welfare.

All increase of knowledge and our mastery of the forces of nature imply, of course, a greater responsibility, but we must trust that the present serious challenge will be met in the proper spirit, and we greet especially the initiative for promoting international cooperation in the field of atomic energy as a most important step in this direction.

On behalf of the foreign guests may I again express our gratitude for your invitation to participate in this congress and our hope that it may contribute decisively to our great common goal.

The Conference Board

XXVI. THE PHYSICAL BASIS FOR INDUSTRIAL USE OF THE ENERGY OF THE ATOMIC NUCLEUS

DET FYSISKE GRUNDLAG FOR INDUSTRIEL
UDNYTTELSE AF ATOMKERNE-ENERGIEN
Tidsskrift for Industri (No. 7–8, 1955) 168–179

Lecture at the National Meeting of Industry 17 March 1955

TEXT AND TRANSLATION

See Introduction to Part II, p. [364].

THE PHYSICAL BASIS FOR INDUSTRIAL USE
OF THE ENERGY OF THE ATOMIC NUCLEUS (1955)

Versions published in Danish

A   Tidsskrift for Industri (No. 7–8, 1955) 168–179
B   *Udnyttelse af atomenergien*, Tidens Stemme **11** (March 1956) 1

B contains only the last paragraph of *A*.

Det fysiske grundlag for industriel udnyttelse af atomkerne-energien

Foredrag af professor, dr. NIELS BOHR på landsindustrimødet den 17. marts 1955

Som følge af den tekniske udvikling, der så gennemgribende har omformet samfundslivet, spiller spørgsmålet om kraftkilder en afgørende rolle. De naturlige kul- og olieforekomster, der hidtil i stedse større omfang er blevet udnyttet for dette formål, vil imidlertid være opbrugt indenfor en overskuelig fremtid, og betydningen af at finde nye kraftkilder har derfor allerede i længere tid tiltrukket sig opmærksomheden i vide kredse.

Alle véd, at man i flere af de større lande, der har haft adgang til de her først og fremmest i betragtning kommende uranmineraler og besiddet de nødvendige store tekniske hjælpemidler, allerede nu planlægger bygningen af atomkraftværker. Selv om det endnu er vanskeligt at bedømme de økonomiske vilkår for sådanne værkers drift, føler man sig overbevist om, at forholdene i løbet af få år vil stille sig langt mere gunstigt som følge af de hastigt voksende erfaringer og tekniske fremskridt på dette nye område. En deltagelse i denne udvikling er selvsagt af største betydning for vort land, der er ganske uden kul- eller olieforekomster, og som heller ikke besidder vandkraft i nogen større udstrækning.

Imidlertid har også manglen på uranmineraler hidtil hindret os i aktive forberedelser på løsningen af denne opgave, men forholdene har i denne henseende ændret sig væsentligt som følge af det af De Forenede Nationer støttede initiativ til fremme af det internationale samarbejde på den fredelige udnyttelse af atomenergien. At vi herhjemme straks kunne tage fat på forberedelserne til deltagelse i et sådant samarbejde skyldes tillige et initiativ fra *Thomas B. Thriges fond*, der i anledning af Thrige-fabrikernes 50-års jubilæum tilstillede *Akademiet for de tekniske Videnskaber* en betydelig understøttelse til igangsættelse af sådanne forberedelser. Akademiet nedsatte straks et udvalg til at arbejde med sagen og især til at indhente oplysninger om undersøgelser foretaget i udlandet.

På grundlag af resultaterne af udvalgets arbejde skønnedes det, at sagen havde et så realistisk perspektiv, at større udgifter til opgavens fremme var forsvarlige og at videre kredse burde inddrages i overvejelserne vedrørende dens videreførelse. Disse synspunkter, der deltes af regeringen og repræsentanter for de lovgivende myndigheder, foranledigede statsministeren til at træffe bestemmelse om at nedsætte *Atomenergikommissionen* og søge en foreløbig bevilling til dens arbejde med sigte på udformningen af forslag til statsmyndighedernes orientering. Foruden Akademiets arbejdsudvalg og fysikere fra landets universiteter og højere læreanstalter fandt man det formålstjenligt at opfordre et antal repræsentanter for det industrielle erhvervsliv med særlig teknisk og organisatorisk sagkundskab til at indtræde i kommissionen. Det er naturligvis klart, at der udenfor en kommission med et begrænset antal medlemmer og et så vidtspændende arbejdsfelt vil findes megen kundskab og indsigt, som det vil være værdifuldt, ja nødvendigt at drage nytte af, og det er derfor ved de forhandlinger, der har fundet sted før kommissionens oprettelse, også udtrykkeligt blevet fremhævet, at kommissionen er berettiget til at indbyde sådanne sagkyndige til at deltage i dens møder i det omfang, det skønnes nødvendigt.

Kommissionens første møde vil blive afholdt umiddelbart efter dette store industrimøde, og den vil bestræbe sig for at fremme de tekniske og organisatoriske forberedelser så hurtigt, som forholdene tillader det.

Før dens forhandlinger er videre fremskredet, vil det naturligvis ikke være muligt at redegøre nærmere for de planer, som det vil være kommissionens opgave at udarbejde, og jeg skal derfor idag indskrænke mig til i nogle hovedtræk at give Dem et indtryk af det fysiske grundlag for spørgsmålet om atomenergiens industrielle anvendelse.

Jeg skal springe lige ind i emnet og først minde Dem om, at forestillingerne om, at alle stoffer er bygget op af *atomer*, går langt tilbage i tiden, og at disse forestillinger viste sig mere og mere frugtbare og uundværlige for at forklare stoffernes fysiske og kemiske egenskaber. Indtil begyndelsen af dette århundrede var det imidlertid en almindelig anskuelse, at der blot var tale om en hypotese, idet man troede, at det ikke, på grund af vore sansers grovhed, ville være muligt nogensinde at undersøge og iagttage virkningen af enkelte atomer. Forholdene har jo imidlertid udviklet sig helt anderledes som følge af den vidunderlige forfinelse af den fysiske eksperimentalteknik og på grund af de mange store opdagelser, som fandt sted omkring århundredskiftet. En af disse opdagelser, der skulle få særlig betydning, og som også var den første, der belærte os om eksistensen af nye energikilder, var opdagelsen af de *radioaktive* stoffer.

Jeg vil gerne benytte et sådant stof til at vise en demonstration af, hvorledes man kan iagttage og registrere enkelte atomer. Når jeg nu tager et radiumpræparat frem fra den svære blyafskærmning, bag hvilken det opbevares, og holder det hen i nærheden af en tælleropstilling, hører De nogle klik i højttaleren. Det, der sker, er, at der fra radiumpræparatet udsendes gammastråling, der er af samme art som røntgenstråling, og som ved at ramme tælleren udløser enkelte *elektroner* i denne. Den svage udladning, der derved startes i tælleren, sendes ind i en forstærker, hvorved impulserne bliver forstærket så meget, at de kan registreres på en almindelig telefontæller, og det er klikkene fra denne, De hører i højttaleren. Den her anvendte tæller repræsenterer en temmelig raffineret art af *elektronik*, idet den i øjeblikket er indstillet til ikke at tælle hver enkelt elektron, men kun hver tyvende, hvilket er gjort, fordi det her anvendte radiumpræparat af hensyn til en senere demonstration er så kraftigt, at tællerstødene

ellers ville komme alt for hurtigt efter hinanden til at kunne skelnes.

Vi har her et eksempel på vore dages højt udviklede elektronik, der er baseret på elektronernes opdagelse for et halvt århundrede siden, som dengang blev betragtet som den yderste grænse for videnskaben. Sådan elektronik benyttes jo også i den forfinede automatik, der finder stadig mere udstrakt anvendelse i industrien. Man må her være forberedt på, at en meget stor del af det kontrollerende arbejde, der i fabrikerne nu udføres af mennesker, inden alt for mange år vil blive overtaget af automater, hvilket naturligvis vil rejse nye store problemer for samfundet om fordelingen af arbejde og forøget specialuddannelse.

Jeg vil derefter gå over til at illustrere nogle hovedtræk af atomfysikens udvikling ved hjælp af et antal lysbilleder. På det første (fig. 1) ser De et eksempel på en velkendt metode indenfor den fysiske eksperimentalteknik. Billedet er et tågekammerfotografi, som viser banerne af de såkaldte alfa-partikler, der udsendes fra det radioaktive stof polonium, der ligesom selve radium blev opdaget af madame *Curie* og derfor er opkaldt efter hendes fødeland. I modsætning til elektronerne, der er meget lettere end atomerne, har alfa-partiklerne en masse af samme størrelsesorden som atomerne, og man har desuden ved at opsamle alfa-partikler i et glasrør kunnet påvise, at de danner helium, idet de i virkeligheden er kerner af heliumatomer. En karakteristisk egenskab ved alfa-partiklerne, som tydeligt fremgår af tågekammerfotografiet, er, at de gennemløber retlinede baner på adskillige centimeters længde ved atmosfæretryk i tågekammeret, hvilket hænger sammen med, at de udsendes fra præparatet med uhyre store hastigheder. På deres vej passerer alfa-partiklerne gennem millioner af luftens atomer, som derfor må besidde en meget åben struktur. Under gennemtrængningen vil alfa-partiklerne dog løsrive elektroner fra atomerne, og der dannes langs banen ialt omkring 100 000 ioner. Det er disse ioner, der bevirker, at alfa-partiklernes baner bliver synlige, idet der i tågekammeret findes overmættet damp af vand eller alkohol, som fortættes til små dråber omkring hver af de dannede ioner, således at partiklens bane gennem luften aftegner sig som en tågestribe.

Det næste tågekammerbillede (fig. 2) skal illustrere en af de vigtigste opdagelser, der er gjort i atomfysiken. Man ser, at en af alfa-par-

Fig. 1

Fig. 2

tiklernes baner her har en forgrening, der må være fremkommet ved, at alfa-partiklen har ramt noget, som har været meget tungere end elektronerne, og som er blevet skubbet til højre ved sammenstødet, medens partiklen selv derved er blevet afbøjet til venstre. Det, som alfa-partiklen har stødt sammen med, er en *atomkerne*, og det var iagttagelser af denne art, der førte til *Rutherford's* epokegørende opdagelse af atomkernen i 1911.

Når radioaktive stoffer udsender alfa- eller beta-partikler, kommer disse partikler fra stoffets atomkerner, som derved omdannes til kerner af et nyt stof. Dette illustreres ved det næste lysbillede (fig. 3), som viser de radioaktive omdannelser indenfor den såkaldte radiumfamilie. Der begyndes med en radiumkerne,

$$\overset{4}{\underset{2}{\nearrow}}He \quad \overset{4}{\underset{2}{\nearrow}}He \quad \overset{4}{\underset{2}{\nearrow}}He \quad \overset{0}{\underset{+1}{\nearrow}}e \quad \overset{0}{\underset{+1}{\nearrow}}e$$

$$\underset{88}{\overset{226}{Ra}} \rightarrow \underset{86}{\overset{222}{Em}} \rightarrow \underset{84}{\overset{218}{Ra\,A}} \rightarrow \underset{82}{\overset{214}{Ra\,B}} \rightarrow \underset{83}{\overset{214}{Ra\,C}} \rightarrow$$

Polonium-   Bly-   Vismut-
Isotop.   Isot.   Isotop.

$$\overset{4}{\underset{2}{\nearrow}}He \quad \overset{0}{\underset{+1}{\nearrow}}e \quad \overset{0}{\underset{+1}{\nearrow}}e \quad \overset{4}{\underset{2}{\nearrow}}He$$

$$\underset{84}{\overset{214}{Ra\,C'}} \rightarrow \underset{82}{\overset{210}{Ra\,D}} \rightarrow \underset{83}{\overset{210}{Ra\,E}} \rightarrow \underset{84}{\overset{210}{Ra\,F}} \rightarrow \underset{82}{\overset{206}{Ra\,G}}$$

Polonium-   Bly-   Vismut-   Polonium   Stabilt Bly
Isotop.   Isot.   Isotop.

Fig. 3

som udsender en alfa-partikel, der som allerede nævnt er identisk med en heliumkerne. Derved omdannes radiumkernen til kernen af et nyt stof, radiumemanation, som også udsender en alfa-partikel. Radiums omdannelse foregår temmelig langsomt, idet det varer 1 600 år, før halvdelen af stoffet er omdannet, medens emanationen omdannes meget hurtigere med en halveringstid på 4 døgn. Dernæst kommer i hurtigere rækkefølge en hel serie omdannelser, indtil rækken ender med almindeligt bly, der ikke er radioaktivt. Af de tal, der er anført ved hvert af stofferne, betegner det nederste atomkernens elektriske ladning, medens det øverste angiver kernens masse i forhold til brintkernens masse, altså stoffets kemiske atomvægt. Da helium har kerneladningen 2 og massen 4, vil en atomkernes ladning formindskes med 2 og dens masse med 4 hver gang en alfa-partikel udsendes, således som det fremgår af tallene på fig. 3. Anderledes går det, når der udsendes beta-partikler, som simpelthen består af elektroner, hvis masse er så forsvindende i sammenligning med kernernes, at den kan sættes lig med nul, medens dens elektriske ladning er en negativ elementarladning. Ved betastråleudsendelse bliver atomkernens masse derfor uforandret, medens kerneladningen vokser med en enhed. Gennem den radioaktive omdannelsesrække vil kerneladningen derfor snart aftage og snart vokse igen, hvorfor der i rækken flere gange kan forekomme stoffer med samme kerneladning. Det er imidlertid atomkernens elektriske ladning, der alene er bestemmende for, hvor mange elektroner der bindes til kernen, og dermed for, hvilket grundstof man har med at gøre. De radioaktive omdannelser førte herved til opdagelsen af, at samme grundstof kan optræde i forskellige modifikationer svarende til forskellig kernemasse. I omdannelsesrækken optræder bly og polonium således hver tre gange og vismut to gange. Sådanne modifikationer af samme grundstof, der har identiske kemiske egenskaber, men forskellig atomvægt, har man kaldt *isotoper*, fordi de henhører på samme plads i grundstoffernes system.

I forbindelse med fig. 3 vil jeg gerne fremhæve et punkt af væsentlig betydning for forståelsen af spørgsmålet om atomenergien. Medens de nederste tal, kerneladningen, altid er nøjagtigt hele tal, idet de angiver et antal af en bestemt udelelig enhed, den elektriske elementarladning, er dette ikke helt tilfældet med de øverste tal, massetallene. Nøjagtige målinger har vist, at disse tal ganske vist ligger temmelig nær ved de hele tal, men at der dog er små afvigelser. Dette hænger sammen med, at der ved de radioaktive omdannelser frigøres en stor mængde energi i form af de udsendte partiklers store bevægelsesenergi, og der må til denne energi svare et tab i masse som følge af ækvivalensen mellem masse og energi, der først erkendtes af *Einstein* som en konsekvens af hans relativitetsteori. Ved almindelige energiomsætninger er masseændringen dog så forsvindende lille, at den ikke kan påvises. Hvis f. eks. 1 kg kul forbrænder, vil forbrændingsprodukterne være en smule lettere end de stoffer, der medgår til forbrændingen, men det drejer sig her kun om et massetab på mindre end en tusindedel milligram. Dette er så lille en størrelse, at den på ingen måde kan mærkes ved vejningerne ved en kemisk analyse. Derimod er energiomsætningerne ved radioaktive omdannelser så meget større i forhold til stofmængden, at når 1 kg radium fuldstændig omdannes til bly, vil den samlede masse af alle omdannelsesprodukterne være 1/10 gram mindre, hvorved vi er oppe på størrelser, der falder indenfor målemulighederne.

Det var dog ikke alene for de radioaktive stoffer, at man konstaterede eksistensen af isotoper, men en sådan isotopi fandtes at optræde hos alle de almindelige grundstoffer. Som et eksempel herpå ses på det næste lysbillede (fig. 4) de forskellige isotoper hos zink. Dette billede er fremstillet ved, at man har ladet ioner af zink passere med stor hastighed gennem et magnetfelt, hvorved de afbøjes forskelligt efter deres masse og rammer en fotografisk plade på forskellige steder. På billedet ligger isotoperne smukt ækvidistant, og ved udmåling kan det fastslås, at isotoperne altid har meget nær hele massetal, svarende til et helt tal gange brintatomets masse. Dette behøver derimod ikke at være tilfældet for stoffets kemiske atomvægt, som blot er som et middeltal, hvis størrelse afhænger af blandingsforholdet mellem stoffets forskellige isotoper.

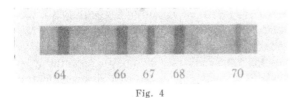

Fig. 4

De radioaktive stoffers omdannelser, hvorved et grundstof forvandles til et andet, er processer, der forløber af sig selv, og som ikke på nogen måde influeres af sædvanlige fysiske eller kemiske indgreb. Man nærede derfor i mange år den anskuelse, at det var umuligt ved eksperimenter at fremkalde forandringer i en atomkerne og derved frembringe en grundstofforvandling. Det lykkedes imidlertid Rutherford i 1919 at om-

Fig. 5

danne en atomkerne til en anden ved beskydning med alfa-partikler. Det næste lysbillede (fig. 5) viser et tågekammerfotografi af alfapartikler, der sendes ind i en kvælstofatmosfære. Forgreningen øverst på billedet hidrører fra et sammenstød mellem en kvælstofkerne og en alfapartikel, der er forløbet på en særlig måde. Ved sædvanlige sammenstød mellem atomkerner (smlgn. fig. 2) vil den ramte kerne kun kunne slynges i mere eller mindre fremadgående retning. Derimod kan den stødende kerne eventuelt kastes tilbage, men vil i så fald have mistet noget af sin hastighed ved sammenstødet. På fig. 5 ses imidlertid en partikel i tilbagegående retning med en bane, der er meget længere end de andre. Det kan derfor umuligt være en partikel af samme slags som de stødende kerner, men det kunne vises at være en brintkerne, der

$$^{14}_{7}\mathrm{N} + {}^{4}_{2}\mathrm{He} \rightarrow \left(\!{}^{18}_{9}\mathrm{F}\!\right) \rightarrow {}^{17}_{8}\mathrm{O} + {}^{1}_{1}\mathrm{H}$$

$$^{9}_{4}\mathrm{Be} + {}^{4}_{2}\mathrm{He} \rightarrow \left(\!{}^{13}_{6}\mathrm{C}\!\right) \rightarrow {}^{12}_{6}\mathrm{C} + {}^{1}_{0}\mathrm{n}$$

Fig. 6

med stor hastighed blev udjaget af den ramte kvælstofkerne.

I den øverste halvdel på fig. 6 er angivet den atomkernereaktion, der her har fundet sted. Kvælstofkernen med ladningen 7 og massen 14 rammes af en heliumkerne med ladningen 2 og massen 4. Hvis disse to kerner ligefrem forenes til en samlet kerne, må denne få ladningen 9 og massen 18, hvilket vil være en isotop af fluor. Af denne kerne udjages imidlertid en brintkerne med ladningen 1 og massen 1, så at der til rest bliver en kerne af en iltisotop med ladningen 8 og massen 17.

Ved disse forsøg havde man for første gang frembragt en forvandling af et grundstof til et andet, og man var dermed, som Rutherford selv har udtrykt det, trådt ind i den nye alkymis tidsalder. Af den kendsgerning, at der ved den omhandlede kerneomdannelse udsendtes en brintkerne, fremgik det endvidere, at denne måtte udgøre en byggesten i atomkernerne, hvorfor man gav brintkernen det særlige navn *proton*.

Fortsatte eksperimenter med beskydning af lette atomkerner med alfa-partikler førte i den følgende tid til en anden grundlæggende opdagelse, for hvilken reaktionsskemaet ses på nederste række i fig. 6. Når en berylliumkerne med ladningen 4 og massen 9 rammes af en heliumkerne, dannes først en kulstofkerne med ladningen 6 og massen 13, der straks udsender en partikel med massen 1, men uden nogen elektrisk ladning, hvorved der til rest bliver den normale kulstofisotop med massen 12.

Den ved denne proces opdagede neutrale partikel, *neutronen,* er den anden af atomkernernes to byggestene, idet vi nu véd, at alle atomkerner simpelthen er opbygget af protoner og neutroner. Da disse to partikler har meget nær samme masse, forstås det, at enhver atomkerne har en masse, der udtrykkes ved det hele tal, der angiver summen af protoner og neutroner i kernen.

I de forskellige atomkerner er disse to slags partikler bundet mere eller mindre stærkt sammen; særlig stærk er bindingen af partiklerne i en heliumkerne, der følgelig er overordentlig stabil og derfor med lethed udsendes som en alfa-partikel fra radioaktive stoffer. En heliumkerne kan principielt dannes ved, at to protoner og to neutroner slutter sig sammen, hvilket vil foregå under en meget stor energifrigørelse. De fleste af Dem har sikkert hørt om, at netop processer af denne art finder sted i solens indre, som følge af den der herskende høje temperatur på mange millioner grader. Det var jo først erkendelsen af disse forhold, der tillod at forstå oprindelsen af den uhyre energi, hvorved solen gennem sin udstråling har kunnet opretholde organisk liv her på jorden gennem så lange tidsrum.

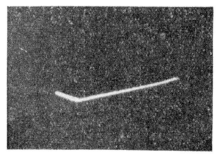

Fig. 7

Opdagelsen af neutronerne var et vigtigt middel til videre fremskridt, idet de viste sig overordentlig effektive til frembringelse af atomkerneomdannelser, hvilket hænger sammen med, at de på grund af deres mangel på elektrisk ladning ikke vil frastødes af atomkernerne, men let trænger ind i disse. Fig. 7 er et tågekammerbillede af en atomkernereaktion frembragt ved sammenstød med en neutron. I modsætning til de tidligere billeder ser man ikke her banen af den stødende partikel, da denne jo er uelektrisk og derfor ikke kan frembringe ioner i luften, hvorom vædskedråberne kan dannes. En neutron kan altså ikke aftegne sin bane i et tågekammer sådan som en alfa-partikel, så det man ser på billedet er kun resultatet af sammenstødet med neutronen, der er kommet fra neden, hvor neutronkilden var anbragt. Tågekammeret var her fyldt med luftarten neon, og fra den ramte neonkerne er der blevet udjaget en heliumkerne, som repræsenteres af det lange banespor, medens det korte er tegnet af den iltkerne, der er blevet til rest. Processen for denne kerneomdannelse er opskrevet i fig. 8.

$$^{20}_{10}\text{Ne} + ^{1}_{0}\text{n} \longrightarrow \left(^{21}_{10}\text{Ne}\right) \longrightarrow ^{17}_{8}\text{O} + ^{4}_{2}\text{He}$$

Fig. 8

At neutronerne er så meget mere effektive end alfa-partiklerne til at frembringe atomkerne-omdannelser hænger endvidere sammen med, at de ikke som alfa-partiklerne stoppes ved at miste energi til ionisering af luftens atomer. Neutronerne stoppes slet ikke ved at passere gennem atomernes åbne struktur, men kan først standses, når de rammer en atomkerne.

En særlig vigtig type af kernereaktioner frembragt ved neutroner er de såkaldte indfangningsprocesser, hvorved neutronen går ind i kernen, uden at denne straks udsender en anden partikel. Kernen får derved sin masse forøget med 1, medens dens ladning forbliver uforandret, så at der ikke sker nogen grundstof-forvandling, men dannes en tungere isotop af det bestrålede stof. Imidlertid vil denne nye atomkerne oftest være ustabil, eller som man kalder det, være gjort kunstig radioaktiv. Et eksempel på en sådan proces ses i fig. 9, hvor det drejer sig om bestråling af sølvisotopen med massen 109, hvorved neutronen indfanges, så at der dannes en radioaktiv sølvisotop med massen 110. Dennes radioaktivitet består i udsendelse af betapartikler, hvorved sølvisotopen med en halveringstid på kun ca. 22 sekunder omdannes til en cadmiumisotop.

På grund af sådanne processers store betydning vil jeg gerne vise Dem endnu en demonstration, og netop af frembringelse af radioaktivitet i sølv. Jeg tager denne lille plade af ganske almindeligt sølv og anbringer den tæt op ad en neutronkilde bag ved blyafskærmningen, hvor den skal bestråles med neutroner i et par minutter. I mellemtiden kan jeg fortælle,

$$^{109}_{47}\text{Ag} + ^{1}_{0}\text{n} \longrightarrow \left(^{110}_{47}\text{Ag}\right) \longrightarrow ^{110}_{47}\text{Ag}^{*}$$

$$^{110}_{47}\text{Ag}^{*} \longrightarrow ^{110}_{48}\text{Cd} + ^{0}_{-1}\text{e}$$

(22 s.)

Fig. 9

at neutronkilden består af et kraftigt radiumpræparat, der er blandet med berylliumpulver, hvorved berylliumkernerne udsættes for et kraftigt alfastrålebombardement fra radium, således at der dannes neutroner efter processen vist på fig. 6. Disse neutroner udsendes imidlertid med meget store hastigheder, og der har nu vist sig den ejendommelighed, at neutronerne er langt mere virksomme ved indfangningsprocesser, hvis de er langsomme, idet de da så at sige lettere klæber til atomkernerne. For at gøre neutronerne langsomme er neutronkilden anbragt inde i en paraffinblok, idet neutronerne da hyppigt vil støde sammen med kernerne i de mange brintatomer, der findes i paraffinen. Ved hvert af disse sammenstød mister neutronerne en stor del af deres hastighed, så at de efter et vist antal sammenstød kommer ned på de samme hastigheder, som brintkernerne besidder ved almindelig temperatur. Disse såkaldte langsomme eller termiske neutroner vil nu med stor sandsynlighed indfanges af atomkernerne i sølvpladen. Hele opstillingen minder i visse henseender stærkt om den moderne atomreaktor, som jeg senere skal omtale, idet neutronerne i denne også gøres langsomme ved at støde mod atomkernerne af et stof, der kaldes moderatoren, og som blot i stedet for paraffin kan bestå af grafit eller tungt vand.

Nu har sølvpladen vist fået en tilstrækkelig bestråling, og når jeg tager den frem og holder den hen i nærheden af tælleropstillingen, høres det tydeligt på tællerklikkene, at den er blevet radioaktiv, og når De lytter nøje efter, vil De høre, at klikkene kommer langsommere og langsommere, og efter ca. ½ minut hører vi kun halvt så mange som i begyndelsen, svarende til halveringstiden på 22 sekunder. Nu kan vi vist godt stoppe forsøget, da det ikke kan nytte at vente på, at radioaktiviteten helt dør hen; sølvet indeholder nemlig også en anden isotop med massen 107, som ligeledes er blevet radioaktiv ved neutronbestrålingen, men som har en meget længere halveringstid på ca. ½ time.

I forbindelse med denne demonstration vil jeg gerne minde Dem om den store betydning, som sådanne radioaktive isotoper har fået for den videnskabelige forskning og også for store områder af tekniken. Ved at benytte radioaktive isotoper som indikatorer har man f. eks. kunnet studere stofskiftet i levende organismer, idet isotoperne jo gennem alle kemiske og biologiske omdannelser må følge det grundstof, de tilhører, og fordi de så simpelt lader sig påvise og genkende gennem deres stråling og halveringstid.

Neutronen er den vigtigste af de atomare partikler med hensyn til frembringelse af atomkernereaktioner og grundstofforvandlinger. Ved dens hjælp kan man lave guld, hvis man ønsker det, men det ville være altfor kostbar en fremgangsmåde. Derimod er der ganske anderledes værdifulde stoffer, som kan frembringes ved neutroner, nemlig sådanne, der dannes ved bestrålingen af de allertungeste grundstoffer uran og thorium ved processer, der er angivet på fig. 10, og som forløber på ganske lignende måde som ved sølv. Når en neutron indfanges i kernen af den almindelige uranisotop med massen 238, dannes en uranisotop med massen 239, som imidlertid er radioaktiv og med en halveringstid på 23 minutter omdannes under betastråleudsendelse til et nyt grundstof, idet kerneladningen forøges til 93. Dette nye stof, som kaldes neptunium, er ligeledes radioaktivt og omdannes med en halveringstid på 2,3 døgn, idet der igen udsendes betapartikler, så at kerneladningen påny vokser med 1 til 94. Der dannes altså endnu et grundstof, *plutonium*, som ikke tidligere var kendt. På lignende måde vises forneden på fig. 10, at thorium ved bestråling med neutroner kan omdannes til en uranisotop med massen 233, idet der som mellemled optræder radioaktive stoffer. De to stoffer, plutonium og uran 233, har særlig betydning for spørgsmålet om atomenergien.

Men her har jeg allerede foregrebet lidt i udviklingen; det var jo først efter endnu en stor opdagelse, at det blev forstået, hvad sådanne stoffer kunne anvendes til. Denne opdagelse, der

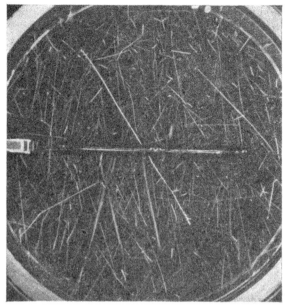

Fig. 11

$$_{92}U^{238} + {_0}n^1 \longrightarrow {_{92}}U^{239} + \gamma$$

$$_{92}U^{239} \xrightarrow{\;23.5\ min\;} {_{-1}}e^0 + {_{93}}Np^{239}$$

$$_{93}Np^{239} \xrightarrow{\;2.3\ day\;} {_{-1}}e^0 + {_{94}}Pu^{239}$$

$$_{90}Th^{232} + {_0}n^1 \longrightarrow {_{90}}Th^{233} + \gamma$$

$$_{90}Th^{233} \xrightarrow{\;23.5\ min\;} {_{91}}Pa^{233} + {_{-1}}e^0$$

$$_{91}Pa^{233} \xrightarrow{\;27.4\ day\;} {_{92}}U^{233} + {_{-1}}e^0$$

Fig. 10

blev gjort af *Hahn* og *Strassmann* i 1939, viste, at der kunne ske en helt ny form for atomkernereaktion, når uran bombarderedes med neutroner, idet man fandt, at der ved bestrålingen dannedes stoffer, som radioaktivt barium, der lå fjernt fra uran i grundstofrækken. Den nærliggende forklaring herpå, at urankernen var blevet spaltet i to omtrent lige tunge brudstykker, blev fremsat af *Lise Meitner*, der havde været Hahns medarbejder, og *Frisch*, der på den tid arbejdede i København, og de foreslog navnet *fission* for denne nye type atomkernespaltning, samtidig med at de gjorde opmærksom på, at processen måtte være ledsaget af en uhyre stor energiudvikling. Det lykkedes kort efter Frisch her i København at måle bevægelsesenergien for de ved spaltningen udslyngede brudstykker, og energifrigørelsen fandtes at være langt større end hidtil iagttaget ved nogen atomproces.

Fissionsprocessen illustreres måske bedst ved følgende smukke tågekammerfotografi (fig. 11), som er optaget af *Bøggild* på Universitetets Institut for teoretisk Fysik. Tværs over kammeret var udspændt en metalstrimmel, hvorpå var anbragt lidt uran, som bestråledes fra en neutronkilde udenfor kammeret. Man ser tydeligt, hvordan der fra den ramte urankerne udgår to meget kraftige baner i modsatte retninger, som er aftegnet af de ved urankernens spaltning dannede to tunge fragmenter, der slynges fra hinanden med stor kraft. De mange andre

$$^{94}_{38}Sr \xrightarrow[2m]{\beta} {}^{94}_{39}Y \xrightarrow[17m]{\beta} {}^{94}_{40}Zr \quad (stabil)$$

$$^{235}_{92}U + {}^{1}_{0}n \to {}^{236}_{92}U$$

$$^{140}_{54}Xe \xrightarrow[16\,sec]{\beta} {}^{140}_{55}Cs \xrightarrow[66\,sec]{\beta} {}^{140}_{56}Ba \xrightarrow[12.8\,d]{\beta} {}^{140}_{57}La \xrightarrow[40h]{\beta} {}^{140}_{58}Ce$$
$$(stabil)$$

Fig. 12

forholdsvis svage banespor hidrører fra andre atomkerner, som er blevet sat i bevægelse ved stød fra neutronerne, især brintkerner fra vanddampen.

Opdagelsen af fissionen og den ledsagende enorme energifrigørelse, der åbnede uanede muligheder for en praktisk udnyttelse af atomenergien, gav anledning til en intensiv udforskning af alle fænomenets enkeltheder. For at spare tid vil jeg nøjes med at vise følgende skematiske billede (fig. 12), der rummer de væsentligste træk ved fissionsprocessen. Man lægger først mærke til, at det ikke er den sædvanlige uranisotop med massen 238, der er tale om ved fissionen, men derimod en isotop med massen 235, som kun udgør mindre end 1 pct. af det i naturen forekommende uran. Når en atomkerne af denne isotop rammes af en langsom neutron, dannes der en urankerne med massen 236, som er så stærkt anslået, at den oftest straks deler sig i to nye kerner, f. eks. en xenonkerne og en strontiumkerne, som på grund af deres indbyrdes frastødning slynges fra hinanden med stor energi. Disse to brudstykker er i høj grad radioaktive, hvilket kommer af, at kernerne har alt for stort et overskud af neutroner, idet der skal være en vis balance mellem en atomkernes protoner og neutroner, for at den kan være stabil. Dette fører til, at stofferne gennem en række radioaktive omdannelser ved betastråleudsendelse får protonantallet til at vokse, indtil en balance atter er opnået mellem kernepartiklerne, så at de to omdannelsesrækker ender med hver sit stabile stof, der her henholdsvis er zirkon og cerium. Blandt de mange radioaktive stoffer ser vi i den nederste række netop den bariumisotop, der gav anledning til opdagelsen af fissionen.

På fig. 12 er anført endnu et meget væsentligt træk ved fissionsprocessen, nemlig at der samtidig med kernens spaltning, eller i hvert fald meget hurtigt efter, udsendes neutroner, der så at sige fordamper fra de stærkt anslåede kerner.

Det er disse neutroner, der skaber muligheden for iværksættelsen af de såkaldte *kædeprocesser*, idet de er i stand til at ramme andre kerner af uran 235, som så igen vil spaltes og udsender nye neutroner, der atter kan frembringe fission i uran, o. s. v.

Det på fig. 12 viste reaktionsskema er ikke den eneste fissionsmulighed, idet urankernen kan spaltes på mange forskellige måder, der dog alle ligner hinanden derved, at ladningstallene for brudstykkerne ligger i nærheden af henholdsvis 38 og 54. Ved uranfission dannes der derfor mange flere forskellige radioaktive stoffer, end der er vist på figuren. Her må også nævnes en omstændighed, som ikke er vist på fig. 12, nemlig at det kan hænde, at nogle af de radioaktive kerner, der er dannet ved fissionen, foruden en beta-partikel udsender en neutron. Disse neutroner kommer derfor lidt senere end de, der udsendes ved selve fissionsprocessen, og kaldes derfor forsinkede neutroner, der, som vi skal se, spiller en afgørende rolle for muligheden for at holde kædeprocesserne under kontrol.

Endelig må det fremhæves at foruden ved uran 235 kan kædeprocesser opretholdes ved plutonium og uran 233, hvis kerner ligeledes kan spaltes af langsomme neutroner under udsendelse af sekundære neutroner.

Muligheden for i enorm målestok at frigøre atomenergien ved fissionsprocesser blev jo bragt til alles kundskab gennem fremstillingen af atombomben, hvis princip er vist i fig. 13. Sprængstoffet i denne, der består af rent uran

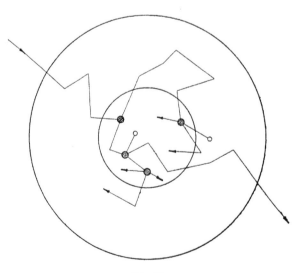

Fig. 13

235 eller plutonium, er anbragt indenfor en kappe der delvis hindrer de neutroner, som opretholder kædeprocesserne, i at undslippe. Som vist på figuren vil blot en eneste neutron, der trænger ind i bombens indre, være tilstrækkelig til at frembringe eksplosion, idet der kan startes en fission med samtidig udsendelse af neutroner, der frembringer nye spaltninger ledsaget af nye neutroner, så at det hele vokser op som en lavine i løbet af uhyre kort tid.

I modsætning til sædvanlige eksplosioner, hvor den kemiske reaktion forplanter sig fra atom til naboatom, gennemløber neutronerne i atombomben betydelige vejlængder, før de rammer en ny atomkerne. Dette medfører, at bomben skal have en vis størrelse for at kunne virke; er stofmængden for lille, vil for mange af neutronerne undslippe, før de når at ramme en atomkerne, og kædeprocessen går i stå. Vi skal imidlertid ikke komme nærmere ind på atombombernes konstruktion og mulighederne for at forstærke deres virkninger ved de såkaldte temonucleare processer.

Her ønsker jeg blot at fremhæve, at det drejer sig om et forøget herredømme over naturens kræfter, der har givet menneskene uanede ødelæggelsesmidler i hænde, og, som enhver anden forøgelse af vor viden og kunnen, rummes heri nye ansvar. I dette tilfælde må alvoren af situationen være så indlysende for alle, at vi tør sætte vor lid til, at der vil findes nye veje til at bilægge stridigheder og muliggøre folkeslagene i fællesskab at indfri de rige løfter om fremgang i menneskelig velfærd, som videnskabens fremskridt på så mange områder holder os for øje. Netop for atomfysikens vedkommende er disse løfter begrundet i muligheden af at kontrollere energiudviklingen ved atomkerneprocesser på en sådan måde, at den kan anvendes til praktiske fredelige formål.

Kontrollerede kædeprocesser kan iværksættes ved anvendelse af almindeligt uran, således som det først blev påvist af *Fermi* i 1942, fra hvilket år man derfor ofte regner atomalderens begyndelse. På fig. 14 er givet en skematisk fremstilling af en såkaldt atomreaktor, og billedet illustrerer nogle af de vigtigste træk ved en sådan reaktors virkemåde. De små cirkler angiver tværsnit af uranstænger, der er omgivet af et metalhylster, som beskytter uranmetallet mod kemiske angreb og hindrer, at de radioaktive fissionsprodukter spredes rundt i reaktoren. De større cirkler viser, at de indkapslede uran-

stænger er anbragt i rør, hvorigennem man sender en luftart eller en vædske, som har til formål at køle uranstængerne, idet langt den overvejende del af den ved fissionen frigjorte energi opstår som varme i disse. Imellem kølerørene er anbragt et stof, der kan bremse neutronerne og bringe deres hastighed ned til termiske hastigheder. Dette stof benævnes moderatoren og bestod i de første reaktorkonstruktioner af kul-

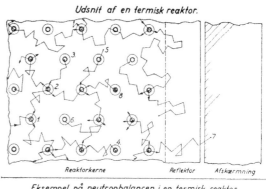

Udsnit af en termisk reaktor.

Reaktorkerne   Reflektor   Afskærmning

Eksempel på neutronbalancen i en termisk reaktor
(Beregnet pr. fission af U^{235}.)

Produktion af neutroner:		Forbrug af neutroner:	
1. Termisk fission af U^{235}	2,5	1. Termisk fission af U^{235}	1,00
2. Hurtig fission af U^{238}	0,06	3. Indfangning i U^{235}	0,2
		4. " " U^{238}	0,9
		5. " " moderator	0,3
		6. " " andre materialer	0,05
		7. Tab til omgivelserne	0,09
		8. Reguleringsoverskud	0,02
	2,56		2,56

Fig. 14

stof i form af grafit. Moderatoren med uranstængerne omgives af en reflektor, der kan være af samme stof som moderatoren, og som skal hjælpe til at hindre, at neutronerne løber ud af reaktoren. Uden om den egentlige reaktor anbringes en betonmur til beskyttelse af omgivelserne mod den kraftige radioaktive stråling. Denne afskærmning er vist yderst til højre på figuren.

Når vi f. eks. betragter uranstangen yderst til venstre i anden række fra neden mærket med tallet 1, ser vi, at en neutron, som efter en række sammenstød med atomkerner i moderatoren, hvilket antydes ved knækkene i banerne, har givet anledning til en fission af uranisotopen 235, hvorved der er frigjort to nye neutroner. Den ene er vist gående nedefter og løber efter en række sammenstød med moderatorens atomkerner ind i en anden uranstang og fremkalder heri en ny fission, hvorved neutroner igen frigøres. Den anden, der går opefter, frembringer også, i en uranstang mærket med tallet 2, en fission af isotopen 235, men, som antydet på figuren, frembringer een af de herved frigjorte neutroner inden den slipper ud af uranstangen, endnu en kernefission. Dette skal minde os om, at der er en ikke ubetydelig chance for, at de hurtige neutroner, der frigøres ved fissionen, også kan fremkalde fission af isotopen 238, hvoraf det sædvanlige uran jo langt overvejende består.

De hidtil omtalte processer er neutronproducerende, men der er også en række neutronforbrugende processer at tage hensyn til. Ud for tallene 3 og 4 ser vi, at en neutron ender inde i en uranstang uden at der sker fission. Medens der i første tilfælde henvises til muligheden for en simpel indfangning i uran 235, er der i andet tilfælde tale om en indfangning i uran 238, hvorved der som slutresultat dannes plutonium (smlgn. fig. 10). Ud for tallet 5 er en neutron forsvundet i moderatoren ved indfangning i atomkerner stammende fra de uundgåelige urenheder i denne. Det er også antydet, hvorledes reflektoren så at sige opfanger de neutroner, der har retning ud af reaktoren og giver dem lejlighed til at vende tilbage, men at enkelte alligevel undslipper og indfanges i atomkernerne i afskærmningen.

Nederst på figuren er der vist en oversigt over de processer, der influerer neutronbalancen i reaktoren. Under den egentlige drift af reaktoren gælder det om at holde balance mellem produktion og forbrug af neutroner, men for at få antallet af neutroner til at vokse op til den store mængde, som er nødvendig for at få en betydelig energiudvikling i reaktoren, må der til at begynde med være et produktionsoverskud, som imidlertid skal kunne elimineres, når energiudviklingen er blevet så stor som ønsket. Dette ordnes i praksis ved hjælp af såkaldte kontrolstænger, som er udført af materialer, der har stor evne til at indfange neutroner, Den sorte plet ud for tallet 8 på figuren skal antyde, at der her i stedet for en uranstang er indlagt en kontrolstang, og vi ser også, at een af neutronerne er blevet indfanget heri. Under driften er disse stænger skudt netop så langt ind i reaktoren, at der bliver balance mellem produktion og forbrug af neutroner, men ved at trække dem lidt ud eller skubbe dem lidt længere ind kan man få neutronantallet til at vokse eller falde efter ønske. I virkeligheden sker fissionsprocesserne overordentlig hurtigt, idet selv de langsomme neutroner går mere end 1 km i sekundet, og den ene generation af neutroner følger derfor hurtigt efter

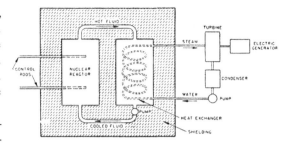

Fig. 15

den anden. Hvis vi kun havde med de ved selve fissionen udsendte neutroner at gøre, ville det således næppe være muligt at bevæge kontrolstængerne tilstrækkelig hurtigt til at regulere systemet. At dette kan opnås, hjælpes imidlertid afgørende ved, at en lille procentdel af neutronerne, som allerede omtalt, er efternølere, der udsendes nogle sekunder efter fissionen.

Princippet for udnyttelsen af den i reaktoren udnyttede energi er forholdsvis simpelt og er antydet på billedet, fig. 15. Yderst til venstre er vist en reaktor, hvori uranet »forbrændes«, og hvorigennem der sendes et eller andet stof, som skal køle uran-brændstoffet og samtidig bortføre den udviklede energi. Kølemediets varme overføres videre til en dampgenerator, hvori der udvikles damp, der på sædvanlig måde ledes igennem en turbine. Kølemediet bliver naturligvis uhyre radioaktivt ved bestrålingen i reaktoren, hvorfor det må holdes inden for en effektiv afskærmning.

Det egentlige brændstof i en atomreaktor, af en sådan type, er jo den mængde uran 235, den indeholder, hvilket som oftest nævnt kun er mindre end 1 % af hele uranmængden i reaktoren. En simpel regning viser imidlertid, at den energi, som på denne måde kan frigøres fra den

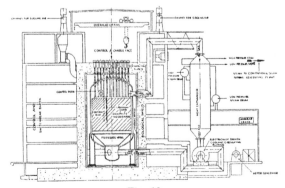

Fig. 16

mængde uran, der er nogenlunde let tilgængelig
på jorden, kun vil kunne dække verdensforbru-
get i forholdsvis få år. Den mulighed, at man af
den hyppige uranisotop 238 kan frembringe et
nyt aktivt stof, plutonium, der virker på samme
måde som den sjældne uranisotop 235, stiller
imidlertid mere end 100 gange så store energi-
reserver i udsigt for menneskene. Ved at ud-
sætte thorium for neutronbestråling i reaktorer
kan man også heraf få dannet et nyt aktivt stof
i form af uranisotopen 233 (smlgn. fig. 10).

De sidste billeder, jeg skal vise, er eksempler
på mere udformede reaktortyper. Billedet, fig. 16,
er gengivet fra den engelske regerings offent-
liggørelse af princippet for de store reaktorer,
der for tiden opføres i England, og som man
regner med vil være i virksomhed i løbet af
nogle få år. Det er ganske det samme almindelige princip som det forrige billede, men her er
henvist til forskellige nye enkeltheder. Reaktoren
køles af en strøm af kulsyre, der holdes ved
7 atm. tryk, og denne overfører varmen til en
dampgenerator. Ved andre forsøg på at få energien ud af en reaktor er man gået så vidt, at
man som kølevædske eller varmeoverførings-
medium tænker på at benytte smeltet natrium,
og fig. 17 er en skitse af, hvorledes dette fore-
slås realiseret i en amerikansk reaktor, man på-
tænker at bygge. Billedet, fig. 18, er en mere omfattende skitse til et fremtidigt anlæg, hvor man
tænker sig at tage energien bort med en strøm
af flydende natrium. Dette bliver så radioaktivt,
at der må indskydes endnu et natriumkredsløb,
inden varmen overføres til dampkedelanlægget.

* * *

Jeg vil gerne, før jeg slutter, sige nogle mere
almindelige ord. For det første, at det er et
meget stort emne både videnskabeligt og teknisk,
idet vi jo er inde i et helt nyt erfaringsområde,
som kræver indgående udforskning. Når vi har
at gøre med så store intensiteter af neutronstrå-
ling, som der her er tale om, sker der jo store
forandringer i stofferne. Ikke alene omdannes
uranet ved fissionen, men mange af de andre
stoffer ændrer deres egenskaber, idet kemiske
forbindelser, der er udsat for sådan stråling, kan
blive opløst i deres komponenter og mange faste
stoffer undergå betydelige ændringer, idet atomerne ved de mange stød fra neutroner bringes ud af deres stilling i krystalgitrene. I »Ingeniøren« (nr. 11/1955) er der af nogle medarbejdere på ingeniør *Topsøe's* virksomhed givet en

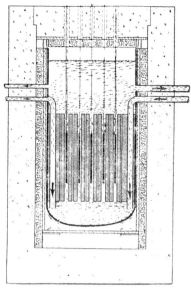

Fig. 17

mere teknisk betonet oversigt over forskellige
muligheder for reaktorer, og over nogle af de
teknologiske problemer, der er forbundet hermed.

Den store mængde radioaktivitet, der dannes
ved driften af atomreaktorer, har givet anledning til ængstelse i vide kredse. Det er klart,
at der ved alt arbejde med reaktorer må træffes
effektive forsigtighedsregler for de mennesker,
der deltager heri, og for dem der opholder sig i
nærheden. En del af de radioaktive stoffer vil
utvivlsomt blive overordentlig nyttige for forskellig art af videnskabelig og teknisk forskning
og indenfor mange grene af industrien. Men for
at undgå skadelige virkninger ved en spredning i
atmosfæren eller grundvandet vil særlige sikkerhedsforanstaltninger være påkrævet. Hele dette
spørgsmål undersøges indgående af fysikere og

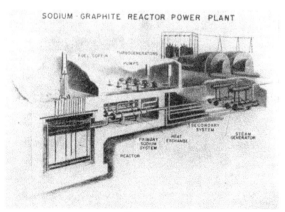

Fig. 18

biologer, og der er, selv om det drejer sig om vanskelige tekniske problemer, næppe grund til at tro, at en ansvarlig og tilfredsstillende løsning ikke vil kunne findes. Naturligvis vil også Atomenergikommissionen fra begyndelsen have opmærksomheden henvendt herpå og holde sig i nøje forbindelse med den biologiske og medicinske sagkundskab.

Jeg har været taknemmelig for lejligheden til at tale ved dette landsindustrimøde, idet en virksom deltagelse fra dansk side i udnyttelsen af atomenergien til samfundets tarv vil kræve interesse og støtte fra mange sider og ikke mindst må drage nytte af den erfaring og de produktionsmuligheder, som forefindes i landets forskelligartede industrielle virksomheder. Det er en sag, som kalder på hele befolkningens med-

virken, og hvis fremme vil forlange kostbare foranstaltninger fra statsmyndighedernes side, såvel som overvindelse af vanskelighederne ved i rette tid at kunne råde over en stor og højt uddannet medarbejderstab til løsning af de foreliggende opgaver. For at vi herhjemme skal kunne følge med i den udvikling af industriens hjælpekilder, som fremskridtene i videnskab og teknik på dette som på så mange andre områder stiller i udsigt, er det ikke alene nødvendigt at samfundet støtter bestræbelserne for at opretholde og videreføre vore traditioner med hensyn til undervisning og forskning, men tillige at der gennem oplysning og vejledning i hjem og skole sikres den til sådanne studier fornødne stedse større tilgang af unge, på hvis evner og friske kræfter vort lands fremtid må bygges.

TRANSLATION

The Physical Basis for Industrial Use of the Energy of the Atomic Nucleus

Lecture by Professor, Dr. NIELS BOHR, at the National Meeting of Industry, 17 March 1955

As a result of the technical development which has so thoroughly transformed life in our society, the question of energy sources has come to play a decisive role. The natural coal and oil deposits which until now have increasingly been used for this purpose, will, however, be depleted within the foreseeable future, and the importance of finding new energy sources has therefore for a long time already been attracting widespread attention.

Everyone knows that in several of the large countries with access to the uranium minerals primarily considered here, and with the necessary large technical resources, there are already now plans to build atomic power stations. Although it is still difficult to estimate the economic conditions for the running of such stations, it seems certain that, within a few years, circumstances will prove to be far more favourable as a result of the rapidly growing experience and technical progress in this new field. Taking part in this development is naturally of the greatest importance for our country, which has no coal or oil deposits at all, and which does not have water power of any significance either.

However, the lack of uranium minerals has also so far prevented us from actively preparing for the solution of this task, but conditions in this respect have changed considerably as a result of the initiative supported by the United Nations to advance international cooperation for the peaceful use of atomic energy. The fact that we in Denmark were immediately able to begin preparations to take part in such cooperation is, in addition, due to an initiative by the *Thomas B. Thrige Foundation* which, on the occasion of the 50th anniversary of the Thrige factories, made a considerable grant to the *Academy of the Technical Sciences* in order to initiate such preparations. The Academy immediately set up a committee to work on the matter and in particular to obtain information about investigations carried out abroad.[1]

On the basis of the results of the committee's work it was judged that the perspectives were so realistic that substantial expenses were justified for the advancement of the task, and that wider circles ought to be involved in the deliberations concerning its continuation. These views, which were shared by the Government and representatives of the Legislature, led the Prime Minister[2] to make the decision to appoint the [Danish] *Atomic Energy Commission* and to apply for a preliminary appropriation for its work with the view of formulating a proposal for the information of the State authorities. In addition to the working committee of the Academy and physicists from our country's universities and other institutions of higher learning, it was found expedient to invite a number of representatives from industry with special technical and organizational knowledge to join the Commission. It is of course obvious that outside a Commission with a limited number of members and such a wide-ranging field of activity, there will be a wealth of knowledge and insight, which it will be valuable, indeed indispensable, to draw on. In the negotiations which took place before the Commission was set up, it was therefore explicitly stressed that the Commission is entitled to invite such experts to attend its meetings as is deemed necessary.

The Commission's first meeting will be held immediately following this large meeting of industry, and it will strive to proceed with the technical and organizational preparations as quickly as circumstances permit.

It will not be possible, of course, to account in greater detail for the plans, which it will be the Commission's task to prepare, until its negotiations have progressed further. Today I shall therefore confine myself to giving you, in outline, an impression of the physical basis for the question of the industrial application of atomic energy.

I shall come straight to the point and first remind you that the conceptions that all matter is made up of *atoms* go back a long time, and that these conceptions proved to be increasingly fruitful and indispensable in explaining the physical and chemical properties of matter. Until the beginning of this century, however, the general view was that this was only a hypothesis, since it was believed that, because of the coarseness of our senses, it would never be possible to examine and observe the effect of individual atoms. However, the

---

[1] The Atomic Energy Committee. See the introduction to Part II, p. [360].
[2] Hans Hedtoft (1903–1955) was Prime Minister of Denmark from 1947 to 1950 and from 1953 until his death.

situation has developed quite differently as a result of the wonderful refinement of the experimental technique of physics and because of the many great discoveries made around the turn of the century. One of these discoveries, which was to gain special importance, and which was also the first to teach us about the existence of new sources of energy, was the discovery of the *radioactive* substances.

I would like to use such a substance to make a demonstration of how single atoms can be observed and registered. When I now take a radium preparation[3] out from the heavy lead shield behind which it is kept, and hold it close to a counter arrangement, you will hear some clicks in the loudspeaker. What is happening is that the radium preparation emits gamma rays, which are of the same kind as X-rays, and which by hitting the counter release a few *electrons* in it. The weak discharge that is thereby started in the counter is sent into an amplifier which amplifies the impulses so much that they can be registered by an ordinary telephone register, and it is the clicks from this that you hear in the loudspeaker. The counter employed here represents a rather refined type of *electronics*, as at the moment it is adjusted not to count every single electron, but only every twentieth. This is done because, in view of a later demonstration, the radium preparation used is so strong that the impulses from the counter would otherwise follow much too rapidly upon one another to be distinguishable.

Here we have an example of modern highly-developed electronics based on the discovery of the electron half a century ago, which was then regarded as the ultimate limit of science. Such electronics is also employed in the refined automation applied more and more extensively in industry. We must here be prepared that a very large part of the control work in the factories now done by people, within not too many years will be taken over by automatic machines. This will of course create new great problems for society concerning distribution of labour and increased special training.

I shall proceed to illustrate some main features of the development of atomic physics by means of a number of slides. On the first one (Fig. 1) you see an example of a well-known method within the experimental technique of physics. The picture is a cloud-chamber photograph showing the tracks of the so-called alpha particles emitted from the radioactive substance polonium which, like radium itself, was discovered by Madame *Curie* and is therefore named after her native country. In contrast to electrons, which are much lighter than atoms,

---

[3] Since radium does not exist in a sufficiently pure form in nature, it has to be prepared by chemical means.

Fig. 1

alpha particles have a mass of the same order of magnitude as atoms, and, by collecting alpha particles in a glass tube, it has furthermore been possible to show that they form helium, as they are actually nuclei of helium atoms. It is a characteristic feature of alpha particles, which appears clearly from the cloud-chamber photograph, that they traverse straight tracks, several centimetres long, at atmospheric pressure in the cloud chamber, a circumstance related to their being emitted from the preparation at extremely high velocities. On their way the alpha particles pass through millions of atoms in the air, which must therefore have a very open structure. As they pass through, the alpha particles will nevertheless break away electrons from the atoms, and about 100,000 ions will be formed along the path. It is these ions that make the paths of the alpha particles visible, as the cloud chamber contains supersaturated vapour of water or alcohol, which condenses into small drops around each of the ions formed, so that the path of the particle through the air stands out like a trail of fog.

The next cloud-chamber photograph (Fig. 2) is meant to illustrate one of the most important discoveries that has been made in atomic physics. You see that one of the paths of the alpha particles has a branching, which must have come about when the alpha particle hit something which was much heavier than the electrons, and which was pushed to the right at the impact, whereas the particle itself was thereby deflected to the left. What the alpha particle collided with is

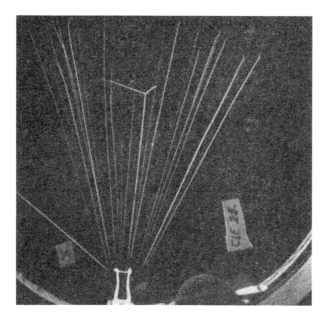

Fig. 2

an *atomic nucleus*, and it was observations of this nature that led to *Rutherford's* epoch-making discovery of the atomic nucleus in 1911.

When radioactive substances emit alpha or beta particles, these particles come from the atomic nuclei of the substance, which are thereby transformed into nuclei of a new substance. This is illustrated by the next slide (Fig. 3) which shows the radioactive transformations within the so-called radium family. We start with a radium nucleus emitting an alpha particle which, as already mentioned, is identical with a helium nucleus. The radium nucleus is thereby transformed into the nucleus of a new substance, radium emanation,[4] which also emits an alpha particle. The transformation of radium occurs quite slowly, in that it takes 1600 years until half of the substance is transformed, whereas the emanation is transformed much faster with a half-life of 4 days. Then follows, in more rapid succession, a whole series of transformations until the chain ends with ordinary lead, which is not radioactive. Of the numbers next to each of the substances the lower one indicates the electric charge of the atomic

---

[4] Radium emanation was discovered in 1900. In 1920, the name radon was adopted for element number 86, and radium emanation was identified as a radon isotope with atomic weight 222. However, the term radium emanation continued to be used.

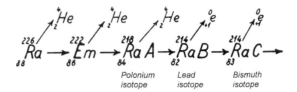

Fig. 3

nucleus, whereas the upper numbers indicate the mass of the nucleus in relation to the mass of the hydrogen nucleus, that is, the chemical atomic weight of the substance. As helium has nuclear charge 2 and mass 4, the charge of an atomic nucleus will be reduced by 2 and its mass by 4 each time an alpha particle is emitted, as is shown by the numbers in Fig. 3. The situation is different for the emission of beta particles, which simply consist of electrons, whose mass is so negligible as compared to the mass of nuclei that it can be put equal to zero, whereas its electric charge is the negative elementary charge. Thus, the mass of the atomic nucleus remains unchanged in beta emission, whereas the nuclear charge increases by one unit. Through the chain of radioactive transformations the nuclear charge will therefore now decrease and now increase again, which is why substances with the same nuclear charge may appear several times in the chain. It is, however, the electric charge of the atomic nucleus alone which determines how many electrons are bound to the nucleus, and thus which element we have to do with. Thereby, the radioactive transformations led to the discovery that the same element can appear in different modifications corresponding to different nuclear mass. Thus, in the chain of transformations, lead and polonium each appear three times and bismuth twice. Such modifications of the same element, which have identical chemical properties but different atomic weight, have been called *isotopes* because they belong in the same place in the system of elements.

In connection with Fig. 3, I would like to emphasize a point of considerable importance for understanding the question of atomic energy. Whereas the lower numbers, the nuclear charge, are always exactly integers since they indicate the

[590]

number of a certain indivisible unit, the electric elementary charge, this is not quite the case with the upper numbers, the mass numbers. Exact measurements have shown that these numbers are indeed quite close to integers, but that there are nevertheless small deviations. This is related to the fact that in radioactive transformations a large amount of energy is released in the form of the great kinetic energy of the emitted particles, and corresponding to this energy there must be a loss of mass as a result of the equivalence between mass and energy, which was first recognized by *Einstein* as a consequence of his theory of relativity. In ordinary energy conversions, however, the change of mass is so negligible that it cannot be detected. If, for instance, one kilogramme of coal burns, the combustion products will be somewhat lighter than the substances consumed in the combustion, but here the loss in mass is less than one thousandth of a milligramme. This is such a small quantity that it can in no way be detected by the weighing involved in a chemical analysis. In radioactive transformations, on the other hand, the energy conversions are so much larger, relative to the amount of substance, that when one kilogramme of radium is completely transformed into lead, the total mass of all the transformation products will be one tenth of a gramme less, whereby we have reached quantities large enough to fall within the possibilities for measurement.

The existence of isotopes, however, was established not only for radioactive substances; rather, such isotopy was found to occur in all the ordinary elements. As an example of this the next slide (Fig. 4) shows the various isotopes of zinc. This picture was obtained by letting zinc ions pass through a magnetic field at high velocity, whereby they are deflected differently according to their mass and hit a photographic plate in different places. In the picture the isotopes are placed beautifully equidistant, and by measurement it can be established that the isotopes always have very nearly whole mass numbers, corresponding to an integer multiplied by the mass of the hydrogen atom. This, however, is not necessarily the case for the chemical atomic weight of the substance, which is merely a mean value whose magnitude depends on the relative abundance of the various isotopes of the substance.

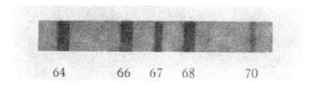

Fig. 4

The transformations of radioactive substances, whereby one element is transmuted into another, are processes that occur by themselves and are in no way influenced by ordinary physical or chemical intervention. For many years the view was therefore held that it would be impossible experimentally to bring about changes in an atomic nucleus and thus produce a transmutation of an element. In 1919, however, Rutherford succeeded in transforming one atomic nucleus into another by alpha particle bombardment. The next slide (Fig. 5) shows a cloud-chamber photograph of alpha particles sent into a nitrogen atmosphere. The branching on the top of the picture results from a collision between a nitrogen nucleus and an alpha particle, which has proceeded in a special way. In ordinary collisions between atomic nuclei (cf. Fig. 2), the target nucleus can only be thrown into a more or less forward direction. The incident nucleus, on the other hand, may possibly be thrown backward, but will in that case have lost some of its velocity in the collision. In Fig. 5, however, we

Fig. 5

$$^{14}_{7}N + ^{4}_{2}He \rightarrow \left(^{18}_{9}F\right) \rightarrow ^{17}_{8}O + ^{1}_{1}H$$

$$^{9}_{4}Be + ^{4}_{2}He \rightarrow \left(^{13}_{6}C\right) \rightarrow ^{12}_{6}C + ^{1}_{0}n$$

Fig. 6

see a particle moving in a backward direction with a path much longer than the others. Therefore, it cannot possibly be a particle of the same kind as the incident nuclei, but could be shown to be a hydrogen nucleus ejected from the target nitrogen nucleus at great velocity.

The upper half of Fig. 6 shows the nuclear reaction that has taken place here. The nitrogen nucleus of charge 7 and mass 14 is hit by a helium nucleus of charge 2 and mass 4. If these two nuclei simply combine into a compound nucleus, this nucleus must get the charge 9 and the mass 18, which is an isotope of fluorine. From this nucleus, however, a hydrogen nucleus of charge 1 and mass 1 is ejected, so that a nucleus of an oxygen isotope of charge 8 and mass 17 remains.

By these experiments one had for the first time produced a transmutation of one element into another, and we had thus, as Rutherford has expressed it himself, entered the age of the new alchemy. From the fact that in this nuclear transformation a hydrogen nucleus was emitted, it furthermore followed that this nucleus had to be a constituent of atomic nuclei, and therefore the hydrogen nucleus was given the special name *proton*.

In the time that followed, continued experiments with bombardment of light atomic nuclei with alpha particles led to another fundamental discovery, whose reaction scheme is shown in the bottom row of Fig. 6. When a beryllium nucleus of charge 4 and mass 9 is hit by a helium nucleus, a carbon nucleus of charge 6 and mass 13 will first be formed, which immediately emits a particle of mass 1, but with no electric charge, whereby the ordinary carbon isotope of mass 12 remains.

The neutral particle discovered in this process, the *neutron*, is the second of the two constituents of atomic nuclei, since we now know that all atomic nuclei are simply made up of protons and neutrons. Because these two particles have very nearly the same mass, it is understood that every atomic nucleus has a mass expressed by the integer indicating the sum of protons and neutrons in the nucleus.

In the various atomic nuclei these two kinds of particles are bound together more or less strongly; the binding of the particles is particularly strong in a helium nucleus, which is consequently extremely stable and is thus easily emitted from radioactive substances as an alpha particle. In principle, a helium nucleus can be formed when two protons and two neutrons combine, which will occur with the release of a very large amount of energy. Most of you have undoubtedly heard that precisely such processes take place in the interior of the sun, as a result of the high temperature of many million degrees prevailing there. It was only the recognition of these circumstances that allowed us to understand the origin of the vast amount of energy whereby the sun, through its radiation, has been able to sustain organic life here on Earth for such long periods of time.

The discovery of the neutrons was an important means for further progress in that they proved extremely effective in producing nuclear transformations. This is related to the fact that, because of their lack of electric charge, they will not be repelled by atomic nuclei, but will penetrate them easily. Fig. 7 is a cloud-chamber picture of a nuclear reaction produced by collision with a neutron. Unlike in the previous pictures, here we do not see the path of the incident particle, as this is non-electric and therefore cannot produce ions in the air around which the liquid drops may form. Thus, a neutron cannot leave an impression of its path in a cloud chamber, as an alpha particle can, so what we see in the picture is only the result of the collision with the neutron, which has come from below where the neutron source was placed. In this case the cloud chamber was filled with neon gas, and from the target neon nucleus a helium nucleus has been ejected, represented by the long track, whereas the short track is left by the remaining oxygen nucleus. The process of this nuclear transformation is written down in Fig. 8.

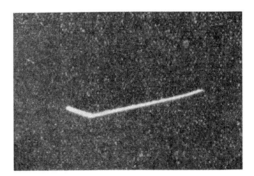

Fig. 7

$$^{20}_{10}\text{Ne} + ^1_0\text{n} \longrightarrow \left(^{21}_{10}\text{Ne}\right) \longrightarrow ^{17}_{8}\text{O} + ^4_2\text{He}$$

Fig. 8

The reason neutrons are so much more effective than alpha particles in producing nuclear transformations is moreover related to the fact that, unlike the alpha particles, they are not stopped by losing energy to the ionization of atoms in the air. The neutrons are not stopped at all on passing through the open structure of atoms, but can only be halted when they hit an atomic nucleus.

A particularly important type of nuclear reactions produced by neutrons is the so-called capture processes, whereby the neutron enters the nucleus without causing an immediate emission of another particle. The nucleus thereby has its mass increased by 1, whereas its charge remains unchanged so that no transmutation of an element takes place, but a heavier isotope is formed out of the irradiated substance. However, this new atomic nucleus will most often be unstable, or be made artificially radioactive, as it is called. An example of such a process is seen in Fig. 9, which shows the irradiation of a silver isotope of mass 109, whereby the neutron is captured, so that a radioactive silver isotope of mass 110 is formed. The radioactivity of this isotope consists of the emission of beta particles, whereby the silver isotope, with a half-life of only about 22 seconds, is transformed into a cadmium isotope.

Because of the great significance of such processes, I should like to show you yet another demonstration which concerns precisely how radioactivity is induced in silver. I take this small plate of quite ordinary silver and place it

$$^{109}_{47}\text{Ag} + ^1_0\text{n} \longrightarrow \left(^{110}_{47}\text{Ag}\right) \longrightarrow ^{110}_{47}\text{Ag}^*$$

$$^{110}_{47}\text{Ag}^* \longrightarrow ^{110}_{48}\text{Cd} + ^0_{-1}\text{e}$$

(22 s.)

Fig. 9

close to a neutron source behind the lead shield where it will be irradiated by neutrons for a couple of minutes. In the meantime, I can tell you that the neutron source consists of a strong radium preparation mixed with beryllium powder, whereby the beryllium nuclei are exposed to a heavy bombardment of alpha rays from radium, so that neutrons are formed according to the process shown in Fig. 6. These neutrons, however, are emitted at very high velocities, and the peculiarity has now become apparent that the neutrons are far more effective during capture processes if they are slow, as they then stick more easily, so to speak, to the atomic nuclei. In order to slow down the neutrons, the neutron source has been placed inside a block of paraffin, as the neutrons will then frequently collide with the nuclei of the many hydrogen atoms found in the paraffin. In each of these collisions the neutrons lose a great deal of their velocity, so that after a certain number of collisions they come down to the same velocities as those of hydrogen atoms at ordinary temperatures. These so-called slow, or thermal, neutrons will now with great probability be captured by the atomic nuclei in the silver plate. In some respects, the entire arrangement bears a strong resemblance to the modern atomic reactor which I shall describe later, in that the neutrons in such a reactor are also slowed down by colliding with the atomic nuclei of a substance called the moderator, with the slight difference that it may consist of graphite or heavy water instead of paraffin.

Now the silver plate should have been irradiated sufficiently, and when I take it out and hold it close to the counter arrangement, the clicks from the counter clearly indicate that it has become radioactive, and when you listen carefully you will hear that the clicks come more and more slowly, and after about half a minute we hear only half as many clicks as in the beginning, corresponding to a half-life of 22 seconds. I think we might as well stop the experiment now, since it is no use waiting for the radioactivity to die out completely; in fact, the silver also contains another isotope of mass 107, which likewise has become radioactive during the neutron irradiation, but which has a much longer half-life of about half an hour.

In connection with this demonstration, I should like to remind you of the great significance that such radioactive isotopes have obtained for scientific research, and for large areas of technology, as well. By using radioactive isotopes as indicators it has been possible to study, for example, the metabolism of living organisms, because the isotopes must follow the elements they belong to through all chemical and biological transformations, and because they are so easily detected and identified by way of their radiation and half-life.

The neutron is the most important of the atomic particles as regards the production of nuclear reactions and transmutations of elements. With its help, you can make gold, if you wish, but this would be much too costly a proce-

$$_{92}U^{238} + {}_{0}n^{1} \longrightarrow {}_{92}U^{239} + \gamma$$

$$_{92}U^{239} \xrightarrow{\text{23.5 min}} {}_{-1}e^{0} + {}_{93}Np^{239}$$

$$_{93}Np^{239} \xrightarrow{\text{2.3 day}} {}_{-1}e^{0} + {}_{94}Pu^{239}$$

$$_{90}Th^{232} + {}_{0}n^{1} \longrightarrow {}_{90}Th^{233} + \gamma$$

$$_{90}Th^{233} \xrightarrow{\text{23.5 min}} {}_{91}Pa^{233} + {}_{-1}e^{0}$$

$$_{91}Pa^{233} \xrightarrow{\text{27.4 day}} {}_{92}U^{233} + {}_{-1}e^{0}$$

Fig. 10

dure. There are, however, far more valuable substances that can be produced by means of neutrons, namely those which are formed by irradiation of the heaviest elements of all, uranium and thorium, in processes shown in Fig. 10, which proceed in much the same way as is the case with silver. When a neutron is captured in the nucleus of the ordinary uranium isotope of mass 238, a uranium isotope of mass 239 is formed. However, this isotope is radioactive, and with a half-life of 23 minutes is transformed by beta emission into a new element, since the nuclear charge increases to 93. This new substance, which is called neptunium, is also radioactive and is transformed with a half-life of 2.3 days, again with the emission of beta particles, so that once more the nuclear charge increases by 1 to 94. Thus yet another element, *plutonium*, is formed, which was previously unknown. Similarly, at the bottom of Fig. 10, it is shown that thorium can be transformed, by irradiation with neutrons, into a uranium isotope of mass 233, with radioactive substances appearing as intermediaries. The two substances, plutonium and uranium 233, are of special relevance for the question of atomic energy.

But here I have already anticipated the development a little; only after yet another great discovery did we understand what such substances could be used for. This discovery, which was made by *Hahn*[5] and *Strassmann*[6] in 1939,[7] showed that a completely new form of nuclear reaction could occur when uranium was bombarded with neutrons, as it was found that substances, such as

---

[5] Otto Hahn (1879–1968).
[6] Fritz Strassmann (1902–1980).
[7] The discovery was actually made in late 1938.

radioactive barium, which were far from uranium in the table of elements, were produced by the irradiation. The natural explanation of this, that the uranium nucleus had been split into two almost equally heavy fragments, was offered by *Lise Meitner*,[8] who had been Hahn's collaborator, and *Frisch*,[9] who was working in Copenhagen at the time. They suggested the term *fission* for this new type of splitting of atomic nuclei, at the same time pointing out that the process had to be accompanied by an extremely large energy generation. Shortly afterwards, here in Copenhagen, Frisch succeeded in measuring the kinetic energy of the fragments ejected by the splitting, and the release of energy was found to be much larger than hitherto observed in any atomic process.

The fission process is perhaps best illustrated by the following beautiful cloud-chamber photograph (Fig. 11), taken by *Bøggild*[10] at the University

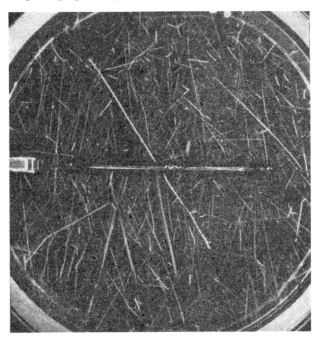

Fig. 11

[8] (1878–1968).

[9] Otto Robert Frisch (1904–1979), Lise Meitner's nephew.

[10] Jørgen Kruse Bøggild (1903–1982), who worked as a physicist at Bohr's institute from 1940 until his retirement in 1970, built and used the first cloud chambers in Copenhagen for the experimental study of fission fragments.

Institute for Theoretical Physics. A metal strip was mounted across the chamber and a little uranium placed on the strip. The uranium was irradiated from a neutron source outside the chamber. You can clearly see how two very strong paths branch off from the target uranium nucleus in opposite directions. These are left by the two heavy fragments produced by fission of the uranium nucleus, which fling apart with great force. The many other relatively weak paths are due to other atomic nuclei that have been set in motion by the impact of neutrons, particularly hydrogen nuclei from the water vapour.

The discovery of fission and the accompanying enormous release of energy, which opened up unsuspected possibilities for practical utilization of atomic energy, gave rise to an intensive exploration of all the details of the phenomenon. In order to save time, I shall only show you the following schematic picture (Fig. 12), which contains the most essential features of the fission process. First, we notice that it is not the usual uranium isotope of mass 238 we are dealing with in the fission process, but rather an isotope of mass 235, which only accounts for less than 1 per cent of the uranium found in nature. When an atomic nucleus of this isotope is struck by a slow neutron, a uranium nucleus of mass 236 is formed, which is so highly excited that as a rule it will immediately split into two new nuclei, for example a xenon nucleus and a strontium nucleus, which because of their mutual repulsion will fling apart from each other with great energy. These two fragments are highly radioactive, which is due to the fact that the nuclei have much too large an excess of neutrons, as there must be a certain balance between the protons and neutrons of an atomic nucleus in order for it to be stable. As a result, the substances make the number of protons increase through a chain of radioactive transformations with the emission of beta rays, until once again a balance has been reached between the nuclear particles, so that the two chains of transformations end up with a stable substance each, in the present case, zirconium and cerium, respectively. Among the many radioactive substances, we see in the bottom row the very barium isotope that led to the discovery of fission.

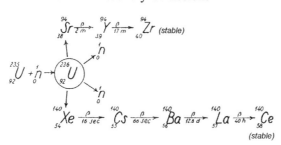

Fig. 12

In Fig. 12, another very essential feature of the fission process is shown, namely that, coincident with the splitting of the nucleus, or at least very quickly afterwards, neutrons are emitted which so to speak evaporate from the highly excited nuclei. It is these neutrons that make the initiation of the so-called *chain processes* possible, as they are able to hit other nuclei of uranium 235, which then again will be split and emit new neutrons, which in turn can produce fission in uranium, and so on.

The reaction scheme shown in Fig. 12 is not the only fission possibility, as the uranium nucleus can be split in many different ways, which are nevertheless all similar in that the charge numbers for the fragments are near to 38 and 54, respectively. Thus, many more different radioactive substances are formed by uranium fission than are shown in the figure. Here should also be mentioned a fact not shown in Fig. 12, namely, that some of the radioactive nuclei formed by fission may, in addition to a beta particle, emit a neutron. These neutrons therefore appear a little later than those emitted in the fission process itself, and are therefore called delayed neutrons which, as we shall see, play a decisive role for the possibility of keeping the chain reactions under control.

Finally, it must be emphasized that chain reactions can be maintained not only by uranium 235, but also by plutonium and uranium 233, whose nuclei can likewise be split by slow neutrons with the emission of secondary neutrons.

The possibility of releasing atomic energy on an enormous scale by means of fission processes was brought to everyone's attention through the production of the atomic bomb, the principle of which is shown in Fig. 13. The explosive substance in the bomb, which consists of pure uranium 235 or plutonium, has been placed inside a shield which partially prevents the neutrons that maintain the chain reactions from escaping. As shown in the figure, the penetration into the interior of the bomb of just one single neutron will be sufficient to produce an explosion, as fission can be set off with simultaneous emission of neutrons which produce new splittings accompanied by new neutrons, so that it all builds up like an avalanche in the course of an extremely short time.

In contrast to ordinary explosions, in which the chemical reaction is transmitted from atom to neighbouring atom, the neutrons in the atomic bomb traverse considerable distances before hitting another atomic nucleus. This means that the bomb must have a certain size in order to work; if the amount of substance is too small, too many of the neutrons will escape before they are able to hit an atomic nucleus, and the chain reaction will stop. We shall not, however, discuss in greater detail the construction of atomic bombs and the prospects of amplifying their effect by means of the so-called thermonuclear processes.

At this point I just wish to emphasize that we are dealing with an increased

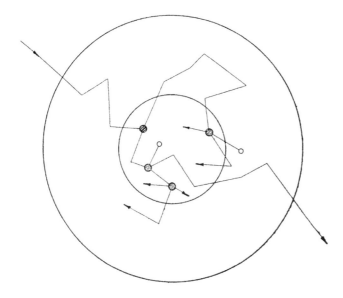

Fig. 13

mastery of the forces of nature, which has provided humanity with unsuspected means of destruction, and, like any other increase of our knowledge and ability, this implies new responsibilities. In this case, the gravity of the situation must be so obvious to everyone that we dare trust that new ways to settle disputes will be found and enable the peoples of the world jointly to fulfil the rich promises of progress for the welfare of mankind that the advance of science foreshadows in so many fields. In the case of atomic physics in particular, these promises are justified by the possibility of controlling the energy generation within nuclear processes in such a way that it can be used for practical peaceful purposes.

Controlled chain reactions can be initiated using ordinary uranium, as first demonstrated by *Fermi*[11] in 1942, the year often reckoned to be the beginning of the Atomic Age. Fig. 14 shows a schematic representation of a so-called atomic reactor, and the picture illustrates some of the most important features

---

[11] Enrico Fermi (1901–1954), Italian physicist, was one of several refugees from fascism whom Bohr helped to obtain work abroad; the Fermi family was enabled to prepare its escape through Sweden by Bohr telling Fermi in advance that he would receive the Nobel Prize in physics for 1938. Fermi's demonstration in Chicago in December 1942 that a chain reaction could be obtained in a nuclear reactor was a crucial step on the way to the making of the atomic bomb.

Cross section of a thermal reactor.

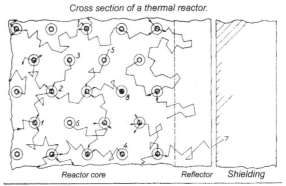

Reactor core   Reflector   Shielding

Example of the neutron balance in a thermal reactor
(Calculated per fission of U^{235})

Production of neutrons:		Consumption of neutrons:	
1. Thermal fission of U^{235}	2.5	1. Thermal fission of U^{235}	1.00
2. Fast fission of U^{238}	0.06	3. Capture in U^{235}	0.2
		4. Capture in U^{238}	0.9
		5. Capture in moderator	0.3
		6. Capture in other materials	0.05
		7. Loss to the surroundings	0.09
		8. Regulating surplus	0.02
	2.56		2.56

Fig. 14

of the mode of operation of such a reactor. The small circles indicate cross sections of uranium rods surrounded by a metal jacket protecting the uranium metal against chemical corrosion and preventing the radioactive fission products from spreading in the reactor. The large circles show that the encapsulated uranium rods are placed in tubes through which a gas or a liquid flows for the purpose of cooling down the uranium rods, as most of the energy released during fission appears as heat in the rods. Between the cooling tubes is placed a substance that can slow down the neutrons and reduce their speed to thermal velocities. This substance is called the moderator and consisted, in the first reactor constructions, of carbon in the form of graphite. The moderator with the uranium rods is surrounded by a reflector, which may be of the same substance as the moderator and is intended to help preventing the neutrons from escaping out of the reactor. A concrete wall is placed around the reactor itself to protect the surroundings from the strong radioactive radiation. This shielding is shown at the extreme right of the figure.

When we look, for example, at the uranium rod farthest left in the second row from below, marked number 1, we see that a neutron, after a number of collisions with atomic nuclei in the moderator, indicated by the breaks in the paths, has led to a fission of the uranium isotope 235, whereby two new neutrons are released. One of these seems to be moving downwards and, after a number of collisions with the atomic nuclei of the moderator, enters another uranium rod in which it produces a new fission, whereby neutrons are again released. The other neutron, moving upwards, also produces a fission of the isotope 235 in a uranium rod marked number 2, but, as shown in the figure, one of the neutrons thus released produces yet another nuclear fission before it escapes from the uranium rod. This is intended to remind us that there is a chance, which is not negligible, that the fast neutrons released by the fission also can produce fission of the isotope 238, of which ordinary uranium mainly consists.

The processes mentioned so far are neutron producing, but there are also several neutron-consuming processes to be considered. At numbers 3 and 4 we see that a neutron ends up inside a uranium rod without causing fission. While the first case is explained by the possibility of a simple capture in uranium 235, the second case concerns capture in uranium 238, with the formation of plutonium as the end result (cf. Fig. 10). At number 5, a neutron has disappeared into the moderator, captured by the atomic nuclei arising from the unavoidable impurities present there. It is also shown how the reflector, so to speak, intercepts the neutrons that are on their way out of the reactor and gives them an opportunity to return, but how a few escape nevertheless and are captured in the atomic nuclei of the shielding.

At the bottom of the figure a survey of the processes influencing the neutron balance in the reactor is shown. During the actual operation of the reactor the object is to keep a balance between production and consumption of neutrons, but in order to increase the number of neutrons to the large quantity necessary to obtain a considerable energy generation in the reactor, there must at first be a surplus production which, however, it must be possible to eliminate when the energy generation has reached the desired level. This is done in practice by means of so-called control rods, which are made of materials with a high capacity for capturing neutrons. The black spot at number 8 in the figure is intended to show that instead of a uranium rod, a control rod has been placed here, and we can also see that one of the neutrons has been captured in it. During operation, these rods are pushed just so far into the reactor that a balance between the production and the consumption of neutrons is achieved. But by pulling them out a little or pushing them a little further in, it is possible to increase or decrease the number of neutrons at will. In reality, fission processes happen extremely fast, since even slow neutrons travel more than 1 kilometre per second, so that one generation of neutrons follows rapidly

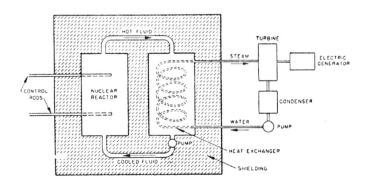

Fig. 15

after the other. Thus, if we were dealing only with the neutrons emitted in the actual fission, it would hardly be possible to move the control rods fast enough to regulate the system. The achievement of this, however, is helped decisively by the fact that a small percentage of the neutrons, as already mentioned, are stragglers, emitted a few seconds after the fission.

The principle for the use of the energy generated in the reactor is relatively simple, as indicated in the picture, Fig. 15. To the far left is shown a reactor in which uranium is "burnt", and through which some substance or other is sent for the purpose of cooling the uranium fuel and at the same time leading away the energy generated. The heat from the cooling medium is transferred on to a steam generator developing steam which is directed through a turbine in the usual way. Naturally, the cooling medium becomes exceedingly radioactive when irradiated in the reactor, and must therefore be kept inside an effective shielding.

The actual fuel in a nuclear reactor of this type is the amount of uranium 235 it contains, which, as often mentioned, is only less than one per cent of the total amount of uranium in the reactor. A simple calculation shows, however, that the energy which can be released in this way from the amount of uranium fairly easily accessible on Earth, will only cover the world's consumption for relatively few years. However, the possibility that a new active substance, plutonium, which reacts in the same way as the rare uranium isotope 235, can be produced from the abundant uranium isotope 238, holds out prospects for humanity of energy reserves more than a hundred times greater. By subjecting thorium to neutron irradiation in reactors we can also make a new active substance in the form of the uranium isotope 233 (cf. Fig. 10).

The last pictures that I am going to show you are examples of more advanced types of reactors. The picture, Fig. 16, is reproduced from the British

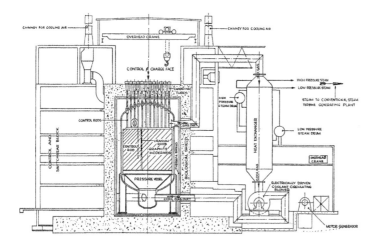

Fig. 16

Government's publication of the principle of the large reactors presently under construction in England, which are expected to be operational within a few years. It is quite the same general principle as in the previous picture, but various new details are now indicated. The reactor is cooled by a flow of carbon dioxide at a pressure of 7 atmospheres, which transfers the heat to a steam generator. In other attempts to extract the energy from a reactor one has gone as far as to consider the use of molten sodium as coolant or heat transmitting medium; Fig. 17 is a sketch of a proposed implementation in a projected American reactor. The picture, Fig. 18, shows a more comprehensive sketch of a future plant in which the removal of the energy by a flow of liquid sodium is being considered. This will become so radioactive that another sodium cycle has to be inserted before the heat is transferred to the steam boiler plant.

* * *

Before closing I should like to make a few more general remarks. In the first place, this is a very broad topic, scientifically as well as technically, as we are after all entering a completely new area of experience requiring intensive exploration. When we are dealing with so high intensities of neutron radiation as is the case here, great changes take place in the substances. Not only is uranium transformed during the fission process, but many of the other substances change their properties because chemical compounds exposed to such radiation can disintegrate into their components, and many solids can

[605]

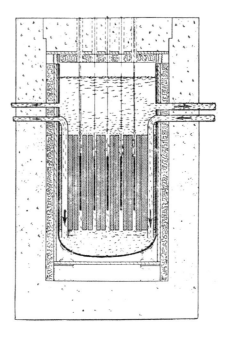

Fig. 17

undergo considerable changes because atoms are displaced from their positions in the crystal lattices by the many impacts of neutrons. In "Ingeniøren"[12] (no. 11/1955) some of the collaborators at engineer *Topsøe's*[13] firm have given a more technically oriented survey of various reactor possibilities and of some of the related technological problems.

The large amount of radioactivity formed by the operation of atomic reactors has given rise to anxiety in wide circles. It is obvious that for all work with reactors, effective precautions must be taken for those people participating in the work as well as for those present nearby. Some of the radioactive substances will undoubtedly become extremely useful for various kinds of scientific and technical research and within many branches of industry. But in order to avoid harmful effects from a leakage into the atmosphere or the ground water, special precautionary measures will be required. This whole issue is being thoroughly investigated by physicists and biologists, and although difficult technical problems are involved, there is hardly reason to believe that it will not be possible

---

[12] The journal of the Danish Association of Engineers.

[13] Haldor Topsøe (1913–), prominent Danish industrial chemist and entrepreneur who pioneered industrial research and development in Denmark.

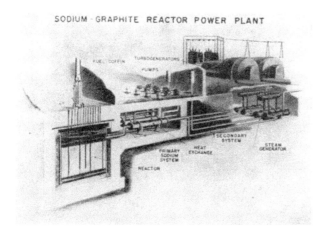

SODIUM-GRAPHITE REACTOR POWER PLANT

Fig. 18

to find a responsible and satisfactory solution. From the outset, also the Atomic Energy Commission will, of course, be directing its attention to this issue and keep in close touch with biological and medical expertise.

I am grateful for the opportunity to speak at this national meeting of industry, since an active Danish participation in the use of atomic energy for the benefit of society will require the interest and support from many quarters, and will not least derive advantage from the experience and the production possibilities found in the diverse industrial establishments in our country. This is an issue which calls for the participation of the whole population and its advancement will require costly measures on the part of the state authorities. It will also be necessary to overcome the difficulties of having at our disposal, at the right time, a large and well-educated staff to solve the existing problems. In order that we in this country may be able to keep up with the development of the industrial resources, which scientific and technical advances in this as well as in many other fields promise us, it is necessary not only that society supports endeavours to maintain and continue our traditions as regards education and research, but also that, through information and guidance at home and in school, we ensure for such studies the necessary ever larger recruitment of young people, upon whose abilities and fresh potential the future of our country must be built.

[607]

XXVII. ATOMS AND SOCIETY

ATOMERNE OG SAMFUNDET
"Den liberale venstrealmanak,"
ASAs Forlag, Copenhagen 1956, pp. 25–32

Address at the Nordic Interparliamentary Meeting,
Copenhagen, 29 June 1955.

TEXT AND TRANSLATION

See Introduction to Part II, p. [365].

Niels Bohr

ATOMERNE
og
SAMFUNDET

★ ★ ★

Foredrag holdt ved Det
Nordiske Interparlamen-
tariske møde i København
den 12. juni 1955

Spørgsmålet om atomenergiens anvendelse som kraftkilde i samfundets tjeneste rummer fremtidsperspektiver, der må påkalde alles interesse. Det er derfor med glæde, at jeg har modtaget opfordringen til, sammen med de indbudte sagkyndige fra de andre nordiske lande, ved dette interparlamentariske møde at sige nogle ord om den videnskabelige baggrund for denne eventyrlige udvikling og om bestræbelserne for herhjemme, ligesom andetsteds, at følge med i udviklingen og blive delagtiggjort i de løfter, som den indebærer.

Før jeg går over til dette emne, vil jeg dog gerne fremhæve, at det store fremskridt i vor beherskelse af naturens kræfter, der er tale om, ligesom enhver anden forøgelse af vor viden og kunnen, selvfølgelig er forbundet med et større ansvar. I dette tilfælde drejer det sig om den alvorligste prøve for hele vor kultur, idet vi på den ene side står overfor muligheden for vidtgående forbedringer af levevilkårene overalt på jorden, medens på den anden side de forfærdende ødelæggelsesmidler, der er kommet menneskene i hænde, vil kunne bringe civilisationens beståen i fare, ja endda føre til udslettelse af alt organisk liv på vor klode, medmindre der findes veje til at bilægge mellemfolkelige stridigheder og til i fæl-

lesskab at indfri de løfter, som videnskabens fremskridt holder os for øje. Netop fordi denne situation må stå klart for enhver, er der ikke grund til at tabe modet, men langt snarere til at samle alle kræfter om virkeliggørelsen af de drømme om et samliv mellem folkeslagene i fred og tryghed, der aldrig har forladt menneskenes sind, og hvis opfyldelse trods alle vanskeligheder turde være nærmere end nogensinde før i kulturens historie. For vor stræben efter dette mål rummes måske den største opmuntring i det af De Forenede Nationer med så stor tilslutning hilste initiativ til fremme af det internationale samarbejde på atomenergiens fredelige udnyttelse.

Det indblik i atomernes opbygning og egenskaber, som vi i vore dage har fået, og som har stillet os så store opgaver, er resultatet af en lang opdagelsesrejse i ukendte egne, hvorved der skridt for skridt, ofte med lange mellemrum, vandtes stedse sikrere fodfæste og åbnedes videre udblik. De første vage forestillinger om atomernes verden går jo helt tilbage til oldtidens tænkere, der forstod, at kun antagelsen af en begrænset delelighed af alt stof kunne forklare den ejendommelige bestandighed, som stoffernes egenskaber udviser trods naturfænomenernes mangfoldighed. Jeg skal her blot erindre om, hvor vanskeligt det ville være at forstå, at vand ved opvarmning kan overgå i dampform, og vanddampen igen ved afkøling kan fortætte sig til en vædske med de oprindelige egenskaber, eller at denne vædske ved afkøling kan blive fast is, der igen kan smelte til sædvanligt vand, hvis vi ved sådanne tilstandsændringer havde at gøre med en gennemgribende forvandling af stoffet. Hvis vi derimod går ud fra, at vandmolekylerne forbliver uforandrede ved fordampningen eller frysningen, men blot i vanddampen er fjernet længere fra hverandre og derfor bevæger sig mere frit, og i isen er ordnet på mere regelbunden måde og derved giver stoffet en fastere struktur, ligger sagen jo ganske anderledes, og vi forstår, at alle vandets egenskaber må genfindes, når molekylerne ved dampens fortætning atter bringes hinanden nærmere, eller deres indbyrdes ordning ved isens smeltning bringes til ophør.

Det var dog først den store opblomstring af naturvidenskaben, der fulgte efter renaissancetiden, som skulle give atomforestillingerne fastere grundlag og videre udvikling. Jeg tænker især på de store fremskridt inden for kemien i det 18. århundrede, der medførte erkendelsen af den væsentlige forskel mellem grundstofferne og de kemiske forbindelser. Den kendsgerning, at vægten af de reagerende stoffer ved kemiske omsætninger inden for målenøjagtigheden er lig med vægten af de dannede produkter, førte endvidere sammen med opdagelsen af de simple lovmæssigheder, der gælder for de vægtmængder af et grundstof, som indgår i dets forskellige forbindelser med andre grundstoffer, til udvikling af enkle anskuelser om molekylernes atomare opbygning. For eksempel forstod man, at vandet er en kemisk forbindelse af grundstofferne hydrogen og oxygen, og at ethvert vandmolekyle netop består af et oxygen- og to hydrogenatomer. Når jeg her, som overalt i det følgende, benytter de latiniserede internationalt aftalte grundstofnavne som hydrogen og oxygen i stedet for de danske betegnelser brint og ilt, eller de norske og svenske vannstof og surstof, väte og syre, er det blot for at undgå de forvekslinger, der så hyppigt kan optræde selv i samkvem med frænder, der taler så nært beslægtede sprog.

I forbindelse med molekylforestillingernes udvikling skal jeg også kort erindre om erkendelsen af, at vi ved kulstoffets forbrænding, der så længe har været en hovedkilde til frembringelse af varme og energi for de forskelligste samfundsbehov, har at gøre med dannelsen af en kemisk forbindelse af grundstofferne carbon og oxygen. I kulsyren eller carbondioxyd, der sædvanligvis dannes ved forbrændingen, består hvert molekyle af et carbon-atom og to oxygen-atomer, medens der ved en mindre fuldstændig forbrænding kan dannes en anden forbindelse, carbon-monoxyd, hvor hvert carbon-atom er simpelt forbundet med et oxygen-atom. Til disse anskuelsers grundfæstelse bidrog også studiet af ændringer i stoffernes egenskaber under forskellige fysiske forhold som temperatur og tryk og elektriske påvirkninger. Således førtes man igennem udviklingen af den mekaniske varmeteori til forståelse af de simple love for forøgelsen af molekylernes bevægelse ved temperaturstigning, ligesom elektriske ladningers atomistiske struktur fremgik af de lovmæssigheder, der gælder for elektrolyse, og hvorefter den udskilte stofmængde altid står i et simpelt forhold til de pågældende atomers eller molekylers relative vægte.

Trods alle disse fremskridt og den følgende tids stedse større og mere frugtbare anvendelse af atomforestillingerne, blev dog lige op til vort århundrede antagelsen af atomernes eksistens almindeligvis betragtet som en hypotese, for hvilken der aldrig ville kunne føres endeligt bevis, idet man antog, at vore måleredskaber og sanseorganer, der jo selv består af utallige atomer, var for grove til direkte iagttagelse af stoffernes mindste dele. Ved forsøgsteknikkens vidunderlige forfinelse skulle det imidlertid blive muligt ikke alene at registrere virkninger af enkelte atomer, men endda at påvise, at atomerne trods grundstoffernes store bestandighed ikke er udelelige, og at opnå vidtgående oplysninger om deres struktur. Jeg tænker jo her på den række af opdagelser, der førte til erkendelsen af, at ethvert grundstofs atomer indeholder en kerne, der trods sin ringe størrelse i forhold til hele atomets udstrækning besidder næsten hele atomets masse. Til atomkernen, der tillige har en positiv elektrisk ladning, er bundet lettere negativt ladede elektroner i et antal, der almindeligvis gør hele atomet elektrisk neutralt. Antallet af positive elementarladninger i kernen svarer netop til det pågældende grundstofs nummer i det såkaldte naturlige system, inden for hvilket stoffernes fysiske og kemiske egenskaber som bekendt varierer på ejendommelig periodisk måde.

Dette simple billede af atomet, der i visse henseender minder om et solsystem i uhyre lille målestok, tillader os umiddelbart at forklare mange af stoffernes mest karakteristiske egenskaber. Således beror grundstoffernes bestandighed på atomkernens upåvirkelighed ved sædvanlige fysiske og kemiske processer, hvor det blot drejer sig om ændringer i elektronernes bindinger til kernerne. Endvidere forstår vi, at der på samme plads i grundstoffernes system, betegnet ved det såkaldte atomnummer, kan optræde isotoper, der trods vidtgående lighed i kemiske og fysiske egenskaber, som alene bestemmes af atomkernens elektriske ladning, kan have forskellig atomvægt svarende til en forskellig sammensætning af kernen. Erkendelsen af disse simple forhold stillede os direkte over for den opgave, der længe havde stået som ideal for naturbeskrivelsen, at føre redegørelsen for stoffernes egenskaber tilbage til en betragtning af rene tal. Denne opgaves løsning viste sig dog at kræve en gennemgribende revision af grundlaget for anvendelsen af de klassiske mekaniske forestillinger, der så fuldstændig beskriver planetbevægelserne omkring solen og funktionen af de mangfoldige maskiner, der benyttes i det praktiske liv. I den så umådelig meget mindre målestok, hvori de enkelte atomprocesser udspilles, møder vi nemlig helt nye lovmæssigheder, der er ansvarlige for atomstrukturernes ejendommelige stabilitet, og som forlanger vidtgående afkald på tilvante forklaringskrav. Som resultat af et intensivt samarbejde af en hel generation af fysikere er det lykkedes skridt for skridt at udvikle almene atommekaniske metoder og principper, hvoraf den klassiske mekanik fremtræder som et grænsetilfælde, og hvis anvendelse har givet os en stedse mere udtømmende beskrivelse af de fysiske og kemiske egenskaber, som stofferne udviser under sædvanlige forhold.

Den seneste udvikling af atomfysikken, der har ledt os på sporet af den i atomkernerne bundne store energi og af mulighederne for dens udvinding, tog imidlertid sit udgangspunkt i opdagelsen af de ejendommelige radioaktive omdannelser, som nogle af de i naturen forekommende grundstoffer undergår, og i påvisningen af grundstofforvandlinger frembragt ved særlig kraftige påvirkninger af atomkernen. De i forhold til energiomsætningerne ved kemiske processer så overordentlig store energi-bindinger og -frigørelser, der ledsager atomkerneomdannelserne, har deres direkte grund deri, at de elementarpartikler, hvoraf kernerne er sammensat, er langt tungere end elektronerne og i meget små afstande påvirker hverandre med stærke kræfter af en ganske anden art end de fra vore sædvanlige erfaringer kendte elektriske tiltræknings- og frastødningskræfter. Medens ingen påviselige vægtændringer finder sted ved kemiske processer, er energiomsætningerne ved atomkerneomdannelser så enorme, at der, i overensstemmelse med den i relativitetsteorien erkendte ækvivalens mellem energi og masse, er en direkte konstaterbar forskel imellem de samlede vægte af atomerne før og

efter de omhandlede processer. I denne forbindelse skal vi også erindre om, at det efter isotopernes opdagelse blev klart, at de sædvanlige kemiske atomvægte, der varierer på tilsyneladende uregelmæssig måde, blot fremkommer som middelværdier af atomvægtene for den i naturen forekommende isotopblanding af hvert grundstof. For adskilte isotoper gælder der langt simplere forhold, idet atomvægtene, regnet med hydrogenets som enhed, kun udviser små afvigelser fra hele tal, svarende til, at kernedelene, de såkaldte nukleoner, alle har nær den samme vægt, der kun i mindre grad ændres ved nukleonernes binding i atomkernen.

Som bekendt optræder nukleonerne i to former, protonen og neutronen, hvoraf den første, bortset fra fortegn, har samme elektriske ladning som elektronen, medens neutronen, som navnet angiver, er elektrisk neutral. Omend der af protoner og neutroner kan dannes stabile kerner, er en neutron i fri tilstand ikke selv stabil, men omdannes efter en middel-levetid på ca. 20 minutter til en proton og en elektron. Selv om forholdene i atomkernerne således i mange henseender frembyder væsentlig nye træk, har det, ved anvendelse af de samme atommekaniske principper, som behersker elektronernes binding i atomerne, været muligt at opnå et stedse klarere overblik over nukleonbindingen i atomkernerne og vidtgående gøre rede for disses karakteristiske egenskaber. Det vil imidlertid føre alt for langt at gå nærmere ind på denne interessante og lovende udvikling, men for at give baggrund for det følgende skal jeg blot her minde om sammensætningen af nogle af de simplest byggede kerner og de først undersøgte kerneomdannelser.

Den enkleste kerne har man jo i sædvanligt hydrogen, hvor den simpelthen består af en enkelt proton; derimod er kernen i den sjældne, dobbelt så tunge hydrogen-isotop, deuterium, sammensat af en proton og en neutron. Ved særlige atomkerneprocesser dannes endvidere en hydrogen-isotop med tredobbelt vægt, det såkaldte tritium, hvis kerne består af en proton og to neutroner. I modsætning til det stabile deuterium har tritium kun en begrænset levetid, idet kernen i løbet af nogle år under elektron-udsendelse omdannes til en heliumkerne bestående af to protoner og en neutron. Til sammenligning skal vi erindre om, at kernen i den almindelige helium-isotop med atomvægt 4 består af to protoner og to neutroner, som indgår en særlig fast forbindelse, ved hvis dannelse store energimængder frigøres. Ved de i solens indre stadig foregående atomkerneprocesser er jo netop opbygningen af sådanne heliumkerner kilden til den enorme varmeudstråling fra solens overflade, der igennem milliarder af år har opretholdt det organiske liv på jorden, men hvis oprindelse man stod ganske uforstående overfor indtil for få år siden.

Det næste stof i grundstoffernes system, lithium, optræder i to isotoper med atomvægtene 6 og 7, hvis kerner indeholder tre protoner og henholdsvis tre og fire neutroner. Den sidstnævnte og hyppigst forekommende lithiumkerne kan ved at rammes af en proton omdannes til to heliumkerner, der ved adskillelsen opnår en bevægelsesenergi mange gange større end den energi, det er nødvendigt at give protonen for at fremkalde processen. Medens man tidligere kun havde kunnet frembringe kerneomdannelser ved hjælp af de fra naturligt radioaktive stoffer udsendte energirige partikler, var netop lithium-omdannelsen det første eksempel på en i laboratoriet ved kunstigt accelererede partikler iværksat kernespaltning og gav derfor anledning til de første alvorlige diskussioner om muligheden for atomenergiens praktiske udnyttelse. På dette tidspunkt var sådan udnyttelse dog udelukket, idet de dannede heliumkerner har altfor ringe lejlighed til at frembringe nye kerneomdannelser, førend de har mistet deres energi ved vekselvirkning med elektronerne i de stoffer, de gennemtrænger. Grunden til, at kerneprocesserne i solen kan fortsættes på kædelignende måde, er jo den høje temperatur i solens indre og den deraf betingede stærke varmebevægelse af alle partiklerne.

Ved at gå videre i grundstoffernes række kommer vi dernæst til beryllium, der kun har en isotop med atomvægt 9, og hvis kerne består af fire protoner og fem neutroner. En af disse sidste kan imidlertid udjages ved sammenstød med en fra radium udsendt hurtig heliumkerne, hvorved der bliver seks protoner og seks neutroner tilbage, der danner en karbon-kerne med meget fast binding. Det var netop denne proces, der førte til opdagelsen af neutronen og i mange år var den vigtigste kilde til frembringelse af de neutroner, der blev et så effektivt middel til iværksættelse af kerneomdannelser af alle mulige grundstoffer, idet de ved ikke at besidde elektrisk ladning hverken afgiver deres energi til elektronerne eller møder nogen frastødning fra atomkernerne og derfor let indfanges i disse. Ved sådanne indfangningsprocesser lykkedes det for mange grundstoffers vedkommende at fremstille radioaktive isotoper, der på grund af den simple konstatering af deres tilstedeværelse igennem registreringen af de udsendte elektroner fandt rig anvendelse som indikatorer for de tilsvarende stabile grundstoffers deltagelse i kemiske omsætninger og stofskiftet hos levende organismer.

En ny epoke i udviklingen, der skulle få de største konsekvenser, indledtes imidlertid i året før udbrudet af den sidste verdenskrig ved den iagttagelse, at der ved neutronbestråling af uran, som med atomnummeret 92 indtil da var det sidste i grundstoffernes række, ikke alene dannes uranisotoper, der ved radioaktive omdannelser overgår til grundstoffer med endnu højere atomnumre, men tillige fremkommer radioaktive isotoper af grundstoffer med langt mindre atomnumre og atomvægte. Det blev klart, at vi her har at gøre med en spaltning af urankernen i to dele, der hver især indeholder omkring halvdelen af de forhåndenværende protoner og neutroner. Som følge af deres store elektriske ladning får disse kernefragmenter ved deres indbyrdes frastødning en langt større bevægelsesenergi end tilfældet er ved omdannelsen af lettere kerner. Den afgørende betydning af opdagelsen af uranspaltningen, der almindeligt betegnes som atomkernefission, var imidlertid ikke alene den store energiudvikling ved de enkelte spaltningsprocesser, men først og fremmest, at sådan fission simpelt kunne fortsættes ved kædeprocesser. Dette beror på, at der ved hver kernefission som følge af den store indre bevægelse i kernefragmenterne så at sige ved fordampning fra hver af disse udsendes nye neutroner, der igen kan frembringe fissionsprocesser i andre urankerner. Det blev derved for første gang muligt under laboratorieforhold at vedligeholde atomkerneomdannelser ved kædeprocesser og derved frigøre en betydelig del af den i kernerne oplagrede store energi.

For de perspektiver, der herved åbnedes, var det tillige af største vigtighed, at man hurtigt blev klar over de meget forskellige roller, som de to i naturen forekommende uranisotoper med atomvægtene 235 og 238 spiller for de omhandlede indfangnings- og fissionsprocesser. Medens den tungere og langt hyppigere uranisotop kun vanskeligt spaltes ved neutronbestråling, vil derimod den lettere isotop, der udgør mindre end 1 procent af det i naturen forekommende uran, spaltes selv af neutroner med ringe hastighed. Denne forskel beror på, at der ved neutronens oprindelige indfangning i uran-235 — i modsætning til i uran-238 — dannes en urankerne, der indeholder et lige antal neutroner. Efter de almindelige regler, der gælder for nukleonbindingen i atomkernerne, kan disse neutroner derfor blive så fast bundne til hverandre, at der ved indfangningen frigøres en energi tilstrækkelig til at sætte spaltningsprocessen i gang. Klarlæggelsen af disse forhold blev udgangspunktet for konstruktionen af atombomber, idet man i rent uran-235, fremstillet i kæmpemæssige isotop-separationsanlæg, kan fremkalde lavineagtige eksplosionsprocesser med enorme virkninger i sammenligning med sædvanlige kemiske eksplosivers.

Som bekendt lykkedes det imidlertid allerede under krigen ved benyttelse af naturligt uran at opretholde kontrollerbare atomkerne-kædeprocesser under særlige betingelser, der sikrer den nødvendige neutronbalance, som kræver, at der ved fission af uran-235 leveres ligeså mange nye neutroner, som der bruges til fissionen og tabes ved indfangning i såvel det tilstedeværende uran-238 som i de andre for uranreaktorens kon-

To faser i en atombombeeksplosion   Nevadaørkenen i 1952

struktion nødvendige materialer. Denne neutron-
økonomi opnås ved, at uranstykker i metallisk
form på passende måde fordeles i en såkaldt moderator, der består af stoffer som grafit eller
tungt vand, hvis atomkerner ikke indfanger neutronerne, men blot fra disse ved de stadige sammenstød borttager bevægelsesenergi, indtil neutronernes hastigheder i middel svarer til molekyl-
bevægelsen ved den pågældende temperatur. Under sådanne forhold vil nemlig neutronernes fis-
sionsvirkninger på uran-235-kernerne i så høj
grad begunstiges i forhold til indfangningerne i de
langt hyppigere uran-238-kerner, at der forbliver
et overskud af neutroner til at opretholde kædeprocessen. For at sikre, at mængden af de neutroner, der gennemkrydser reaktorens indre, holdes på et passende niveau og ikke stiger på uøn-
sket måde, er der endvidere i reaktoren anbragt
kontrolstænger af neutron-absorberende materialer, der automatisk kan skydes mere eller mindre
dybt ind i reaktoren. At det er muligt at operere
tilstrækkeligt hurtigt med sådanne kontrolstænger, beror derpå, at frigørelsen af neutroner vel
for størstedelens vedkommende følger umiddelbart efter fissionen, men at der tillige fra nogle af
de stærkt radioaktive spaltningsprodukter udsendes såkaldte forsinkede neutroner, der, til
trods for deres forholdsvis ringe antal, spiller en
afgørende rolle for ligevægtens opretholdelse.

Ved de første under krigen konstruerede reaktorer, der holdtes på en lav temperatur ved
kraftig afkøling, drejede det sig dog ikke i første
linie om energifrembringelse. Hovedformålet var
derimod ved neutronbestråling i uran-238 at fremstille plutonium, hvis atomkerne dannes ved successive elektronudsendelser fra den radioaktive
oranisotop-239, og som i kemisk ren tilstand på
lignende måde som uran-235 kan anvendes som
eksplosiv i atombomber. I denne sammenhæng
skal jeg også blot lige berøre, at det som bekendt
ved benyttelse af sådanne eksplosiver som tænd-
sats i de så meget omtalte hydrogen-bomber har
været muligt at skabe forhold, under hvilke enorme mængder af energi og radioaktivitet frembringes ved såkaldte termonukleare processer. For
undgåelsen af de farer for civilisationen, som anvendelsen af sådanne forfærdende ødelæggelsesmidler indebærer, må vi jo som allerede fremhævet sætte vor lid til udbredelsen af forståelse af
det fælles ansvar. For det mellemfolkelige samarbejde på atomenergiens fredelige udnyttelse, til
hvilket der netop i sådan henseende må stilles
store forhåbninger, har imidlertid selv de mere
militærtekniske undersøgelser ydet betydningsfulde bidrag. Således giver plutonium-fremstil-
lingen mulighed for til energiudvinding at udnytte ikke blot den naturlige uran-235, der med
det nuværende energiforbrug kun ville slå til i et

[615]

meget begrænset antal år, men for efterhånden at omdanne hele uranforekomsten på jorden til såkaldt fissilt materiale egnet til brændstof i atomkraftværker. Endvidere er det muligt yderligere betydeligt at forøge det tilgængelige atombrændstof ved i uran- og plutonium-reaktorer at bestråle thorium med neutroner og derved omdanne dette til en uranisotop med atomvægt 233, der igen har lignende fissile egenskaber som uran-235 og plutonium.

På dette grundlag og med disse udsigter er der i de lande, der har været førende i den tekniske udvikling, gjort et energisk arbejde på atomenergiens anvendelse, og som bekendt er større atomkraftværker allerede under udførelse. For at blive i stand til at overse den mest økonomiske og praktiske konstruktion af sådanne kraftværker må man imidlertid endnu foretage omfattende videnskabelige og teknologiske undersøgelser. Det drejer sig jo her om forhold, der afviger overmåde meget fra dem, hvorunder vi hidtil har vundet erfaringer om stoffernes i teknik og industri anvendte egenskaber, idet især det store antal neutroner, der stadig gennemkrydser atomkraftværkerne, vil have mange slags særlige virkninger. Ikke alene frembringes ved kerneomdannelserne store mængder af radioaktive stoffer, hvis behandling kræver den yderste forsigtighed, men ved deres sammenstød med atomkernerne vil neutronerne tillige i stort mål fremkalde ændringer i de kemiske forbindelser, ja selv de i faste materialer, der er nødvendige for alle tekniske konstruktioner, idet atomerne så at sige stadig vil blive puffet bort fra deres ligevægtsstillinger i krystalgitrene og derved give anledning til metallurgiske forandringer af hidtil lidet studeret karakter. Under dette forskningsarbejde har benyttelsen af fissile stoffer som uran-235 og plutonium i mere eller mindre ren tilstand været af allerstørste betydning, idet deres anvendelse tillader under mindre komplicerede og mere overskuelige forhold at frembringe forsøgsbetingelser, der svarer til forholdene i store fremtidige atomkraftværker.

Selv om man jo uden for de store lande hidtil har måttet nøjes med at være mere eller mindre fjerntstående tilskuere til denne intensive tekniske atomforskning, har man dog naturligvis også i de nordiske lande søgt på enhver tilgængelig måde at deltage i forskningsarbejdet og selv at erhverve sådan uddannelse og erfaring, der er nødvendig for at udvide denne deltagelse. I Norge har man jo gennem mange år som biprodukt, ved den på vandkraft baserede store produktion af hydrogen og oxygen ved elektrolyse, fremstillet anseelige mængder af tungt vand og derved forsynet laboratorier verden over med dette for atomundersøgelser så værdifulde hjælpemiddel.

Få år efter krigens afslutning blev der da også, på grundlag af tungt vand fra Norsk Hydro og en betydelig mængde uran, der kort før krigsudbrudet var erhvervet af laboratoriet i Leiden, og som det lykkedes at holde skjult under den tyske besættelse, indledt et norsk-hollandsk samarbejde, der har ført til opbygningen af en atomreaktor i Kjeller i nærheden af Oslo, som har været i drift siden 1952. Også i Sverige er som bekendt en reaktor i drift, og flere under planlæggelse, med benyttelse af det uran, som findes i de store svenske uranholdige skiferaflejringer, og med hvis teknisk vanskelige udvinding man gik i gang umiddelbart efter krigen.

Her i Danmark, hvor man har været afskåret fra tilgang af de nødvendige udgangsmaterialer, har vi hidtil ikke kunnet bygge atomreaktorer og har været henvist til efter bedste evne at følge med i den rent videnskabelige udvikling på atomkerneforskningens område. Imidlertid er situationen i vort tilfælde, som i mange andre ligestillede nationers, blevet gennemgribende ændret ved det allerede omtalte initiativ til fremme af det internationale videnskabelige samarbejde på den fredelige udnyttelse af atomenergien og især ved tilbud fra Amerika og England om til dette formål at stille betydelige mængder af beriget uran til rådighed. Som det har været meddelt, er der således fornylig imellem USA og Danmark truffet en aftale, hvorefter den amerikanske atomenergikommission vil holde os forsynet med en i forhold til det naturlige uran stærkt beriget isotopblanding indeholdende 6 kg uran-235 til driften af en forsøgsreaktor, der vil blive konstrueret under ledelse af den nyoprettede danske atomenergikommission, og som vi regner med vil være færdig til brug i løbet af et års tid. Endvidere er der med statsmyndighedernes samtykke truffet aftale mellem den britiske Atomic Energy Authority og den danske atomenergikommission om et samarbejde, i forbindelse med hvilket der her i landet vil blive bygget en forsøgs-reaktor af samme type som den high-flux reactor, hvis opførelse er påbegyndt i det britiske Atomic Energy Research Establishment i Harwell. Til denne reaktor, som det vil tage mellem to og tre år at bygge, og som er konstrueret til særlige teknologiske undersøgelser med direkte sigte på at vinde erfaringer for konstruktionen af atomkraftværker, vil vi fra engelsk side blive forsynet med fissilt materiale i form af højt beriget uran, medens den til driften nødvendige mængde af tungt vand vil blive leveret fra Amerika, hvor man har udviklet nye metoder til dets fremstilling i industriel målestok.

*) Halvøen Risø i Roskilde fjord (red.)

[616]

Sådan bistand til fremme af det internationale samarbejde vil også blive ydet mange andre lande i tilsvarende situation, og det er nylig blevet meddelt, at USA allerede har sluttet aftale med 16 lande i Europa og andre verdensdele. For Danmarks vedkommende vil gennemførelsen af de omhandlede planer tillige give os mulighed for at udveksle erfaringer med vore nordiske nabolande, ligesom vi jo håber stadig at kunne uddybe det længe bestående nøje samarbejde inden for den teoretiske atomforskning. Alt i alt forventer vi, at der inden for de nordiske lande vil kunne gives frugtbare bidrag til atomkraftens anvendelse for samfundets tarv. Hvornår man her vil kunne skride til at bygge atomkraftværker til elektricitetsforsyning, er det vanskeligt i øjeblikket at forudse, men i Danmark ligesom i andre lande vil man begynde derpå, såsnart man ved hjælp af de indhøstede erfaringer og på baggrund af de herskende økonomiske forhold kan skønne, at sådanne værker allerede vil være rentable, eller at prøvelsen af deres drift vil være nødvendig for at sikre en fremtidig rentabilitet.

Til overvejelse for de danske statsmyndigheder er atomenergikommissionen i færd med at udarbejde forslag til oprettelse af en atomenergiforsøgsstation. På denne, der vil blive anbragt på en passende isoleret grund i ikke for stor afstand fra København*), vil der foruden de omtalte atomreaktorer blive opført de for driften nødvendige fysiske og kemiske laboratorier og mekaniske og elektrotekniske værksteder. — Ligesom i de store lande vil foretagender af denne art stille vidtgående krav til sagkundskab såvel i videnskabelig som teknisk henseende og vil forlange et meget betydeligt uddannelsesprogram. Værdifuld hjælp er her tilbudt fra de specielle reaktorskoler, der er indrettet såvel i Amerika som England, men desuden vil det være nødvendigt i Danmark som andetsteds stadig at udvide undervisning og forskning ved universiteterne og de tekniske højskoler med henblik på de store krav, som den stadig voksende industrialisering på dette som på så mange andre områder stiller.

*F*ør jeg slutter, skal jeg endnu kort minde om de særlige problemer, som er knyttet til al forskning og teknisk udvikling på atomenergiens område på grund af de farer, der er forbundet med den i reaktorerne frembragte stråling og behandlingen af de i disse dannede radioaktive stoffer. I modsætning til, hvad der gælder for virkningerne af atombombeeksplosioner, turde imidlertid en betryggende beskyttelse af personalet ved atomkraftværker og atomenergiforsøgsstationer og den omkringboende befolkning kunne opnås ved passende sikkerhedsforanstaltninger, udarbejdet med fuld hensyntagen til den medicinske sagkundskab. Også i sådanne henseender forventer vi jo at få værdifulde oplysninger gennem den af De Forenede Nationer i Genève i august arrangerede kongres, hvor ikke alene de med atomenergiens udnyttelse forbundne fysiske og tekniske spørgsmål vil blive indgående diskuteret, men hvor også den hele samfundsmæssige baggrund vil blive alsidigt belyst. Først og fremmest håber vi imidlertid, at denne kongres på afgørende måde vil bidrage til det mellemfolkelige samarbejde på atomenergiens fredelige anvendelse til gavn for hvert enkelt samfund, og gennem fjernelsen af skranker for gensidig oplysning vil fremme den forståelse af folkeslagenes fælles interesser, der er betingelsen for et harmonisk samliv til højnelse af menneskekulturen i alle dens afskygninger.

TRANSLATION

Niels Bohr

ATOMS
and
SOCIETY

* * *

*Address held
at the Nordic
Interparliamentary
Meeting in
Copenhagen
12 June 1955[1]*

The question of the application of atomic energy as a source of power in the service of society holds perspectives for the future which must invoke the interest of everyone. It is therefore with pleasure that I have received the request to say a few words at this interparliamentary meeting, together with the invited experts from the other Nordic countries,[2] about the scientific background of this marvellous development and about the endeavours in our country, just as elsewhere, to keep up with the development and to be given a share in the promises it implies.

[1] The correct date is 29 June 1955.
[2] Speeches were also given by Gunnar Randers (1914–1992) from Norway and Rickard Johannes Sandler (1884–1964) from Sweden. See the introduction to Part II, p. [365].

[618]

Before I turn to this topic, I should, however, like to emphasize that the great advance in our mastery of the forces of nature in question, just as any other increase in our knowledge and ability, is naturally bound up with a greater responsibility. In this case what is involved is the most serious test of our whole culture in that, on the one hand, we face the possibility of far-reaching improvements of the living conditions everywhere on Earth, while, on the other hand, the terrible means of destruction, which have come into the hands of mankind, will be able to endanger the existence of civilization, indeed even lead to annihilation of all organic life on our globe, unless ways are found to settle international conflicts and to fulfil together the promises that the progress of science holds. Precisely because this situation must be obvious to everybody, there is no reason to lose courage, but much rather to concentrate all strength on the realization of the dreams of coexistence between all peoples in peace and security that have never disappeared from people's minds and whose fulfilment, despite all difficulties, ought to be closer than ever before in the history of civilization. The greatest encouragement for our striving after this goal is perhaps found in the initiative by the United Nations for the advancement of international cooperation for the peaceful use of atomic energy, which has received such great approval.[3]

The insight we have achieved in our time into the structure and properties of atoms, and which has presented us with such great tasks, is the result of a long voyage of discovery in unfamiliar territory whereby, step by step and often with long intervals, ever more secure foothold was gained and wider vistas were opened. The first vague ideas about the world of atoms go all the way back to the thinkers of Antiquity who understood that only the assumption of a limited divisibility of all matter could explain the peculiar stability displayed by the properties of substances despite the diversity of natural phenomena. I shall here only remind you how difficult it would be to understand that water, on being heated, can change into the state of vapour and that the vapour, on being cooled, can again condense into a liquid with the original properties, or that this liquid, on being cooled, can become solid ice which can again melt into ordinary water, if with such changes of state we had to do with a fundamental transformation of the substance. However, if we assume that the water molecules remain unchanged on the evaporation or the freezing, but

---

[3] Bohr here refers to the developments set in motion by President Eisenhower's proposal of his Atoms for Peace programme. See the introduction to Part II, p. [360].

in the water vapour are only removed farther from one another and therefore move more freely, and in the ice are arranged in a more orderly fashion and thereby give the substance a more solid structure, the matter is quite different, and we understand that all the properties of water must reappear when, on the condensation of the vapour, the molecules are again brought closer to one another, or their mutual ordering comes to an end on the melting of the ice.

It was, however, only the great flowering of natural science following upon the Renaissance that should give the ideas of the atom a more solid basis and further development. I think especially of the great advances within chemistry in the 18th century, which led to the recognition of the essential difference between the elements and the chemical compounds. The fact that, within the accuracy of measurement, the weight of the reacting substances on chemical conversions is equal to the weight of the products formed, led furthermore, together with the discovery of the simple regularities holding for the quantities of weight of an element entering into its various combinations with other elements, to the development of simple ideas of the atomic structure of molecules. For example, it was understood that water is a chemical compound of the elements hydrogen and oxygen and that each water molecule consists precisely of one oxygen atom and two hydrogen atoms. When here, as everywhere in the following, I use the latinized names of the elements agreed upon internationally, such as hydrogen and oxygen, instead of the Danish terms "brint" and "ilt" or the Norwegian and Swedish "vannstof" and "surstof", "väte" and "syre", it is only in order to avoid the confusion which can so frequently occur even in communication between kinsmen who speak so closely related languages.

In connection with the development of the conceptions of molecules, I shall also briefly remind you of the recognition that on the combustion of carbon, which for so long has been a major source for the production of heat and energy for the most diverse needs of society, we have to do with the formation of a chemical compound of the elements carbon and oxygen. In carbonic acid, or carbon dioxide, which is usually formed on the combustion, each molecule consists of one carbon atom and two oxygen atoms, while on a less complete combustion another compound, carbon monoxide, can be formed, where each carbon atom is simply combined with one oxygen atom. The study of the changes in the properties of substances under different physical conditions, such as temperature and pressure and electrical influences, also contributed to the consolidation of these ideas. Thus one was led, through the development of the mechanical theory of heat, to an understanding of the simple laws for the increase in the movement of the molecules on a rise of temperature, just as the atomic structure of electrical charges became apparent from the regularities holding for electrolysis, and according to which the quantity of material separated always stands in a simple ratio to the relative weights of the atoms or molecules in question.

Despite all these advances and the ever greater and more fruitful application of the conceptions of the atom in the following time, the assumption of the existence of atoms was nevertheless generally regarded right up to our century as a hypothesis, of which it would never be possible to produce final proof, since it was assumed that our measuring instruments and sense organs, themselves made up of innumerable atoms, were too coarse to be able to observe the smallest parts of matter directly. By the marvellous refinement of experimental technique it should, however, become possible not only to register effects of the individual atoms, but even to demonstrate that, despite the great stability of the elements, atoms are not indivisible and to obtain far-reaching information about their structure. Here I am referring to the series of discoveries that led to the recognition that the atoms of any element contain a nucleus that, despite its small size compared with the whole extension of the atom, holds nearly the whole mass of the atom. To the atomic nucleus, which furthermore has a positive electric charge, are bound lighter negatively charged electrons in such a number as generally to make the whole atom electrically neutral. The number of positive elementary charges in the nucleus corresponds precisely to the number of the element in question in the so-called natural system, in which, as is well known, the physical and chemical properties vary in a remarkable periodic manner.

This simple picture of the atom, which in certain respects is reminiscent of a solar system on an extremely small scale, allows us to explain straight away many of the most characteristic properties of the substances. Thus the stability of the elements is due to the immutability of the atomic nucleus in ordinary physical and chemical processes, where only changes in the bindings of the electrons to the nuclei are involved. Furthermore, we understand that, at the same place in the system of the elements, designated by the so-called atomic number, there can appear isotopes which, despite far-reaching similarity in chemical and physical properties determined alone by the electric charge of the atomic nucleus, can have different atomic weights corresponding to a different composition of the nucleus. The recognition of these simple circumstances confronted us directly with the problem, which had long stood as the ideal for the description of nature, of bringing the explanation of the properties of substances back to a consideration of pure numbers. However, the solution of this problem proved to require a thorough revision of the basis for the application of the classical mechanical ideas, which so fully describe the planet

[621]

movements around the sun and the function of the numerous machines that are used in practical life. In the so extremely much smaller scale where the individual atomic processes take place, we meet, in fact, quite new regularities which are responsible for the remarkable stability of the atomic structures and which require far-reaching renunciations of customary requirements for explanation. As a result of an intensive cooperation by a whole generation of physicists, it has been possible to develop, step by step, general atomic-mechanical methods and principles, of which classical mechanics appears as a borderline case, and whose application has given us an ever more exhaustive description of the physical and chemical properties displayed by substances under ordinary conditions.

The latest development in atomic physics, which has led us onto the track of the great energy tied up in the atomic nuclei and of the possibilities for extracting it, had its starting point, however, in the discovery of the remarkable radioactive transformations undergone by some of the elements occurring in nature, and in the demonstration of the transmutations of elements induced by especially strong influences on the atomic nucleus. The energy bindings and releases – so enormously large in comparison with the energy conversions in chemical processes – which accompany the transformations of the atomic nucleus, have their direct basis in the fact that the elementary particles constituting the nuclei are much heavier than the electrons and, at very small distances, affect each other with strong forces of a quite different kind than the electric forces of attraction and repulsion known from our ordinary experience. While no demonstrable changes in weight take place on chemical processes, the energy conversions on transformations of the atomic nuclei are so enormous that, in accordance with the equivalence between energy and mass recognized in the theory of relativity, there is a directly discernable difference between the total weight of the atoms before and after the processes in question. In this connection we should also remember that it became clear, after the discovery of isotopes, that the ordinary chemical atomic weights, which vary in a seemingly irregular manner, only appear as mean values of the atomic weights of the isotopic mixture of each element occurring in nature. Far simpler conditions hold for separated isotopes, as the atomic weights, calculated with that of hydrogen as the unit, only show small deviations from whole numbers, corresponding to the circumstance that the nuclear constituents, the so-called nucleons, all have nearly the same weight, which is only changed to a small degree on the binding of the nucleons in the atomic nucleus.

As is well known, nucleons appear in two forms, the proton and the neutron, of which the first, apart from the sign, has the same electric charge as the electron, while the neutron, as the name indicates, is electrically neutral. Even though stable nuclei can be formed from protons and neutrons, a neutron in a free state is not itself stable but is transformed to a proton and an electron after a mean life of about 20 minutes. Although the conditions in the atomic nuclei thus in many ways present essentially new features, it has been possible, by application of the same atomic-mechanical principles that control the binding of the electrons in the atoms, to achieve an ever clearer comprehension of the nucleon binding in atomic nuclei and to give a far-reaching explanation of the characteristic properties of these. It would, however, go much too far to enter into more detail about this interesting and promising development, but in order to give the background for what follows I will here just remind you of the composition of some of the most simply constructed nuclei and the nuclear transformations examined first.

The most simple nucleus is found in ordinary hydrogen, of course, where it simply consists of a single proton; on the other hand, the nucleus in deuterium, the rare isotope of hydrogen which is twice as heavy, is composed of one proton and one neutron. Moreover, on special nuclear processes an isotope of hydrogen with three times the weight, the so-called tritium, whose nucleus consists of one proton and two neutrons, is formed. In contrast to the stable deuterium, tritium has only a limited life, as the nucleus in the course of some years with electron emission is transformed into a helium nucleus consisting of two protons and one neutron. For comparison, we should mention that the nucleus in the ordinary helium isotope with atomic weight 4 consists of two protons and two neutrons, which enter into a particularly firm combination, on whose creation great amounts of energy are released. In the nuclear processes constantly taking place in the interior of the sun, it is precisely the formation of such helium nuclei that is the source of the enormous heat radiation from the sun's surface, which throughout billions of years has sustained organic life on Earth, but whose origin was quite impossible to understand until a few years ago.

The next substance in the system of elements, lithium, appears in two isotopes with atomic weights 6 and 7, whose nuclei contain three protons and three and four neutrons, respectively. The latter and most frequently occurring lithium nucleus can, on being hit by a proton, be transformed into two helium nuclei, which on the separation obtain a kinetic energy many times greater than the energy that it is necessary to give to the proton in order to bring about the process. While previously it was only possible to induce nuclear transformations with the help of high-energy particles emitted from naturally radioactive substances, it was precisely the lithium transformation which was the first example of a nuclear fission brought about in the laboratory using artificially accelerated particles and which therefore gave rise to the first serious discussions about the possibility of the practical application of atomic energy. At that time such an application was impossible, however, since the helium nuclei formed have much too small a chance of inducing new nuclear transformations before they have lost their energy through interaction with the electrons in the substances they penetrate. The reason that the nuclear processes inside the sun can be continued in a chain-like fashion is of course the high temperature in the interior of the sun and the strong thermal movement of all the particles caused by this.

On going further in the series of elements, we meet next beryllium, which has only one isotope of atomic weight 9, and whose nucleus consists of four protons and five neutrons. One of the latter can, however, be expelled by collision with a rapid helium nucleus emitted from radium, whereby six protons and six neutrons remain forming a carbon nucleus with very firm binding. It was precisely this process that led to the discovery of the neutron and for many years was the most important source for the production of the neutrons that became such an efficient means for bringing about nuclear transformations of all kinds of elements, as, by not possessing an electric charge, they neither transfer their energy to the electrons nor encounter any repulsion from the atomic nuclei and can therefore be easily captured in these. By such capture processes it proved possible in the case of many elements to produce radioactive isotopes which, because of the simple recognition of their presence through the registration of the emitted electrons, found rich use as indicators of the involvement of the corresponding stable elements in chemical processes and the metabolism of living organisms.

A new epoch in the development, which was to have the greatest consequences, was begun, however, in the year before the outbreak of the last world war by the observation that, on neutron irradiation of uranium, which with the atomic number 92 was until then the last in the series of elements, not only uranium isotopes are formed, which by radioactive transformations turn into elements of even higher atomic numbers, but, in addition, radioactive isotopes of elements with much lower atomic numbers and atomic weights appear. It became clear that here we have to do with a splitting of the uranium nucleus into two parts, each of which containing about half of the protons and neutrons present. As a result of their great electric charge, these nuclear fragments obtain, on their mutual repulsion, a much greater kinetic energy than is the case with the transformation of lighter nuclei. The decisive significance of the

Two phases of an atomic bomb explosion in the Nevada Desert in 1952.

discovery of uranium splitting, which is usually termed nuclear fission, was, however, not alone the great generation of energy on the individual splitting processes, but first and foremost that such fission could be continued in a simple manner by chain processes. The reason for this is that on each nuclear fission, as a result of the great internal movement within the nuclear fragments so to say on evaporation from each of these, new neutrons are emitted, which in turn can induce fission processes in other uranium nuclei. It thereby became possible for the first time under laboratory conditions to sustain nuclear transformations by means of chain processes and thereby release a considerable part of the great energy stored in the nuclei.

For the perspectives hereby revealed it was, in addition, of the greatest importance that one quickly recognized the very different roles that the two uranium isotopes found in nature with atomic weights 235 and 238 play for the capture and fission processes described here. While the heavier and much more frequently occurring uranium isotope can only be split with difficulty by means of neutron irradiation, the lighter isotope, which makes up less than 1 percent of the uranium found in nature, is, on the other hand,

[625]

split even by neutrons with low velocity. This difference is due to the fact that on the neutron's original capture in uranium 235 – as opposed to in uranium 238 – a uranium nucleus is formed containing an even number of neutrons. According to the general rules valid for the binding of nucleons in atomic nuclei, these neutrons can therefore become so firmly bound to each other that, on the capture, sufficient energy is released to set off the fission process. The clarification of these circumstances became the starting point for the construction of atomic bombs, as it was possible, in pure uranium 235 produced in huge isotope separation plants, to develop avalanche-like explosion processes with enormous effects as compared to ordinary chemical explosives.

As is well known, it proved possible, however, already during the war, to sustain, with the use of natural uranium, controllable nuclear chain processes under special conditions that ensure the necessary neutron balance, which requires that, on the fission of uranium 235, just as many new neutrons are supplied as are used for the fission and lost by capture in the uranium 238 present, as well as in the other materials necessary for the construction of the uranium reactor. This neutron economy is achieved by pieces of uranium in metallic form being distributed in a suitable way in a so-called moderator, which consists of substances such as graphite or heavy water, whose nuclei do not capture the neutrons, but on the constant collisions only remove kinetic energy from them, until the velocities of the neutrons on average correspond to the molecular movement at the temperature in question. Under such conditions the fission effects of the neutrons on the uranium 235 nuclei will in fact be favoured to such a high degree in relation to the captures in the much more prevalent uranium 238 nuclei that there remains a surplus of neutrons to sustain the chain reaction. To ensure that the quantity of the neutrons traversing the inside of the reactor is kept at a suitable level and does not increase in an undesirable way, control rods of neutron-absorbing materials are, moreover, placed in the reactor, which can automatically be slid more or less deeply into the reactor. The reason that it is possible to operate sufficiently quickly with such control rods is that although the release of neutrons, certainly for the greatest part, follows immediately after the fission, in addition, so-called delayed neutrons are emitted from some of the strongly radioactive fission products, which, despite their relatively small number, play a decisive role for maintaining of the equilibrium.

For the first reactors constructed during the war, which were kept at a low temperature by strong refrigeration, the production of energy was not of the first importance, however. On the contrary, the main objective was, by means of neutron irradiation of uranium 238, to produce plutonium, whose atomic nucleus is formed by successive electron emissions from the radioactive uranium isotope 239, and which in a chemically pure state can be used as an explosive in atomic bombs in a similar way to uranium 235. In this connection, I will only just mention that, as is well known, it has been possible, with the use of such explosives as detonators in the so much discussed hydrogen bombs, to create conditions under which enormous amounts of energy and radioactivity are produced by so-called thermonuclear processes. In order to avoid the dangers for civilization which the use of such terrifying means of destruction implies, we must, of course, as already emphasized, put our trust in the spreading of an understanding of the shared responsibility. However, even the investigations directed primarily towards military technology have provided important contributions as regards international cooperation for the peaceful use of atomic energy, in which there, precisely in this respect, must be placed great hopes. Thus the production of plutonium opens the possibility to utilize, for the extraction of energy, not only the naturally occurring uranium 235, which with the present energy consumption would only last for a very limited number of years, but, as time passes, to transform all uranium found on Earth to so-called fissile material suitable for fuel in atomic power stations. Furthermore, it is possible to increase the available atomic fuel even more significantly by irradiating thorium with neutrons in uranium and plutonium reactors and thereby transform it to a uranium isotope with atomic weight 233, which again has fissile properties similar to those of uranium 235 and plutonium.

On this basis, and with these prospects, intensive work regarding the use of atomic energy has been carried out in the countries that have been leading in the technical development, and, as is well known, large atomic power stations are already under construction. In order to be able to assess the most economical and practical construction of such power stations, however, extensive scientific and technological investigations must still be made. What is involved here are of course conditions extremely different from those under which we hitherto have gained experience about the properties of the substances used in technology and industry, as especially the large number of neutrons constantly traversing the atomic power plants will have many kinds of peculiar effects. Not only do the nuclear transformations produce large amounts of radioactive substances, whose handling requires the utmost care, but on their collisions with the atomic nuclei, the neutrons will also, to a great extent, bring about changes in the chemical compounds, indeed even those in the solid materials necessary for all technical constructions, as the atoms will so to say constantly be pushed away from their equilibrium positions in the crystal lattices and thereby give rise to metallurgical changes of a character hitherto subject to little study. During this research work the use of fissile materials, such as uranium 235 and plutonium in a more or less pure state, has been of the very greatest importance, as their application allows, under less complicated and more comprehensible circumstances, the creation of experimental conditions corresponding to the circumstances in future large atomic power stations.

Although outside the big countries one has hitherto had to be content with being a more or less remote observer to this intensive technical atomic research, one has naturally also in the Nordic countries tried in every way available to take part in the research work and even to acquire such education and experience as is necessary to increase this participation. In Norway, of course, considerable amounts of heavy water have been produced for many years as a by-product of the large production of hydrogen and oxygen by electrolysis based on hydropower,[4] and laboratories all over the world have thus been supplied with this resource which is so valuable for atomic research. A few years after the end of the war, a Norwegian–Dutch cooperation was thus initiated on the basis of heavy water from Norsk Hydro and a considerable amount of uranium, which had been acquired by a laboratory in Leyden shortly before the war broke out and which it had been possible to keep hidden during the German occupation.[5] This cooperation has led to the construction of an atomic reactor at Kjeller near Oslo, which has been in operation since 1952.[6] In Sweden too, as is well known, there is a reactor in operation[7] and several being planned, with utilization of the uranium which is found in the large Swedish slate deposits containing uranium, and whose technically difficult extraction was started immediately after the war.

Here in Denmark, where we have been cut off from supplies of the necessary basic materials, we have hitherto not been able to build atomic reactors and have been relegated to keeping up, to the best of our ability, with the purely scientific development in the field of nuclear research. However, in our case the situation, as for so many nations in a similar position, has been radically

[4] The production of heavy water was carried out by the Norwegian plant Norsk Hydro, established in 1905 in the Norwegian town Rjukan.

[5] The Norwegian–Dutch cooperation originated in a meeting between Gunnar Randers and Hendrik Anthony Kramers during the winter of 1950.

[6] This work was led by Gunnar Randers. The Norwegian reactor went critical on 30 July 1951.

[7] The first Swedish reactor, built at the Royal Technical University in Stockholm, went critical in July 1954.

altered by the initiative, already mentioned, for the advancement of international scientific cooperation for the peaceful use of atomic energy and especially by offers from America and England to provide considerable amounts of enriched uranium for this purpose. As has been announced, an agreement has thus recently been made between the U.S.A. and Denmark, according to which the American Atomic Energy Commission will keep us supplied with an isotopic mixture, greatly enriched as compared to natural uranium, containing 6 kilogrammes of uranium 235, for the operation of a test reactor which will be constructed under the leadership of the newly established Danish Atomic Energy Commission, and which we expect to be ready for use within a year. Furthermore, an agreement has been made, with the approval of the Danish State authorities, between the British Atomic Energy Authority and the Danish Atomic Energy Commission, about a collaboration, in connection with which a test reactor will be built in this country of the same type as the high-flux reactor that is under construction at the British Atomic Energy Research Establishment at Harwell. For this reactor, which will take two to three years to build, and which is constructed for special technological investigations with the direct aim of gaining experience for the construction of atomic power stations, we will be supplied with fissile material from England in the form of greatly enriched uranium, while the amount of heavy water necessary for the operation will be delivered from America, where new methods for its production on an industrial scale have been developed.

Such support for the advancement of international cooperation will also be given to many other countries in a similar situation, and it has recently been announced that the U.S.A. has already entered agreements with 16 countries in Europe and other parts of the world.[8] As far as Denmark is concerned, the implementation of the plans in question will additionally give us the opportunity to exchange experience with our neighbouring Nordic countries, just as we of course still hope to be able to deepen the long-standing close cooperation within theoretical atomic research. All in all, we expect that it will be possible, within the Nordic countries, to give fruitful contributions to the utilization of

* The Risø peninsula in Roskilde Fjord (ed.).

---

[8] At the time of Bohr's speech, four countries had formally signed bilateral agreements with the United States for "cooperation concerning civil uses of atomic energy". These countries were, in the order that the agreements were signed, Turkey, Belgium, Canada and the United Kingdom. Bohr, no doubt, had information that agreements were about to be signed with Israel, Taiwan, Lebanon, Colombia, Portugal, Venezuela, the Philippines, Argentina, Brazil, Greece, Chile and Pakistan. Denmark signed the agreement on 25 July 1955, four days after Venezuela and three days before the Philippines.

atomic power for the benefit of society. It is difficult at present to foresee when it will be possible to proceed here to build atomic power stations for electricity supply, but in Denmark, just as in other countries, a start will be made on this as soon as it can be estimated, by means of the experience gained and on the basis of the prevailing economic conditions, whether such stations will be profitable already or whether a trial of their operation will be necessary in order to ensure a future profitability.

The Atomic Energy Commission is in the process of preparing a proposal for the establishment of an atomic energy research station, for consideration by the Danish Government. At this station, which is to be placed on a suitable isolated plot of land not too far from Copenhagen*, there will be built, in addition to the atomic reactor mentioned, the physical and chemical laboratories, as well as the mechanical and electro–technical workshops, necessary for its operation. Just as in the big countries, enterprises of this kind will make extensive demands on specialized knowledge as regards both science and technology and will require a very significant educational programme. Valuable help has been offered in this regard by the special reactor schools[9] that have been set up in America as well as in England, but, in addition, it will be necessary in Denmark, as elsewhere, to constantly increase education and research at the universities and technical colleges with a view to the great demands that the steadily growing industrialization makes in this field as in so many others.

Before I close, I shall again briefly mention the special problems connected with all research and technical development in the field of atomic energy because of the dangers related to the radiation produced in the reactors and the handling of the radioactive substances formed in them. In contrast to what is the case for the effects of atomic bomb explosions, however, a reassuring protection of the staff at atomic power stations and atomic energy research stations and the neighbouring population ought to be achieved by suitable safety measures, drawn up with full attention to medical expertise. Also in such regards we expect of course to receive valuable information through the congress in Geneva in August arranged by the United Nations, where not only the physical and technical questions connected with the utilization of atomic energy will be thoroughly discussed, but where also the whole social background

---

[9] The reactor physicist Povl L. Ølgaard was the first Dane to attend the Argonne School of Nuclear Science and Engineering in 1956. Ølgaard relates that there was also Danish attendance at a similar school at Oak Ridge and that others learned from witnessing the construction of reactors in the United States similar to those to be built at Risø.

will be illuminated from all sides. We hope first and foremost, however, that this congress will contribute in a decisive way to the international cooperation for the peaceful use of atomic energy to the benefit of every single society and, through the removal of barriers to mutual information, will advance that understanding of the interests common to all peoples, which is the precondition for a harmonious co-existence for the heightening of human culture in all its shapes and forms.[10]

[10] On the Geneva conference and Bohr's contribution there, see the introduction to Part II, p. [366]. Bohr's lecture at the conference, *Physical Science and Man's Position*, Ingeniøren **64** (1955) 810–814, is reproduced in Vol. 10, pp. [102]–[106].

XXVIII. GREETING TO THE EXHIBITION
FROM PROFESSOR NIELS BOHR

HILSEN TIL UDSTILLINGEN FRA PROFESSOR NIELS BOHR
Elektroteknikeren **53** (1957) 363

One of several greetings from prominent Danes
to a national exhibition organized by the Danish electrical industry

TEXT AND TRANSLATION

See Introduction to Part II, p. [366].

[633]

Hilsen til udstillingen fra professor Niels Bohr

Professor Niels Bohr.

Elektroteknikeren har på anmodning modtaget nedenstående udtalelse af professor, dr. phil. & scient. & techn. Niels Bohr.

I vort århundrede, hvor der er åbnet så store udsigter til at nyttiggøre naturvidenskabens og teknologiens fremskridt i samfundslivet, er det blevet af stedse større betydning at kunne råde over tilstrækkelige energikilder. På et tidspunkt, hvor udnyttelsen af vandkraft og organisk brændsel ikke længere slår til for menneskehedens behov, har derfor muligheden af at frigøre en del af den i atomernes kerner oplagrede uhyre energimængde åbnet nye lovende perspektiver. Vi står her overfor store opgaver for teknik og industri, til hvis løsning alle folkeslag efter evner og muligheder må stræbe at bidrage, og jeg vil gerne udtrykke håbet om, at denne udstilling må tjene til i vide kredse at udbrede kendskab til og forståelse af disse opgaver og deres betydning for vort samfund.

7. OKTOBER 1957

363

[635]

TRANSLATION

Greeting to the exhibition from Professor Niels Bohr

Upon request Elektroteknikeren has received the following statement from Professor Niels Bohr, D.Phil. & Sc. & Tech.

In our century, when such great prospects have been opened for utilizing the advances of science and technology in society, it has become of ever greater importance to have sufficient sources of energy at disposal. At a time when the exploitation of water power and organic fuels no longer meets the needs of humanity, the possibility of releasing some of the enormous amounts of energy stored in the nuclei of atoms has therefore opened new promising perspectives. Here we are confronted with great tasks for technology and industry, towards whose solution all peoples, according to their abilities and opportunities, must strive to contribute, and I would like to express the hope that this exhibition may serve to spread in wide circles the knowledge and understanding of these tasks and their importance for our society.

Niels Bohr

XXIX. THE PRESENTATION OF
THE FIRST ATOMS FOR PEACE AWARD
TO NIELS HENRIK DAVID BOHR,
OCTOBER 24, 1957

Pamphlet printed by the National Academy of Sciences
reporting the award ceremony, pp. 1, 5, 18–22

See Introduction to Part II, p. [367].

[637]

The Presentation

of the first

ATOMS FOR PEACE AWARD

to

NIELS HENRIK DAVID BOHR

at the

NATIONAL ACADEMY OF SCIENCES

Washington, D.C.

October 24, 1957

Contents

[639]

NIELS HENRIK DAVID BOHR

[640]

The Citation

Niels Henrik David Bohr, in your chosen field of physics you have explored the structure of the atom and unlocked many of Nature's other secrets. You have given men the basis for greater understanding of matter and energy. You have made contributions to the practical uses of this knowledge. At your Institute for Theoretical Physics at Copenhagen, which has served as an intellectual and spiritual center for scientists, you have given scholars from all parts of the world an opportunity to extend man's knowledge of nuclear phenomena. These scholars have taken from your Institute not only enlarged scientific understanding but also a humane spirit of active concern for the proper utilization of scientific knowledge.

In your public pronouncements and through your world contacts, you have exerted great moral force in behalf of the utilization of atomic energy for peaceful purposes.

In your profession, in your teaching, in your public life, you have shown that the domain of science and the domain of the humanities are in reality of single realm. In all your career you have exemplified the humility, the wisdom, the humaneness, the intellectual splendor which the Atoms for Peace Award would recognize.

It is fitting, therefore, that you receive this, the first presentation of the Atoms for Peace Award, and it is deemed a great honor by the Trustees of the Award to present it to you. On behalf of the Trustees of the Award, I present to you this medal, symbolizing the Award. May this token signify the importance to all men of your contributions "for the benefit of mankind."

THE ATOMS FOR PEACE AWARD MEDAL

Response

NIELS HENRIK DAVID BOHR

It is a great privilege to me in this distinguished assembly, honored by the presence of the President of the United States, to express my deep gratitude for having been selected as recipient of the Atoms for Peace Award, which is dedicated to a cause embraced by the hopes of humanity and is associated with the memory of two great Americans, who contributed so much to developing modern society.

It has been my good fortune at close hand to follow how new vast fields of knowledge, hitherto beyond the reach of man, have been opened through a most intensive cooperation of scientists from all over the globe.

In science, where we build on the achievements of preceding generations and strive to enrich the heritage, the individual worker can only add a brick or shape a column to a great edifice — but to see it rising by common effort is a most inspiring adventure.

The exploration of the world of atoms, which has so greatly advanced our insight into the structure of matter and revealed novel aspects of our position as observers of nature, has at the same time provided mankind with unprecedented opportunities.

However, every increase in knowledge and potentialities entails greater responsibilities. Indeed, the rapid advance of science and technology in our age, which involves such bright promises and grave dangers, presents civilization with a most serious challenge. To meet this challenge, which calls upon the highest human aspirations, the road is indicated by that world-wide cooperation which has manifested itself through the ages in the development of science.

XXX. PROFESSOR NIELS BOHR ON RISØ

PROFESSOR NIELS BOHR OM RISØ
Elektroteknikeren **54** (1958) 238–239

TEXT AND TRANSLATION

See Introduction to Part II, p. [369].

[645]

Professor Niels Bohr om Risø

Ved indvielsen af Risø den 6. juni holdt atomenergikommissionens formand, professor N i e l s B o h r, denne tale, som kort, klart og populært skildrer baggrunden for forsøgsarbejdet:

Professor *Niels Bohr.*

Ved indvielsen af atomenergi-forsøgsstationen her på Risø ved den skønne Roskildefjord, hvor der for blot få år siden lå bondegårde med kornmarker og græsenge, som vi stadig ser det i det omgivende landskab, mødes fortid og fremtid på harmonisk måde. Vi er jo vidne til det første skridt til muliggørelsen af vort lands deltagelse i en ny teknisk udvikling, der rummer store muligheder for samfundslivets fremme.

Ligesom tidligere tiders fremgang på agerbrugets og industriens område er denne udvikling frugt af den menesker iboende trang til at lære den os omgivende natur at kende, uden at vide hvorhen det måtte føre os. Her har det endda drejet sig om noget os tilsyneladende så fjerntliggende som de smådele, hvoraf alle stoffer er opbyggede, og hvis størrelse i forhold til vore egne legemer og redskaber er lige så ringe som disses forhold til hele jordkloden.

Selvom, som bekendt, atomforestillingerne i deres oprindelse går tilbage til oldtidens tænkere, og i århundredernes forløb viste sig stadig mere frugtbare for forklaringen af stoffernes fysiske og kemiske egenskaber, er det helt naturligt, at det for blot få menneskealdre siden var en almindelig antagelse, at det altid ville ligge uden for vore erfaringers rækkevidde at iagttage virkninger af de enkelte atomer.

Den fysiske experimenterkunsts vidunderlige udvikling i de mange grundlæggende opdagelser omkring sidste århundredskifte skulle imidlertid ikke alene give os et i enkeltheder gående indblik i atomernes opbygning af endnu mere elementære bestanddele, men tillige afsløre uanede muligheder for at frigøre den i atomernes indre bundne energi. Som vi nu ved, overgår denne energi i stort mål den, som for exempel ved forbrænding af kul kan vindes ved atomernes omlejring i de kemiske molekyler. Ja, vi har endda erkendt, at atomenergien er selve kilden til den enorme lys- og varmestråling fra solen, der igennem tiderne har opretholdt alt liv her på jorden.

Det første fingerpeg om tilstedeværelsen af sådanne energikilder fik man for omkring 60 år siden ved de til navnene *Becquerel* og *Curie* knyttede opdagelser af de i naturen forekommende radioaktive stoffer. Det afgørende skridt til forståelsen af disse stoffers egenskaber var *Rutherfords* opdagelse i 1911 af atomets kerne, hvis uforanderlighed under sædvanlige fysiske og kemiske forhold forklarer de almindelige grundstoffers bestandighed, men som i de radioaktive stoffer undergår sønderdelinger uden ydre påvirkninger. Ved sin efterfølgende skelsættende påvisning af, at også andre grundstoffers kerner, når de udsættes for bombardement af de fra radioaktive atomkerner udsendte energirige partikler, kunne undergå omdannelser, skabte Rutherford et helt nyt forskningsområde, den såkaldte atomkernefysik, eller, som han selv malende udtrykte det: moderne alkymi.

Selvfølgelig skal jeg ikke ved denne lejlighed forsøge nærmere at skildre atomkernefysikkens æventyrlige udvikling, men blot kort minde om nogle af de for atomenergiens udnyttelse mest afgørende fremskridt. I de nærmest følgende år skyldtes sådan fremgang først og fremmest Rutherford selv og hans medarbejdere ved det berømte Cavendish laboratorium i Cambridge. Der udviklede *Cockcroft* nye hjælpemidler til at give protoner en energi tilstrækkelig til at fremkalde kunstige atomkerneomdannelser, og forsøgene hermed viste så store energifrigørelser, at spørgsmålet om atomenergiens praktiske udvinding for første gang kom i søgelyset. Et videre skridt, der skulle vise sig afgørende for at fremme sådanne muligheder, var *Chadwicks* opdagelse af en hidtil ukendt kernebestanddel — neutronen — der, løsrevet fra atomkernerne, skulle blive et så virksomt middel til at frembringe nye kerneomdannelser, hvis studium fra år til år øgede kendskabet til atomkernernes opbygning og egenskaber.

Udviklingen indtrådte som bekendt i et nyt stadium ved *Hahns* opsigtsvækkende opdagelse i Berlin 1939 af spaltningen af urankerner ved neutronbestråling. Forståelsen af denne såkaldte fissionsproces uddybedes af Hahns mangeårige medarbejder, *Lise Meitner*, der på den tid havde måttet søge tilflugt i Sverige, og hendes nevø, *Otto Frisch*, der i de år arbejdede på Universitetets institut for teoretisk fysik i København, hvor han foretog de første målinger af den overordentlige energifrigørelse ved uranspaltningen. For udnyttelsen af atomkerneenergien i stor målestok skabtes få måneder derefter grundlaget ved de af *Joliot* og *Fermi* i henholdsvis Paris og New York udførte forsøg, der viste, at der som følge af den voldsomme indre bevægelse af ved fissionsprocessen udslyngede kernebrudstykker fra disse udsendes neutroner, der igen kan spalte urankerner og derved sætte kædereaktioner i gang. Som alle ved, lykkedes det Fermi i december 1942 at fuldføre konstruktionen af en på sådanne kædeprocesser baseret atomreaktor, en bedrift, som man ofte betegner som indledningen til atomalderen, i hvis 16de år vi efter sådan tidsregning skulle befinde os.

Det er her hverken stedet eller tiden til at gå nærmere ind på den påfølgende begivenhedsrige udvikling, hvis hovedtræk er enhver bekendt, og som har givet menneskeheden midler i hænde, hvis anvendelse rummer så rige løfter, men samtidig så store farer. Det er jo klart, at enhver forøgelse af vor viden og kunnen medfører et forøget ansvar, og den nye udvikling betyder i denne henseende intet mindre end den alvorligste prøvelse for hele vor civilisation, der kun kan bestå ved et samarbejde mellem alle folkeslag i gensidig tillid. På denne baggrund skabte præsident *Eisenhower*, gennem sin tale i 1953 til De Forenede Nationer, nye håbefulde udsigter gennem det initiativ til mellemfolkelig bistand til atom-energiens fredelige udnyttelse, der fandt så stor tilslutning.

Spørgsmålet om også herhjemme at deltage i udviklingen på dette område indtrådte hermed i en ny fase. Hidtil havde vi jo, på grund af mangelen på de nødvendige udgangsmaterialer, været afskåret fra at iværksætte sådanne teknologiske undersøgelser, der allerede var vidt fremskredne i de store lande, og måtte indskrænke os til efter bedste evne at følge med i den rent videnskabelige forskning på atomfysikkens område. Selv dette krævede betydelige midler, der med forståelse og offervilje stilledes til rådighed såvel fra statsmyndighedernes side som fra Carlsberg- og Thrigefondet. Da de nye udveje åbnedes, var det også en stor hjælp, at et forberedende arbejde straks kunne iværksættes herhjemme ved en gave fra Thrigefondet til Akademiet for de tekniske Videnskaber.

Professor *Robert Henriksen,* som vi i dag så dybt savner, var dengang præsident for Akademiet, og ved sin deltagelse i det af dette nedsatte arbejdsudvalg og senere som atomenergikommissionens næstformand udøvede han med sin store indsigt i alle spørgsmål vedrørende elektricitetsværker og sin levende forståelse af det så hastigt voksende behov for industrielle kraftkilder en indflydelse af største værdi for sagens fremme.

Naturligvis måtte det være forbundet med vanskeligheder at opbygge et foretagende på et sådant for os helt nyt område, og især måtte man nære ængstelse for muligheden af at imødekomme kravet om en tilstrækkelig stor og velkvalificeret medarbejderstab. Det lykkedes imidlertid til virksomheden fra første færd at knytte nogle af vore atomfysikere med særlig erfaring såvel med hensyn til forskning som til uddannelse og tillige et antal kemikere, biologer og ingeniører med den fornødne indsigt til at lede arbejdet på de mange forskellige felter, som Forsøgsstationen omfatter. Lykkeligvis medførte også den i vide kredse vågne interesse for det nye foretagende, at et større og større antal evnerige unge med videnskabelig og teknisk uddannelse har meldt sig som medarbejdere.

Det var fra begyndelsen klart, at foretagendets omfattende karakter med dets mange såvel hjemlige som internationale forbindelser ville nødvendiggøre en let-arbejdende og effektivt virkende organisation, ved hvis ledelse og sammensætning det har været muligt at drage nytte af den indsigt og erfaring, der findes i landets centraladministration og større industrivirksomheder.

Som enhver forskningsvirksomhed, der tilsigter at bidrage til fælles menneskelig viden, har vi ved opbygningen af Forsøgsstationen og tilrettelæggelsen af arbejdet i videst muligt omfang støttet os til erfaringer og resultater vundet i andre lande. I denne henseende har de kollegiale og venskabelige forbindelser rundt om i verden, som var knyttet under den mangeårige danske deltagelse i det internationale samarbejde på atomfysikkens område, været os til største nytte. Ikke mindst tænker vi med taknemmelighed på de værdifulde råd og den stedse beredvillige hjælp, som på alle stadier blev ydet os af *Sir John Cockcroft,* som det er en særlig glæde at have iblandt os i dag, og til hvis dybe indsigt og rige erfaring vi håber også i fremtiden at kunne støtte os.

Om atomenergikommissionens lovformelige oprettelse og statsmyndighedernes medvirken til Forsøgsstationens skabelse har finansminister *Kampmann* selv gjort rede, og på atomenergikommissionens vegne vil jeg gerne rette en varm tak til regeringen og de bevilgende myndigheder for den store tillid, der er blevet vist os, og som vi har stræbt ikke at svigte.

Det har været vor overbevisning, at iværksættelsen af et forskningsarbejde som det, der her er begyndt, var nødvendigt for herhjemme at vinde den erfaring og skabe de uddannelsesmuligheder, uden hvilke de af hele befolkningen delte forhåbninger ikke vil kunne virkeliggøres. Alene herved vil vi jo kunne bevare en til vore traditioner svarende stilling blandt andre nationer og med disse deltage i den udforskning af videregående perspektiver, der rummer løfter om at skabe uudtømmelige energikilder for menneskehedens behov ved, omend i beskedneste mål, at efterligne de forhold, der råder i naturens eget vældige kraftværk på solen.

Jeg vil gerne slutte med at give udtryk for de inderligste ønsker om, at der på den atomenergi-forsøgsstation, som indvies og fremvises i dag, må blive udført et arbejde til gavn for det danske samfund og til styrkelse af vort samarbejde med andre folkeslag på løsningen af tidens store opgaver.

TRANSLATION

Professor Niels Bohr on Risø

At the inauguration of Risø on 6 June, Niels Bohr, chairman of the Atomic Energy Commission, gave the following speech, which in a brief, clear and popular way describes the background for the research:

With the inauguration of the atomic energy research station here at Risø beside the beautiful Roskilde fjord, where only a few years ago farms with cornfields and pastures lay as can still be seen in the surrounding countryside, the past and the future meet in a harmonious fashion. We are witnessing the first step towards the realization of participation by our country in a new technical development which holds great opportunities for the benefit of society.

Just like the progress in previous times in the domain of agriculture and industry, this development is the fruit of the innate urge of humans to learn about the nature around us, without knowing where this might lead. In this case, it has even been a question of something seemingly so remote from us as the small parts of which all matter is made, and whose size in relation to our own bodies and tools is just as minute as these are in relation to the whole Earth.

Even though, as is well known, ideas about atoms in their origin go back to the thinkers of Antiquity, and in the course of the centuries proved to be increasingly fertile for the explanation of the physical and chemical properties of matter, it is quite natural that only a few generations ago it was generally assumed that the observation of the effects of the individual atoms would always lie beyond the reach of our experience.

The wonderful development of the art of experimental physics and the many fundamental discoveries made around the turn of this century should, however, not only give insight in detail into how atoms are built up from even more elementary constituents, but also reveal unimagined possibilities for releasing the energy bound inside the atoms. As we now know, this energy greatly exceeds that which can be gained, for example, by the rearrangement of the atoms in the chemical molecules by the burning of coal. Indeed, we have even recognized that atomic energy is the very source of the enormous radiation of light and heat from the sun which through the ages has sustained all life here on Earth.

The first hint of the presence of such energy resources was given about sixty years ago by the discoveries, linked with the names *Becquerel*[1] and *Curie*,[2] of the radioactive substances occurring in nature. The decisive step towards the understanding of the properties of these substances was the discovery by *Rutherford* in 1911 of the nucleus of the atom whose immutability under normal physical and chemical conditions explains the stability of the ordinary elements, but which in the radioactive substances undergoes disintegration without outside influence. With his subsequent epoch-making demonstration that also the nuclei of other elements could undergo transformations when subjected to bombardment of the energy-rich particles emitted from the radioactive atomic nuclei, Rutherford created a whole new branch of research, so-called nuclear physics or, as he himself vividly expressed it: modern alchemy.

Of course, I shall not on this occasion attempt to describe in detail the fantastic development of nuclear physics, but just briefly mention some of the most decisive advances as regards the utilization of atomic energy. During the following years such progress was first and foremost due to Rutherford himself and his collaborators at the famous Cavendish Laboratory in Cambridge. There *Cockcroft*[3] developed new tools to give protons an energy sufficient to produce artificial nuclear transformations, and these experiments displayed such great releases of energy that the question of the practical generation of atomic energy came into focus for the first time. A further step which should prove decisive for advancing such possibilities was *Chadwick's*[4] discovery of a hitherto unknown nuclear constituent – the neutron – which, detached from the atomic nuclei, should become such an effective means for producing new nuclear transformations, the study of which, year by year, increased the knowledge of the constitution and properties of the atomic nuclei.

As is well known, the development entered a new stage with *Hahn's*[5] sensational discovery in Berlin in 1939 of the splitting of uranium nuclei by means of neutron irradiation. The understanding of this so-called fission process was clarified by Hahn's collaborator for many years, *Lise Meitner*,[6] who at that time had had to seek refuge in Sweden, and her nephew *Otto Frisch*,[7] who in those years was working at the University Institute for Theoretical Physics

---

[1] Henri Becquerel (1852–1908).
[2] See p. [515], ref. 20.
[3] John Douglas Cockcroft (1897–1967).
[4] James Chadwick (1891–1974).
[5] See p. [597], ref. 5.
[6] See p. [598], ref. 8.
[7] See p. [598], ref. 9.

in Copenhagen, where he made the first measurements of the enormous release of energy from uranium fission. The basis for the utilization of nuclear energy on a large scale was created a few months later with experiments conducted by *Joliot*[8] and *Fermi*,[9] in Paris and New York, respectively, which showed that, as a result of the violent internal movement of the nuclear fragments slung out in the fission process, neutrons are emitted from these, which can in turn split uranium nuclei and thereby set off chain reactions. As is generally known, in December 1942 Fermi succeeded in completing the construction of an atomic reactor based on such chain processes,[10] an achievement which is often called the beginning of the Atomic Age, in whose 16th year we should now be, according to such a calendar.

This is neither the time nor the place to describe in more detail the subsequent eventful development, whose main features are familiar to everyone, and which has given humanity possession of resources whose application holds such rich promises but at the same time such great dangers. It is of course obvious that any increase of our knowledge and ability brings with it increased responsibility, and in this respect the new development means nothing less than the most serious challenge for our entire civilization, which can only survive through cooperation between all peoples in mutual trust. Against this background, President *Eisenhower* created, with his speech in 1953 to the United Nations, new hopeful prospects through the initiative for international support for the peaceful uses of atomic energy, which found such great accord.

Thus, the question of whether also we in this country should take part in the development entered a new phase. Due to the lack of the necessary basic materials, we had hitherto been cut off from implementing such technological investigations, which were already far advanced in the large countries, and had to limit ourselves to keeping up, to the best of our ability, with the purely scientific research in the field of atomic physics. Even this required considerable resources which, with understanding and generosity, were made available by the State authorities as well as the Carlsberg and Thrige Foundations. When the new possibilities opened up it was also of great help that preparatory work could start immediately in this country by means of a gift from the Thrige Foundation to the Academy of Technical Sciences.[11]

---

[8] See p. [515], ref. 22.

[9] See p. [601], ref. 11.

[10] See p. [601], ref. 11.

[11] The gift to the Academy of Technical Sciences, which was founded in 1937, amounted to 200,000 Danish kroner. See the introduction to Part II, p. [360].

Professor *Robert Henriksen*,[12] whom we miss so deeply today, was President of the Academy at that time, and by his participation in the working group set up by the Academy and later as Vice-Chairman of the Atomic Energy Commission, he exercised an influence of the greatest value for the advancement of the matter with his great insight in all questions regarding electricity plants and his vivid understanding of the so rapidly growing demand for industrial power sources.

Naturally, there must be difficulties in connection with the establishment of a venture in a field totally new to us, and especially there must be fears as regards the possibility of meeting the requirements for a sufficiently large and qualified staff. However, there has been success from the very start in attaching to the venture some of our atomic physicists with special experience in research as well as teaching and, in addition, a number of chemists, biologists and engineers with the insight required to lead the work in the many different fields covered by the research station. Happily, the lively interest in wide circles for the new enterprise also resulted in a steadily increasing number of gifted young people with scientific and technical education joining the staff.

It was clear from the start that the comprehensive character of the enterprise, with its many ramifications both at home and abroad, would necessitate a flexible and efficient organization, with whose management and composition it has been possible to make use of the insight and experience found in the Government administration and large industrial companies in this country.

Like any research enterprise aiming to contribute to common human knowledge, we have, by the establishment of the research station and the organization of the work, to the greatest possible extent made use of experience and results gained in other countries. In this regard the professional and friendly connections around the world, welded during the Danish participation over many years in the international cooperation in the field of atomic physics, have been of the greatest use. Not least do we remember with gratitude the valuable advice and the always willing assistance offered to us at all stages by *Sir John Cockcroft*, whom it is a special pleasure to have in our midst today, and whose profound insight and rich experience we hope to be able to rely on in the future too.

Minister of Finance *Kampmann*[13] himself has described the atomic energy commission's establishment by legislation and the State's participation in the

---

[12] (1890–1958). Henriksen had died on 17 May. On Henriksen, see also the introduction to Part II, pp. [360] and [364].

[13] Viggo Kampmann (1910–1976). See the introduction to Part II, p. [363]. Kampmann would be become Danish Prime Minister from 1960 to 1962.

creation of the research station,[14] and on behalf of the Atomic Energy Commission I would like to direct sincere thanks to the Government and the authorizing bodies for the great trust that we have been shown and which we have striven not to betray.

It has been our conviction that the implementation of research work, like that started here, was necessary in order to gain the experience and to create the opportunities for education here at home without which it would not be possible to realize the hopes shared by the whole population. Only in this way will we be able to maintain a position among other nations corresponding to our traditions, and together with them take part in the investigation of further perspectives which hold promises for creating inexhaustible resources of energy for the needs of humanity by imitating, even though on the most modest scale, the conditions that prevail in nature's own enormous power station in the sun.

I would like to close by expressing the most heartfelt wishes that work may be carried out at the atomic energy research station, inaugurated and presented today, for the benefit of Danish society and for the strengthening of our cooperation with other peoples for the solution of the great tasks of the time.

[14] Bohr here refers to Kampmann's lecture on the same occasion. Only Bohr's lecture, however, was published in Elektroteknikeren.

Niels Bohr followed closely the building of the Risø Research Establishment of the Atomic Energy Commission. Here he is seen on the site together with the physicists J.C. Jacobsen and Ebbe Rasmussen.

6. CONTACT WITH MATHEMATICIANS

XXXI. ON THE PROBLEM OF MEASUREMENT IN ATOMIC PHYSICS

OM MAALINGSPROBLEMET I ATOMFYSIKKEN
"Festskrift til N.E. Nørlund
i Anledning af hans 60 Aars Fødselsdag den 26. Oktober 1945
fra danske Matematikere, Astronomer og Geodæter, Anden Del",
Ejnar Munksgaard, Copenhagen 1946, pp. 163–167

TEXT AND TRANSLATION

See Introduction to Part II, p. [370].

Niels Erik Nørlund and Niels Bohr, 1947.

Om Maalingsproblemet i Atomfysikken.

Af Niels Bohr.

De eksakte Videnskabers Opgave er i videste Omfang at finde talmæssige Forbindelser mellem de Størrelser, som indgaar i Beskrivelsen af Naturfænomenerne. Bestræbelserne for at opnaa den størst mulige Nøjagtighed i Maalingen af saadanne Størrelser har derfor ofte været afgørende for Videnskabens Fremskridt, og det behøves næppe at erindre om det Forbillede, som Geodæsi og Astronomi i denne Henseende har frembudt i Tidernes Løb. Ved det store Gennembrud paa saa mange af Naturvidenskabens Omraader, der indleder den nyere Udvikling, kom ogsaa Maalingers Betydning i Forgrunden, saaledes som det blev udtrykt i Galileis bekendte Program om overalt at efterforske kvantitative Relationer, selv ved Naturfænomener, der ikke umiddelbart frembyder talmæssige Kendetegn.

I den efterfølgende Tid skabtes der ikke alene gennem Newtons Storværk et fast Grundlag for Mekanikken, der tillod saa omfattende en Udnyttelse af de astronomiske og geodætiske Maalinger, men ogsaa andre Omraader af Naturvidenskaben skulde snart blive Genstand for en Behandling efter eksakte Metoder. Ligesom det var den paa nøjagtige Vægtbestemmelser grundede kvantitative kemiske Analyse, der især gennem Lavoisiers Indsats skulde blive afgørende for Erkendelsen af Kemiens Grundlove, var det jo Udviklingen af Metoder til at maale Temperaturer og Varmemængder, som efterhaanden skabte Grundlaget for Formuleringen af Termodynamikkens Hovedsætninger.

Nye Perspektiver aabnede sig atter, da de elektriske og magnetiske Fænomener, der allerede fra Oldtiden havde tiltrukket sig Opmærksomheden, men hvis dybere Sammenhæng først kom for Dagen efter de store Opdagelser i Begyndelsen af det forrige Aarhundrede, blev Genstand for kvantitativ Udforskning. Gennem Udviklingen af Metoder til nøjagtig Maaling af elektromagnetiske Kræfter og elektriske Strømme skabtes Grundlaget for en Elektrodynamik, der, hvad Konsekvens og Rækkevidde angaar, kunde maale sig med den Newtonske Mekanik og endda viste Veje udover denne, der sluttelig førte til den Revision af selve Grundlaget for Rum-Tids-Koordinationen, som fandt sit Udtryk i Ein-

steins Relativitetsteori og som har givet den klassiske Fysiks Begrebsbygning en saa harmonisk Afrunding.

Den seneste Udvikling af Fysikken, der har givet os et dybt
Indblik i Atomernes Verden og har medført yderligere Fremskridt
i Erkendelsen af Vilkaarene for videnskabelig Analyse og Syntese,
har ogsaa fra første Færd været betinget af Bestræbelserne for
Maalingernes stadige Forfinelse. Netop paa dette Omraade var
der jo særlige Vanskeligheder at overvinde, og man behøver næppe
at erindre om, at Atomteorien, hvis Oprindelse gaar tilbage til
de allerældste Bestræbelser for at opnaa et almindeligt filosofisk
Grundlag for Forstaaelsen af de Lovmæssigheder, som Naturfænomenerne trods deres Rigdom og Foranderlighed ved nærmere Betragtning udviser, længe maatte anses for en Arbejds-
hypotese, for hvilken intet umiddelbart Bevis kunde gives, idet
man gik ud fra, at vore Sansers Grovhed for bestandig vilde
udelukke os fra at iagttage Virkninger af individuelle Atomer.

Imidlertid skulde man efterhaanden lære en Række kvantitative
Lovmæssigheder at kende, der tillod at drage sikre Slutninger
om Stoffernes Opbygning af Atomer og endda gav Oplysninger
om de enkelte Atomers Egenskaber. Det første store Skridt paa
denne Vej var Daltons Forklaring af Loven om de multiple Pro-
portioner, der behersker Masseforholdene ved kemiske Reaktioner.
Det var endvidere den konsekvente Videreførelse af samme Tanke-
gang, der danner Grundlag for den Tolkning af Faradays elektro-
lytiske Ækvivalenslove, der skulde føre til Erkendelsen af Elektri-
citetens atomistiske Karakter. Ydermere gav Spektralanalysens
saa overordentlig forfinede Maaleteknik ikke alene talmæssige
Kendetegn paa Grundstofferne, der tillader os at fastslaa deres
Tilstedeværelse selv paa fjerne Stjerner, men aabenbarede tillige
karakteristiske Regelmæssigheder vedrørende den fra Atomerne udsendte Straaling, der skulde faa største Betydning for den senere
Udvikling.

Et nyt Stadium i vort Kendskab til Stoffernes Byggestene
indlededes omkring Aarhundredeskiftet gennem Undersøgelsen af
helt nye Fænomener som Katodestraaler og Røntgenstraaler. Især
skulde Opdagelsen af Elektronen som en fælles Bestanddel af alle
Stoffernes Atomer og de stadig nøjere Bestemmelser af dens
Masse og Ladning hurtigt bringe Fysikerne paa Spor efter kvantitative Forbindelser mellem mange hidtil ukorrelerede Fænome-

ner. Fremfor alt skulde dog Undersøgelsen af den naturlige Radioaktivitet hos de tungeste Grundstoffer give os Midler i Hænde til en direkte Udforskning af Atomernes Indre, der gennem Rutherfords Paavisning af Atomkernens Eksistens og af dens Omdannelsesmuligheder i saa høj Grad skulde fuldstændiggøre Forestillingerne om Atomernes Bygning og tillige gennem de nøjagtige Maalinger af de ved Kerneomdannelserne frigjorte store Energimængder give en afgørende Bekræftelse af den i Relativitetsteorien forudsagte almindelige Forbindelse mellem Masse og Energi.

Samtidig med at Kendskabet til Atomernes Struktur skabte Klarhed over mange Spørgsmaal vedrørende Grundstoffernes Slægtskabsforhold, afslørede det paa slaaende Maade de klassiske mekaniske og elektrodynamiske Forestillings Utilstrækkelighed for Forklaringen af den ejendommelige Stabilitet hos Atombygningen, der behersker Grundstoffernes specifike Egenskaber, ja endda betinger selve Eksistensen af de Maaleredskaber, hvorpaa alle Fænomeners kvantitative Udforskning i sidste Instans hviler. En Nøgle til Opklaringen af disse Forhold var imidlertid givet ved Opdagelsen af Virkningskvantet, hvortil Planck var ført gennem den sindrige Analyse af de Love for Varmestraalingsfænomenerne, som stadig mere nøjagtige Maalinger havde bragt for Dagen. Denne skelsættende Opdagelse bragte jo Erkendelsen af, at selv de mest elementære mekaniske Begreber, til Trods for at de tillader saa indgaaende en Beskrivelse af udstrakte Erfaringsomraader, kun er Idealisationer, hvis Anvendelighed er betinget af, at de i Betragtning kommende Virkninger er tilstrækkelig store i Forhold til det universelle Virkningskvantum.

Ved Redegørelsen for Atomernes Egenskaber er man under disse Omstændigheder henvist til en principielt statistisk Beskrivelsesmaade, hvis Maal er at bestemme Sandsynlighederne for de ikke nærmere analysebare individuelle Kvanteeffekter. Efter forskellige Tilløb er det paa dette Grundlag lykkedes ved Indførelsen af en egnet matematisk Formalisme at udvikle en konsekvent Atommekanik, der er at betragte som en rationel Almindeliggørelse af den Newtonske Mekaniks Begrebsbygning. Herved skabtes en Ramme for Udredningen af kvantitative Forbindelser mellem Maaleresultater, som er blevet indsamlet paa de mest forskellige Omraader og som, hvad Mangfoldighed angaar, taaler

[659]

Sammenligning med hele det Erfaringsmateriale, den klassiske Fysik hviler paa.

Samtidig har Erkendelsen af, at Virkningskvantets Eksistens sætter en nedre Grænse for den ved enhver Iagttagelse uundgaaelige Vekselvirkning mellem Objekterne for Undersøgelsen og Maaleinstrumenterne, stillet selve Maaleproblemet i en væsentlig ny Belysning. Medens det hidtil har været en Forudsætning, at alle Fænomener uanset den af det valgte Henførelsessystem for Rum og Tid betingede Relativitet kan føres tilbage til en af Iagttagelsen uafhængig Opførsel af Objekterne, er man ved Studiet af de individuelle Kvanteeffekter efter Sagens Art afskaaret fra skarpt at skelne mellem en selvstændig Opførsel af atomare Objekter og deres Vekselvirkning med de Maalemidler, der er uundværlige for Fænomenernes Definition.

Uden at der for saa vidt sættes nogen absolut Grænse for Nøjagtigheden af de enkelte Maalinger, betyder denne Omstændighed som bekendt, at Lokaliseringen i Tid og Rum af atomare Objekter og Kendskabet til deres Bevægelsesmængde og kinetiske Energi er underkastet en ejendommelig, ved de Heisenbergske Ubestemthedsrelationer udtrykt reciprok Begrænsning. Denne Erkendelse bragte den fuldstændige Opklaring af de Paradokser, som vi møder ved Spørgsmaalet om atomare Objekters Natur, og som stammer fra, at der i Redegørelsen for saavel Fotoners som Elektroners Egenskaber maa benyttes tilsyneladende uforenelige Billeder som Korpuskel og Bølgefelt. Det drejer sig faktisk her om en helt ny Situation vedrørende Erfaringernes Analyse og Syntese, idet Fænomener iagttagne under forskellige Forsøgsanordninger kan staa i et saadant gensidigt Udelukkelsesforhold, at de ikke kan sammenfattes i et enkelt anskueligt Billede, men at Maaleresultaterne ikke desto mindre maa betragtes som komplementære i den Forstand, at de først tilsammen tillader en fuldstændig Beskrivelse af Objekternes Opførsel under hvilkesomhelst veldefinerede Omstændigheder.

Hvor store Vanskeligheder der end maatte være at overvinde paa Atomfysikkens Omraade for at opnaa en rationel Indordning af de mange nye Fænomener, som stadig kommer inden for vore Erfaringers Kreds, kan der paa ingen Maade være Tale om at vende tilbage til en Beskrivelse, der imødekommer vore fra Dagliglivet tilvante Krav til Anskuelighed, men langt snarere om en bestandig videre Udvikling af en matematisk Formalisme, der er tilpasset til de paa de nye Iagttagelsesfelter opnaaelige Maaleresultater. Der er al Grund til at tro, at saadanne Bestræbelser stadig vil vise sig frugtbare, og at det vil lykkes paa konsekvent Maade at tilvejebringe flere og flere talmæssige Forbindelser inden for vort hurtigt voksende Kundskabsomraade.

I denne Sammenhæng maa det paa det stærkeste understreges, at hverken Atommekanikkens statistiske Beskrivelsesmaade eller de for de atomare Fænomeners Analyse gældende Indskrænkninger indeholder noget Afkald paa de eksakte Videnskabers Metode. Tværtimod drejer det sig ved det saakaldte Komplementaritetssynspunkt om en Almindeliggørelse af Kausalitetsidealet, tilstrækkelig vid til at tage Hensyn til de Træk af Individualitet hos Atomprocesserne, der er betinget af Virkningskvantets Eksistens, og muliggøre en modsigelsesfri Syntese af Fænomener, der ikke kan finde Plads inden for den klassiske Fysiks Rammer, men som i saa væsentlig Grad har forøget vor Viden om Naturens Lovmæssigheder.

TRANSLATION

On the Problem of Measurement in Atomic Physics.

By Niels Bohr.

———

The aim of the exact sciences is to establish, as far as possible, numerical relationships between the quantities that are part of the description of natural phenomena. The efforts to achieve the highest possible accuracy in the measurement of such quantities have therefore often been decisive for the advance of science, and it is hardly necessary to call to mind the example that geodesy and astronomy have provided in this respect in the course of time. With the great break-through in so many fields of science inaugurating the recent development, also the importance of measurements came into the foreground, as expressed in Galilei's[1] well-known programme to investigate quantitative relations everywhere, even with regard to natural phenomena that do not directly show numerical characteristics.

During the following period not only was there created, through Newton's[2] great work, a solid basis for mechanics which allowed such a comprehensive use of the astronomical and geodetical measurements, but also other fields of the natural sciences would soon become the object of treatment based on exact methods. Just as it was the quantitative chemical analysis based on accurate measurements of weight that, especially through Lavoisier's contributions, would be decisive for the recognition of the fundamental laws of chemistry, so it was the development of methods for the measurement of temperatures and quantities of heat that gradually created the basis for the formulation of the laws of thermodynamics.

New perspectives opened once again when the electrical and magnetical phenomena, which already from ancient times had attracted attention but whose deeper relationship came to light only after the great discoveries in the beginning of the last century, became the object of quantitative investigation.

---

[1] Galileo Galilei (1564–1642).
[2] Isaac Newton (1642–1727).

Through the development of methods for the accurate measurement of electro-magnetic forces and electric currents, the basis was created for a formulation of electrodynamics which, as regards consistency and scope, could compare with Newton's mechanics and, in addition, pointed to paths beyond it, finally leading to that revision of the very basis for the space–time coordination, which found its expression in Einstein's theory of relativity and has given the conceptual structure of classical physics such a harmonious completion.

The most recent development in physics, which has given us a deep insight into the world of the atoms and has led to further advances in the recognition of the conditions for scientific analysis and synthesis, has also from the very beginning depended on the efforts towards still further refinement of measurements. Precisely in this field special difficulties had to be overcome, and it is hardly necessary to recall how, for a long time, the atomic theory – whose origin goes back to the very earliest efforts to obtain a general philosophical basis for the understanding of the regularities that natural phenomena, despite their richness and variability, show on closer inspection – had to be considered as a working hypothesis for which no direct proof could be obtained, as it was assumed that the coarseness of our senses would forever exclude us from observing effects of individual atoms.

Gradually, however, a number of quantitative regularities were disclosed, from which definite conclusions with regard to how substances are built up from atoms could be drawn and which even provided information about the properties of individual atoms. The first major step on this road was Dalton's[3] explanation of the law of multiple proportions, which governs the mass ratios in chemical reactions. It was, moreover, the consistent development of this line of thought, which forms the basis for the interpretation of Faraday's[4] laws of electrolytical equivalence, that would lead to the recognition of the atomic character of electricity. Furthermore, the exceedingly refined measurement technique of spectral analysis not only gave the elements numerical characteristics, allowing us to determine their presence even in distant stars, but also revealed characteristic regularities concerning the radiation emitted from atoms which should become of the greatest significance for the later development.

A new stage in our knowledge of the building stones of the elements began around the turn of the century through the study of entirely new phenomena such as cathode rays and X-rays. In particular, the discovery of the electron as a common constituent of the atoms of all substances, and the steadily more

---

[3] John Dalton (1766–1844).
[4] Michael Faraday (1791–1867).

accurate determinations of its mass and charge, would soon bring the physicists on the track of quantitative relationships between many hitherto uncorrelated phenomena. Above all, however, the study of the natural radioactivity of the heaviest elements would give us the means for a direct investigation of the interior of atoms which, through Rutherford's demonstration of the existence of the atomic nucleus and of the possibilities for transforming it, to such a high degree would complete the ideas of the constitution of the atoms and would also, by way of the accurate measurements of the large amounts of energy released by nuclear transformations, yield a decisive confirmation of the general connection between mass and energy predicted by the theory of relativity.

At the same time as the knowledge of the structure of atoms clarified many questions about the relationships between the elements, it revealed, in a startling way, the inadequacy of the classical mechanical and electrodynamical ideas to explain the peculiar stability of the atomic constitution governing the specific properties of the elements and which is indeed a condition for the very existence of the measuring instruments upon which the quantitative investigation of all phenomena ultimately rests. A key to the clarification of these circumstances was provided, however, through the discovery of the quantum of action, to which Planck was led by his ingenious analysis of the laws governing the phenomena of heat radiation, which had been brought to light through steadily more accurate measurements. This epoch-making discovery led of course to the recognition that even the most elementary mechanical concepts, although they allow such a detailed description of broad realms of experience, are only idealizations whose applicability depends upon the effects under consideration being large enough in relation to the universal quantum of action.

In these circumstances any account of the properties of atoms must rely upon a mode of description which is statistical in principle and whose aim is to determine the probabilities of individual quantum effects that do not permit closer analysis. After various attempts it has become possible on this basis, by the introduction of a suitable mathematical formalism, to develop a consistent atomic mechanics which is to be regarded as a rational generalization of the edifice of Newtonian mechanics. Hereby a framework was created for the explanation of quantitative relationships between measurement results which have been collected in the most diverse fields and which, with regard to variety, are comparable to the entire empirical material upon which classical physics rests.

At the same time, the very problem of measurement has been placed in an essentially new light by the recognition that the existence of the quantum of action sets a lower limit for the interaction – unavoidable in any observation – between the objects of the investigation and the measuring instruments. While until now it has been assumed that all phenomena – irrespective of the relativity contingent on the chosen frame of reference for space and time – can be traced back to a behaviour of the objects independent of observation, in the study of the individual quantum effects we are, by the nature of the case, prevented from distinguishing sharply between an independent behaviour of atomic objects and their interaction with the means of observation indispensable for the definition of the phenomena.

Without really setting any absolute limit for the precision of individual measurements, this circumstance means, as is well known, that the localization in space and time of atomic objects and the knowledge of their momentum and kinetic energy are subject to a peculiar reciprocal limitation expressed by Heisenberg's indeterminacy relations. This recognition led to the complete resolution of the paradoxes we meet as regards the question of the nature of atomic objects and which originate from the necessity of using apparently incompatible pictures, such as corpuscle and wave field, in the account of the properties of photons as well as of electrons. Here we have in fact to do with an entirely new situation with regard to the analysis and synthesis of experience, as phenomena observed under different experimental arrangements can stand in such a state of mutual exclusion that they cannot be brought together in a single visualizable picture, but that the measurement results must nevertheless be regarded as complementary in the sense that only together do they allow a complete description of the behaviour of the objects under any well-defined conditions.

However great might be the difficulties to overcome in the field of atomic physics in order to achieve a rational arrangement of the numerous new phenomena still entering our domain of experience, there can be no question of reverting to a description meeting the demands for visualizability to which we are accustomed from everyday life, but rather of a constant further development of a mathematical formalism which is adapted to the measurement results obtainable in the new fields of observation. There is every reason to believe that such efforts will continue to be fruitful and that we shall succeed, in a consistent manner, in obtaining more and more numerical relationships within our rapidly growing area of knowledge.

In this connection it must be emphasized as strongly as possible that neither the statistical mode of description of atomic mechanics nor the limitations prevailing in the analysis of atomic phenomena involve any renunciation of the method of exact science. On the contrary, in the so-called complementarity approach we are concerned with a generalization of the ideal of causality, sufficiently broad to take into account the features of individuality in the atomic processes owing to the existence of the quantum of action, and to make pos-

[665]

sible an unambiguous synthesis of phenomena which cannot be placed within the framework of classical physics but which to such an essential extent has increased our knowledge of the regularities of nature.

XXXII. MATHEMATICS AND NATURAL PHILOSOPHY

The Scientific Monthly **82** (1956) 85–88

Dedication of the Institute of Mathematical Sciences
at New York University, 29 November 1954

See Introduction to Part II, p. [370].

[667]

MATHEMATICS AND NATURAL PHILOSOPHY (1956)

Versions published in English

A   The Scientific Monthly **82** (1956) 85–88
B   Technion Yearbook 1957, New York 1957, pp. 54–56, 139–140

B is identical to *A*.

Niels Bohr and Richard Courant, New York, 1958.

Made in United State of America
Reprinted from THE SCIENTIFIC MONTHLY
Vol. 82, No. 2, February, 1956

Mathematics and Natural Philosophy

NIELS BOHR

Dr. Bohr, professor of physics at the University of Copenhagen, is director of the Institute for Theoretical Physics and president of the Royal Danish Academy of Sciences and Letters. This article is based on an address given 29 November 1954 at the dedication of the Institute of Mathematical Sciences at New York University.

TO every one of us, the first revelation of the power of mathematical reasoning in its simplest form was the acquaintance with numbers and their use. From remembrances of our own childhood and the teaching of our children, we have all learned how playful counting is gradually replaced by a more conscious appreciation of the powerful tool for ordering manifolds of things and events that is represented by the rules of addition, subtraction, multiplication, and division. Recalling our education in elementary mathematics, we are likewise reminded of the wonderful experience in our early youth when we learned to estimate distances and heights of trees by the simple geometric constructions that were used by the ancient Egyptians and Mesopotamians with such skill in both geodesy and astronomy.

The significance of the study of mathematics for the development of logical thinking surely cannot be overestimated, and we must realize that every student in his own mind, although in far easier circumstances and with correspondingly greater speed, must travel step by step along the same ever more lofty paths that mankind has wandered on and paved through the ages. A milestone on this climb was reached in ancient Greece, where, at the same time when art flourished in unsurpassed measure, the endeavors to base mathematical science on clearly stated logical principles succeeded to a degree that evokes our admiration and presents us with an everlasting challenge.

I need not stress the invaluable exercise in stringent argumentation that Euclid's elements still offer or how much we learned by the profound exploration of geometric proportions that led Eudoxus to the distinction between so-called rational and irrational numbers, which was basic for ever wider mathematical generalizations. The awareness in the minds of the Greek philosophers of the paradoxes encountered in problems involving infinite sequences such as that in the humorous tale about the race between Achilles and the tortoise sharpened the demands that were made on mathematical proofs. An instructive illustration in this respect is

Archimedes' reluctance to rely on the methods, akin to modern infinitesimal calculus, by which he first derived his famous formulas for the volume of pyramids and spheres.

Appreciation of mathematics as a guide in natural philosophy also dates back to the days of the ancient Greeks. We all know how Pythagoras emphasized the importance of simple numerical relationships for musical harmonies as well as cosmology and how much significance the study of regular polyhedrons had for Plato's ideals of beauty and perfection. Among the lasting contributions of Greek mathematicians to physical science, we think especially of the laws of equilibrium of supported and floating bodies that Archimedes, with unfailing intuition, founded on simple arguments of symmetry and balance. In the treatment of dynamic problems, however, great difficulties long stood in the way of eliminating arguments that were motivated by the feeling of exertion experienced in our own movements and by the purposes behind our actions.

Liberation from the Aristotelian approach to dynamics was, as is well known, first accomplished at the time of the Renaissance when Galileo recognized the elementary character of uniform motion and the restricted application of the idea of forces to the alterations of such motion. On this basis, Newton built the marvelous edifice of classical mechanics that—both because of its immense power and scope and because of its adaptation to mathematical calculation—came to stand as an ideal model for scientific explanation and led to the so-called mechanistic conception of nature. Besides, in the analytic geometry of Descartes, the appropriate mathematical tools were found in differential calculus to which Newton himself, equally eminent as a physicist and as a mathematician, contributed so fundamentally.

This revolutionary development initiated an extremely intimate correlation between physical and mathematical research; discoveries in physics presented mathematicians with new challenges, and, in turn, mathematical abstractions and generalizations furthered the clarification of physical problems. As a typical example, we may recall how the studies of heat conduction inspired Fourier to develop the harmonic analysis that up to now has remained an important branch of pure mathematical research and has proved more and more indispensable to innumerable domains of physics. We may also mention the interplay between Faraday's fundamental research in electricity and magnetism and Maxwell's theory of electromagnetic fields, which inspired the development of mathematical disciplines such as vector and tensor analysis that have been so fruitful in many fields of physical science.

A very impressive account of the powerful tools that physicists now possess—thanks to the ingenious work of mathematicians in the later centuries—is given in the masterly treatise by Courant and Hilbert on the methods of mathematical physics. In this work, invaluable to every student, a lucid exposition is given of logical generalizations that not only have proved to be of extreme fertility in the explorations of multifarious problems within the domain of classical physics but have also shown themselves to be equally inspiring for the elucidation of the novel problems with which modern developments in physical science have confronted us.

The new and broader approach to the description and comprehension of natural phenomena originated in the recognition of the limited scope of the very ideas of absolute space, time, and causality on which the mechanical conception of nature rested. The first hint came, as is well known, from refined optical measurements that demonstrated the absence of the expected influence of the motion of the earth around the sun and revealed that observers moving relatively to each other at large velocities will coordinate phenomena differently. In fact, not only may such observers take different views of shapes and positions of rigid bodies, but events at separate points of space may to one person appear to be simultaneous and may by another be judged to occur at different times.

Far from entailing confusion and complication, the recognition of the extent to which description of physical phenomena depends on the standpoint of the observer proved to be a powerful inspiration to establish general physical laws common to all observers. I need hardly recall Einstein's discovery of the universal relationship between energy and mass or how, by stressing the equivalence from the observational standpoint of the effect of gravitational fields and accelerated frames of reference, he deeply remolded Newton's ideas. The general theory of relativity, which broadened our horizon and gave our world picture a unity surpassing all

previous imagination, is surely one of the greatest triumphs of rational human thinking.

For my theme, it is of principal interest that mathematical generalizations that were developed without reference to practical applications but merely in the pursuit of logical harmony have offered adequate tools for the realization of Einstein's great program. Abandoning not only the ideas of absolute space and time but even Euclidean geometry as the foundation, Einstein took recourse to a curved four-dimensional Riemannian spacetime metric that automatically accounts for gravitational effects and for the singular role of the speed of light that represents an upper limit for any consistent use of the physical concept of velocity. Mathematicians had indeed gradually become familiar with abstractions of this kind through a development of non-Euclidean geometry and its various models.

Despite all the new features, it was possible in relativity theory to retain and even to refine the deterministic description that is characteristic of classical physics. An inherent limitation of the very idea of causality has, however, been revealed in the last decades by the exploration of the atomic constitution of matter, which was made possible by modern developments in experimental technique. It is interesting to note that, although the assumption of a limited divisibility of substances goes back to antiquity, it was up until quite recent times considered to be a hypothesis for which no direct verification could be obtained. With the great progress in chemistry and physics in the last centuries, atomic ideas proved yet more fruitful, and in particular it was found possible to develop statistical mathematical methods of treating the average behavior of systems consisting of a large number of particles and in this way to account for the empirical laws of thermodynamics. A decisive step was the elucidation by Boltzmann of the general relationship between the concept of entropy and the probability of the degree of order of the state of such systems.

This great accomplishment was indeed the clue to the analysis of the regularities of thermal radiation that, in the first year of this century, led Planck to his epoch-making discovery of the universal quantum of action. In pointing to a feature of indivisibility in physical processes that is quite foreign to the mechanistic conception of nature, Planck's discovery revealed that the laws of classical physics are idealizations that are applicable only to the description of phenomena where the actions involved are sufficiently large to permit neglect of the quantum. Whereas this condition is amply fulfilled in phenomena on the ordinary scale, we meet in atomic processes regularities of quite a new kind, regularities that defy the deterministic pictorial description. Very striking illustrations are afforded by the well-known dilemmas regarding the properties of electromagnetic radiation as well as of material corpuscles, evidenced by the circumstance that in both cases contrasting pictures as waves and particles appear equally indispensable for the full account of experimental evidence.

Here we are clearly in a situation where it is no longer possible to define unambiguously attributes of physical objects independently of the way in which the phenomena are observed, and especially to ignore the interaction between objects and measuring instruments, the disregard of which is characteristic of the mechanistic conception of nature. This situation has demanded a new revision of the foundations for the description and comprehension of physical experience. On the one hand, we must realize that however far the phenomena transcend the scope of classical physics, it is obviously necessary to describe the experimental arrangement and the observations in plain language suitably supplemented with technical physical terminology. On the other hand, just the necessity of accounting for the functioning of the measuring agencies on classical lines excludes in principle in proper quantum phenomena an accurate control of the reaction of the measuring instruments on the atomic objects.

This circumstance prevents in particular the unrestricted combination of space-time coordination and the dynamic conservation laws on which deterministic description in classical physics rests. In fact, any unambiguous use of the concepts of space and time refers to an experimental arrangement that involves a transfer of momentum and energy, uncontrollable in principle, to instruments such as measuring rods and synchronized clocks, which are required for fixing the reference frame. Conversely, an account of phenomena governed by conservation of momentum and energy involves in principle a renunciation of detailed space-time coordination.

The essential indivisibility of proper quantum phenomena finds logical expression in the circumstance that any attempt at a well-defined subdivision would require a change in the experimental arrangement that precludes the appearance of the phenomenon itself. Under these conditions, it is not surprising that phenomena observed with different experimental arrangements appear to be contradictory when it is attempted to combine them in a single picture. Such phenomena may appropriately be termed complementary in the sense that they represent equally important aspects of the knowledge obtainable regarding the atomic objects and only together exhaust this knowledge. The notion of complementarity implies no arbitrary renunciation of our accustomed demands on physical explanation but simply refers to our position as observers in this new domain of experience.

Actually, it has been possible by the concerted efforts of a whole generation of theoretical physicists to develop a rational generalization of classical mechanics that permits a complete account of a wide field of experience along the lines of the complementary mode of description. In this quantum mechanical formalism, the ordinary kinematic and dynamic variables are replaced by operators, subject to a noncommutative algorism involving Planck's constant. We are here again dealing with mathematical abstractions already widely explored; for example, it was early realized that the composition of rotational movements of rigid bodies as a sequence of rotations around different axes is dependent on the order in which such operations are performed.

In the quantum mechanical terminology, the noncommutabilty of the symbolic operators directly reflects the mutual exclusion of experimental arrangements that permit the accurate definition of the corresponding physical quantities. Moreover, the reciprocally restricted applicability of kinematic and dynamic variables in the quantum mechanical description of the state of a physical system finds quantitative expression in Heisenberg's indeterminacy relationships which proved to be of fundamental importance for the clarification of the situation, especially with regard to the limits of the customary ideal of causality.

In conformity with the circumstance that several individual quantum processes may take place in a given experimental arrangement, the predictions of the formalism concerning observations are of an essentially statistical character. It must be realized, however, that in this respect we are presented, not with an analog to the use of probability considerations in the account of the behavior of complicated mechanical systems, but with the impossibilty of defining any directive for the course of individual processes beyond those afforded by the self-consistent generalization of deterministic mechanics.

For anyone who through the years has been concerned with the difficulties and paradoxes in quantum physics, it is indeed a deep satisfaction that logical order should be attained to such degree by means of the subtle methods offered by mathematical science. Truly, it has been a wonderful experience to witness how an immense amount of experimental evidence regarding atomic and molecular spectra, chemical bonds, and radioactive processes in the course of a few years was accounted for in detail and brought into unique connection with simple data regarding inertial masses and electric charges of the particles of which all atoms are composed.

We are confronted here with regularities fundamental to the properties of matter that, although they are quite beyond the scope of the mechanical principles so fruitful in diverse fields of technology, nevertheless lend themselves to mathematical formulation and numerical calculation. In this connection, it is also important that the existence of high-speed computers—such as the wonderful UNIVAC installed in the Institute of Mathematical Sciences at New York University—which have initiated great advances in the treatment of many problems within the domain of classical physics, holds similar promises for the exploration of atomic problems.

The general lesson of the role that mathematics has played through the ages in natural philosophy is the recognition that no relationship can be defined without a logical frame and that any apparent disharmony in the description of experiences can be eliminated only by an appropriate widening of the conceptual framework. This lesson, familiar to mathematicians, and conspicuous in studies of the foundations of their science, has been enforced by the development of physics in a way that has a bearing on many other fields of human knowledge and interest in which we meet with similar situations in the analysis and synthesis of experience.

Opening in 1937 of the new radium station at the Finsen Institute in Copenhagen. The event took place in the new X-ray Hall. Seated closest to the camera are the Danish King and Queen, Christian X and Alexandrine. The first, third and fifth person on the first row are H.M. Hansen, Niels Bohr and J.E. Meulengracht, who would become chairman of the National Association for Combating Cancer the following year.

7. ON FOOTBALL, CANCER,
ISRAEL AND THE DANISH RESISTANCE

XXXIII. GREETING FROM NIELS BOHR

NIELS BOHR'S HILSEN
"Akademisk Boldklub 1939–1949",
Copenhagen 1949, pp. 7–8

Greeting recorded on a gramophone record
during Bohr's stay in the U.S.A. in 1939

TEXT AND TRANSLATION

See Introduction to Part II, p. [372].

The Mittweida Ball Club from Germany visiting the Academic Ball Club in Copenhagen, 1905. Niels Bohr is in the second row to the extreme right, with his brother, Harald, standing beside him.

NIELS BOHR'S HILSEN

I anledning af A.B.s 50 års jubilæum i 1939 udtalte professor dr. phil. Niels Bohr følgende hilsen, indtalt på grammofonplade under hans ophold i Amerika 1939:

Niels Bohr.

Det er en stor glæde for mig, trods jeg i øjeblikket befinder mig så langt borte fra Danmark, alligevel gennem radioen med nogle få ord at få lejlighed til at være med til at fejre Akademisk Boldklubs 50-års jubilæum. I disse dage vil mange med mig mindes de uforglemmelige sorgløse og forfriskende timer, som vi har tilbragt sammen på boldbanen som medspillende eller tilskuere.

Ikke mindst for den studerende ungdom er jo sporten en uvurderlig kilde til udhviling og forfriskelse efter det stillesiddende liv og de åndelige anstrengelser, og alle har vi jo ikke alene følt den fornyede kraft, hvormed vi efter vor fri tumlen på den grønne plæne har været i stand til at tage fat på arbejdet igen, men vil heller aldrig glemme den umiddelbare glæde, det friske samvær med kammeraterne under og efter kappestriden var for os. Den legemlige opdragelse og den selvdisciplin, som deltagelse i sportslivet har givet os, har vel endda de fleste af os først rigtig lært at skatte i den senere alder, hvor vi var nødt til at holde mere hus med kræfterne og ikke længere tålte de anstrengelser, vi som yngre kunne byde os.

Vi tænker alle med taknemlighed på, at der også i Danmark på så tidligt et tidspunkt fandtes en forståelse for sportens betydning for studielivet, og vi mindes med tak først og fremmest de mænd, der med så stort et fremsyn og så stor en begejstring for 50 år siden tog de første vanskelige skridt til skabelsen af Akademisk Boldklub.

De beskedne forhold, under hvilke spillet dengang fandt sted, kender jo kun vi ældre, der, selv om vi ikke var med fra begyndelsen, levende mindes, hvorledes vi som drenge i den nystiftede junior-afdeling sammen med seniorerne, som vi så op til med så megen beundring og respekt, havde til huse i de små værelser i baggården ved Blegdamsvejen, hvorfra vi hver gang måtte bringe mål eller gærder ud på fælleden og stille dem op på den bedste ledige plads, vi kunne finde. At forholdene for spillets udøvelse nu er så meget bedre og rammerne for klublivet så udvidede og forskønnede, skylder vi jo den forstående og velvillige interesse, der efter den første indsats voksede op i stedse større kredse. Om alt dette vil dog andre med større indsigt og beføjelse tale om i disse dage, og jeg vil derfor blot gerne have lov til at udtale min personlige tak til den gamle klub for alle de mange glæder og minder, som den har givet mig, og udtale mine allervarmeste ønsker for dens fortsatte trivsel til styrkelse og opdragelse for den studerende ungdom i Danmark.

TRANSLATION

GREETING FROM NIELS BOHR

On the occasion of A.B.'s 50-year jubilee in 1939, Professor
Niels Bohr, D.Phil., gave the following greeting, recorded on
a gramophone record during his stay in America in 1939:

Although being so far away from Denmark at present,[1] it is a great pleasure
for me to have the opportunity nevertheless of taking part in the celebration
of the 50-year jubilee for Akademisk Boldklub [Academic Ball Club] with a
few words over the radio. At this time, many will join me in remembering the
unforgettable carefree and refreshing hours that we have spent together on the
football pitch as players or spectators.

Sport is, of course, not least for the student generation, an invaluable source
of relaxation and refreshment after the sedentary life and the intellectual effort,
and all of us have not only felt the renewed energy with which, after our
untrammelled play on the greensward, we have been able to take up work
again, but neither will we forget the instant pleasure the refreshing meeting
with comrades during and after the game was for us. The physical training and
the self-discipline participation in sport has given us, most of us have probably
really learned to value only at the later age when we found it more necessary
to husband our powers and could no longer overcome the exertions we could
bid ourselves when younger.

We all remember with gratitude that in Denmark, too, at so early a time,
there was understanding for the importance of sport in the life of study, and we
remember with thanks first and foremost those men who, 50 years ago, with
such great foresight and such great enthusiasm, took the first difficult steps
towards the establishment of Akademisk Boldklub.[2] The modest conditions
under which the sport was then practised are of course only familiar to us
veterans who, even though we were not there from the beginning, vividly
remember how, as boys in the newly established junior branch, we shared our
premises with the seniors, to whom we looked up with much admiration and

---

[1] Bohr was visiting the United States. See the introduction to Part II, p. [340].

[2] Bohr's father, Christian Bohr (1855–1911), was involved in the establishment of the Club in
1889.

respect, in the small rooms in the back yard near Blegdamsvej, from where every time we had to bring goals or fences out on to the common and put them on the best available spot not already in use. That the conditions for playing the game are now so much better, and the framework for life in the club so extended and adorned, we owe of course to the sympathetic and generous interest which grew in ever larger circles after the first efforts. However, others with greater insight and entitlement will speak about all this at this time and I therefore only want to be allowed to express my personal thanks to the old club for all the many pleasures and memories it has given me, and express my most heartfelt wishes for its continued welfare for the benefit and the upbringing of the student generation in Denmark.

XXXIV. ADDRESS BROADCAST ON
THE DANISH NATIONAL RADIO, 16 OCTOBER 1953

TALE VED STATSRADIOFONIENS UDSENDELSE
DEN 16. OKTOBER 1953
"Det kongelige Teater, Forestillingen lørdag den 17. oktober 1953", pp. 4–6

Reproduced in the Royal Danish Theatre's programme
for the performance of "Pygmalion" on 17 October 1953,
held in support of the National Association for Combating Cancer

TEXT AND TRANSLATION

See Introduction to Part II, p. [373].

Tale ved Statsradiofoniens udsendelse
den 16. oktober 1953

NIELS BOHR
Præsident for den danske
Kræftkomité

Næppe nogen sag kalder stærkere på vor deltagelse og hjælp end kampen mod de snigende kræftsygdomme, der medfører så store lidelser og kræver så mange menneskeliv. Det gælder jo om at bringe lindring og helbredelse til medmennesker, som endnu i mange år kunne leve til glæde og støtte for deres nærmeste og fortsat tjene samfundet. Især må det ligge alle på sinde at fjerne den trusel, der hænger over yngre mødre netop i de år, hvor livet stiller dem de største opgaver, og hvor sygdom og død foruden sorgen skaber ulykke og savn i deres hjem og for dem, de bragte til verden og værnede om med den kærlighed, der er familiens grundvold.

Bestræbelserne for at bekæmpe kræftsygdommene med støtte af et initiativ, der direkte henvender sig til enhvers offervilje, går her hjemme som i mange andre lande tilbage til århundredets første år, hvor, samtidig med at livmoderkræftens foruroligende udbredelse kom til almindelig kundskab, opdagelsen af de vidunderlige stråler, der kan trænge dybt ind i det menneskelige legeme, skabte nye muligheder for at angribe de ondartede svulster. I 1905 oprettedes »Den almindelige danske lægeforenings Cancer-komité«, der i begyndelsen virkede særligt ved at udbrede forståelse af situationens alvor og mane til energiske bestræbelser for forbedring af vilkårene for kræftsygdommenes udforskning og behandling.

På denne baggrund stiftedes i 1912 »Radiumfondet«, hvis midler, indsamlet til minde om kong FREDERIK VIII, muliggjorde det første danske indkøb af radium og tillod indretningen, først i København og kort derefter i Århus og Odense, af behandlingssteder – de såkaldte radiumstationer – til undersøgelse og bestråling af kræftsvulster. Trods de små midler, som stod til rådighed, og de meget beskedne forhold, under hvilke behandlingen fandt sted, arbejdedes der med en iver og målbevidsthed, der måtte gøre et dybt indtryk på enhver, som kom i berøring med den gryende virksomhed. Mange af os vil altid mindes den ildhu, der besjælede de kyndige og varmhjertede personligheder, som ledede arbejdet og forberedte den næste landsindsamling i 1921, hvis rige resultat viste, hvor stærkt hele befolkningen allerede dengang var grebet af den for alle så vigtige sag.

På det ved indsamlingen skabte nye økonomiske grundlag muliggjordes store fremskridt for Radiumfondets virksomhed. Langt stærkere radiumpræparater kunne nu anskaffes og fordeles mellem radiumstationerne, der samtidig i høj grad forbedredes og udvidedes, således at der foruden den ambulante behandling kunne indrettes sengepladser til hårdt angrebne patienter. Det viste sig imidlertid hurtigt, at der krævedes endnu større anstrengelser for at kampen mod kræften kunne føres på en måde, der tog fuldt hensyn til sygdommens udbredelse og tillod udnyttelsen af alle muligheder for dens behandling. Med de forhåndenværende midler kunne radiumstationernes daglige drift kun opretholdes med vanskelighed, og der kunne ikke være tale om at imødekomme de krav til deres yderligere udvidelse og forbedringen af deres udstyr, der stedse stærkere meldte sig. Navnlig blev det mere og mere klart, at det var nødvendigt i større omfang at iværksætte videnskabelige undersøgelser for at kunne følge med i udviklingen og udnytte alle veje, som forskningen måtte anvise.

I erkendelse af nødvendigheden af at skabe et bredere økonomisk og organisatorisk grundlag for kampen mod kræften indbød Lægeforeningens Cancerkomité repræsentanter for alle interesserede kredse til et møde på Christiansborg den 23. oktober 1928. På dette møde, der skulle få skelsættende betydning, vedtoges det at stifte »Landsforeningen til Kræftens Bekæmpelse« og samtidig at udvide rammerne for Cancerkomitéen, der omdannedes til et rådgivende og kontrollerende organ med navnet »Den danske Kræftkomité«. Kong CHRISTIAN X, der lige fra Radiumfondets oprettelse havde omfattet sagen med varmeste interesse, indvilgede i at være Landsforeningens protektor og som bekendt har også vor nuværende konge på denne måde ydet Landsforeningen til Kræftens Bekæmpelse sin betydningsfulde støtte.

Tanken om at kalde på hele befolkningens medvirken viste sig at være den helt rigtige, og takket være den lykkelige hånd, hvormed sagen blev grebet an, vandt Landsforeningen hurtigt stor tilslutning. Allerede det første år indmeldte sig over 40.000 medlemmer, og i dag ligger medlemstallet omkring 200.000. Det virksomme initiativ og det fremsyn, som Landsforeningens ledende mænd fra første færd har lagt for dagen, er også i stedse højere grad blevet mødt med forståelse og offervilje. Foruden medlemsbidragene og indtægterne fra de årlige Kræftdage og statstelegrafens festblanketter er der i årenes løb tilgået foreningen mange både mindre og større gaver, der har bidraget til at styrke grundlaget for virksomheden.

Det er umuligt i den korte tid, der står til min rådighed, blot i hovedtræk at skildre alt hvad det er lykkedes Landsforeningen til Kræftens Bekæmpelse i de forløbne 25 år at udrette. Det hele føjer sig sammen til et sandt og gribende eventyr, der trods den alvorlige baggrund rummer en opmuntring for os alle af sjælden art. De store og smukke radiumstationer i Århus, København og Odense, udstyrede med de nyeste hjælpemidler, står i dag rede til at modtage de syge og give dem en behandling baseret på videnskabens seneste fremskridt, hvortil bidrag ydes af det bredt anlagte forsknings-arbejde, som Landsforeningen har iværksat. Samtidig er der over hele landet ved artik-ler, foredrag og radioudsendelser gjort et stort oplysningsarbejde med det vigtige formål at tilskynde enhver, der mærker de første symptomer på kræftlidelser, til straks at søge lægekyndig bistand, idet trods alle fremskridt muligheden for helbredelse altid vil være større, jo tidligere behandlingen kan begynde.

Den store samfundssag kræver imidlertid stadig fornyede anstrengelser, og selv om staten i de sidste år, hvor udgifterne ved hospitalsvirksomhed er steget så overordentligt, har bevilget meget betydelige tilskud til radiumstationernes drift, er der endnu mange vigtige opgaver, som venter på deres løsning. Ikke mindst er det en betingelse for op-retholdelsen af det høje stade på kræftforskningens område, hvor nogle af de berømteste foregangsmænd netop har virket her i landet, at der skaffes denne forskning stedse bedre arbejdskår. Endvidere er der en stadig voksende trang til, at hårdt angrebne kræftpatienter efter behandlingen kan få adgang til rekreation for at genvinde kræfter og mod til at genoptage deres gerning i livet og samfundet.

I taknemmelighed for hvad der er nået og i fælles ønske om fortsat fremgang må vi alle af hjertet dele håbet om, at den jubilæumsindsamling, som Landsforeningen i an-ledning af sin 25-årige beståen foranstalter i disse dage, vil finde en sådan tilslutning, at vi ikke alene kan bevare værdifulde traditioner, men at alt, hvad vi formår, vil kunne sættes ind på at bekæmpe kræftsygdommenes trusel og hjælpe lidende mennesker til helbredelse.

NIELS BOHR

TRANSLATION

Address broadcast on Danish National Radio
16 October 1953

NIELS BOHR
President of the Danish Cancer Committee

Hardly any matter calls more strongly for our participation and help than the battle against the insidious cancer illnesses which cause such great suffering and claim so many human lives. What is important, of course, is to bring relief and recovery to fellow human beings so that they can live for many more years to the joy and support of their family and friends and continue to serve society. In particular, it must be a matter of importance for everyone to remove the threat which hangs over young mothers in precisely those years when life offers them the biggest tasks and when illness and death cause, apart from the sorrow, unhappiness and loss in their home and for those they brought into the world and shielded with the love that is the foundation of the family.

The endeavours to combat cancer with the support of an initiative appealing directly to the individual's readiness to give go back, here at home as in so many other countries, to the first years of the century when, at the same time as the disturbing spread of uterine cancer became generally known, the discovery of the wonderful rays that can penetrate deeply into the human body created new possibilities for attacking the malignant tumours. In 1905, "The Danish Medical Association's Cancer Committee"[1] was established, which in the beginning worked especially to spread understanding of the seriousness of the situation and to encourage vigorous endeavours for the improvement of the conditions for investigation and treatment of cancer.

On this background, the "Radium Foundation" was established in 1912, whose means, collected in memory of King FREDERIK VIII,[2] made possible the first Danish purchase of radium and permitted the establishment, first in Copenhagen and shortly thereafter in Århus and Odense, of places of treatment – the so-called radium stations – for the investigation and irradiation of cancer

---

[1] Bohr was President of the Cancer Committee from 1935 until his death. See the introduction to Part II, p. [373].

[2] (1843–1912). He had acceded to the throne in 1906.

tumours.[3] Despite the small means available and the very modest conditions under which the treatment took place, work was done with a zeal and a singleness of purpose which had to make a great impression on anyone who came into contact with the budding activity. Many of us will always remember the enthusiasm which inspired the capable and warm-hearted personalities who led the work and prepared the next national subscription in 1921, whose rich result showed how strongly the whole population already then was gripped by the cause so important for all.

Great advances for the Radium Foundation's activity were made possible on the new financial basis created by the subscription. Far stronger radium preparations could now be acquired and distributed among the radium stations, which at the same time were improved and extended to a great degree, so that in addition to ambulant treatment, hospital beds could also be furnished for seriously affected patients. However, it rapidly became evident that even greater efforts were required in order to be able to conduct the fight against cancer in a way that took full account of the spreading of the disease and allowed application of all possibilities for its treatment. With the means at hand it was possible only with difficulty to maintain the day-to-day operation of the radium stations, and there was no possibility of meeting the demands, which were made ever more strongly, for their further extension and the improvement of their equipment. In particular, it became more and more obvious that it was necessary to a greater extent to initiate scientific investigations in order to be able to keep up with the development and utilize all paths that research might indicate.

In recognition of the necessity to create a broader financial and organizational basis for the fight against cancer, the Cancer Committee of the Danish Medical Association invited representatives from all interested circles to a meeting at Christiansborg on 23 October 1928. At this meeting, which was to have epoch-making importance, it was voted to establish "The National Association for Combating Cancer" and at the same time extend the framework for the Cancer Committee, which was transformed into a consulting and controlling body with the name "The Danish Cancer Committee". King CHRISTIAN X, who already from the establishment of the Radium Foundation had embraced the

---

[3] The first radium station in Denmark was established in Copenhagen in 1913, the ones in Aarhus and Odense the year after. They were all run by the Radium Foundation. In 1922 the Copenhagen radium station moved to the Finsen Institute (see p. [516], ref. 25) after the second national subscription for the cancer cause the year before. In 1928 the newly established National Association for Combating Cancer took over the responsibility for running the radium stations until they became institutions of the Danish State in 1963, the year after Bohr's death.

cause with the warmest interest, agreed to be Patron of the National Association and, as is well known, our present king, too, has in this way given the National Association for Combating Cancer his important support.

The idea of appealing for the participation of the whole population turned out to be just the right one, and thanks to the fortunate way the matter was tackled, the National Association rapidly won great support. Already in the first year more than 40,000 members joined, and now the membership numbers around 200,000. The active initiative and the foresightedness which the leading men of the National Association displayed from the very start, have also, to an ever increasing degree, been met with understanding and readiness to give. In addition to contributions from the members and income from the annual Cancer Days[4] and the State telegraph service's greetings telegram forms,[5] over the years many gifts, both large and small, have reached the Association, which have contributed to strengthening the basis for the enterprise.

It is impossible in the short time at my disposal to describe even in outline all that the National Association for Combating Cancer has succeeded in achieving in the past 25 years. It all combines into a true and gripping adventure which despite its serious background holds encouragement of a rare kind for us all. The large and beautiful radium stations at Århus, Copenhagen and Odense, equipped with the newest facilities, stand today prepared to receive the sick and give them a treatment based on the most recent advances of science, to which contributions are made by the broadly based research work that the National Association has initiated. At the same time, great educational work, using newspaper articles, lectures and radio broadcasts, has been carried out all over the country, with the important purpose of urging anyone who notices the first symptoms of a cancer illness immediately to seek medical advice, as despite all advances, the possibility of recovery will always be greater the sooner the treatment can start.

The great social issue requires, however, ever renewed efforts, and although the State in recent years, when the expenses for running hospitals have risen so excessively, has allotted very significant grants to the operation of the radium stations, there are still many important tasks awaiting their solution. It is not least a prerequisite for the maintenance of the high level in the field of cancer research, where some of the most famous pioneers have worked precisely in this country, that ever improved working conditions are obtained for this

[4] The annual Cancer Day, when the whole Danish population was encouraged to contribute to the cause, was instituted in 1942.

[5] The National Association started receiving this income in 1934.

research. Furthermore, there is a constantly growing need for seriously affected cancer patients to be able to have access to convalescence after treatment in order to regain strength and courage to resume their activity in life and society.

In gratitude for what has been achieved and in a common wish for continued progress, may we all from our hearts share the hope that the jubilee appeal that the National Association is organizing at this time, on the occasion of its 25 years of existence, will find such great support that we not only can hold on to valuable traditions but that all our resources can be applied to combat the threat of cancer and to help suffering people to recovery.

NIELS BOHR

XXXV. THE REBUILDING OF ISRAEL: A REMARKABLE KIND OF ADVENTURE

ISRAELS GENOPBYGING: ET ÆVENTYR AF EJENDOMMELIG ART
Israel **7** (No. 2, 1954) 14–17

Talk on Danish national radio 2 February 1954

TEXT AND TRANSLATION

See Introduction to Part II, p. [376].

Israels genopbygning.
Et æventyr af ejendommelig art.

Af Prof., Dr. phil. NIELS BOHR

Skabelsen af et hjem for mennesker af jødisk afstamning på den gamle historiske grund, hvor Israels folk i sin tid bidrog så meget til at udbygge det etiske grundlag, hvorpå senere slægters liv har hvilet, er et eventyr af helt ejendommelig art. Det drejer sig jo her ikke om en mere eller mindre jævn fremgang i levekår og kultur, som vi kender det fra andre landes historie, men om en tilpasning, som trods alle vanskeligheder foregår med en hastighed, der næppe har noget sidestykke. Mennesker fra nære og fjerne dele af verden, hvor de har deltaget i nationernes liv — omend på grund af fordomme de ofte, ja helt op til vor tid, har været udsat for forfærdende forfølgelser — søger, grebet af en fælles samlende idé, i Israel at udnytte de kundskaber og eraringer, som de hver især har medbragt, til at opbygge et nyt samfund og udvikle landets naturlige hjælpemidler for at skabe livsbetingelser for en befolkning mange gange større end den, de små landstrækninger på Lilleasiens kyst i de senere århundreder har ernæret.

En sådan situation kræver naturligvis den største anspændelse og personlige opofrelse af alle, både af dem, der med nærmere kendskab til forholdene må tage ledende del i arbejdet, og de mange, der stadig strømmer til, og for hvem der må skaffes den efter deres højst forskellige uddannelsestrin mest egnede beskæftigelse. Overalt i Israel ser man lejre med boliger af meget primitiv art, hvori man anbringer indvandrere, der straks må sættes til at deltage i den opdyrkning af landet, som ofte kræver store forberedende drænings- og reguleringsarbejder. Disse lejre omdannes efterhånden til smukke landsbyer omgivet af marker og frugtplantager, der i det varme klima giver rigt udbytte, så snart de rette betingelser er til stede. Foruden de store olivenskove, der især i Galilæas bjergkløfter på grund af de gunstige vandingsforhold har holdt sig fra oldtiden af, ser man allerede nu mange steder i Israel sydfrugter af enhver art og store strækninger beplantet med majs og bomuld.

For hele denne opdyrkning såvel som for opbygningen af den for landets økonomi stadig mere nødvendige industri gælder det om at udnytte al kundskab, der kan tilvejebringes. I denne forbindelse har netop Chaim Weizmanns lykkelige fremsyn som statsmand og videnskabsmand spillet en ganske afgørende rolle. Fra første begyndelse forstod han, at det nye samfunds trivsel ville være betinget af de mest energiske bestræbelser for at fremme videnskabelig forskning, og han tænkte her ikke alene på den umiddelbare hjælp, som videnskaben måtte kunne yde til udnyttelsen af alle landets i så lang tid forsømte

Chaim Weizmann — videnskabsmand,

rfigdomskilder, men tog samtidig sigte på betydningen af det nye samfunds deltagelse i den fælles menneskelige søgen efter kundskaber, der måske mere end noget andet binder folkeslagene sammen.

Som følge af de gamle traditioner og den respekt for lærdom og forskning, der igennem tiderne har været levende i de over verden spredte jødiske folkegrupper, er der i Israel en udbredt forståelse af videnskabelige studiers betydning. Allerede få år efter den første verdenskrig stiftedes med udstrakt international støtte universitetet i Jerusalem, der indrettedes i de imponerende bygninger på Scopus-bjerget lige øst for den gamle by med udsigt over Jordandalen med dens rige minder. Efter den våbenstilstand, hvormed opgøret i landet i årene efter den sidste verdenskrig sluttedes, ligger imidlertid Scopus-bjerget på den arabiske side af grænsen mellem Israel og Jordan, og de store universitetsbygninger med det værdifulde bibliotek står derfor ikke længere til rådighed for de mange videnskabsmænd og studenter i Jerusalem. Indtil de nye bygninger, som påtænkes opført i den vestlige udkant af byen, kan tages i brug, er de alle henvist til at arbejde under de mest sammentrængte forhold i midlertidige lokaler, hvilket dog ikke hindrer udfoldelsen af en flid og begejstring, der gør dybt indtryk på enhver besøgende.

En ganske anden tilblivelseshistorie har Weizmann-Instituttet i Rehovoth haft. Her, midt i det nuværende Israel, hvor der længe kun havde været en ørken med lave sandbanker, tilsyneladende uegnet til beboelse og opdyrkning, opførtes for omkring 20 år siden med støtte fra Weizmanntaknemmelige venner nogle mindre laboratoriebygninger, der skulle blive den første begyndelse til det nu så store og betydningsfulde forskningscentrum. På toppen af den højeste banke indrettede Weizmann sit eget hjem, og ørkenen blev ved hjælp af overrisling tilplantet og består nu af ap-

. . . og politiker.

[691]

pelsin- og citronplantager, der ind-
rammer institutbygningerne på
skønneste måde. For at hædre Weizmanns minde har den israelske stat
besluttet at gøre hele arealet omkring instituttet til en national
park, der rummer Weizmanns grav
og hans bolig, der med tiden vil
blive indrettet til et museum for
minderne om Israels første store
præsident.

Jeg traf selv Chaim Weizmann
første gang i Manchester, da jeg arbejdede hos Rutherford under den
første verdenskrig. Han var dengang professor i organisk kemi ved
Manchester Universitetet og havde
netop udarbejdet nye metoder til
fremstilling af de for krigsførelsen
nødvendige sprængstoffer, hvorved
han kom til at yde de allierede magter uvurderlige tjenester. Dog ikke
alene hans ry som videnskabsmand
omgav ham med respekt, men alle,
der kom i berøring med ham, fik et
dybt indtryk af hans kraftfulde og
ædle personlighed. Sidste gang jeg
så ham var ved et besøg i hans
hjem i New York få dage før Israels uafhængighedserklæring. Han
var da syg og sengeliggende, men
med en indre ild fortalte han beta-
gende om de forhandlinger, han i
de gamle dage havde ført med kong
Feisal af Arabien. Disse forhandlinger fandt sted omkring et bål i ørkenen med den berømte Lawrence
som tolk og førtes i en tillids- og
venskabsånd, som man må håbe vil
komme tilbage.

Mindeparken i Rehovot blev
med deltaagelse af Israels regering
og talrige repræsentanter fra hele
landet indviet i begyndelsen af november på årsdagen for Weizmanns
død, og den nedlæggelse af grund-
stenen til et fysisk institut, som vi
var indbudt til at deltage i, var led
i højtidelighederne, der strakte sig
over en hel uge og endte med afhol-
delsen af et internationalt videnskabeligt symposium til diskussion

af de problemer, der særlig er genstand for forskningen ved Weizmann-Instituttet. Denne forskning,
der til at begynde med naturligt
koncentreredes om de områder inden for den organiske kemi, til hvilke Weizmann selv havde givet så
betydningsfulde bidrag, er med tiden stadig vokset i omfang, og talrige fremragende videnskabsmænd
er beskæftiget med forskningspro-
blemer både af almen art og tagende
sigte på anvendelsen i samfundets
tjeneste. Til de sidste hører ikke
mindst bestræbelserne for at skabe
levevilkår for en betydelig del af
befolkningen i den nu næsten øde
Negev-ørken i det sydlige Israel og
udnytte de mineralske forekomster
omkring Det døde Hav.

Under højtidelighederne indvie-
des også et af en ven af Israel skænket moderne udrustet, biologisk institut, hvor man især vil beskæftige
sig med spørgsmål af betydning for
jordbundens frugtbargørelse og bekæmpelse af de sygdomme, der truer menneskenes og husdyrenes
sundhed. Det gælder jo her om en
deltagelse, tilpasset Israels forhold,
i en forskning inden for hvilken
der for tiden gøres så stor en indsats overalt i verden. Fuldendelsen
af det planlagte fysiske institut vil

*Chaim Weizmann og Emir Feisal, arabernes
anfører i oprøret mod tyrkerne. Mødet i ørkenen
gav stødet til livsvarigt venskab.*

betyde et videre skridt til virksom deltagelse i aktuel international forskning, og i det hele taget peger arbejdet på Weizmann-Instituttet langt ud over Israels umiddelbart foreliggende praktiske problemer og tager sigte på landets samhørighed med andre nationer i menneskehedens stræben efter kundskabsforøgelse.

Det er derfor en overordentlig smuk tanke til mindet om Chaim Weizmann at knytte bestræbelserne for at støtte det institut, der bærer hans navn. For dette formål er der oprettet en international komité med deltagelse fra hele verden, og desuden er der i mange lande dannet særlige nationale komitéer for at støtte arbejdet. Herhjemme har man navnlig sat sig som mål at fremme det internationale videnskabelige samarbejde ved at muliggøre både deltagelse fra viden-skabsmænd fra vort land i forskningen på Weizmann-Instituttet og studiebesøg af de til dette knyttede videnskabsmænd på vore forskningsinstitutioner.

Når man i Danmark har ønsket på denne måde at være med til at hjælpe i en sag af en fælles menneskelig karakter til gavn for alle, mindes vi også de stærke venskabsbånd, der forbinder vort gamle land med det nye samfund. Dette venskab, der har en særlig baggrund i den overalt i Israel med taknemmelighed omfattede redningsdåd i besættelsens tid, hvorved det lykkedes at bringe så mange danske borgere af jødisk afstamning i sikkerhed, finder et synligt udtryk i det højt i Judæabjergene så smukt beliggende hospital for tuberkulostlidende børn, som er indrettet med dansk hjælp og som bærer Kong Christian den Tiendes navn.

TRANSLATION

The rebuilding of Israel.
A remarkable kind of adventure.

By Prof. NIELS BOHR, D.Phil.

The creation of a home for people of Jewish extraction on the old historic ground where the people of Israel long ago contributed so much to developing the ethical basis upon which the life of later generations has rested, is a quite remarkable kind of adventure. What is involved here is not a more or less smooth advance in living conditions and culture, as we know it from the history of other countries, but an adaptation which, despite all difficulties, is taking place at a pace which hardly has any parallel. People from far and near parts of the world, where they have taken part in the life of the nations – even though because of prejudice they often, indeed even right up to our time, have been subject to terrifying persecutions – seek in Israel, gripped by a common uniting idea, to make use of the skills and experience that they each in particular have brought with them to build up a new society and develop the country's natural resources in order to create living conditions for a population many times larger than the one the small tracts of land on the coast of Asia Minor have supported in recent centuries.

Such a situation obviously requires the greatest exertion and personal sacrifice from everybody, both from those who, with detailed knowledge of the circumstances, have to take a leading part in the work, and the many who come in a steady stream, and for whom there must be found the most suitable employment according to their widely differing levels of education. Camps with dwellings of a very primitive kind are to be seen everywhere in Israel, giving shelter to immigrants who must immediately be directed to take part in the cultivation of the country, which often requires great preparatory drainage and regulation. These camps are in time converted to beautiful villages surrounded by fields and fruit orchards, which in the hot climate give a rich yield as soon as the right conditions are in place. In addition to the large olive groves, which, especially in the ravines of Galilee, have survived from Antiquity because of the favourable watering conditions, exotic fruits of every kind and large tracts planted with maize and cotton can be seen already now many places in Israel.

Chaim Weizmann – scientist,

For all this cultivation, as well as for the building-up of industry ever more necessary for the economy of the country, it is important to utilize all the knowledge that can be found. In this connection, precisely Chaim Weizmann's[1] happy foresight as statesman and scientist has played a quite decisive role. From the very beginning he understood that the well-being of the new society would be contingent on the most vigorous endeavours to advance scientific research, and here he not only considered the immediate assistance that science should be able to offer for the utilization of all the sources of wealth in the country, so long neglected, but at the same time had in view the significance of the participation of the new society in the common human search for knowledge, which perhaps more than anything else ties peoples together.

As a result of the old traditions and the respect for learning and research that throughout the ages have been alive in the groups of Jewish people spread all over the world, there is in Israel wide-spread sympathy for the importance of scientific studies. Already a few years after the First World War, the university in Jerusalem was founded with wide international support.[2] It was established in the impressive buildings on Mount Scopus, just east of the old city, looking

[1] (1874–1952). See the introduction to Part II, p. [376].
[2] The Hebrew University of Jerusalem was inaugurated on 1 April 1925.

[695]

... and politician.

over the Jordan Valley with its rich memories. Since the ceasefire whereby the conflict in the country in the years after the last World War was terminated, Mount Scopus, however, has been on the Arabian side of the border between Israel and Jordan, and the large university buildings with the valuable library are therefore no longer available for the many scientists and students in Jerusalem.[3] Until the new buildings that are planned to be built in the western fringe of the city can be put to use, they are all obliged to work under the most crowded conditions in temporary quarters, which nevertheless does not prevent the display of an assiduity and enthusiasm that makes a profound impression on any visitor.[4]

The Weizmann Institute in Rehovot has had a quite different genesis. Here, in the middle of the present Israel, where for a long time there has only been a desert with low sandhills seemingly unsuitable for habitation and cultivation, about 20 years ago, with support from grateful friends of Weizmann, some small laboratory buildings were erected, which were to be the first beginning of the now so large and important research centre. On the top of the highest

[3] The separation occurred with the Arab–Israeli war of 1948–1949.
[4] Construction of a new campus was started in 1953.

hill, Weizmann established his own home, and with the help of irrigation, the desert was planted and now consists of orange and lemon orchards which frame the institute buildings most beautifully. To honour Weizmann's memory, the State of Israel has decided to make the whole area around the institute into a national park containing Weizmann's grave and his dwelling, which in time will be established as a museum for the mementos of Israel's great first president.[5]

I personally met Chaim Weizmann for the first time in Manchester when I worked with Rutherford during the First World War. At that time he was professor of organic chemistry at the University of Manchester and had just developed new methods for the manufacture of explosives required for the warfare, whereby he extended invaluable services to the Allied powers. It was, however, not only his reputation as a scientist which made him respected, but everyone who came into contact with him received a profound impression of his powerful and noble personality. The last time I saw him was on a visit to his home in New York a few days before Israel's declaration of independence.[6] At that time he was ill in bed, but with an inner fire he told movingly of the negotiations he had had in the old days with King Faisal of Arabia.[7] These negotiations took place around a campfire in the desert with the famous Lawrence[8] as interpreter and were held in a spirit of trust and friendship, which one must hope will return.[9]

The memorial park at Rehovot was inaugurated in the presence of the Government of Israel and numerous representatives from all parts of the country at the beginning of November, on the anniversary of Weizmann's death, and the laying of the foundation stone for the physics institute, in which we were invited to take part, was part of the festivities which lasted for a whole week and concluded with the holding of an international scientific symposium for discussing the problems which are particularly subjected to research at the Weizmann Institute. This research, which in the beginning was naturally concentrated on the fields in organic chemistry to which Weizmann himself had contributed so importantly, has with time steadily grown in extent, and numerous eminent scientists are occupied with research problems both of a general nature and aiming at application in the service of society. To the latter

---

[5] Weizmann's home was opened to the public a few years after the death in 1968 of his widow, Vera, who had continued to live there. It was reopened in 1990 after a substantial restoration.

[6] The meeting probably took place during the first half of May 1948, when Bohr was visiting the United States as described in the introduction to Part I, p. [69].

[7] Emir Faisal ibn Husseini (1885–1933), who from 1921 until his death was King Faisal I of Iraq.

[8] Thomas Edward Lawrence (1888–1935).

[9] The Faisal–Weizmann Agreement was signed on 3 January 1919.

Chaim Weizmann and Emir Faisal, the Arabs' leader in the rebellion against the Turks. The meeting in the desert gave rise to a lifelong friendship.

belong not least the endeavours to create living conditions for a considerable proportion of the population in the now nearly desolate Negev desert in the southern part of Israel and to utilize the mineral deposits around the Dead Sea.

During the festivities a modernly equipped biological institute donated by a friend of Israel[10] was also inaugurated, where work will be done in particular on questions of importance for soil fertilization and the combating of the diseases threatening the health of humans and domestic animals. What is important here, of course, is a participation, adapted to Israel's situation, in research within which at the present time such a great effort is being made all over the world. The completion of the planned physics institute will signify a further step towards active participation in current international research, and the work at the Weizmann Institute points in general far beyond Israel's immediate practical problems and aims at the country's concord with other nations in the striving of humanity for the increase of knowledge.

It is therefore an exceedingly beautiful thought to link to the memory of Chaim Weizmann the endeavours to support the institute bearing his name. For this purpose an international committee has been set up with worldwide participation,[11] and, in addition, special national committees have been estab-

[10] Isaac Wolfson (1897–1992), British businessman and philanthropist, gave name to the Building for Experimental Biology, which was dedicated on 3 November 1953.

[11] This was the Board of Governors, of which Bohr served as an active member. By 1960, when Bohr was still on the board, it counted 85 members, including Bohr's close physicist colleagues Felix Bloch (1905–1983), J. Robert Oppenheimer (1904–1967) and Victor Weisskopf (1908–2002).

lished in many countries to support the work.[12] Here at home the goal has been set in particular to advance international scientific cooperation by enabling both participation by scientists from our country in the study at the Weizmann Institute and study visits at our research institutes by scientists from there.

When in Denmark there has been a wish to take part in this way towards helping in a cause of common human character for the benefit of all, we are also reminded of the strong ties of friendship binding our old country to the new society. This friendship, which has a special background in the rescue action at the time of the occupation whereby it was possible to bring so many Danish citizens of Jewish descent to safety, which is regarded everywhere in Israel with gratitude, finds a visible expression in the hospital for children suffering from tuberculosis, so beautifully situated high in the mountains of Judea, which is established with Danish assistance and which bears the name of King Christian X.[13]

---

[12] After Bohr's death the Danish Committee continued under the leadership of Stefan Rozental (1903–1994), Bohr's assistant who had accompanied him on his trip to Israel. It is still in existence today.

[13] The King Christian X Hospital was established in 1950 by Den Danske Israelindsamling (The Danish Subscription Fund for the Benefit of Israel). It is still in existence today as a mental hospital.

XXXVI. THE GOAL OF THE FIGHT:
THAT WE IN FREEDOM MAY LOOK FORWARD
TO A BRIGHTER FUTURE

KAMPENS MÅL:
AT VI I FRIHED KAN SE HEN TIL EN LYSERE FREMTID
"Ti år efter" ["Ten Years After"],
Kammeraternes Hjælpefond, Copenhagen 1955

Memorial booklet prepared by the
Comrades' Assistance Fund
for previous concentration camp prisoners
including the programme for a reunion 19–20 March 1955

TEXT AND TRANSLATION

See Introduction to Part I, p. [376].

[701]

KAMPENS MÅL

At vi i frihed kan se hen
til en lysere fremtid

Af professor NIELS BOHR

Det er mig en glæde med nogle ord at bidrage til det mindehæfte der i tiåret for Danmarks befrielse udsendes af Kammeraternes Hjælpefond. Selv var jeg jo ikke blandt de mange, der under frihedskampen indespærredes i tyske fangelejre, men efter at være blevet advaret om min forestående fængsling undslap jeg ved modstandsbevægelsens hjælp til Sverige, hvorfra jeg på den britiske regerings indbydelse bragtes til Skotland og kort efter med den engelske videnskabelige mission kom til Amerika for at deltage i det store atomenergiprojekt. Som enhver der var langt borte fra vort land i de år, hvor kampen for Danmarks selvstændighed under Frihedsrådets ledelse samlede hele befolkningen, fik jeg imidlertid det stærkeste indtryk af den beundring hvormed man overalt i den frie verden fulgte begivenhederne i det besatte Danmark, og af den betydning modstanden mod voldsmagten havde for folkets anseelse og landets stilling da befrielsen kom. Iværksættelsen af det arbejde, hvori jeg tog beskeden del, var jo påkrævet for at undgå skæbnesvangre overraskelser fra mulige tilsvarende forberedelser fra fjendtlig side, men for alle der deltog deri var det tillige forbundet med forvisningen om, at den videnskabelige og tekniske udvikling ville få afgørende betydning for virkeliggørelsen af menneskenes gamle drøm om et fredeligt samarbejde mellem folkeslagene på kulturens fremme. Det er jo klart at ethvert fremskridt af vor kundskab og af vor beherskelse af naturens kræfter medfører et forøget ansvar, og trods alle hidtidige skuffelser må vi sætte vor lid til at alvoren af den situation menneskeheden nu er stillet overfor vil finde så udbredt forståelse at nye veje vil åbnes til bilæggelse af stridigheder og muliggøre at folkene i fællesskab kan indfri de rige løfter om fremgang i velfærd, som videnskaben på så mange områder holder frem for os. Det er med dette store mål for øje at vi føler, at det mod og den opofrelse der i de mørke tider blev udvist ikke har været forgæves, og vor sorg over dem der faldt i kampen fog vor medfølelse med dem der led varigt tab af helbred får en baggrund af dyb taknemmelighed for hver dåd, der bidrog til at sikre at vi i frihed kan se hen til en lysere fremtid.

TRANSLATION

THE GOAL OF THE FIGHT
That we in freedom can look forward to a brighter future
By Professor NIELS BOHR

It is a pleasure for me to contribute with a few words to the memorial issue published by the Comrades' Assistance Fund on the tenth anniversary of Denmark's liberation. I myself was not, of course, among the many who were confined in German prison camps during the fight for freedom, but, after having been warned of my imminent imprisonment, I escaped with the help of the resistance movement to Sweden, from where I was brought to Scotland at the invitation of the British Government and, shortly thereafter, came to America with the English scientific mission in order to take part in the great atomic energy project. Like anyone far away from our country during those years, when the fight for Denmark's independence under the leadership of the Danish Liberation Council[1] united the whole population, I got the strongest impression, nonetheless, of the admiration with which the events in occupied Denmark were followed all over the free world and of the importance the resistance against tyranny had for the respect for the people and the standing of the country when the liberation came. The implementation of the work in which I took a modest part was of course necessary in order to avoid fateful surprises from possible corresponding preparations on the part of the enemy, but for all who participated, it was also bound up with the conviction that the scientific and technical development would have decisive importance for the realization of humanity's old dream of peaceful cooperation between the peoples for the advancement of culture. It is obvious, of course, that every advance of our knowledge and of our mastery of the forces of nature brings with it increased responsibility, and, despite all previous disappointments, we

---

[1] *Danmarks Frihedsråd* was established on 16 September 1943 as a coordinating body for the Danish resistance. One of its leading members was Mogens Fog, who mediated Bohr's contact with Yakov Terletzkii after the war. The Soviet government accepted the Council as the rightful Danish government upon a visit by the Council's representative Thomas Døssing in 1944. Bohr's postwar contacts with Terletzkii and Døssing are described in the introduction to Part I on pp. [57] and [60], respectively.

[703]

must have faith that the seriousness of the situation now facing humanity will find such wide understanding that new paths will be opened for the resolution of disputes and allow that the peoples together can fulfil the rich promises of progress in welfare that science holds out for us in so many fields. It is with this great goal in view we feel that the courage and the sacrifice demonstrated in the dark times have not been in vain, and our grief for those who fell in the fight and our sympathy with those who suffered permanent loss of health get a background of deep gratitude for every deed that contributed to ensuring that we in freedom can look forward to a brighter future.

Niels Bohr

APPENDIX. SELECTED CORRESPONDENCE

TEXTS AND TRANSLATION

This appendix contains the entire documents that are partially quoted in the introduction to Part II.

All of them being letters to or from Bohr, they are arranged in alphabetical order according to correspondents. All letters were originally written in English except for one in Danish which is followed by a translation.

It has been attempted to make the layout of letterheads etc. correspond as closely as possible to that of the original letters. As a help for the reader to place them in context, the list overleaf includes references to the pages in the introduction to Part II where the letters are quoted. In addition, explanatory footnotes are provided in the letters themselves in order to identify particular persons, institutions and events.

The correspondence with Courant is microfilmed in BSC (28.1), whereas the other letters belong to the Bohr General Correspondence at the NBA.

[705]

LIST OF DOCUMENTS

ALSING ANDERSEN

ANDERSEN TO BOHR, 17 May 1955
[Typewritten]

DANSK INTERPARLAMENTARISK
GRUPPE   *København, den* 17. maj 1955.
Folketinget, CHRISTIANSBORG

———

Kære professor Bohr!

I fortsættelse af vore telefonsamtaler vil jeg gerne gennem disse linier forklare Dem situationen:

Hvert andet år holdes i et af de nordiske lande et nordisk interparlamentarisk delegeretmøde, det vil sige et møde af nordiske parlamentsmedlemmer af alle partier. I år holdes dette møde på Christiansborg i København i dagene 28/6 – 29/6. Mødet vil omfatte henved 100 deltagere, og Det nordiske interparlamentariske Råd har besluttet at sætte spørgsmålet om atomkraftens anvendelse på mødets dagsorden den 29/6. Samtidig vedtog mødet eenstemmigt at opfordre Dem til at være indledende taler om dette emne. Jeg tilføjer, at det naturligvis først og fremmest er mulighederne for atomenergiens praktiske anvendelse i det civile liv og mulighederne for et nordisk samarbejde på dette område, som mødets deltagere er interesseret i.

På Det interparlamentariske Råds vegne bringer jeg Dem en oprigtig tak for Deres tilsagn om at deltage i dette møde som indledende taler og i den efterfølgende diskussion. Jeg ved, at det vil vække tilfredshed og glæde hos alle mødets deltagere.

Da det drejer sig om et særligt emne af vanskelig karakter, ville det være til stor nytte for de mødedeltagere, der ikke har dansk som modersmål, dersom De ville have mulighed for at overlade mig Deres manuskript, således at vi kunne lade Deres tale eller et resumé deraf trykke og omdele til mødets deltagere.

Jeg har ingen mulighed for at bedømme, hvor lang taletid, De vil have brug for; men af hensyn til forhandlingernes gennemførelse på rimelig tid, ville det være hensigtsmæssigt, dersom Deres foredrag kunne holdes indenfor 45 minutter.

Det træffer sig desværre så uheldigt, at jeg er nødt til at tage min ferie fra i morgen onsdag den 18. maj; men jeg er ikke længere borte end i Liseleje,[1] og jeg vil tillade mig at telefonere til Dem en dag for at tale med Dem om sagen; og hvis De ønsker det, vil jeg kunne komme ud til Dem en dag, når jeg kommer til byen.

Venlig hilsen,
Deres hengivne

Alsing Andersen[2]

Hr. professor, dr.phil. & sc. & techn. Niels Bohr,

Gl Carlsbergvej,

Valby.

Translation

DANISH INTERPARLIAMENTARY GROUP
Folketinget, CHRISTIANSBORG

Copenhagen, 17 Maj 1955.

Dear Professor Bohr!

In continuation of our telephone conversations I would like to explain the situation to you with these lines:

Every second year, a Nordic interparliamentary delegate meeting, that is, a meeting of Nordic members of parliament from all parties, is held in one of the Nordic countries. This year this meeting is to be held at Christiansborg in Copenhagen in the days 28/6 – 29/6. The meeting will comprise about 100 participants, and the Nordic Interparliamentary Council has decided to put the question of the application of atomic power on the agenda for the meeting on 29/6. At the same time the meeting decided unanimously to invite you to be the introductory speaker on this topic. I would like to add that it is of course first and foremost the possibilities for the practical application of atomic energy in civilian life and the possibilities for Nordic cooperation in this field, that the participants in the meeting are interested in.

---

[1] Danish fishing village and summer resort on northern Zealand.
[2] (1893–1962). See the introduction to Part II, p. [364].

On behalf of the Nordic Interparliamentary Council I offer you sincere thanks for your acceptance to take part in this meeting as the introductory speaker and in the following discussion. I know that this will give satisfaction and pleasure to all participants in the meeting.

As a special topic of a difficult character is involved, it would be very useful for the participants who do not have Danish as their mother tongue if you should have the possibility of giving your manuscript to me, so that we could have your talk, or a summary of it, printed and handed out to the participants in the meeting.

I have no way of judging how much time you require for your talk; but with a view to the completion of the negotiations within a reasonable time, it would be suitable if your talk could be kept within 45 minutes.

Unfortunately it so happens that I am obliged to take my holidays from tomorrow Wednesday 18 May, but I am no further away than in Liseleje,[1] and I will allow myself to ring to you one day and talk about the matter; and if you wish, I could come out and visit you one day when I am in town.

Yours sincerely,

Alsing Andersen[2]

Professor Niels Bohr, D.Phil., D.Sc., D.Tech.Sc.

Gl Carlsbergvej,

Valby.

RICHARD COURANT

COURANT TO BOHR, 13 May 1954
[Typewritten]

NEW YORK UNIVERSITY
INSTITUTE OF MATHEMATICAL SCIENCES
25 WAVERLY PLACE, NEW YORK 3, N.Y.

TELEPHONE: ORegon 4-0734

May 13, 1954

AIR MAIL.

Professor Niels Bohr
Gl. Carlsberg
Valby
Copenhagen, Denmark

Dear Niels:

I was under so much pressure all these last months that I always delayed writing to you about many things that are on my mind. In the meantime, so much has happened that writing again is difficult. The case of Robert O.[3] certainly must have been disquieting for you as it is for all of us here. It has so many sides that I personally cannot accept the prevailing simplified version and attitude. Did you get complete transcripts of the public statements?[4] I may send a copy to Aage. At any rate, I would welcome very much an opportunity of talking to you personally about this and other matters.

Probably I shall go to Europe with Nina[5] this summer. My schedule will be very tight, but I would be happy if I could meet you one way or the other,

---

[3] In December 1953, President Eisenhower ordered that J. Robert Oppenheimer's security clearance be revoked. In turn, the Atomic Energy Commission named a three-man board to hear the case. The hearing took place from the middle of April to early May 1954.

[4] The massive transcripts of the hearings were promptly printed by the United States Government Printing Office in several thousand copies, and it may have been one of these that Courant offered to send. The transcripts were published as a book: *In the Matter of J. Robert Oppenheimer: Transcript of Hearing before Personnel Security Board and Texts of Principal Documents and Letters*, MIT Press, Cambridge, Massachusetts, and London, England, 1970.

[5] Nerina Runge Courant (1891–1991), Courant's wife.

maybe some time during July. I probably can get away from other obligations for a few days during that period.

As I understand, you are coming here for the Columbia celebration.[6] In this connection, please don't feel under pressure when I submit to you the following question: As you will see from the enclosed pamphlet, our Institute now is being realized. In some ways I am sure it will play an important role. Now we are planning this fall to have some kind of inauguration. We would not want a big affair, but still I want to ask you whether it would be at all possible on this occasion for you to give a talk. The date of this inauguration is still flexible and could quite possibly be adjusted to your convenience, within certain limits.[7] An honorarium of course would be paid, which could presumably help a little bit in supplementing Columbia and Princeton.

Please, if you have no time, tell Aage to answer this letter, possibly pretty soon.

Everybody in my family is quite well. Ernest's[8] work is proceeding satisfactorily. As a matter of fact, we are doing some very interesting calculations on our UNIVAC computing machine[9] for the stability problem of the supersynchrotron, and the results are definitely encouraging.

Hans got a small Fulbright grant for Paris and will go there next fall to work with the cosmic ray group.[10]

I hope everything is well with Margarete and you and your family. With most cordial regards,

As ever yours,

Richard C.

RC: er   R. Courant

---

[6] Bohr received an honorary Doctor of Science degree at Columbia University, New York, at the end of October 1954. See the introduction to Part II, p. [370].

[7] As noted in the introduction to Part II, p. [370], Bohr gave his talk at the inauguration on 29 November 1954. The published version, *Mathematics and Natural Philosophy*, The Scientific Monthly **82** (1956) 85–88, is reproduced on pp. [667] ff.

[8] Ernest Courant (1920–), Richard Courant's son, was a distinguished physicist, working at the Brookhaven National Laboratory, U.S.A.

[9] The UNIVAC (UNIVersal Automatic Computer) had been placed at the New York University in 1953 by the Atomic Energy Commission and became part of Courant's institute. It was the first university-based mainframe computer.

[10] Like his brother Ernest, Hans Courant (1924–), was a physicist. He spent his Fulbright grant at the École Polytechnique, where Louis Leprince–Ringuet (1901–2000) had developed a strong research programme in cosmic ray research.

BOHR TO COURANT, 28 May 1954
[Carbon copy]

May 28, 1954.

Professor R. Courant,
Institute of Mathematical Sciences,
New York University,
25 Waverly Place,
New York 3, N.Y.

Dear Richard,

I thank you for your kind letter of May 13, 1954. I would certainly be happy if we could meet during your visit to Europe this summer and it would give Margrethe and me great pleasure if you and Nina could visit us in Tisvilde[11] any time after the middle of July. I am, of course, anxious to hear about developments and surely we have been very worried about Robert's troubles. If it can fit in with my obligations in Columbia and Princeton, it shall certainly be a great pleasure to me to give a talk at the inauguration of your Institute where I might perhaps speak about the inherent connection between the appropriate mathematical tools and the epistemological problems in physical science. But perhaps we can talk it over when we meet.

It was a great pleasure to hear the progress of Ernest's splendid work and also that Hans is going to Paris to take part in the very successful researches on cosmic rays there.

With heartiest greetings to you and Nina from Margrethe and

Yours ever

(Niels Bohr).

---

[11] Summer resort on northern Zealand, east of Liseleje, ref. 1. Bohr acquired a summer house there in 1924, where he spent the summers with his family for the rest of his life. The property also included a pavilion, where Bohr often withdrew with fellow physicists for study and discussion.

COURANT TO BOHR, 6 April 1955
[Typewritten]

NEW YORK UNIVERSITY
INSTITUTE OF MATHEMATICAL SCIENCES
25 WAVERLY PLACE, NEW YORK 3, N.Y.

TELEPHONE: ORegon 4-0734

April 6, 1955

Professor Niels Bohr
Gl. Carlsberg
Valby
Copenhagen, DENMARK

Dear Niels:

I am sending, with apologies for the delay, a slightly revised manuscript of your talk with a copy of the original which shows what changes have been proposed.[12]

Incidentally, my suggestion would be to publish it in Science.[13]

I was very happy that I had at least a brief chance of seeing you in Europe. I hope that in the meantime the CERN situation[14] has been cleared up and that you have had a little rest.

Yesterday I met Lewis Strauss[15] by chance in von Neumann's[16] office. Lewis

---

[12] The manuscripts and proofs for Bohr's talk and article, ref. 7, are in BMSS (10.4), AHQP.

[13] The article was ultimately published in The Scientific Monthly, ref. 7.

[14] Bohr was strongly involved in the establishment of CERN, the European Organization for Nuclear Research. By this time the site of Geneva had long since been decided. Denmark signed the CERN Convention on 23 December 1953 and deposited its instruments of ratification on 5 April 1954.

[15] (1896–1974). Strauss had been appointed chairman of the U.S. Atomic Energy Commission in 1953.

[16] John von Neumann (1903–1957), Hungarian-born mathematician who moved to the United States in 1932. Ranging broadly in his work, he contributed substantially to the theoretical understanding of computers as well as to the foundations of quantum theory. He was also a prominent adviser to the U.S. government.

is very happy about his forthcoming trip to Europe and about your invitation which he deeply appreciates.[17]

Many cordial regards and wishes to all of you,

As ever yours,
Richard

RC:js   R. Courant
Enclosures

P.S.: Incidentally, I hope I told you that Jakob Goldschmidt[18] was rather sick with a heart condition during the months of January and February. He is now much better. He always apologizes to me that he has not written to you but you know anyway how much your visit meant to him.

BOHR TO COURANT, 24 May 1955
[Carbon copy]

May 24, 1955.

Professor R. Courant,
Institute of Mathematical Sciences,
New York University,
25 Waverly Place,
New York 3, N.Y.

Dear Richard,

It was a great pleasure to see you on my journey to Geneva where you so kindly visited me in the train between Hannover and Göttingen.[19] I was also

---

[17] Strauss visited Copenhagen in May 1955.

[18] Jakob Goldschmidt (1882–1955) was Bohr's host in New York when Bohr gave his talk at Courant's institute.

[19] Bohr attended one of his innumerable meetings at CERN on 24 and 25 February 1955 when Courant was visiting Europe.

very grateful shortly after my return to Copenhagen to receive your helpful corrections to the manuscript of my address at the dedication of your institute and I have a very bad conscience not earlier to have returned it. In the last months, however, I have been very occupied with administrative duties and hoped all the time to be able to improve the address materially by the elaboration of various points. As you will see, however, from the manuscript which I send under separate cover, I have now decided only to introduce a few smaller corrections and additions and I hope that you will not be too dissatisfied with it.[12] As I told you, I have been approached from various sides as regards its publication, but I have not made any arrangements and shall leave it all to your better judgement and kind help. If you think I ought to see a proof, I shall take care that it is returned without delay. I may add that I have just received a telegram from Shepard Stone[20] announcing his visit here on the 3rd of June and I need not say that I shall write to you again when I have talked with him.

With warmest greetings and best wishes from home to home

Yours ever

(Niels Bohr).

[20] (1908–1990). In 1953, Stone became head of the Ford Foundation's section on International Affairs (US and Europe), which provided funds to Bohr's institute during the following years to promote international cooperation.

JAMES R. KILLIAN

KILLIAN TO BOHR, 18 July 1956
[Typewritten]

Atoms for Peace Awards, Inc.

A MEMORIAL TO HENRY FORD AND EDSEL FORD

77 MASSACHUSETTS AVENUE, CAMBRIDGE 39, MASSACHUSETTS

July 18, 1956

Professor Niels Bohr
Institute for Theoretical Physics
University of Copenhagen
Copenhagen, Denmark

My dear Professor Bohr:

The trustees of Atoms for Peace Awards, Inc. are most desirous of having before them, when they select the recipients of the Award, the names of all persons who have made significant contributions to the peaceful use of atomic energy. To this end, they are asking some of the leading workers in the theoretical and applied fields concerned with atomic and nuclear energy to submit nominations.

I am enclosing a copy of the booklet describing the organization of Atoms for Peace Awards and its objectives.[21] Also enclosed is a statement of the criteria upon which selection of the Award recipient will be made and of the nomination procedure.

I sincerely hope that you will wish to assist the trustees in their decision by submitting nominations for people known by you to have made contributions which will qualify them for consideration.

Sincerely yours,

James R. Killian, Jr.[22]
President

---

[21] These items are held together with the present letter at the NBA.

[22] (1904–1988). Killian was President of the Massachusetts Institute of Technology (MIT) and would serve as the first Presidential Science Adviser from 1957 to 1959.

JOHN S. SINCLAIR

SINCLAIR TO BOHR, 3 June 1954
[Typewritten]

NATIONAL INDUSTRIAL CONFERENCE BOARD INC
LET THERE BE LIGHT
FOUNDED 1916

NATIONAL INDUSTRIAL CONFERENCE BOARD
INCORPORATED
247 PARK AVENUE, NEW YORK 17, N.Y.

JOHN S. SINCLAIR
   PRESIDENT

June 3, 1954

Dr. Niels Bohr, Director
Institute for Theoretical Physics
Blegdamsvej 15
Copenhagen, Denmark

Dear Dr. Bohr:

The National Industrial Conference Board[23] sponsors an annual conference devoted to "The Peaceful Uses of Atomic Energy." The conference is designed to bring leaders of industry, science and government together for discussions of how to achieve maximum peacetime benefits. In the opinion of many American government officials and industrial leaders, these meetings are an outstanding medium for the exchange of ideas and information on the nonmilitary phases of atomic energy developments.

In recognition of the world-wide economic and social implications of this new energy source, and in keeping with President Eisenhower's policy of encouraging international cooperation, we are inviting the official heads and

---

[23] This institution was established in 1916 and is still in existence as the Conference Board.

outstanding scientists of atomic energy programs in Argentina, Australia, Belgium, Brazil, Canada, Denmark, France, Great Britain, India, Italy, Mexico, The Netherlands, Norway, Spain, Sweden, Switzerland and the Union of South Africa to participate in our Third Annual Atomic Energy Conference from October 13 through 16 here in New York City.

On behalf of the Trustees and Associates of The Conference Board, may I have the honor of inviting you to address one of the sessions of this important conference. We sincerely hope that you will be able to accept our invitation and review the activities and objectives of your country's peacetime atomic energy program for the benefit of the hundreds of scientists, businessmen and government officials who will attend.[24]

In addition to the world-wide review of the progress being made in developing peaceful applications of atomic energy, the conference includes sessions on the following subjects: (1) The Current Outlook on Atomic Power Costs, (2) Legal Problems of Peacetime Atomic Developments, (3) Uses of Radioisotopes in Agriculture and Food Products, (4) How to Get "Know-How" in Atomic Energy, (5) Use of Radioisotopes in Medicine and Pharmaceuticals, (6) Uses of Low-Power and Small Atomic Reactors, (7) Industrial Uses of Radioisotopes in Process Industries, (8) Public Safety and Reactor Operations, (9) Economic Factors in Locating Atomic Power Reactors, (10) Peacetime Utilization of Atomic Fission Products, (11) Industrial Uses of Radioisotopes in Metal and Metal Fabricating Industries, and (12) Atomic Energy Planning by American and Canadian Companies. Government officials, industrial leaders, scientists, and elected representatives of the United States Congress will address these sessions.

Enclosed are agenda of our 1952 and 1953 conferences and a copy of the published transcript of last year's meeting. Also enclosed is a copy of the Board's Annual Report for 1952 which will acquaint you with The Conference Board in the event that you are not already familiar with it.[25] The Conference Board was established almost forty years ago to conduct objective research on economic and business administration problems and to act as a forum for the exchange of ideas and information in these areas. The Board is supported by the voluntary contributions of more than 3,000 Associates, most of which are

---

[24] Bohr spoke at lunch on the first day of the conference. His talk was published as *Greater International Cooperation is Needed for Peace and Survival* in *Atomic Energy in Industry: Minutes of 3rd Conference October 13–15, 1954*, National Industrial Conference Board, Inc., New York 1955, pp. 18–26. It is reproduced on pp. [561] ff. See the introduction to Part II, p. [361].

[25] These items are not held together with the letter in the NBA.

the leading business and industrial firms of America. However, its supporters also include the leading labor unions and educational institutions of the country.

We sincerely hope that you can participate in this meeting and would very much appreciate hearing from you at your earliest convenience in order that plans for the final program can be completed at an early date.

Yours sincerely,

John S. Sinclair[26]

JSS:RED   President
Enclosures

BOHR TO SINCLAIR, 19 June 1954
[Carbon copy]

June 19, 1954

Dear President Sinclair,

I thank you for your kind letter of June 3 with the invitation of the National Industrial Conference Board to participate in your third annual Atomic Energy Conference. It shall be a welcomed experience to me to accept this invitation and to give, at one of the sessions, a brief account of the preparations in Denmark to join, according to our modest possibilities, in the great and promising development. I am coming to U.S. for some months next fall, and when I arrive

---

[26] Sinclair (1897–1972) was President of the Conference Board from 1949 to 1963.

in New York at the end of September I shall take pleasure in calling upon you or your associates to learn about arrangements of your Conference.

Yours sincerely

Niels Bohr

Mr. John S. Sinclair,
President
National Industrial Conference Board Inc.
247 Park Avenue,
New York 17,
N.Y.
U.S.A.

JEROME B. WIESNER

WIESNER TO BOHR, 16 July 1956
[Typewritten]

MASSACHUSETTS INSTITUTE OF TECHNOLOGY
RESEARCH LABORATORY OF ELECTRONICS
CAMBRIDGE 39, MASS.

July 16, 1956

Dr. Niels Bohr, Director
Institute for Theoretical Physics
Copenhagen, Danmark

Dear Dr. Bohr:

As you no doubt know, Henry Ford II[27] has established a prize to be known as the "Atoms for Peace Award" to be given to the person who, in the opinion of a board of judges, has made the major contribution to the peaceful use of atomic energy during the previous year.

---

[27] (1917–1987).

As a consequence of his role in organizing the Geneva Conference[28] and his leadership in the initiation of the CERN Laboratory, I.I. Rabi[29] is being considered for the prize, and I have been asked to collect material documenting the part that he played in these activities. Since you also participated in the early planning of these activities, I am hopeful that you will be willing to write a letter giving your view of the part Rabi played and endorsing our effort to have the prize awarded to him.[30]

If you feel that you are in a position to write such a letter, would you please address it to me at M.I.T.

Sincerely yours,

Jerome B. Wiesner[31]

Jerome B. Wiesner, Director
Research Laboratory of Electronics

---

[28] See Bohr's response, below.

[29] Isidor Isaac Rabi (1898–1988), prominent American physicist spending his career at Columbia University, New York. At the fifth General Conference of UNESCO in Paris in June 1950, Rabi, as a first-time member of the U.S. delegation, introduced a resolution promoting European collaboration in nuclear physics, which would prove crucial for the establishment of CERN.

[30] In the end, it was Bohr himself who was awarded the first Atoms for Peace Award. The pertinent information, including Bohr's acceptance speech, contained in the pamphlet *The Presentation of the first Atoms for Peace Award to Niels Henrik David Bohr, October 24, 1957*, National Academy of Sciences, Washington, D.C. 1957, is reproduced on pp. [637] ff.

[31] (1915–1994). Like Killian, Wiesner worked at MIT, where he too became President in 1971. Also like Killian, he served as Presidential Science Adviser (the third, from 1961 to 1964).

BOHR TO WIESNER, 10 August 1956
[Carbon copy]

August 10, 1956.

Jerome B. Wiesner, Director
Research Laboratory of Electronics
M.I.T.
Cambridge 39, Mass.

Dear Dr. Wiesner:

Thank you for your letter of July 16, which I received on my return from a journey to Yugoslavia to take part in the Tesla Centennary Celebrations.[32]

I can most heartily support the proposal that the "Atoms for Peace Award" be given to Professor I.I. Rabi who as chairman of the Organizing Committee played so helpful a part in making the Geneva Conference on the Peaceful Uses of Atomic Energy such a great success and a decisive contribution to the promotion of genuine international cooperation in the field of atomic science and its peaceful applications.

I also heartily concur in the gratitude which we all owe to Professor Rabi for initiating the endeavours leading to the establishment of the CERN Organization which already has created fruitful cooperation among European scientists as well as with their colleagues in other countries.[29]

Hoping that this letter may be helpful for your purpose, and with kindest regards,

Sincerely yours,

Niels Bohr.

---

[32] At these celebrations Bohr presented a tribute to the physicist Nikola Tesla which was subsequently published in *Centenary of the Birth of Nikola Tesla 1856–1956*, Belgrade 1959, pp. 46–47. The published version is reproduced in Vol. 12.

INVENTORY OF RELEVANT MANUSCRIPTS
IN THE NIELS BOHR ARCHIVE

INTRODUCTION

The following is a list of folders in the microfilmed Bohr MSS containing manuscripts by Bohr of special relevance for the present volume. The list is arranged chronologically and covers the years 1927–1960. As in earlier volumes, the list does not contain documents from other, not microfilmed, collections. For example, folders from the BPP, which constitutes an essential basis for Part I of the present volume, are not included.

In the first line of each listed item, the title of the folder has been assigned by the cataloguers, as has any date in square brackets. Unbracketed dates are taken from the manuscripts. A number in the margin indicates the page number on which a particular document has either been reproduced (R) or cited (C) or on which the publication (P) resulting from the manuscript is printed.

The next line indicates whether the documents in the folder are typewritten, carbon copies or handwritten, as well as the language used. Finally, the number of the microfilm on which the folder can be found is provided (e.g., "mf. 11").

Other relevant information about the folders – such as the contents of individual manuscripts and the relation of manuscripts among themselves or to documents reproduced in the present volume – is given in explanatory notes in small print.

[725]

1 *Foreningen Norden*   13 December 1927

Carbon copy, 1 p., Danish, mf. 11.

Statement for a pamphlet about "Foreningen Norden" (The Nordic Association). Argument for the importance of Scandinavian cooperation, as subsequently expressed in the publication on p. [501].

2 *Tale ved Kræftkomiteens Møde*   14 September 1929

Carbon copy, 4 pp., Danish, mf. 12.

Speech at the meeting of the Danish Cancer Committee in the Copenhagen Town Hall, 14 September 1929. First known manuscript relating to activities described in the publication on p. [681]. Bohr points to the medical importance of the discoveries in physics of X-rays and radioactivity as well as to the Manchester Hospital as a model for the Copenhagen Radium Station.

3 *Lord Mayor of London*   5 September 1930

Typewritten, carbon copy, 2 pp., English, mf. 12.

Draft of speech on the occasion of a visit by the Lord Mayor of London, Sir William Waterlow, to Bohr's institute on 5 September 1930. The visit included a guided tour of the institute.

4 *Om Stoffets og Lysets Natur*   [22 April 1931]

Typewritten with handwritten corrections and additions [S. Lauritzen], 15 pp., Danish, mf. 12.

(On the Nature of Matter and Light) Lecture delivered before the Society for the Dissemination of Natural Science (Selskabet for Naturlærens Udbredelse), Copenhagen 22 April 1931.

5 *Overdragelse af Æresbolig*   11 December 1931

Typewritten, carbon copy, 2 pp., Danish, mf. 12.

Speech delivered before the Royal Danish Academy of Sciences and Letters on 11 December 1931 on the occasion of the awarding by the Academy of the honorary residence at Carlsberg to Niels Bohr.

6 *Speech to Dr. Vincent*   April 1932

Carbon copy, 2 pp., English, mf. 13.

Draft of speech to George Edgar Vincent (1864–1941), the Rockefeller Foundation, given in April 1932. The speech is formulated as a response to Vincent's address, emphasizing Rockefeller philanthropy's importance for international scientific cooperation during difficult times.

7   *Indvielse af Matematisk Institut*   8 February 1934

Typewritten, carbon copy, 2 pp., Danish, mf. 13.

Title: "Tale ved Indvielsen af Universitetets Mathematiske Institut, 8.2.1934" (Speech at the inauguration of the Mathematical Institute of the University of Copenhagen, 8 February 1934). The institute, which was directed by Bohr's brother, Harald Bohr, was built adjacent to Bohr's institute, and the speech was held in Auditorium A of the latter institution.

8   *Landsforeningen til Kræftens Bekæmpelse*   [18 March 1935]

Carbon copy, 4 pp., Danish, mf. 14.

Manuscript of address on medicine and science given at the meeting of the National Association for Combating Cancer, in which Bohr argues for the unity of science.

9   *Det moderne Verdensbillede*   24 June 1935

Typewritten, with handwritten corrections [N. Bohr], 8 pp., Danish, mf. 14.

Title: "Det moderne verdensbillede" (The modern world view). Manuscript of lecture on the general implications of the new physics given at the Scandinavian Students' Meeting on 24 June 1935.

[501] P

10   *Tale ved Banket, Skandinaviske Naturforskermøde*   14 August 1936

Typewritten, carbon copy, 3 pp., Danish, mf. 14.

Speech at the banquet of the 19th Scandinavian Meeting of Natural Scientists in Helsinki 11 to 15 August 1936. The manuscript is not quite identical to the final publication.

[509] R

11   *Indvielse af Institutets Højspændingsanlæg*   5 April 1938

Carbon copy, 8 pp., Danish, mf. 15.

Report, probably transcribed from notes in shorthand, of a speech given at the inauguration of the high-voltage plant at Bohr's institute.

[527] P

12 *Radio Talk to America*   5 April 1938

Typewritten, carbon copy, with handwritten corrections, 7 pp., English, mf. 15.

Radio talk to America on the occasion of the inauguration of the high-voltage plant at Bohr's institute. Exact manuscript for the reproduced publication, plus one partly handwritten page.

13 *Nytårstale i Radioen*   31 December 1938

Typewritten, carbon copy, 2 pp., Danish, mf. 15.

(New Year Speech on the Radio) Draft and manuscript of speech given on the radio. Cf. folder 34.

[379] P

14 *Bohr Præsident for Videnskabernes Selskab*   [20 October 1939]

Typewritten, carbon copy, with handwritten corrections [N. Bohr], 5 pp., Danish, mf. 16.

Manuscript of Bohr's contributions to the meeting of the Royal Danish Academy of Sciences and Letters, for the first time in his capacity as President of the Academy. This is the complete manuscript for Bohr's talk which is partly paraphrased in the published version.

15 *Tale i Videnskabernes Selskab*   1 December 1939

Handwritten [Erik Bohr], 3 pp., Danish, mf. 16.

Draft for report of meeting of the Royal Danish Academy of Sciences and Letters, containing a speech given at the supper following the meeting. Bohr states his concern for the fate of Finland, which he would also express in the publication on p. [385].

16 *Christian X's 70-Års Dag*   26 September 1940

Typewritten, carbon copy, 5 pp., Danish, mf. 16.

Draft and manuscript for a short contribution in a broadcast on Danish National Radio celebrating King Christian X's 70th birthday. The contribution describes the day-to-day work at Bohr's institute.

17 *Landsforeningen til Kræftens Bekæmpelse* [25 November] 1940

Carbon copy, 3 pp., Danish, mf. 16.

Manuscript of opening and closing addresses, given at the annual meeting of the Danish Cancer Committee.

[469] P

18 *K. Linderstrøm–Lang, Ørsted-Medaillen* 19 March 1941

Carbon copy, 3 pp., Danish, mf. 16.

Manuscript of speech given at the meeting of the Society for the Dissemination of Natural Science (Selskabet for Naturlærens Udbredelse). The manuscript is practically identical to the published version.

[533] P

19 *Universitetet og Forskningen* 1941

Carbon copy, 6 pp., Danish, mf. 16.

Speech given at the University of Copenhagen. The manuscript has handwritten corrections, some of which are taken into account in the published version.

[493] P

20 *Mindeaften for Kirstine Meyer* 14 October 1942

Carbon copy, 7 pp., Danish, mf. 16.

The manuscript is identical to the published version, except for an introductory paragraph in memory of the physicist and inventor Valdemar Poulsen (1869–1942), who had recently died, and a final paragraph expressing thanks to Mogens Pihl for his contribution.

[397] P

21 *Videnskabernes Selskabs 200-Aarsdag* 13 November 1942

Carbon copy, 9 pp., Danish, mf. 16.

Title: "Det Kongelige Danske Videnskabernes Selskabs 200-Aarsdag 13.11.1942" (The 200th anniversary of the Royal Danish Society of Sciences and Letters 13 November 1942). The manuscript is identical to the published version except for brief passages that are only paraphrased in the latter.

22 *Politiets Sprogforening* [5 April 1943]
Typewritten, 1 p., Danish, mf. 16.

A few words written in the visitor's book of the Language Club of the Danish Police on the occasion of its visit to Bohr's institute. Bohr stresses the importance of international cooperation and expresses his gratitude for the readiness of the police authorities throughout the years to allow visits by young foreign scientists.

23   *Speech in England*   [October 1943]

Typewritten, carbon copy, with handwritten corrections [Aage Bohr], English and Danish, 6 pp., mf. 16.

Speech prepared in Stockholm in October 1943, before Bohr went to England. The two typed versions include handwritten amendments. The intention to give the speech in England was not realized due to the secrecy of Bohr's journey.

[26] C

24   *Royal Society Club Dinner*   4 November 1943

Handwritten [Aage Bohr], 2 pp., English, mf. 16.

Outline of speech, given at the Royal Society Club Dinner. The outline contains some elements of the manuscript in folder 23.

25   *Notes for Sir Edward Appleton*   [1945]

Carbon copy, 5 pp., English, mf. 16.

Notes on the physics of the atomic bomb prepared on the request of Sir Edward Appleton for Appleton's personal use.

26   *Udtalelse i Radio ved Hjemkomst*   25 August 1945

Typewritten, 1 p., Danish, mf. 16.

Manuscript of a short announcement on the radio on Bohr's return to Denmark.

[136] R

27   *Udtalelse i Pressens Radioavis*   26 August 1945

Stencil, carbon copy, 9 pp., Danish, mf. 16.

Announcement contained in the news on the occasion of Bohr's return to Denmark. The carbon copy is slightly different from the stencil used in the reproduction.

28   *Støtte til Hollandshjælpen*   31 August 1945

Typewritten, carbon copy, 1 p., Danish, mf. 16.

Title: "Udtalelse i Dagspressen i Anledning af Hollandshjælpen 31.8.1945" (Announcement to the press concerning the Aid to Holland Foundation, 31 August 1945). Bohr's statement, which was published in the Danish press, emphasized the need to help those hardest hit by the war.

29   *Fond til Fædrelandets Vel*   9 September 1945

Carbon copy, 1 p., Danish, mf. 16.

Title: "Udtalelse som Led i Radioudsendelsen 9.9.1945 for Tilslutning til Stiftelsen af 'Fondet til Fædrelandets Vel' i Anledning af Kong Christian X's 75-Aars Fødselsdag 26.9.1945" (Statement made as a part of the radio broadcast on 9

September 1945 in support of the establishment of the "Foundation for the Benefit of the Country" on the occasion of King Christian X's 75th birthday on 26 September 1945).

[294] R

30 *Frigørelse af Kerneenergi*   3 October 1945
Typewritten, 24 pp. Danish, mf. 16.

(Release of nuclear energy) Manuscript of lecture delivered before the Danish Association of Engineers (Ingeniørforeningen) as well as a one-page press release the following day. Only the latter has been reproduced.

[145] R

31 *Studenternes Fakkeltog ved 60-Aars Dagen*   7 October 1945
Stencil, 2 pp., Danish, mf. 16.

Speech given at Carlsberg on the occasion of the torchlight procession celebrating Bohr's 60th birthday.

[125] P

32 *A Challenge to Civilization*   12 October 1945
Typewritten, carbon copy, 4 pp., English and Danish, mf. 16.

Manuscripts for the version in "Science", reproduced in this volume, as well as for the Danish version, referred to on p. [126]

33 *Landsforeningen til Kræftens Bekæmpelse*   30 November 1945
Carbon copy, handwritten [Aage Bohr], 2 pp., Danish, mf. 16.

Outline of lecture on "Atomic Physics and Radiology", given at the annual meeting of the National Association for Combating Cancer. The outline, consisting of nine separate points, delineates the history of nuclear physics and its increasing importance for medical science.

[149] P

34 *Nytårstale i Radioen*   31 December 1945
Typewritten, carbon copy and handwritten [Margrethe Bohr, N. Bohr, Aage Bohr and S. Rozental], 90 pp., Danish, mf. 16.

(New Year speech on the radio) Extensive dated drafts, as well as the manuscript, of speech given on Danish National Radio on 31 December 1945.

[655] P

35 *Om Maalingsproblemet i Atomfysikken*   9 March 1946
Typewritten, carbon copies and handwritten [N. Bohr, Aage Bohr, L. Rosenfeld and S. Rozental], 84 pp., Danish, mf. 17.

(On the problem of measurement in atomic physics) Drafts and manuscripts. Almost all pages are supplied with dates from 13 September 1945 to 9 March 1946.

36   *Welcome to International Astronomical Union*   March 1946

Typewritten, 1 p., English, mf. 17.

Words of welcome on behalf of the Royal Danish Academy of Sciences and Letters to the meeting of the International Astronomical Union held in Copenhagen.

[131] P   37   *Atomic Physics and International Cooperation*   [21 October 1946]

Typewritten, carbon copy, with handwritten corrections [Aage Bohr], 4 pp., English, mf. 17.

Two versions of a manuscript that is practically identical to the published version.

38   *Danmarks Amerikanske Selskab*   2 December 1946

Typewritten, carbon copy, 4 pp., Danish, mf. 17.

Title: "Tale ved Mødet d. 2.12.1946 arrangeret af Danmarks Amerikanske Selskab" (Speech at the meeting 2 December 1946 arranged by the Danish American Society). Bohr stresses the importance of giving Danish students the opportunity of visiting the United States.

[419] P   39   *Mindeord om Kong Christian X*   25 April 1947

Typewritten, 2 pp., Danish, mf. 17.

Memorial speech for King Christian X at the meeting of the Royal Danish Academy of Sciences and Letters. The manuscript is identical to the published version except for an additional last paragraph.

[425] P   40   *Tale til Kong Frederik IX*   17 October 1947

Carbon copy, 3 pp., Danish, mf. 17.

Speech to King Frederik IX at the meeting of the Royal Danish Academy of Sciences and Letters. The manuscript is identical to the published version except for a few paraphrases in the latter and a minor change at the end.

41   *Welcome to Council of Scientific Unions*   13 September 1949

Typewritten, carbon copy, 1 p., English, mf. 19.

Words of welcome on behalf of the Royal Danish Academy of Sciences and Letters to the delegates at the meeting of the International Council of Scientific Unions, held in Copenhagen.

42   *Speech to Lord Boyd Orr*   14 December 1949
Carbon copy, 4 pp., English, mf. 19.

Two slightly different versions of a speech to Lord (John) Boyd Orr (1880–1971), the recipient of the 1949 Nobel Peace Prize.

43   *Indvielse ved Lunds Universitet*   31 May 1951
Typewritten, carbon copy, 2 pp., Danish, mf. 19.

Title: "Professor Niels Bohrs tale ved indvielsen af de nye fysiske og matematiske institutioner ved Lunds universitet 31.5.1951" (Professor Niels Bohr's speech at the inauguration of the new physical and mathematical institutes at the University of Lund, 31 May 1951).

44   *Lecture at the Conference at the Niels Bohr Institute*   6 July 1951
Handwritten [Aage Bohr], 7 pp., English, mf. 19.

Title: "Lecture Friday July 6th" (the manuscript is dated 5 July 1951). Lecture given at the opening of conference at Bohr's institute from 6 to 10 July 1951. The outline consists of 23 points which indicate that Bohr introduced the conference in light of the earlier development of theoretical physics, particularly in relation to his institute.

[463] P   45   *Statens almindelige videnskabsfond*   4 February 1952
Carbon copy, 3 pp., Danish, mf. 20.

(National Science Foundation) Statement on Danish National Radio. The manuscript is identical to the published version.

[477] P   46   *Uddelelse af H.C. Ørsted Medaljen*   3 December 1952
Carbon copy, 3 pp., Danish, mf. 20.

(Presentation of the H.C. Ørsted Medal) Speech and concluding remarks at the meeting of the Society for the Dissemination of Natural Science. Preliminary manuscript for the published version.

47   *American Scandinavian Foundation*   30 May 1953
Typewritten, 4 pp., English, mf. 20.

Title: "Speech by Professor Niels Bohr, recorded on May 30th 1953, for a radio programme for the American–Scandinavian Foundation series". The draft, which is less complete than the manuscript in the same folder, is dated 20 May 1953.

[681] P   **48**   *Landsforeningen til Kræftens Bekæmpelse*   16 October 1953

Carbon copy, 5 pp., Danish, mf. 20.

(National Association for Combating Cancer) Title: "Tale ved Statsradiofoniens udsendelse d. 16.10.1953" (Address broadcast on Danish National Radio, 16 October 1953). On the occasion of the 25th anniversary of the founding of the Danish Cancer Society. The manuscript is identical to the published version.

[689] P   **49**   *Det nye Israel og videnskaben*   2 February 1954

Typewritten, 6 pp., Danish, mf. 20.

(The new Israel and science) Lecture given in the radio programme "Information", 2 February 1954. The manuscript is practically identical to the published version.

[561] P   **50**   *International Cooperation for Peace and Survival*   [13 October 1954]

Carbon copy, with handwritten corrections [A. Petersen], 10 pp., English, mf. 20.

Note (S. Hellmann): "Tale holdt 13.10.1954 in U.S.A." (Speech given on 13 October 1954 in the U.S.A.). Given at the conference on "Peaceful Uses of Atomic Energy" arranged by the National Industrial Conference Board, 13 to 15 October 1954. Preliminary manuscript for the published version.

[667] P   **51**   *Mathematics and Natural Philosophy*   29 November 1954

Typewritten, with handwritten corrections [R. Courant, A. Petersen and S. Hellmann], 37 pp., English, mf. 20.

Drafts and manuscript of talk given at the dedication of the Institute of Mathematical Sciences, New York University, as well as manuscript and proof of published article based on the talk.

[701] P   **52**   *Kammeraternes Hjælpefond*   [19–20 March 1955]

Carbon copy, 1 p., Danish, mf. 21.

Contribution to the pamphlet, edited by "Kammeraternes Hjælpefond" (The Comrades' Assistance Fund) in connection with a reunion of prisoners-of-war ten years after the liberation of Denmark. The manuscript is identical to the published version.

[609] P   **53**   *Atomerne og samfundet*   29 June 1955

Typewritten, carbon copy, 10 pp., Danish, mf. 21.

(Atoms and Society) Notes for lecture at the meeting of the Nordic Interparliamentary Council, Copenhagen. The material is noted as typed on 8 June 1955. The notes are very different from the published version.

54   *Dansk Idræts Forbund*   [14 February 1956]
Carbon copy, 1 p., Danish, mf. 22.

Speech on Danish National Radio on the occasion of the 60th anniversary of Dansk Idræts Forbund (Danish Sports Union).

55   *Nordens Dag, Foreningen Norden*   30 October 1956
Carbon copy, 2 pp., Danish, mf. 22.

Title: "Tale af professor Niels Bohr på 'Nordens Dag' 30.10.1956" (Speech by Professor Niels Bohr on "Nordic Day" 30 October 1956).

56   *Tale i B.B.C.*   [17] May 1957
Typewritten, 1 p., Danish, mf. 22.

Speech recorded 11 May 1959 and broadcast by the British Broadcasting Corporation on 17 May 1957 in connection with the visit by Elisabeth II to Denmark.

57   *Lykke Lotteriet*   2 September 1957
Typewritten, 2 pp., Danish, mf. 22.

Statement on the radio in support of "Lykke-Lotteriet" (Lucky Lottery). Note: "Indtalt 2.9.1957" (Recorded 2 September 1957). Lykke-Lotteriet was a lottery organized by four Danish charities.

[637] P

58   *Atoms for Peace Award*   [24 October 1957]
Typewritten, carbon copy, 2 pp., English, mf. 22.

Speech on receiving the Atoms for Peace Award in Washington. The manuscript is identical to Bohr's "Response" in the publication.

59   *En åben verden, seminar ved M.I.T.*   15 November [1957]
Handwritten [A. Petersen], 2 pp., Danish, mf. 22.

Notes for Bohr's contribution to the discussion at the Open World Seminar at the Massachusetts Institute of Technology, 15 November 1957.

[735]

60   *3 taler, 1959*   6 February 1959, 17 September 1959, 1 December 1959

Typewritten, with handwritten notes [N. Bohr], 3 pp., Danish and English, mf. 23.

(3 speeches) 1 page of typescript of radio talk on the Greenland Foundation, 6 February 1959; 1 page of typescript of talk on receiving (in Copenhagen) the diploma of membership of the Academy of Sciences of the U.S.S.R., 17 September 1959; 1 page of handwritten notes of talk in Randers to the Danish Association of Engineers, 1 December 1959.

61   *Science and Technology*   25 June 1959

Typewritten, with handwritten corrections [N. Bohr and Margrethe Bohr], 20 pp., English, mf. 23.

Various typescripts in preparation for an address delivered at the dedication of the John Jay Hopkins Laboratory for Pure and Applied Science, San Diego, California.

62   *Taler, 1960.*   [5 February, 6 September, 1 October 1960]

Typewritten, with handwritten corrections, 7 pp., Danish and English, mf. 24.

(Speeches, 1960) Drafts for the following talks: (a) Speech delivered on the occasion of the inauguration of the CERN proton synchrotron, 5 February 1960; (b) Address of welcome at a conference on the use of Radioisotopes in Physical Sciences and Industry, Copenhagen, 6 September 1960; (c) Remarks for 140th anniversary of the establishment of the Danish Student Association and the 50th anniversary of the dedication of its building, 1 October 1960.

INDEX

Subjects which appear throughout the volume, such as "politics" and "physics", are not listed. Some particularly relevant subjects, notably "atomic bomb", "atomic energy" and "international cooperation", have been included, even though they appear often. There are index terms for some of Bohr's characteristic expressions, such as "atomic bomb as challenge to civilization", "harmony", "international cooperation", "natural phenomena", "responsibility" and "unity/concord". Names of publications are generally not indexed, with the exception of Bohr's "Open Letter to the United Nations" as well as three of his most important "publishable" writings – the "Memorandum" and "Addendum" to Roosevelt, and the "Comments" to Marshall – have been indexed.

All persons (other than Niels Bohr) and most institutions in the running text are listed, whereas places are only indexed selectively. When a term on a page is found only in a footnote, the page number in the index is followed by the letter n. Picture captions are fully indexed, with page numbers followed by the letter p. Subjects in the front material are only indexed if they also appear in the main body of the volume. It is hoped that the cross references will help the reader identify subjects that are expressed in two or more ways in the text.

.

Printed and bound by CPI Group (UK) Ltd, Croydon, CR0 4YY

03/10/2024

01040329-0018